THE BIOLOGY OF
TERRESTRIAL MOLLUSCS

THE BIOLOGY OF
TERRESTRIAL MOLLUSCS

Edited by

G.M. Barker

Landcare Research
Hamilton
New Zealand

CABI *Publishing*

CABI Publishing is a division of CAB International

CABI Publishing
CAB International
Wallingford
Oxon OX10 8DE
UK

Tel: +44 (0)1491 832111
Fax: +44 (0)1491 833508
Email: cabi@cabi.org

CABI Publishing
10 E 40th Street
Suite 3203
New York, NY 10016
USA

Tel: +1 212 481 7018
Fax: +1 212 686 7993
Email: cabi-nao@cabi.org

A catalogue record for this book is available from the British Library, London, UK.

Library of Congress Cataloging-in-Publication Data
The biology of terrestrial molluscs/edited by G.M. Barker.
        p.   cm.
    Includes bibliographical references.
    ISBN 0-85199-318-4 (alk. paper)
        1. Mollusks.   I. Barker, G.M.

    QL407 .B56 2001
    594--dc21

00-065708

ISBN 0 85199 318 4

Typeset by AMA DataSet Ltd, UK.
Printed and bound in the UK by Cromwell Press, Trowbridge.

Contents

Contributors

T. Backeljau, Royal Belgian Institute of Natural Sciences, Vautierstraat 29, B-1000 Brussels, Belgium

G.M. Barker, Landcare Research, Private Bag 3127, Hamilton, New Zealand

A. Baur, Department of Integrative Biology, Section of Conservation Biology (NLU), University of Basel, St Johanns-Vorstadt 10, CH-4056 Basel, Switzerland

B. Baur, Department of Integrative Biology, Section of Conservation Biology (NLU), University of Basel, St Johanns-Vorstadt 10, CH-4056 Basel, Switzerland

B. Berger, Institut für Zoologie und Limnologie, Abteilung Ökophysiologie, Universität Innsbruck, Technikerstraße 25, A-6020 Innsbruck, Austria

R. Chase, Department of Biology, McGill University, 1205 Av. Docteur Penfield, Montréal, Québec, Canada H3A 1B1

A. Cook, School of Environmental Studies, University of Ulster, Coleraine, Northern Ireland BT52 1SA, UK

R. Dallinger, Institut für Zoologie und Limnologie, Abteilung Ökophysiologie, Universität Innsbruck, Technikerstraße 25, A-6020 Innsbruck, Austria

I. Deyrup-Olsen, Department of Zoology, University of Washington, Seattle, Washington, USA

V.K. Dimitriadis, School of Biology, Faculty of Sciences, Aristotle University of Thessaloniki, Thessaloniki, Greece 54006

E. Furuta, Department of Histology, Dokkyo University School of Medicine, Mibu, Tochigi 321-0293, Japan

B.J. Gómez, Departamento de Zoología y Dinámica Celular Animal, Facultad de Farmacia, Universidad del País Vasco, Paseo de la Universidad 7, 01006 Vitoria, Spain

A. Gomot de Vaufleury, Laboratoire de Biologie et Ecophysiologie, Faculté des Sciences et des Techniques, Université de Franche-Comté, Place Leclerc, 25030 Besançon Cedex, France

J.M. Healy, Centre for Marine Studies, The University of Queensland, Brisbane, Queensland 4072, Australia

J. Heller, Department of Evolution, Systematics and Ecology, The Hebrew University, Jerusalem 91904, Israel

H. Köhler, Animal Physiological Ecology, Zoological Institute, Konrad-Adenauer-Strasse 20, D-72072 Tübingen, Germany

D.L. Luchtel, Department of Environmental Health, School of Public Health and Community Medicine, University of Washington, Box 357234, Seattle, WA 98195-7234, USA

U. Mackenstedt, Institut für Zoologie, Fachgebiet Parasitologie, Universität Hohenheim, Emil-Wolff-Str. 34, D-70599 Stuttgart, Germany

K. Märkel, Lehrstuhl für Spezielle Zoologie, Ruhr-Universität Bochum, Universitätstraße 150, D-44780 Bochum, Germany

B. Speiser, Research Institute of Organic Agriculture (FiBL), Ackerstrasse, CH-5070 Frick, Switzerland

R. Triebskorn-Köhler, Steinbeis-Transfer Center Ecotoxicology and Ecophysiology, Kreuzlinger Strasse 1, D-72108 Rottenburg, Germany *and* Animal Physiological Ecology, Zoological Institute, Konrad-Adenauer-Strasse 20, D-72072 Tübingen, Germany

K. Yamaguchi, Institute of Medical Science, Dokkyo University School of Medicine, Mibu, Tochigi 321-0293, Japan

Preface

With an estimated 35,000 species, terrestrial gastropod molluscs are one of the most successful and diverse animal groups in land-based eco-systems. These animals have long been of importance to human societies as food, medicine, crop pests, vectors of parasites, and as tools, personal ornamentation and currency in trade. A relatively small fraction of the global terrestrial gastropod diversity – in the order of 100 species – has proved to be highly adaptive to environmental change brought about by human activity, and often become a highly abundant and characteristic component of invertebrate faunas in modified habitats. These species generally exhibit high propensities for both passive dispersal associated with human trading activities and invasiveness when introduced to new areas. They have become increasingly important as crop pests in agriculture and as vectors of helminth parasites in humans and domestic livestock. Because of their increasing ubiquity and economic importance, and their utility as laboratory animals, these same species are among the most intensively studied invertebrates. They have proved to be excellent model systems for studies in, for instance, neurophysiology, behavioural ecology and population genetics. By virtue of their capacity to accumulate metal in their tissues, these animals are also of increasing utility as bioindicators of environmental pollution. Many of these areas of investi-gation show that research on terrestrial gastropods can contribute to the wider fields of research endeavour encompassing physiology, ecology, evolution and conservation biology.

The greater part of the terrestrial gastropod biodiversity, however, goes unnoticed by people – they are mostly small creatures, seeking out a living as detritivores, contributing significantly to nutrient cycling through facilitation of decomposition and return of plant litter to the soil. In many parts of the world, terrestrial gastropod communities are being greatly perturbed or lost through human-induced habitat degradation and loss.

This book presents a synthesis of current knowledge and research on the biology of terrestrial gastropod molluscs. In a series of peer-reviewed chapters, it provides authoritative coverage of the topics of morphology, phylogeny and systematics, structure and function of the various organ systems, feeding behaviour, life history strategies, behavioural ecology, population and conservation genetics, and soil biology and ecotoxicology. This book is for both students and professionals concerned with terrestrial Mollusca.

Gary M. Barker
Landcare Research
Hamilton
New Zealand

Acronyms

ABARE	Australian Bureau of Agricultural and Resource Economics
ACIAR	Australian Council for International Agricultural Research
ACSAD	Arab Centre for Studies of Arid Zones and Dry Lands
ANZECC	Australian and New Zealand Environment and Conservation Council
AOAD	Arab Organization for Agricultural Development
ATO	African Timber Organization
BD	Bilateral donors
CAI	Current annual increment
CATIE	Tropical Agricultural Research and Higher Education Centre
CBD	Convention on Biological Diversity
CCAB-AP	Central American Council on Forests and Protected Areas
CCAD	Central American Commission on Environment and Development
CCC	Convention on Climate Change
CCD	Convention to Combat Desertification
CCFM	Canadian Council of Forest Ministers
CFTT	Centre for Forest Tree Technology
C&I	Criteria and indicators
CIFOR	Center for International Forestry Research
CPOM	Coarse particulate organic matter
CRPF	Centre Régional de la Propriété Forestière
CSA	Canadian Standards Association
CSCE	Conference on Security and Cooperation
CSD	Commission on Sustainable Development
CWD	Coarse woody debris
DEST	Department of Environment, Sport and Tourism (Aust.)
DFID	Department for International Development (UK)

DNRE	Department of Natural Resources and Environment (Vic.)
DPCSD	Department for Policy Coordination and Sustainable Development (UN)
ECE	Economic Commission for Europe
EFI	European Forestry Institute
EMS	Environmental management system
ENGO	Environmental non-government organization
ESFM	Ecologically sustainable forest management
EU	European Union
Fa	Facilitator
FAO	Food and Agriculture Organization
FLORES	Forest Land Oriented Resource Envisioning System
FMU	Forest management unit
FRA 2000	Forest Resources Assessment Programme
FSC	Forest Stewardship Council
GIS	Geographic Information System
GTZ	Deutsche Gesellschaft für Technische
HIID	Harvard Institute for International Development
IBFRA	International Boreal Forest Research Association
IBI	Index of Biotic Integrity
ICRAF	International Centre for Research in Agroforestry
IFF	Intergovernmental Forum on Forests
IFN	National Forest Inventory
IFPS	Integrated Forest Planning System
IIED	International Institute for Environment and Development
ILO	International Labor Organization
InBio	National Biodiversity Institute (Costa Rica)
IPCC	Intergovernmental Panel on Climate Change
IPF	Intergovernmental Panel on Forests
Ir	Infiltration rate
ISO	International Organization for Standardization
ITFF	Inter-Agency Task Force on Forests
ITTA	International Tropical Timber Agreement
ITTO	International Tropical Timber Organization
IUCN	International Union for the Conservation of Nature
IUFRO	International Union of Forestry Research Organizations
I&V	Indicators and verifiers
Ks	Saturated hydraulic conductivity
LR	Long run data
MAF	Ministry of Agriculture (New Zealand)
MAI	Mean annual increment
MAS	Multi-actor systems
MCDA	Multi-criteria decision aid
MCFFA	Ministerial Council on Forestry, Fisheries and Aquaculture (Aust.)

MCPEE	Ministerial Conference on the Protection of Forests in Europe
MDF	Medium-density fibre board
MRI	Multiple resource inventory
MTR	Mean Trophic Ranking
NEFD	National Exotic Forest Description
NFI	National Forest Inventory
NGO	Non-government organization
NZFOA	New Zealand Forest Owners Association
ODA	Overseas Development Agency (UK)
OECD	Organization for Economic Cooperation and Development
ONF	Office National des Forêts
PDP	Parallel distribution processing
PE	Partial equilibrium
PHABSIM	Physical Habitat Simulation
RBD	Regional Development Banks
RECOFT	Regional Community Forestry Training Centre
RHS	River Habitat Survey
RIVPACS	River Invertebrate Prediction and Classification System
SAF	Society of American Foresters
SCOPE	Scientific Committee on Problems on the Environment
SFM	Sustainable forest management
SFRI	Statewide Forest Resource Inventory
SG4	Structural Group of 4
SIAR	Security of inter-generational access to resources
SOM	Soil organic matter
SS	Soil strength
TBT	Technical Barriers to Trade
THS	Timber Harvesting Committee
TPA	Totally protected area
UN	United Nations
UN CSD	United Nations Commission on Sustainable Development
UN/ECE	United Nations Economic Commission for Europe
UNCED	United Nations Conference on Environment and Development
UNCTAD	United Nations Conference on Trade and Development
UNDP	United Nations Development Programme
UNEP	United Nations Environmental Programme
UNESCO	United Nations Educational, Scientific and Cultural Organization
UNIDO	United Nations Industrial Development Organization
UNSO	United Nations Special Office to Combat Desertification and Drought
WBG	World Bank Group
WCED	World Commission on Environment and Development
WCFSD	World Commission on Forests and Sustainable Development

WCMC   World Conservation Monitoring Center
WIPO   World Intellectual Property Organization
WRI   World Resources Institute
WTO   World Trade Organization
WWF   World Wide Fund for Nature

1 Gastropods on Land: Phylogeny, Diversity and Adaptive Morphology

G.M. BARKER

Landcare Research, Private Bag 3127, Hamilton, New Zealand

The Mollusca is a very old monophyletic invertebrate lineage, dating from before the Cambrian. Molluscs are in appearance, anatomy, ecology and physiology a highly diverse group, for which the phylogenetic pathways and higher classification have been controversial since the very beginning of comparative investigation (e.g. Morton, 1963; Salvini-Plawén, 1972, 1980, 1984, 1990; Lauterbach, 1984; Ivanov, 1996; Salvini-Plawén and Steiner, 1996, and references therein). Within the Mollusca, many problems in our understanding of systematic relationships arise among the ecologically disparate Gastropoda (Haszprunar, 1988b). Ponder and Lindberg (1997) summarize the history of evolutionary and classificatory work on gastropods over the past two centuries, leading to the renewed interest in gastropod phylogeny, at taxonomic levels varying from ordinal to species-group level, over the past 15–20 years. This upsurge in interest has been contingent upon the application of new techniques, such as electron microscopy for investigation of anatomical characters, molecular methods for characterizing the genome, and cladistic methodologies for formulating phylogenetic hypotheses, the latter aided by computer algorithms for more rigorous and replicable development and testing of phylogenetic reconstructions.

In this modern era of gastropod systematics, Haszprunar (1985c, 1987, 1988a,b) has been most influential in establishing a phylogenetic framework (Fig. 1.1) based on cladistic principles, albeit with some critism of the methodologies employed to construct the trees manually (Bieler, 1990; Haszprunar, 1990). Ponder and Lindberg (1996) presented a preliminary computer-generated parsimony analysis based on 25 superfamily/family taxa and 95 morphological characters. Subsequently, these workers (Ponder and Lindberg, 1997) published the results of an array of parsimony analyses for an expanded data set comprising 40 taxa and 117 morphological characters. The analysis by Ponder and Lindberg

©CAB *International* 2001. *The Biology of Terrestrial Molluscs*
(ed. G.M. Barker)

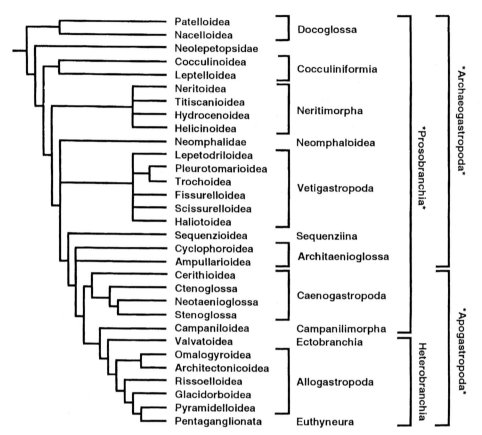

Fig. 1.1. The phylogeny and classification of gastropods according to Haszprunar (1987, 1988a,b). Neolepetopsidae identified in Haszprunar's cladograms as 'Hot-vent-C'. *Taxon* indicates an orthophyletic group.

(1997) provides the most robust phylogenetic reconstruction for the major groups in the Gastropoda published to date (Fig. 1.2). Their analysis indicates a primary division of the monophyletic Gastropoda into two primary groups, the Eogastropoda – for Patellogastropods and ancestors – and the Orthogastropoda – the remainder. Orthogastropoda were shown to comprise several well-defined clades. The Vetigastropoda encompass most of the gastropods previously included in the paraphyletic Archaeogastropoda (Fissurelloidea, Trochoidea, Scissurelloidea and Pleurotomarioidea) (see Haszprunar, 1988a,b, 1993; Hickman, 1988) as well as lepetodriloidean and lepetelloidean limpets and sequenziids. The relationships of the Peltospiridae and Neomphalidae were unresolved. The Cocculiniformia were shown to possibly be paraphyletic, with Lepetelloidea belonging to the vetigastropod clade, and Cocculinoidea to the Neritopsina. The placement of the cocculinoid–neritopsine clade was not well unresolved, but was treated by Ponder and Lindberg (1997) as a sister clade to the rest of the orthogastropods in their consensus tree.

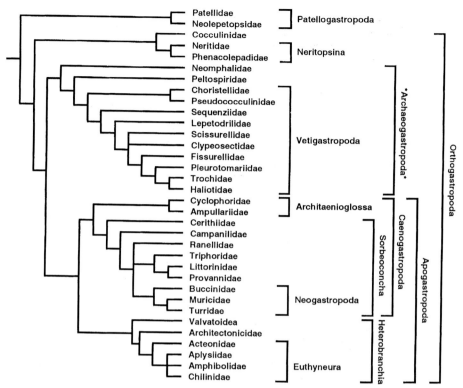

Fig. 1.2. The phylogeny and classification of gastropods according to the parsimony analyses of Ponder and Lindberg (1997), based on 40 taxa and 117 morphological characters. *Taxon* indicates an orthophyletic group.

Caenogastropods and heterobranchs were each shown to be monophyletic, and collectively constitute a clade (Apogastropoda) of sister status to the orthogastropods traditionally grouped as 'Archaeogastropoda'. In Ponder and Lindberg's (1997) analysis, the Architaenioglossa is placed as a basal divergence within the caenogastropod clade. The Euthyneura represent the climax of the heterobranchian radiation.

Terrestrial species occur in the Neritopsina, Architaenioglossa, Sorbeoconcha and Euthyneura. Thus, the phylogeny-based classifications of Haszprunar and of Ponder and Lindberg confirm what has been known for centuries – that terrestriality has been achieved independently in several gastropod lineages. Reflecting their focus on higher relationships within the Gastropoda, however, the Euthyneura were poorly represented among the taxa included in the phylogenetic reconstructions of Haszprunar (1987, 1988a,b) and Ponder and Lindberg (1996, 1997), and thus do not provide for hypotheses on the evolutionary history of the foremost radiation on land, namely that within the Pulmonata. While the monophyly of the Pulmonata has generally been accepted, the phylogenetic relationships of the various adaptive radiations recognizable within these euthyneurans have not been fully resolved. This has, for the most part, been due to differing interpretations of pathways of evolutionary change in conchological and anatomical characters (e.g. Pelseneer, 1901; Thiele, 1929–35; Hubendick, 1945; Pilsbry, 1948; Morton, 1955a,b,c; Baker, 1955, 1956; van Mol, 1967; Delhaye and Bouillon, 1972a,b,c; Minichev and Starobogatov, 1979; Tillier, 1984a,b,c, 1989, Haszprunar, 1985; Golikov and Starobogatov, 1975, 1988; Haszprunar and Huber, 1990; Nordsieck, 1985, 1992; Salvini-Plawén and Steiner, 1996) and the failure to employ autapomorphies as the basis for taxon definition at all levels of classification. As a consequence, taxonomy and nomenclature have been unstable (Fig. 1.3).

Molecular data potentially provide more characters for phylogenetic analysis. However, these methods have to date (e.g. Tillier *et al.*, 1992, 1994, 1996; Rosenberg *et al.*, 1994, 1997; Harasewych *et al.*, 1997; Winnepenninckx *et al.*, 1998; Colgan *et al.*, 2000) generally provided poor resolution of relationships among major groups and only modest congruency with trees derived from analyses of morphological data (Ponder and Lindberg, 1997). Using partial 28S ribosomal (rDNA) and histone H3 sequences, Colgan *et al.* (2000) found support for the monophyly of the Patellogastropoda, the 'higher' vetigastropods and the Euthyneura, and for the polyphyly of the 'Cocculiniformia'. However, there was little support for Ponder and Lindberg's (1997) division of the gastropods into two major clades (Eogastropoda and Orthogastropoda) or for Caenogastropoda and Heterobranchia as monophyletic taxa. The data of Colgan *et al.* (2000) do not clarify the relationships of the Neritopsina or Achitaenioglossa.

Tillier *et al.* (1996) suggested that the expectation that DNA sequences will resolve the higher relationships within the Pulmonata, and in particular the Stylommatophora, may not be realized because rapid radiation may have involved only a low number of substitutions (i.e. a low number of potential synapomorphies), while the comparatively long time after the initial radiation may have led to accumulation of many convergent substitutions. Nevertheless, there remains strong interest in resolving relationships and developing more evolutionary-based classification schemes for groups within the Pulmonata based on morphological and molecular data (e.g. Emberton, 1988, 1991a,b, 1994; Emberton *et al.*, 1990; Pearce, 1990; Roth, 1996; Scott, 1996; Thomaz *et al.*, 1996; Hausdorf, 1998; Cuezzo, 1998; Douris *et al.*, 1998; Wade *et al.*, 1998, 2001; Muratov, 1999). Colgan *et al.* (2000) have estimated slow evolutionary

Fig. 1.3. (Opposite) Examples of the varied phylogenies and classification schemes proposed for Pulmonata. (A) The manually derived cladogram of van Mol (1967), based on assumed polarity in characters of the cerebral ganglia. (B) Relationships inferred from the classification scheme of Tillier (1984b), based on assumed polarity and apomorphic states in characters of various organ systems. (C) The manually derived cladogram of Nordsieck (1985), based on assumed polarity and apomorphic states in characters of various organ systems. (D) Relationships inferred from the classification scheme of Stanisic (1998), based on an interpretative consensus of various phylogenetic hypotheses. In the phylogeny proposed by Nordsieck (1985) and the classification scheme proposed by Stanisic (1998), the Succineidae and Athoracophoridae are treated as part of the stylommatophoran radiation.

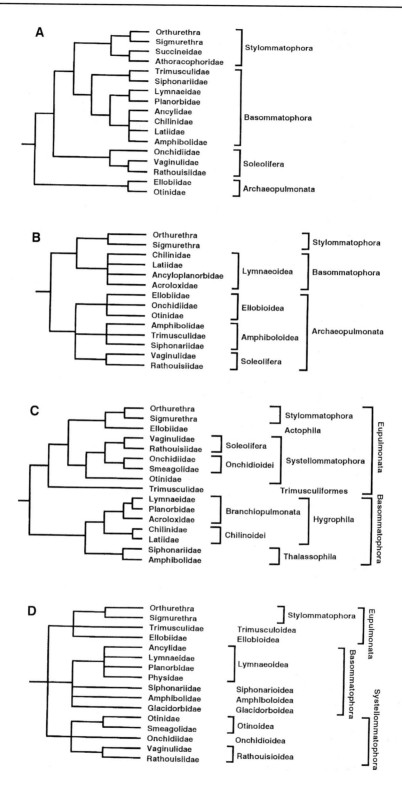

rates in 28S rDNA of Euthyneura, and evolutionary stasis subsequent to the initial divergence, relative to that in Patellogastropoda, Veitgastropoda and basal Heterobranchia.

In this chapter, a new phylogenetic reconstruction for the Gastropoda is presented on the basis of morphological characters. The phylogeny of the terrestrial Pulmonata is analysed in more detail based on morphological and molecular data. These reconstructions are then used as the historical hypotheses for a synoptic discussion of gastropod radiation into terrestrial environments.

New Phylogenetic Hypotheses of Adaptative Radiation in Gastropoda

Figure 1.4 presents the most parsimonious reconstruction of the phylogeny of Gastropoda based on 72 morphological characters (Appendices 1.1 and 1.2). While many of the characters are common to both analyses, the topology of the tree presented in Fig. 1.4 has only moderate congruency with the consensus tree of Ponder and Lindberg (1997) (cf. Fig. 1.2). This reflects differences in scoring of character states, in *a priori* assessments of their evolutionary polarity, and in the choice of taxa for inclusion in the analyses. Consistent with Ponder and Lindberg (1997), the Heterobranchia are hypothesized to have arisen prior to the Caenogastropoda. In the present analysis, the terrestrial radiation in Heterobranchia is monophylogenetic, with Ellobiidae basal to a onchidioid–rathouisioid and stylommatophoran–succineoid divergence. Among the lower

Fig. 1.4. (Opposite) Phylogram for Gastropoda as indicated by the single most-parsimonious reconstruction (tree length 1380; consistency index 0.69; homoplasy index 0.90; retention index 0.81; rescaled consistency index 0.55) generated by PAUP* 4.0 (Swofford, 1998) from 72 morphological characters (Appendices 1.1 and 1.2). The branch lengths are proportional to the numbers of character changes (scale bar = 10 character state changes). Taxon nomenclature generally follows that of Fretter *et al.* (1998), Rudman and Willan (1998) and Smith and Stanisic (1998), except that for this analysis the Sequenzioidea (with single family Sequenziidae) was treated independently of the Vetigastropoda (comprising superfamilies Pleurotomarioidea, Fissurelloidea and Trochoidea), the family name Vaginulidae is used in preference to Veronicellidae, the Planorbidae is taken to include species often assigned to Ancylidae, and Succineidae and Athoracophoridae were treated independently of the Stylommatophora (the latter comprising numerous superfamilies and families). The analysis presupposed that the taxa included are monophyletic, and no attempt was made to substantiate *a priori* their monophyly by investigating their potential autapomorphies. Heuristic searching was employed, with multiple states of characters within taxa interpreted as polymorphism, preference given to reversals over parallelisms using the DELTRAN option, branch swapping by tree bisection–reconnection (TBR), and a random addition sequence with ten replications. While analysis without *a priori* assumptions is desirable, preliminary analyses with unordered character states indicated a number of implausible transformations in some basic molluscan characters. Solutions to this enigma included treating these characters as irreversible or Dollo, or by constructing step matrices or character state trees that weighted heavily against reversals. Only trees compatible with higher taxa relationships inferred from the osphadial ultrastructure data (cilia bottles and Si1, Si2, Si4 cells) of Hazsprunar (1985a,b) were retained. The node numbers are relevant to the information on apomorphic changes in Appendix 1.3. Bootstrap analysis (Felsenstein, 1985), as implemented in PAUP* 4.0, yielded >65% support for nodes 1–11, 19–29 and 33–47.

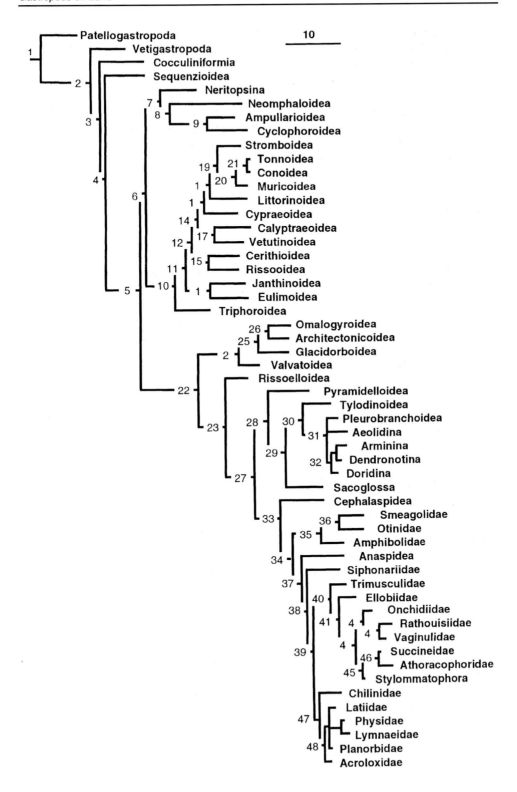

gastropods, the Neritopsina are grouped with the Neomphaloidea and achitaenoglossan Cyclophoroidea and Ampullarioidea, in a clade sister to the Caenogastropoda. Ponder and Lindberg (1997) found support for an evolutionary association of Neritopsina and Neomphaloidea in several of their reconstructions, but Cyclophoroidea and Ampullarioidea were indicated to have a sister relationship to the Caenogastropoda.

Parsimony analysis of the terrestrial radiation in Pulmonata, using the morphological data presented in Appendices 1.4 and 1.5, confirms the monophyly of the Succineoidea and its sister relationship to the Stylommatophora. However, the relationships within the Stylommatophora are generally poorly resolved. While some clades that traditionally have been assumed to be monophyletic are recovered (e.g. Arionidae–Philomycidae; Limacidae–Agriolimacidae), others are not. In particular, the Orthurethra, Limacoidea, Helicioidea and Achatinoidea as monophyletic clades are not well supported. Further, this reconstruction is not well concordant with that indicated by analysis of 28S rDNA sequences (Wade *et al.*, 1998, 2001; C.M. Wade, personal communication). Figure 1.6 presents a maximum parsimony re-analysis of the morphological data in Appendices 4 and 5 with the constraint that retained trees are congruent with the clades well supported in the 28S rDNA analysis of Wade *et al.* (1998, 2001; C.M. Wade, personal communication): Succineoidea, Orthurethra, Limacoidea, Helicioidea and Achatinoidea were taken as monophyletic clades, but no assumptions were made about their inter-relationships. This analysis provides for mapping of morphological characters on to the 28S rDNA phylogeny, and thence for discussion of the adaptive radiation in these gastropods.

Figure 1.7 is a diagramatic summation of the phylogenetic hypotheses developed in Figs 1.4 and 1.6, with illustration of the body plan at four grades of gastropod evolution, namely Vetigastropoda, sorbeoconch Caenogastropoda, ellobioidean Pulmonata, and stylommatophoran Pulmonata. Tables 1.1 and 1.2 briefly review habitat occupancy and feeding modes as the two principal adaptative zones in Gastropoda.

Fig. 1.5.   (Opposite) Phylogram for the terrestrial Pulmonata as indicated by the single most parsimonious reconstruction (tree length 1464; consistency index 0.48; homoplasy index 0.95; retention index 0.78; rescaled consistency index 0.36) generated by PAUP* 4.0 (Swofford, 1998) from 57 morphological characters (Appendices 1.4 and 1.5) without any *a priori* topological constraints imposed. The branch lengths are proportional to the numbers of character changes. Taxon nomenclature generally follows that of Tillier (1989). The analysis pre-supposed that the taxa included are monophyletic, and no attempt was made to substantiate their monophyly by investigating their potential autapomorphies, although the various subfamilies in Arionidae and the two subfamilies in Athoracophoridae were included separately in the analyses to test the ability to recover these clades: the analysis indicated the respective monophyly of the Arionidae and Athoracophoridae. Heuristic searching was employed, with multiple states of characters within taxa interpreted as polymorphism, preference given to reversals over parallelisms using the DELTRAN option, branch swapping by tree bisection–reconnection (TBR), and a random addition sequence with ten replications. Ellobiidae, Onchidiidae and Vaginulidae were treated as outgroups. Preliminary analyses with unordered character states indicated a number of implausible transformations, so various characters were treated as irreversible or Dollo, or by constructing step matrices or character state trees that weighted heavily against reversals.

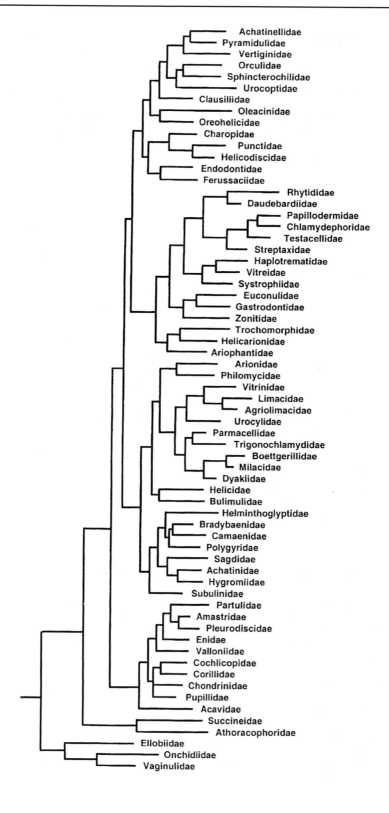

Basic Gastropod Organization

Primitively Gastropoda are marine animals, but several groups in this molluscan class have made the adaptive shift from aquatic to terrestrial existence. To appreciate fully the adaptive radiation into the terrestrial environment, it is necessary first to examine the body plan of the primitive gastropod from which these terrestrial taxa evolved.

In the primitive gastropod, the body is divisible into a head–foot and a visceral hump, interconnected by a neck or waist. The visceral hump, located within a calcareous shell, houses the viscera. It seems that the earliest gastropods had a limpet-like shell and that the asymmetrical shell coiling, typifying the great majority of gastropods, evolved several times in this molluscan class (Haszprunar, 1988b). Some groups have developed secondarily non-coiled, limpet-like shells or have reduced the shell completely. Terrestrial gastropods evolved from clades with coiled shells. This shell coiling is associated with an asymmetrical body plan, with reduction in the size of the organs on the right, inner side of the spiral. Gastropod shells are composed of several different layers – a thin outer organic layer or periostracum consisting of conchiolin overlying much thicker crystalline calcium carbonate layers. The shell is formed by accretionary growth at the mantle edge. The mantle is a thin membrane that extends minutely beyond the shell aperture, and at its edge adds a shell increment to the aperture margin so that each increment copies a configuration of the mantle edge at that time. The shell tube coils in a logarithmic spiral, retaining isometric proportions between length, area and volume parameters as it grows. The inner walls that form the axis of shell coiling is referred to as the columella. The whole of ontogeny is conserved in the shell, as the larval shell (termed the protoconch) is generally retained as the apex to (sometimes concealed within) the post-embryonic shell, termed the teleoconch.

The head–foot is concerned with sensory and locomotor activities, and is protruded from the protective shell during movement and feeding. Head–foot protrusion is effected mainly by blood pressure but is withdrawn by contraction of muscles, a larger right and smaller left, originating on the columella of the shell. A pocket-shaped space, the

Fig. 1.6.   (Opposite) Phylogram for the terrestrial Pulmonata as indicated by the single most parsimonious reconstruction (tree length 1719; consistency index 0.48; homoplasy index 0.96; retention index 0.75; rescaled consistency index 0.37) generated by PAUP* 4.0 (Swofford, 1998) from the same 57 morphological characters as employed in the analysis presented in Fig. 1.5 (Appendices 1.4 and 1.5) but with *a priori* constraint that trees must be compatible with a 28S rDNA sequence phylogeny (Wade *et al.*, 1998, 2001; C.M. Wade, personal communication) in which there is strong support for Orthurethra, Limacoidea, Endodontoidea, Helicoidea, Succineoidea and Achatinoidea as monophyletic lineages. The branch lengths are proportional to the numbers of character changes. The node numbers are relevant to the information on apomorphic changes in Appendix 1.6. Bootstrap analysis (Felsenstein, 1985) as implemented in PAUP* 4.0 yielded >70% support for 1ll labelled nodes. Taxon nomenclature generally follows that of Tillier (1989). The *a priori* treatment of character transformations and conditions for Heuristic searching were as given in Fig. 1.5.

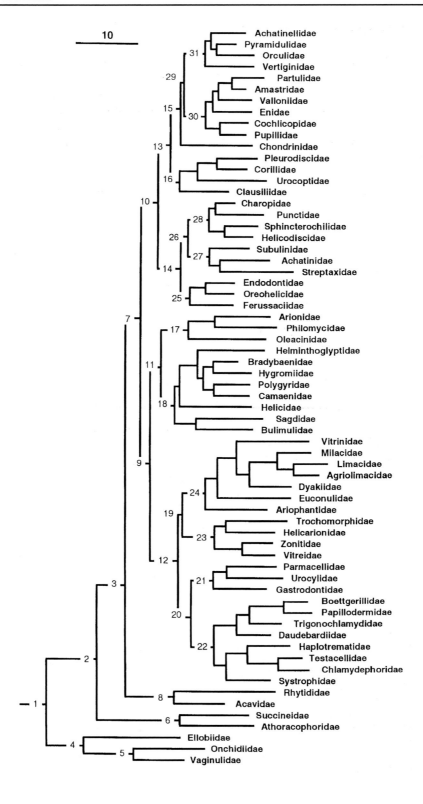

10

31 — Achatinellidae
— Pyramidulidae
— Orculidae
— Vertiginidae

30 — Partulidae
— Amastridae
— Valloniidae
— Enidae
— Cochlicopidae
— Pupillidae
— Chondrinidae

16 — Pleurodiscidae
— Corillidae
— Urocoptidae
— Clausiliidae

28 — Charopidae
— Punctidae
— Sphincterochilidae
— Helicodiscidae

27 — Subulinidae
— Achatinidae
— Streptaxidae

25 — Endodontidae
— Oreohelicidae
— Ferussaciidae

17 — Arionidae
— Philomycidae
— Oleacinidae

18 — Helminthoglyptidae
— Bradybaenidae
— Hygromiidae
— Polygyridae
— Camaenidae
— Helicidae
— Sagdidae
— Bulimulidae

24 — Vitrinidae
— Milacidae
— Limacidae
— Agriolimacidae
— Dyakiidae
— Euconulidae
— Ariophantidae

23 — Trochomorphidae
— Helicarionidae
— Zonitidae
— Vitreidae

21 — Parmacellidae
— Urocylidae
— Gastrodontidae

22 — Boettgerillidae
— Papillodermidae
— Trigonochlamydidae
— Daudebardiidae
— Haplotrematidae
— Testacellidae
— Chlamydephoridae
— Systrophidae

8 — Rhytididae
— Acavidae

6 — Succineidae
— Athoracophoridae

4 — Ellobiidae
5 — Onchidiidae
— Vaginulidae

pallial cavity, occurs above and behind the head, but within the shell. Its floor is formed by the dorsal surface of the head–foot, its roof by the mantle skirt – a thin fold from the anterior face of the visceral hump joined laterally to the head–foot – leaving the cavity open anteriorly. The mantle cavity is filled with water when the animal is active but accommodates the retracted head–foot. Primitively, the head bears dorsally a pair of sensory peducles, the cephalic tentacles, with an eye of an open vesicle or closed cup type located at their base. The ventral surface of the foot, termed the sole, is the locomotary pad and comprises a densely ciliated epithelium overlying an elaborate pedal musculature. At its anterior end and for a short distance down the sides, a deep groove separates a thin fold, the mentum; near the median line, at the bottom of this groove, is the external opening of the pedal mucous gland (suprapedal gland) which functions by laying down a mucus pad over which the animal crawls. Anteriorly, separated from the sole and mentum by a deep groove, is the snout or rostrum, with the mouth near the centre. Locomotion generally is achieved through a series of waves passing over the mesopodial sole, brought about by local contraction and relaxation of the pedal musculature. These waves may be direct, starting at the posterior, or retrograde, starting at the anterior, and may be monotaxic in occupying the whole width of the sole, or ditaxic in comprising two series out of phase with one another and each occupying half the width of the sole. Miller (1974) suggested that retrograde waves are the more primitive of the two wave types, because of their occurrence in chitons and lower gastropods. In that they confer greater agility, ditaxic waves are assumed to be more advanced than monotaxic ones. These basic types of locomotion, based on wave propagation in the pedal musculature, have been subject to various modifications in numerous gastropod lineages (summarized in Fretter *et al.*, 1998). Further, a number of gastropods have adopted ciliary gliding as the means of locomotion. The dorsal surface of the body is covered by a simple, non-ciliate, columnar epithelium, while the sole is covered by a ciliated epithelium. Both usually contain numerous mucus cells. A peripheral zone around the foot is elaborated primitively as an epipodium, often with tentacles. Because of its pedal innervation, the epipodium represents a sensory specialization of the gastropod foot.

Fig. 1.7. (Opposite) Diagrams to illustrate the general organization of the adult animal at different grades of structural evolution in the Gastropoda, with emphasis on pallial, nervous, digestive and reproductive systems. (A–C) Generalized vetigastropod (B and C illustrating variation in coiling of the intestine). (D–E) Generalized sorbeoconch caenogastropod. (F–H) Generalized ellobiidoidean pulmonate. (I–K) Generalized stylommatophoran pulmonate. The broken line denotes the limit of the pallial cavity. ai, anterior loop of intestine; an, anus; au, auricle; b, buccal mass; bc, bursa copulatrix; cg, cerebral ganglion; ct, ctenidium; dg, digestive gland; e, eye; et, epipodial tentacle; f, foot; g, gonad; it, inferior tentacle; n, nephridium; oec, oesophageal crop; oep, oesophageal pouches; oes, oesophagus; op, operculum; os, osphradium; p, penis; pal, pallial gonoduct; pc, pedal cord; pg, pedal ganglion; per, pericardium; pi, posterior loop of intestine; plc, pallial cavity; plg, pleural ganglion; pn, pneumostome; prm, penial retractor muscle; pv, pulmonary veins; r, rectum; rg, renal gonoduct; s, shell; sb, suboesophageal ganglion; sg, seminal groove; slg, salivary gland; sp, supraoesophageal ganglion; st, stomach with spiral caecum; t, cephalic tentacle; u, ureter; v, ventricle; vd, vas deferens.

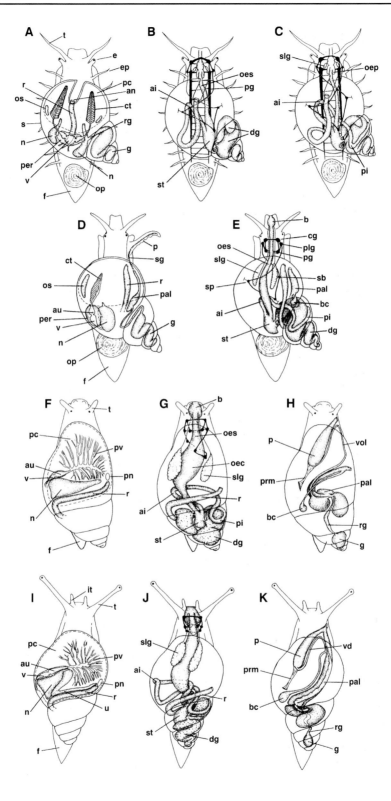

Table 1.1. Habitats occupied by various gastropod clades

	Marine benthic		Marine plankton	Estuarine	Freshwater	Terrestrial		
	Deep sea	Shallow sea				Amphibious	Damp	Arid
Patellogastropoda	*	*		(*)				
Vetigastropoda	*	*	*					
Neritopsina	*	*		*	*		*	(*)
Helicinoidea							**	*
Hydrocenoidea							*	
Cyclophoroidea							**	
Ampullarioidea					**	*		
Sorbeoconcha	*	*	*	*	*	*	*	(*)
Littorinoidea		**		*	*	*	(*)	(*)
Rissooidea	*	**		*	*	*	*	
Neogastropoda	*	*		*	(*)			
Heterobranchia	*	*	*	*	*	*	*	*
Allogastropoda	*	*		*	*			
Opisthobranchia	*	*	*	*	(*)			
Ellobioidea		*		**			*	
Onchidioidea		*		**	*		*	
Rathouisioidea							*	
Succineoidea						*	**	*
Stylommatophora							**	*

**Predominate, *present, (*)rare. Modified from Ponder and Lindberg (1997).

Table 1.2. Feeding modes in various gastropod clades

	Detritivory	Herbivory	Grazing carnivory	Active predation	Parasitic	Suspension
Patellogastropoda	*	**				
Vetigastropoda	**	**		*		*
Neritopsina	**	**				
Helicinoidea	*					
Hydrocenoidea	*					
Cyclophoroidea	*	*				
Ampullarioidea	*	*				
Sorbeoconcha	*	*	*	*	*	*
Littorinoidea	*	**				
Rissooidea	*	**				
Neogastropoda	(*)	(*)	*	**	(*)	
Heterobranchia	*	*	*	*	*	(*)
Allogastropoda	*	*	*		*	
Opisthobranchia		*	*	*		(*)
Ellobioidea	*	*				
Onchidioidea	*	*		(*)		
Rathouisioidea				*		
Succineoidea	**	*				
Stylommatophora	**	**	*	*		

**Predominate, *present, (*)rare. Modified from Ponder and Lindberg (1997).

Gastropod opercula are hard, disc-shaped structures secreted by and carried on the dorsal surface of the foot. They are found during ontogeny in all gastropods with free larval development. The primary function of the operculum is to close the shell aperture, thus providing protection for the head–foot when the larva is retracted inside the shell (Bandel, 1982). This function also occurs in the adult gastropod when the operculum is retained throughout ontogeny. Because the opercula are always attached to the dorsal surface of the foot, are of conchiolin, and are often calcified, Adanson (1757) suggested a strict homology between the operculum and the shell. Accordingly, gastropods and lamellibranchs were both true bivalves and the gastropod columellar muscle could also be said to be homologous with bivalve adductors. While Houssay (1884) and Kessel (1942) showed that the operculum is secreted by an epithelium differentiated on the dorsal side of the foot, homology with the shell was still supported by Fleischmann (1932), Pruvot-Fol (1954) and others. More recently, Checa and Jiménez-Jiménez (1998) presented evidence for derivation of the epithelium that secretes the operculum from the shell-secreting mantle. These authors suggested that the parietal segment of the periostracal groove migrated towards the epipodium and became independent from the rest of the mantle. The concomitant development of an opercular disc allowed the successive turns of the periostracal strip to seal together. The flexiclaudent spiral operculum, evident today in neomphalid and vetigastropod Archaeogastropoda and cerithioidean Caenogastropoda, is thought to be plesiomorphic in Gastropoda. The flexiclaudent operculum is secreted when the animals are partly or wholly extended, and achieves a seal of the aperture in retracted animals by virtue of it flexing at the margins. During the course of gastropod evolution, the rigiclaudent operculum emerged several times from the ancestral flexiclaudent type. The rigiclaudent operculum is secreted when the animal is in the retracted position and the operculum takes on the shape of the aperture.

A further unique aspect of the gastropod is an event called torsion, which occurs early in larval life and involves a shift in body orientation. At first, the larval gastropod has the pallial cavity at the posterior end of the body, but two 90° rotations of the visceral hump occur over a short period of time, which brings the pallial cavity to an anterior position just above and behind the head. In basal Vetigastropoda, a respiratory organ (ctenidium or gill), with skeletal rods and a chemosensory organ (osphradium) running axially along its ventral membrane, lies along each side of the pallial cavity. Towards the posterior of the pallial cavity are paired hypobranchial glands. The anus opens medially, flanked on each side by a nephridial opening. To function as a respiratory space, the pallial cavity is ventilated by cilia on the ctenidial lamellae beating so that water is drawn into the pallial cavity anterolaterally and ventrally and passed over the osphradia and gills. Right and left streams converge medianly over the anal and nephridial openings and leave the pallial cavity anterodorsally. The heart, surrounded by the pericardium, is usually placed asymmetrically and transversely on the left, where visceral hump and pallial cavity meet. It consists of a ventricle flanked on each side by an auricle.

Although coelomatic, the body cavity *sensu stricto* of these gastropods is reduced to the pericardium. As a consequence, the body cavity *sensu lato* is comprised of three major venous sinuses: (i) the cephalopedal sinus of the head–foot region; (ii) the visceral sinus of the visceral hump; (iii) and the subrenal sinus located near the columellar muscle at the base of the visceral mass. There is considerable variation with respect to the circulatory system among primitive gastropods, with nearly every archaeogastropod superfamily having its own type (Fretter and Graham, 1962; Andrews, 1981, 1985; Haszprunar, 1988b; Ponder and Lindberg, 1997). In the primary coiled Vetigastropoda, blood is gathered into main veins leading to the right nephridium, where it is filtered before passing to the mantle skirt and ctenidia. Each auricle receives oxygenated blood from a ctenidium and passes it to the ventricle, from which two aortae distribute it to the sinuses (anterior and posterior, respectively), where aortic branches ultimately discharge to haemocoelic spaces inter-penetrating the various organs. Respiratory exchange occurs mainly across the surfaces of the ctenidial leaflets, although in some intertidal species there is increased vascularization of the mantle skirt permitting respiratory exchange there also (Deshpande, 1957). A blood filtrate enters the pericardial cavity from the auricles; the pericardial gland within the auricle walls, with specialized cells called podocytes, serves as the filtration site. In most vetigastropods, numerous outpouchings of the auricular wall form filtration chambers, but in Fissurellidae the podocytes are confined to areas of each auricle close to the openings of the veins. At systole, when the ventricle is refilled, blood is retained in the pericardial gland chambers and the resulting pressure gradient causes fluid and molecules up to a given size to pass to the pericardium. This primary urine escapes to the nephridia via renopericardial ducts, where it is modified to the final urine to be shed to the pallial cavity.

The excretory organs of basal archaeogastropods are paired, coelomatic ducts, situated wholly within the visceral hump and opening to the innermost end of the pallial cavity. In the plesiomorphic state, it is assumed that the nephridia were simple sacs and were in communication with other coelomic spaces, each being joined to the pericardium by a ciliated renopericardial duct which, on the right side, received the duct from the single gonad (Fretter and Graham, 1962). This arrangement is retained in diotocardian members of the extant vetigastropods, although the paired nephridia have long been known to differ in structure and function (e.g. Perrier, 1889; Cuénot, 1899; Delhaye, 1976; Andrews, 1988). The post-torsional right nephridium is concerned with nitrogen excre-tion, the left with the ionic composition of the blood. The separation of functions probably arose because the right nephridium receives largely deoxygenated blood, rich in waste products from the viscera, whereas the left receives blood low in waste and replete in oxygen on passage through the ctenidium (Andrews, 1985). In vetigastropods, the left nephridium has assumed a pallial location, with a tall, papillate, usually ciliated epithelium surrounding a core of spongy connective tissue penetrated by blood spaces. In Trochoidea the wall of this papillary renal sac, towards the pericardium, has become specialized as a nephridial gland (Andrews, 1985, 1988). This gland comprises a pad of connective tissue penetrated by tubular extensions of the nephridial lumen, blood lacunae and muscle fibres. It offers a large surface in contact with blood on one side and urine on the other, and, therefore, has a function in osmoregulation. The blood in the nephridial gland collects in an efferent vessel, separate from that which collects blood from the rest of the nephridium, to pass directly to the auricle.

The mouth, placed at the end of a pretentacular snout, leads into a buccal cavity to which paired salivary glands discharge. On the roof of the buccal cavity anteriorly are two cuticular jaws, and on its floor lies the odontophore, a protrusible tongue-like organ coated with a radular membrane on which are mounted teeth. The radula is unique to molluscs. In their simplest form, each tooth consists of a basal plate, by which it is attached to the radular membrane, and one or more elevated sharp edges or cusps. The radular membrane is symmetrical, with the teeth arranged in transverse rows. In Vetigastropoda, the radular membrane is flexible (flexoglossate), but a more primitive radula type, with rigid radular membrane (stereoglossate), is to be found in Patellogastropoda. During feeding, complex muscles control movement of the radula and associated support cartilages. In Vetigastropoda and all higher gastropods, the radular membrane is thus rotated partly over the tip of the underlying support when produced through the mouth. When the odontophore is protruded, the teeth on the radular membrane are elevated and can scrape, pierce, cut or tear the object with which they come into contact and produce small bits for swallowing. During feeding, the teeth become worn, with cusps gouged, chipped or ground down. This wear problem is solved by new rows of teeth forming at the posterior end of the radula throughout the life of the animal, while worn teeth at the anterior end are discarded continuously to be swallowed and passed out in the faeces. The membrane gradually grows forward, moving new teeth into the feeding position and worn teeth to the area where the membrane can be resorbed by the animal. The radular membrane is extended beyond the posterior limit of the buccal mass as a radular sac, towards the terminus of which are the odontoblastic cells responsible for formation of the teeth and membranoblasts that furnish the radular membrane; the length of this sac is proportional to the wear and hence to the replacement rate of the radular teeth. In most archaeogastropods there are two pairs of buccal cartilages, which is probably the basic number but which has been enlarged or reduced secondarily in a number of lineages.

The oesophagus arises from the dorsal aspect of the buccal mass and reaches the stomach through the neck linking head–foot and visceral mass, and so is affected by torsion. The oesophagus is compressed dorsoventrally, and is characterized by two strong, longitudinal folds that, together with a dorsomedian ciliary tract, form the dorsal food channel, a pair of lateral, glandular pouches (oesophageal glands), and a ventro-median ciliated tract. Because of torsion, these oesophageal zones rotate 180° counter-clockwise along its midsection. The stomach, complex internally, lies at the base of the visceral mass, on the left. The oesophagus opens to the stomach halfway down the right, columellar side, alongside ducts from a voluminous digestive gland that occupies much of the visceral mass. The middle part of the stomach is equipped with a series of parallel ridges that constitute a sorting area, and adjacent a cuticularized gastric shield. A caecum extends back from the sorting area. From the posterior-most opening of the digestive gland arises a ridge, the major typhlosole, to run into the intestine. A second ridge (minor typhlosole) arises more anteriorly to also run to the intestine. Between them, these typhlosoles divide the outlet from the stomach into two channels. One is the style sac, which leads from the main cavity and receives the stomach string, i.e. indigestible remains of food. The other is the intestinal groove into which ridges from the sorting area extend: it forms a pathway for the liver string, i.e. waste leaving the digestive gland. From the style sac, the intestine follows a rather lengthy course that always includes an anteriorly directed loop. The intestine enters the pericardial cavity, passing through the ventricle, before joining the rectum to open as the anus in the left side of the mantle cavity.

These animals are predominantly grazers and detritivores. Food is passed to the stomach along the ciliated oesophageal channel, where it is mixed with mucus and enzymes from the lateral villi. In the stomach, it meets further enzymes from the digestive gland, passes into, along and out of the caecum, and is then squeezed by the gastric shield. Digested food and some particulate matter pass into the digestive gland for uptake. The residue enters the style sac, where it is rotated, compacted and bounded with mucus into the beginnings of a faecal rod, the protostyle. These gastropods generally produce faecal strings.

In the primitive gastropod, the nerve centres are weakly concentrated in the head–foot, forming a loose ring around the oesophagus. There are three pairs of ganglia. The cerebral ganglia, linked by a cerebral commissure, are placed far forward dorsally over the buccal mass. Each cerebral ganglion sends two connectives ventrally, one each to the pleural and pedal ganglia located ventrally at the level of the anterior border of the foot. The left and right pleural ganglia are linked to the respective left and right pedal ganglia by a short connective. The pedal ganglia comprise long cords that extend posteriorly and are linked by many commissures. This ganglionic arrangement, with the pleural ganglia sited closer to the pedal ganglia than to the cerebral ganglia, is termed hypoathroid. Primi-tively, the visceral loop is represented by a neural cord, but in vetigastro-pods distinct ganglia have developed. Paired visceral ganglia lie at the base of the visceral hump, linked to the pleural ganglia by connectives that traverse the neck and so are affected by torsion. As a consequence the half-loop starting at the right pleural ganglion crosses over the oesophagus to the visceral ganglion on the left, and the half starting at the left pleural ganglion passes under the oesophagus to the visceral ganglion on the right, so giving the crossed condition of the visceral loop known as streptoneury. The dorsal pleural–visceral connective carries medially a supraoesophageal ganglion that innervates the left ctenidium and osphradium. The ventral connective carries a corresponding sub-oesophageal ganglion, innervating the same organs on the right. The osphradium is the primary sensory organ of the pallial cavity and, as in vetigastropods, is generally located on the pallial roof in association with the efferent branchial membranes.

The cerebral ganglia receive tactile and olfactory sensory input from cephalic tentacles on the head and from the lips around the mouth, and visual signals from the eyes; there is a single nerve to each cephalic tentacle. The cerebral ganglia also innervate the buccal muscles. The pedal ganglia control locomotor movements of the foot and receive sensory information from a series of epipodial tentacles. The pleural ganglia innervate the mantle edge, and the visceral ganglia the organs of the visceral mass.

The gonad is of mesodermal origin, arising during organogenesis by migration and multiplication of pericardial cells. In many basal gastropods, the sexes are separate and, therefore, individuals are dioecious with a single gonad located near the apex of the visceral hump. Some are hermaphroditic, however, with the single gonad producing both ova and spermatozoa. Debate continues as to whether gonochorism or hermaphroditism is the primary condition among the Gastropoda (e.g. Haszprunar, 1988b). The gonad communicates with the cavity or duct of the right kidney, whose aperture is thus a urinogenital pore. At the most primitive level, gametes are broadcast into the surrounding water, and fertilization is external. Therefore, generally, there is little or no elaboration of the genital duct or modification for copulation, and the jelly-like material that surrounds the ova arises in large part in the ovaries. In some Vetigastropoda, however, the distal portion of the right kidney duct is modified into a glandular region that produces a mucous layer in which eggs are deposited during spawning. Further, in a number of archaeogastropods, including some vetigastropods, a glandular gonoduct is developed in the pallial roof and fertilization is internal. There is a trend for the cephalic tentacles or anterior cephalic processes to be used as a copulatory organ. Primitively, the eggs are broadcast into the water column, but a number of vetigastropods produce egg strings or benthic egg masses, while others brood eggs in the pallial cavity. Trochophore and veliger stages are present in embryonic life, but the stage at hatching varies from early trochophore to crawling young after metamorphosis. The embryonic shell, that forms the protoconch atop of the adult shell, is produced at once by the shell gland prior to hatching.

Gastropods on Land

Worldwide, terrestrial gastropods have been estimated to number about 35,000 extant species (Solem, 1984; van Bruggen, 1995). The gastropod groups represented in terrestrial environments are Hydrocenoidea, Helicinoidea, Cyclophoroidea, Rissooidea, Littorinoidea, Ellobioidea, Onchidioidea, Rathouisioidea, Succineoidea and all superfamilies in the Stylommatophora. This diversity encompasses approximately 112 gastropod families, but the number of independent invasions of land from the aquatic environment is thought to exceed 10 considerably, as terrestriality has been achieved repeatedly in some families (e.g. Truncatellidae; Rosenberg, 1989, 1996a). The vast majority of terrestrial gastropods are stylommatophoran pulmonates, probably exceeding 30,000 species. In most regions of the world, it is these stylommatophoran pulmonates that dominate the terrestrial faunas. In some regions, however, prosobranchs constitute a significant part of the faunas, the most notable being Central America and the islands of the Caribbean (de la Torre and Bartsch, 1938, 1941; de la Torre *et al.*, 1942; Bartsch and Morrison, 1942; Bartsch, 1946; Solem, 1956) and Madagascar (Fischer-Piette *et al.*, 1993; Emberton, 1995a,b; Emberton and Pearce, 1999). Generally, the prosobranchs are considered the more primitive or ancient element among the terrestrial gastropods, but this is due more to the conservatism of the prosobranch lineages and their retention of primitive aspects of morphology than to the duration of their existence on land.

Archaeogastropod Neritopsina

Neritopsina are thought to have arisen early in the evolutionary history of the gastropods, with a fossil record extending back to the Middle Devonian (Knight *et al.*, 1960a). The extant members of this clade comprise Neritopsidae, Phenacolepadidae and Titiscaniidae in marine environments, Neritidae in marine, brackish water, freshwater, semi-terrestrial estuarine and coastal terrestrial environments, and Helicinidae and Hydrocenidae in terrestrial environments. Fossils from the Carboniferous of North America have been attributed to the Helicinidae (Solem and Yochelson, 1979), but the next oldest record for this family is from the Upper Cretaceous of Europe (Knight *et al.*, 1960a) and North America (Bishop, 1980). Extant Helicinidae are mostly tropical, being distributed in Central America–West Indies and the Indo-Asian–Polynesian regions. A fossil record for the Hydrocenidae extends back only to the Pleistocene (Knight *et al.*, 1960a) but their contemporary widespread occurrence suggests a more ancient origin. Hydrocenids have radiated most extensively in Borneo, and there constitute a significant component of the terrestrial gastropod fauna (Thompson and Dance, 1983).

The family Hydrocenidae frequently is placed in close relationship with the Helicinidae, but the internal anatomy is so divergent from the more primitive anatomical states of the Helicinidae that only a remote relationship can be established on the basis of morphological data (Thiele, 1910; Bourne, 1911; Baker, 1925b; Thompson, 1980). Indeed, Haszprunar (1988b) suggested that the Hydrocenidae and Helicinidae represent two independent invasions of land from aquatic ancestors. This divergence is recognized here in the assignment of Hydrocenidae and Helicinidae to separate superfamilies, respectively Hydrocenoidea and Helicinoidea. Several groups within the Helicinidae have been recognized previously (e.g. Wagner, 1907–11; Baker, 1922, 1923, 1956; Thiele, 1929–35; Boss and Jacobson, 1975; Thompson, 1980), but confirmation of their status and relationships awaits a comprehensive phylogenetic analysis using cladistic methodologies. There has been no systematic review of the Hydrocenidae since Pfeiffer (1876). As a consequence, there currently is little understanding of the hydrocenid evolutionary history and within-family relationships, and indeed little agreement on the status of the various generic names in use (Thompson and Dance, 1983; Solem, 1988). The hydrocenids are a poorly studied group, principally because of their minute size and cryptic behaviour. However, they are often locally abundant (e.g. Berry, 1966).

Neritopsina primitively retain the diotocardian heart, with the ventricle penetrated by the rectum, a rhipidoglossate radula, and both left and right columellar muscles. They are characterized by their primitive bipectinate (with a row of filaments on both sides of the median branchial vein) ctenidia without skeletal rods, by ciliary lateral specialization of the sensory epithelium of the osphradium, and by specialization and hypertrophy of the genital organs in the pallial cavity, with suppression of the (post-torsional) right pallial organs. The circulatory system is essentially as seen in Vetigastropods. Ultrafiltration and production of a primary filtrate occur in the auricle of the heart (Andrews, 1985; Estabrooks *et al.*, 1999). The pallial cavity roof and floor are vascularized, which may have been a pre-adaptation to colonization of land. In the terrestrial families, the Helicinidae and Hydrocenidae, the pallial cavity lacks gills and osphradia. The left kidney is retained as a functional renal organ, the nephridium, with a tubular urinary chamber that opens to the posterior of the pallial cavity via a papilla (Delhaye, 1974b; Andrews, 1981). The nephridium is lacking a nephridial gland (Andrews, 1988). Mucoid cells are generally present in the renal, renopericardial and renal orifice epithelium, whose mucopolysaccharide secretions may have a role in decreasing surface tension in the nephridium and, in terrestrial species, decreasing renal desiccation (Estabrooks *et al.*, 1999). The terrestrial Neritopsina, in contrast to marine gastropods but in common with those inhabiting freshwater, regulate their blood composition osmotically and ionically, and the nephridium is involved in ion resorption. The nephridium produces dilute urine (Little, 1972). Loss of the right auricle of the heart has occurred in most Neritopsina, but two auricles still persist in some, including the two most primitive groups within the Helicinidae: in *Ceres* Gray (Ceresinae), the right auricle is functional and nearly as large as the left; in *Hendersonia* Wagner (Hendersoninae), the right auricle is functional but very much reduced in size. In Helicinidae, both columellar muscles are retained, though the left is reduced in size. In Hydrocenidae, this left muscle is entirely absent.

Neritopsins retain the dioecious condition, but they have attained internal fertilization. The genital duct is divisible into gonadial, renal and pallial sections. No physical connection between the gonoduct and the renal system is to be observed in extant Neritopsina. However, a short duct leaves the renal gonoduct and opens to the pallial cavity, alongside the vestigial right gill in Neritidae. The opening of this duct bears the same relationship to the right gill as that of the left nephridium to the left gill, and appears to be homologous with the opening of the right nephridium of vetigastropods (Fretter, 1965). The short glandular renal gonoduct of the more primitive gastropods has been supplemented by a tubular pallial extension to form a glandular pallial gonoduct. In the females, this is associated with provision of the fertilized ova with nutritive albumen and their enclosure within a capsule. The pallial gonoduct may have arisen through the extension of the glandular renal gonoduct in the anterior pallial vein; the rectum follows the same course (Fretter, 1965). The terminal part of the renal gonoduct, at the back of the pallial cavity, is elaborated to form either a lobate pocket or a blind diverticulum: its function is not known, although Baker (1925b) observed the lumen in female *Hendersonia* to be crowded with spermatozoa (allosperm). The junction of the visceral and pallial portions of the gonoduct is elaborated into a fertilization chamber which bears two acces- sory chambers or diverticula, one functioning as the bursa copulatrix for reception of allosperm (or spermatophores where they are produced) by way of its opening to the posterior of the pallial cavity via a short papilla, the other functioning as a receptaculum seminis as indicated by the presence of crowded spermatozoa arranged radially so that their nuclei associate with cytoplasmic processes of the lumen wall (Baker, 1925b). Spermatophores apparently are not produced in Helicinidae and Hydrocenidae. The bursa copulatrix represents the metamorphosed right nephridium and the papilla its orifice (Thiele, 1902; Baker, 1925b). The glandular renal section of the neritopsinan gonoduct runs to the glandular pallial section near the posterior limit of the pallial cavity, where the first secretory tissues, probably derived from the glandular right nephridium, have assumed an albuminiparous function. The single left hypobranchial gland opens to the posterior of the pallial cavity independently of the reproductive system in helicinid Proserpinellinae, but in Helicininae and Hendersoninae the gland has a secondary duct opening to the papilla of the bursa copulatrix. In more advanced Neritopsina, such as the marine Phenacolepadidae, the hypobranchial gland has been fully incorporated into the proximal part of the pallial gonoduct in the females (Fretter, 1984). The pallial gonoduct and the rectum are intimately associated along their course in the pallial cavity: in females of many Neritopsina, the openings of the rectum and genital duct are linked by a sac that provides particulate material (of rectal origin) for reinforcement of the egg capsule wall. In the Neritidae, the sac is provisioned with calcium carbonate sourced from the digestive gland (Andrews, 1937). In males, the pallial gonoduct is elaborated into several types of prostatic tissues. The males of marine Neritopsina usually have an external phallus of pedal origin on the right side of the head, but this is absent in the terrestrial taxa. In phallate Neritopsina, the pallial gonoduct continues forward to the penis as the spermiduct, either an open groove or a closed vas deferens.

Development in Neritopsina involves both trochophore and veliger stages. However, in these animals, the trochophore stage is passed within the egg capsule. Primarily, the larval phase is short, and feeding does not occur (lecithotrophic). In some species, however, the veliger has a planktonic life and feeds actively (planktotrophic). The freshwater *Theodoxus* shows significant modifications of this bauplan, with the single embryo in each egg capsule feeding on nurse eggs and hatching as a juvenile animal. In the terrestrial Neritopsina, intracapsular development occurs with hatching delayed until metamorphosis of the veliger larva is complete. The eggs of the Malayan *Hydrocena monterosatiana* Godwin Austen & Nevill are enclosed in calcareous capsules that are glued to surfaces such as rocks and vegetation. The capsules comprise two halves, with a suture between the base and lid (Berry, 1965), similar to that seen in neritids (Fretter, 1946).

The digestive system in Neritopsina lacks salivary glands, although secretory crypts embedded in the buccal mass at the origin of the oesophagus (Baker, 1925b) may be homologous to the salivary glands in other gastropods. The radula remains of the rhipidoglossate type, but the central and first lateral teeth are often absent. The jaws are absent. The oesophagus is rather long and, in most neritopsins, the oesophageal glands take on the form of diverticulate pouches. The stomach is large, with the caecum vestigial or absent. The style is absent. The intestine of the Neritopsina remains long, describing a single or double forward loop, linked to the anterior aorta by arterial branches. The rectum extends to the front of the pallial cavity and conveys waste to the exterior. In the marine and freshwater Neritidae, the rectum passes through the ventricle en route to the anus, but in the terrestrial Neritopsina, the rectum does not penetrate the pericardium. In contrast to more primitive gastropods, neritopsins produce faecal pellets.

Like the Vetigastropoda, the Neritopsina possess a hypoathroid central nervous system (CNS), with the pedal ganglia extended posteriorly as cords connected by a principal, anterior commissure, and often with several minor, more posteriad commissures. In helicinids such as *Hendersonia*, the anterior ends of the pedal cords are enlarged into pedal ganglia. Furthermore, several Neritopsina have the pleural ganglia fused to the anterior portion of the pedal ganglia and form a ring-like connection with the suboesophageal ganglion. In the Neritopsina, the pedal cords strongly diverge at about a 60–75° angle. In Helicinidae and Hydrocenidae, the pedal cords are only weakly divergent. In addition to the cerebral commissure, the labial or subcerebral commissure is retained in Neritopsina. The head of Helicinidae bears the usual pair of cephalic tentacles, with an advanced, closed-vesicle eye at the base of each. In Hydrocenidae, the tentacles are absent and the eyes are borne on short stalks.

Hydrocenids are found only in moist regions. Generally they are ground dwelling, although some species occur on tree trunks in rainforests. They possess conical to globose shells, generally less than 10 mm in height, with the umbilicus closed by the callosity of the columellar. The operculum, rigiclaudent and paucispiral or concentric, is always present, with a prominent internal apophysis embedded in the columellar muscle. Helicinidae occupy environments ranging from moist tropical rainforest to xeric scrubland. Most helicinids are terrestrial, but some are partially or wholly arboreal. Many helicinids are able to aestivate during dry weather, withdrawn into their shells and cemented to rocks or tree trunks. The shells of the Helicinidae may be conical, globose or flattened, usually less than 30 mm in diameter, imperforate with the umbilicus closed by the columellar callus. The rigiclaudent operculum, when present, is composed of an inner corneous and an outer paucispiral or concentric calcareous layer: a true apophysis is lacking, but a small ridge provides for columellar muscle attachment. In the Proserpininae and Proserpinellinae, the operculum is absent, with loss of this structure representing a secondary condition in these subfamilies rather than the non-operculate condition being a plesiomorphic one for the Helicinidae as a whole, as suggested by Thompson (1980). The shell aperture in helicinids is often with palatal and parietal plicae and columellar barriers. In common with other neritopsin snails, Hydrocenidae and Helicinidae exhibit progressive resorption of the shell columella and internal partitions as the animal increases in size, resulting in a flatted spheroidal visceral mass that is simply encircled by the last shell whorl. The foot sole is furrowed in Hydrocenidae. In Helicinidae, the sole is usually uniform, but in Proserpininae the sole extends laterally up on to the sides of the foot where it is delimited by a parapodial groove and thus similar to the aulacopodous condition seen in some terrestrial pulmonates. Baker (1928) noted the large sole in *Helicina delicatula* Shuttleworth to be functionally tripartite; locomotion seems to be accomplished mainly by the broad and firm, central zone, which develops definite and numerous retrograde pedal waves. *Helicina* de Lamarck species are rather active animals. In contrast, Baker (1928) observed species of *Schasicheila* Shuttleworth to be relatively slow moving and inactive. The sole of the foot in these animals is elliptical, comparatively short and, as evident in preserved material, divided into two halves by a longitudinal groove. Locomotion is accomplished by irregular retrograde waves, with only one or two on the sole at any one time. Usually, the two sides of the foot move in unison, but they can act separately to a small extent.

Little is known of the biology of the terrestrial Neritopsina. Many helicinids and hydrocenids are known to be most abundant in, and indeed often restricted to, limestone areas. Others exhibit broad ecological tolerance and accordingly are widely distributed. The calcicolous Malayan hydrocenid species *H. monterosatiana* feeds on algae, moss and lichen (Berry, 1961).

Architaenioglossan caenogastropods

The extant Architaenioglossa, as presently recognized, comprises two groups, the freshwater and often amphibious Ampullarioidea and the terrestrial Cyclophoroidea. While the ampullarioidean Ampullariidae and Viviparidae are not strictly terrestrial and therefore not among the gastropods being addressed in this chapter, they do provide a useful reference point when considering the adaptive diversity in Cyclophoroidea. Analyses by Ponder and Lindberg (1996, 1997) suggest that Architaenioglossa is a sister group to the remainder of the caenogastropods. My analyses indicate a closer relationship to Neritopsina and Neomphaloidea. Taxa recognizable as architaenioglossans separated early in the evolution of the caenogastropods, possibly in the Devonian, as the first fossils attributed to this group occur in deposits of Lower Carboniferous age (Knight *et al.*, 1960a; Bandel, 1993) (but see Solem and Yochelson, 1979).

The Viviparidae comprise three recognized subfamilies – the Viviparinae from Europe, Asia and North America; the Campelominae from North America; and the Bellamyinae from South-East Asia, Japan and Australia. These are strictly freshwater animals. The Ampullariidae occur in tropical and subtropical freshwater habitats of Africa (including Madagascar), South-East Asia, and the Americas. Several exhibit amphibious tendencies. Debate continues on the relationships and supraspecific groups within the Ampullariidae (e.g. Michelson, 1961; Berthold, 1989, 1991; Bieler, 1993).

The most comprehensive investigation of the Cyclophoroidea to date is that by Tielecke (1940). On the basis of several conchological and anatomical characters, Tielecke recognized five cyclophoroidean families: Cyclophoridae, Poteriidae, Pupinidae, Cochlostomatidae and Maizaniidae. As noted by Solem (1959) and Thompson (1967), Tielecke's classification probably reflects natural relationships, but it lacked explicit use of apomorphies in recognition of family units and ignored nomenclatural priority among the many suprageneric names available. Thompson (1969, p. 35) noted, 'Recent authors tend to recognize the group as a superfamily with several families, but no one has demonstrated the presence of characters that consistently separate and characterize the families.' Morton (1952), Creek (1953), Berry (1964), Prince (1967), Thompson (1969), Climo (1973), Kasinathan (1975), Girardi (1978), Jonges (1980), Varga (1984) and others have significantly advanced knowledge on the anatomy of cyclophoroideans, but the superfamily awaits revision using modern phylogenetic analysis methods. In a checklist of caenogastropods, Ponder and Warén (1988) recognized

Cyclophoridae, Neocyclotidae (= Poteriidae *sensu* Tielecke), Neopupidae, Craspedopomatidae (= Maizaniidae *sensu* Tielecke), Diplommatinidae, Pupinidae and Aciculidae as extant cyclophoroidean representatives, and their classification is followed here. Cyclophoridae occur in southern Africa, Madagascar and the Indo-Pacific. Neocyclotidae is represented in Central America, South America, the West Indies and the South Pacific. Craspedopomatidae is confined to Africa and Pupinidae to the Indo-Pacific. Diplommatinidae comprise two subfamilies of vicariant distributions: the Cochlostomatinae confined to the European region, and the Diplommatininae mainly to Madagascar, East Asia and the western Pacific, but possibly including the American *Adelopoma* Döring. The New Zealand endemic genera *Cytora* Kobelt & Moellendorff and *Liarea* Pfeiffer have been attributed to a distinct family (Liareidae by Powell, 1946; Cytoridae by Climo, 1970) (Climo, 1975), but are treated as part of the pupinid radiation by Ponder and Warén (1988). The systematic relationships of the Aciculidae have long been recognized as problematic and the family generally attributed to the caenogastropod Littorinoidea in the absence of definitive phylogenetic analysis (e.g. Thiele, 1929–35).

Architaenioglossans retain only one set of pallial organs. The Ampullarioidea retain a monopectinate ctenidium (one row of filaments) in the pallial cavity, while the cyclophoroideans have lost the gill. In both superfamilies, the anterior of the pallial cavity is highly vascular and functions as a lung. The relative sizes of the ctenidium and 'lung' in ampullarioideans, and their relative importance in respiration, vary among species according to the relative lengths of time spent in or out of water, and the degree to which water quality in their habitat affects the respiratory efficacy of the gill (Andrews, 1965). The ctenidium is greatly reduced in the more terrestrial ampullariids, such as members of the genera *Turbinicola* Annandale & Prashad and *Asolene* d'Orbigny, and respiration is effected primarily by the pallial cavity when the animals are not submerged. In Cyclophoridae, the pallial cavity is the primary respiratory site, but it is complemented by respiratory gas exchange in the vascular spaces in the nephridium (Andrews and Little, 1972). With the pallial cavity widely open to the exterior, as it is in their marine ancestors, ventilation in cyclophoroideans is brought about by movement of the head in large animals such as cyclophorids (Andrews, 1965; Kasinathan, 1975) but seems to depend on diffusion in small ones. In minute species, such as aciculids of the genus *Acicula* Hartman, respiratory exchange occurs through the water that fills the pallial cavity (Creek, 1953). In *Acicula*, cilia on the head and hypobranchial gland set up currents in the pallial water that probably facilitate gaseous exchange over the respiratory surface. As in Neritopsina, the nephridium has a primarily visceral location in ampullariid Ampullarioidea, but is wholly pallial in viviparids and Cyclophoroidea. Structurally, the nephridium of *Viviparus* de Montfort, in Viviparidae, differs little from the basic monotocardian plan, the only modification being the development of a pallial ureter. In *Pomacea* Perry, in Ampullariidae, the nephridium is subdivided into two chambers, the posterior part – sited largely in the viscera – being concerned with excretion and water storage, and the anterior part – in the mantle skirt – largely concerned with resorption. The nephridium opens laterally, reflecting partial rotation of the nephridium to the post-torsional right. This trend for subdivision and rotation of the nephridium is elaborated further in cyclophoroideans. In these animals, the nephridium consists of two distinct regions: the main body with lamellate internal structure and primarily excretory in function, and an unfolded region surrounding the renopericardial and external openings and concerned with resorption of water from the urine. Further, the opening of the nephridium to a narrowed part at the posterior of the pallial cavity has been brought about by maintenance of the plesiomorphic position of the orifice but migration with rotation of the nephridium into the mantle cavity. In some cyclophoroideans, the separation of the posterior bay of the pallial cavity from the more anterior, respiratory part is achieved by the apposition of the overhanging nephridium to the rectum and genital duct. In others, a transverse septum has developed, and separation from the pallial cavity is more substantive (Kasinathan, 1975; Andrews and Little, 1982). The partial or near complete separation of the posterior part of the pallial cavity from the rest of the cavity raises the possibility that urine may be retained there long enough to permit further resorption of ions and water through the pallial epithelium (Andrews and Little, 1972, 1982). The nephridium in these gastropods apparently secretes a stream of almost pure water that moistens the surface of the pallial cavity. This aids respiration and keeps the tissues from drying out. The efficiency of these excretory/respiratory systems is marginal, and activity in cyclophoroideans is confined to periods of very high humidity. When the humidity is low, the animals remain retracted within their shells, with the aperture sealed by the operculum to prevent moisture loss. Many have special adaptations of the shell aperture to allow respiration to continue at these times (Rees, 1964).

Architaenioglossans retain the primitive gastropod condition in that urine is formed by ultrafiltration through the wall of the auricle into the pericardial cavity. Filtration chambers, similar in structure to those in vetigastropods, occur extensively in the auricle of Viviparidae (Boer *et al.*, 1973; Andrews, 1979) and less so in Ampullariidae (Andrews, 1965). Cyclophoroidea lack pericardial gland chambers but none the less retain podocytes on the auricular wall (Andrews and Little, 1972). The reduced development of the filtration area in cyclophoroideans, cutting the filtration rate and volume, can be viewed as a water conservation mechanism (Andrews and Little, 1982). The ultrafiltrate enters the nephridium through the renopericardial duct and is modified there by resorption of ions, organic molecules and water, and by the excretion of purines and lipids (Andrews and Little, 1971, 1972, 1982). The nephridial gland is present in the nephridium of viviparids, vestigial in ampullariids, and absent in cyclophoroideans (Andrews and Little, 1982; Andrews, 1988). The involvement of the nephridium in osmoregulation has led to increased excretory activity of the hypobranchial gland. This gland is elaborated variously among families into extensive subepithelial glandular tissue which discharges products into the excretory groove that runs on the right side of the pallial cavity from the nephridial orifice to the mantle collar. The hypobranchial gland has both excretory and secretory functions in these animals, releasing purines and lipids that are combined with the products of the nephridium in the pallial groove to form an excretory string to be carried from the pallial cavity (Andrews and Little, 1972).

Cyclophoroideans are dioecious, with internal fertilization. There is only one gonad, representing the post-torsional right side of the animal, and the renal gonoduct that runs from the gonad is separated from the functional nephridium. There may be a vestige of the gonopericardial duct. In many cyclophoroideans, the renal gonoduct still retains an opening to the pallial cavity. A true bursa copulatrix, for reception of allosperm during mating and resorption of excess gametes, and a receptaculum seminis, for storage of received allosperm, are associated with the distal section of the renal gonoduct. Their sacculate reservoirs invariably are located in the visceral cavity, but their ducts exhibit a range of configurations. In various taxa, the bursa copulatrix and/or the receptaculum seminis has been lost. Unlike the situation in early gastropods, that part of the gonoduct of ectodermal origin in architaenioglossans is extended through the pallial cavity to open towards the mantle collar and is closely associated with the anus. In this respect, the evolution of the reproductive system in architaenioglossans closely parallels that in Neritopsina. In cyclophoroideans, the pallial gonoduct comprises a series of glandular chambers separated by septa. In the females, this is associated with investments of the fertilized ova with nutritive albumen and their enclosure in a capsule. The albumen gland is found either as a widening of the renal gonoduct, as a pouch opening into the renal gonoduct or as a sacculate chamber continuous with the proximal part of the capsule gland, usually located in the visceral cavity. In the primary condition (e.g. still evident in Neopupidae and some Cyclophoridae), the female pallial gonoduct is widely open to the pallial cavity along its ventral axis. In more evolved forms, the pallial gonoduct is a closed tube, opening to the pallial cavity via a narrow orifice at its anterior extremity. Such a change from an open channel to a closed duct has occurred independently in several cyclophoroidean lineages, as evidenced by the ontogeny of many species which have closed ducts in the mature animal but in which, during development, the pallial oviduct first appears as a longitudinal groove in the mantle skirt that later closes over. Morton (1952) regarded closure of the pallial gonoducts as an adaptation to terrestrial existence, but this can hardly be true given that such closure has occurred repeatedly in Architaenioglossa and indeed in many other caenogastropod lineages, in marine, freshwater and terrestrial environments. The glandular cells are confined to the epithelium that lines the lumen (Morton, 1952; Creek, 1953).

In cyclophoroidean males, the renal gonoduct is a simple duct, terminating in the posterior part of the pallial gonoduct. The pallial gonoduct of the males is glandular, functioning to produce prostatic secretions and, in at least some species, spermatophores (Weber, 1924; Kasinathan, 1975; Jonges, 1980). Corresponding to the condition in the females, the neopupid males possess a prostate open to the pallial cavity via a slit along its columellar margin. In the Neopupidae, the male system is continued forward as an open seminal groove along the right side of the head to run to the tip of a penis located behind the right cephalic tentacle. This condition of an open groove running to a penis behind the tentacle is evident in Aciculidae and Maizaniidae, some Cyclophoridae and Cochlostomatinae Diplommatinidae. According to Creek (1953), the seminal groove in *Acicula* assumes the form of a duct during copulation, due to contraction of the circular muscle fibres that lie beneath the epithelium. In some Cyclophoridae and Pupinidae, the spermiduct takes the form of a closed tube, the vas deferens, firstly running a short distance through the anterior of the pallial cavity, then penetrating the anterior cephalic body wall to enter the penis where it runs internally to the tip. The conditions of open spermiduct and closed vas deferens are both represented in Neocyclotidae, but here the penis is located on the dorsal midline of the head, its base located within the pallial cavity. In a number of cyclophoroideans, including craspedopomatid *Maizania* Bourguignat and neocyclotid genera such as *Ostodes* Gould and *Neocyclotus* Fischer & Crosse, the penis bears a terminal flagellum, while in pupinid *Cytora* the penis is equipped with a retractable, terminal intromittent organ. Some Pupinidae and Diplommatinidae are aphallic.

The free-living phase of the trochophore larva, characteristic in the development of Vetigastropoda and other primitive prosobranchs, is absent in Architaenioglossa. The Ampullariidae and Cyclophoroidea generally produce spherical capsules, each enclosing a single embryo that undergoes direct development. The veliger larval stage is suppressed. Some Cyclophoroidea are ovoviviparous or viviparous, and all Viviparidae are ovoviviparous, with the embryos completing development in a 'brood pouch' within the pallial gonoduct. The sparse information available on the reproductive biology of cyclophoroideans relates only to oviparous species. Kasinathan (1975) provides some observations on the reproductive biology of the oviparous cyclophorids *Cyclophorus jerdoni* (Benson), *Micraulax scabra* (Theobald), *Pterocyclus bilabiatus* (Sowerby), *Theobaldius ravidus* (Benson) and *Theobaldius shiplayi* (Pfeiffer) from the Alagarkoil Hills, southern India. The duration of copulation in these species varies greatly (3–4 to 6–10 h), but all involve transfer of spermatophores from the male to the female. Deposition of egg capsules generally commenced within a few days after copulation and, while the eggs pass singly through the pallial gonoduct, females deposited 40–150 eggs over the following few weeks. The eggs of *C. jerdoni*, *M. scabra* and *P. bilabiatus* were found to be spherical, with a white calcareous shell. In the case of the two *Theobaldius* Nevill species, the eggs possessed opaque yellow capsules that were spherical except for a small, flattened area. Van Benthem Jutting (1948) states that *Cyclophorus perdix* (Broderip & Sowerby), of Javanese forests, produces eggs with a calcareous shell, and deposits them in clutches, which in one case comprised ten eggs. Dupuis and Putzeys (1901) describe the eggs of the African *Maizania intermedius* von Martens as 'black, small, and very numerous'. According to Prince (1967), the diplommatinid *Cochlostoma septemspirale* (Razoumowsky), of rocky habitat in southern Europe, produces clutches of 3–8 eggs, each egg heavily invested with albumen within a calcareous shell, and thinly coated in faecal material and other debris. At egg laying, the adult holds each egg by the foot while using its mouth to transfer calcium-rich faecal pellets to the surface of the egg. In the related diplommatinid genus *Obscurella* Clessin, the eggs vary among species – those of *Obscurella obscurum* (Draparnaud) are grey, and have a weak calcareous shell, while those of *Obscurella hidalgoi* (Crosse) are cream coloured, soft and sticky and readily adhere to plant material. Berry (1964) found breeding in the Malayan diplommatinid *Opisthostoma retrovertens* Tomlin to occur throughout the year but with a pronounced peak coinciding with the monsoon rains of September to January. The eggs of this species are provided with a calcareous shell comprising two halves joined by a suture passing around the greatest circumference.

The cyclophoroidean CNS retains the primitive features of the hypo-athroid condition of the cerebropedal ring, the pedal ganglia with cords running the length of the foot and united by numerous commissures, and the crossed or streptoneurous condition of the visceral loop. In the fresh-water architaeniglossan sister group Ampullariidae, the cerebropedal ring is similarly hypoathroid, while in the Viviparidae the hypoathroid condition is restricted to the left side and the right side is epiathroid with close apposition of cerebral and pleural ganglia. The primary hypoathroid condition and only partial epiathroidy strongly suggests architaenio-glossans affinities with the archaeogastropods, rather than with the caenogastropods that are characterized by full epiathroidy; as suggested by Ponder (1991), the epiathroid condition probably arose independently in Architaenioglossa and Caenogastropoda. Various cyclophoroideans exhibit zygoneury with connections between the right pleural and sub-oesophageal ganglia or between the supraoesophageal and the left pleural ganglia, which are thought to impart improved neural coordination. None the less, the CNS of Architaenioglossa is little more concentrated than that in vetigastropods, and the presumptive neurosecretory cells remain widely dispersed through the various ganglia and cords (e.g. Gorf, 1961). The osphradia of the freshwater Ampullarioidea are similar to those of Neritopsina in possessing a sensory epithelium dividing into a central zone and lateral zones with ciliated cells. In Ampullarioidea, the lateral zones are uniquely differentiated to deep grooves (Haszprunar, 1985a). Contrary to statements in the literature, the osphradium is often retained in the cyclophoroidean pallial cavity, albeit reduced in size (e.g. Creek, 1953; Kasinathan, 1975), but the absence of information on its structure currently prevents any useful comparison with that in Ampullarioidea or indeed any other gastropod.

Cyclophoroideans are generally regarded as detritivores, and indeed some are known to feed on decaying leaf material (Prince, 1967). However, many are apparent specialists, feeding predominantly on algae, fungi or mosses (e.g. de la Torre *et al.*, 1942; Morton, 1952; Berry, 1964). The digestive system is known in any detail for only a few cyclophoroideans. That of the pupinid *Cytora pallidum* (Hutton) exhibits many of the basic features seen in archaeogastropods (Morton, 1952), including a buccal mass equipped with paired jaws; a long radular sac; the stomach differentiated into a proximal digestive sac and a distal style sac with mucoid protostyle; and an intestine with a forwardly directed loop. Features in which *C. pallidum* is more advanced than Vetigastropods include the loss of the radular diverticulum; the reduction of the odontophoral cartilages to just one pair; the taenioglossan radula; simplification of the anterior digestive tract with reduction of the oesophageal glands to a ventral glandular area (curves to a dorsal position posteriorly) and loss of the ventral ciliated tract; simplification of the sorting area of the stomach and associated loss of the gastric caecum; loss of the anterior lobe of the digestive gland; the failure of the intestine to penetrate the pericardium; and the anus opening towards the anterior of the mantle cavity. Not all of these advanced features are shared by other cyclophoroideans. For example, some cyclophorids retain the gastric caecum, and most retain the two lobes of the digestive gland. Yet other cyclophoroideans exhibit features more advanced than in *C. pallidum*, with marked simplification of the stomach so that there is no style or gastric shield, and with considerable reduction in the intestinal length so that the forward-directed loop is barely discernable. In the taenioglossate radula, each transverse row comprises just seven teeth – a central tooth flanked on either side by a robust lateral tooth, each in turn flanked by two elongate marginal teeth. The taenioglossate-type radula is common among higher gastropods and, generally assumed to have arisen only once in gastropod evolution, has been regarded as indicative of the monophyly of the caenogastropods and heterobranchs. However, the taenioglossate radula represents a simplification or reduction of the rhipidoglossate radula, and Haszprunar (1988b) and Ponder and Lindberg (1997) are justified in questioning its monophyletic status.

The majority of Cyclophoroidea live on the ground, amongst leaf litter, decaying wood and mosses in forests. Some species, particularly among cyclophorids, remain buried deeply in humus; Kasinathan (1975) noted, for example, that *M. scabra* of the Alagarkoil Hills, southern India, occurred to a depth of 15–20 cm, even during the monsoon season. Some ground-dwelling species, particularly in Diplommatinidae, are also to be found amongst leaf litter and humus accumulated in epiphytes. Many cyclophoroideans are found associated with limestone rock formations, and some are obligate calcicoles. Others live on tree trunks or foliage and can be regarded as truly arboreal. Cyclophoroideans are, as a general rule, strongly dependent on high humidity conditions for activity and, as such, are largely confined to moist woodland and rainforest ecosystems. None the less, there are species that occupy more open habitat, such as certain species of the diplommatinid genus *Obscurella* that occur in an exposed rock rubble habitat in the European mountains. The ecology of archi-taenioglossans is poorly known. Berry (1962, 1964, 1966, 1975) provides some information on abundance and reproductive phenology of *O. retrovertens* on Bukit Chintamani, a limestone hill in Pahang.

There is great variation among cyclophoroidean species in shell size (1–60 mm), shape (discoidal, turbinate, turreted, pupiform; open or closed umbilicus) and orientation of coiling (predominately dextral, but some sinistral), even within a single family. The operculum is rigiclaudent and multispiral. The animal generally has the head produced to a long snout, and carries a pair of slender, cylindrical, highly contractile cephalic tentacles with the advanced, lens-equipped eyes on their outer bases, usually on short peduncles. The sole generally is undivided. *Acicula* progresses by ciliary gliding.

Sorbeoconch caenogastropods

Within the sorbeoconchan caenogastropods occur a diverse assemblage of families. While the primitive marine habitat still predominates, a great many sorbeoconchan lineages are represented in marine littoral and supralittoral, freshwater and terrestrial environments. Of interest in our discussion on land-based gastropods are the littorinimorphan superfamilies Littorinoidea and Rissooidea.

The extant Littorinoidea comprises the mainly marine Littorinidae, the marine Zerotulidae, Skeneopsidae and Pickworthiidae (the latter of uncertain affinity), and the terrestrial Pomatiasidae. They possess moderately large shells, up to 50 mm in height, which are turbinate, trochoidal or conical in shape. Their opercula are rigiclaudent and paucispiral. The Littorinidae can only be traced to the beginning of Tertiary times (Bandel, 1993), with fossils in Palaeocene age deposits of Europe and California (Reid, 1998). According to a model offered by Ponder (1988), the Littorinoidea arose from a common ancestor with the Eatoniellidae (assigned to the Cingulopsoidea by Ponder and de Keyzer, 1998) in the cold water of southern Pangea and reached European waters by the beginning of the Cretaceous. As noted by Bandel (1993), fossil evidence to support this hypothesis is missing so far. According to Reid (1998), the family Littorinidae contains 13 extant genera and about 180 species, grouped in three subfamilies: Lacuninae, Laevilitorininae and Littorininae. This classification is based on a phylogeny of all 33 recognized subgenera, produced by cladistic analysis of morphological data (Reid, 1989). However, Reid's (1989) interpretation of polarity for many of the character states employed in the cladistic analyses is highly questionable, and thus acceptance of the suprageneric relationships must await re-analysis of the data set. There is an extensive literature on the systematics, anatomy and ecology of the marine Littorinidae, but little is known about the remaining littorinoidean families.

Despite their colonization of the littoral fringe, none of the littorinid Littorininae have become truly terrestrial. In fact, one of the developments that permitted their successful colonization of the littoral fringe, i.e. the pelagic egg capsule, has tied them to the marine environment (Reid, 1989); those littorinids living in the high littoral zone generally migrate to the water to spawn. The few examples of non-planktotrophic development in the littoral Littorininae, involving ovoviviparity in *Echininus viviparus* Rosewater, *Tectarius viviparus* (Rosewater) and *Littoraria aberrans* (Philippi), of high littoral fringe cliffs in the former two, and landward fringe of mangrove forests in the latter, are apparently of recent origin and have not produced terrestrial radiations (Reid, 1989). The amphibious *Cremnoconchus* Blanford, in the subfamily Lacuninae, is the only littorinid lineage to be found outside the marine environment. It is represented by two species, *Cremnoconchus syhadrensis* (Blanford) and *Cremnoconchus conicus* Blanford, occupying wet cliff habitats beside streams in the western Ghats of India (Blanford, 1863, 1870). As a sister group of the marine *Bembicium* Philippi and *Risellopsis* Kesteven, with a possible Gondwanaland origin, *Cremnoconchus* is believed to be an ancient littorinid advance on to land (Reid, 1989), which none the less has not produced an extensive radiation.

The Pomatiasidae are, according to Wenz (1938–44) and Hrubesch (1965), present in the Late Cretaceous of southern France. The extant Pomatiasidae occur in the Americas and the West Indies, Europe and the Middle East, to Africa and India. These animals have turreted to planispiral shells, usually less than 10 mm in greatest dimension, but ranging up to 80 mm in the Madagascan *Tropidophora curveriana* (Petit), the largest extant land operculate gastropod. In some genera, the shell of adult animals often is decollated. Slight sexual dimorphism in shell size often is evident, with females a little larger than males. The rigiclaudent operculum is multispiral or paucispiral. These animals generally occur in moist forest, dwelling in leaf litter and rock talus, but with the activity of *Tudora* Gray extending arboreally into the trees and shrubs. Some Pomatiasidae, such as Mexican species of *Choanopoma* Pfeiffer, are abundant in xeric and mesic forests. Particularly diverse pomatiasid radiations have long been recognized among a complex of genera in the Americas (de la Torre and Bartsch, 1938, 1941; Solem, 1961) and within *Tropidophora* Troschel of Madagascar (Fischer-Piette *et al.*, 1993; Emberton, 1995b). The Old World and the Neotropical pomatiasids often have been assigned to separate families (Pomatiasidae and Annulariidae, respectively), principally on radular characters (e.g. Henderson and Bartsch, 1920; Thompson, 1978). Baker (1924) and Solem (1961) considered the basis for separation as different families to be inadequate and to obscure the very close anatomical relationship between the Old World and New World forms. The phylogenetic relationships among pomatiasids have yet to be resolved, and Solem (1961) highlighted the need for a substantive systematic revision of these animals.

Rissooidea is the largest group in Littorinimorpha, and indeed among the Sorbeoconcha; Ponder (1988) recognized 17 families, of which Assimineidae, Pomatiopsidae, Truncatellidae and Hydrobiidae are represented in terrestrial habitats. Rissooideans are small animals, rarely exceeding 10 mm in shell length. They exhibit a large diversity in shell shape, ranging from planispiral to very elongate, and from tightly to loosely coiled or even uncoiled and tusk-like. The shells of males are sometimes slightly smaller than those of females. The shells of adult Truncatellidae are usually decollate through loss of the protoconch and early teleoconch whorls. The rigiclaudent opercula of rissooideans vary from multispiral to concentric, often with a peg arising from the inner surface to embed in the columellar muscle. As pointed out by Ponder (1988) and Ponder and de Keyzer (1998), the superfamily as presently conceived cannot be defined by autapomorphies and is possibly paraphyletic or even polyphyletic. Rissooidea can only be traced to the middle of the Mesozoic, with records from the Jurassic of Italy and New Zealand (Bandel, 1993). The evolutionary relationships within the superfamily have been reviewed by Ponder (1988), but his proposed rissooidean phylogeny is largely rejected here because of the highly questionable polarity he assigned to morphological characters employed in the cladistic analysis. None the less, the family units used here follow Ponder (1988) and Ponder and Warén (1988). Ponder (1988) suggested that the Assimineidae, Pomatiopsidae and Truncatellidae comprised a monophyletic group within the Rissooidea that radiated in the Upper Jurassic of the southern regions of the Pangean continent. Davis (1979, 1982) places the centre of dispersal of the pomatiopsids in what is now Antarctica on that part of the continent that abutted India, Australia and Africa. The group is postulated to then have reached South-East Asia by being rafted on the Indian plate during the Tertiary. Davis (1979) argues that rissooidean fossils from the Upper Cretaceous of India and South Africa are probably pomatiopsids.

Extant Pomatiopsinae occur in South America, Japan, Manchuria, north-west North America, Australia, China, Taiwan, the Philippines, Sulawesi and South Africa, with some eight genera and about 30 species (Davis, 1981). Pomatiopsinae are freshwater aquatic, estuarine or freshwater amphibious, or terrestrial. Japanese *Blanfordia* Adams, North American *Pomatiopsis* Tryon and some members of South African *Tomichia* Benson are terrestrial. Triculinae make up the balance of the pomatiopsid radiation and are restricted to freshwater habitat in Asia.

The occurrence of fossils reputed to be Truncatellidae (Wenz, 1939–44) indicates the assimineid–pomatiopsid–truncatellid radiation to have been present in Europe, Asia, North Africa and Australia during the Palaeocene. Ponder (1988) postulated that the Assimineidae developed on the southern shores of the Tethys Sea. According to Wenz (1939–44), assimineids did not occur in Europe before Tertiary times, but Hrubesch

(1965) found *Turbacmella* Thiele, a genus represented in the extant faunas of South-East Asia, to be present in the Late Cretaceous of France. The first fossils assignable to assimineids in North America are Miocene (Wenz, 1939–44).

The extant Assimineidae comprises two subfamilies. Assimineinae principally occur in temperate and tropical lowland regions, where they are amphibious in marine supralittoral, freshwater, and moist terrestrial environments. Most have been referred to the genera *Assiminea* Fleming and *Paludinella* Pfeiffer but, in the absence of comparative studies, the systematic relationships and distributional limits of these genera cannot be defined. Species of the strictly terrestrial subfamily Omphalotropinae are found in the Indo-Pacific. They occur in a variety of niches, including forest litter (e.g. some *Omphalotropis* Pfeiffer), rock rubble (e.g. *Electrina* Baird), arboreally in tree foliage (e.g. *Pseudocyclotus* Thiele, some *Omphalotropis*) and in wet moss on cliffs (e.g. *Tutuilana* Hubendick). Particularly diverse *Omphalotropis* radiations are known to occur on Mauritius and Rodriguez (Madge, 1939). Despite a number of studies on the anatomy and ecology of selected species (e.g. Marcus and Marcus, 1965a; Fowler, 1980; Solem *et al.*, 1982; Hershler, 1987) and systematic revisions of regional species groups (e.g. Abbott, 1949, 1958; Turner and Clench, 1972; Giusti, 1973; Kershaw, 1983; Fukuda and Mitoki, 1995, 1996a,b), there have been no comprehensive, family-level revisions since Thiele (1927).

The majority of species in Truncatellidae live at or just above the marine high tide line, among mats of cast-up debris and coastal vegetation in warm temperate and tropical areas. These are generally referred to the genus *Truncatella* Risso. These truncatellids are capable of long-distance dispersal in marine shoreline debris, and most species have large geographic ranges. A number of *Truncatella* species occur, however, in more terrestrial habitat, often some distance inland and at elevations up to 600 m. Rosenberg (1989, 1996a) demonstrated that the two *Truncatella* species in Barbados achieved terrestriality independently from supralittoral ancestral stock. Two genera, *Geomelania* Pfeiffer in forested mountain areas of the Caribbean and *Taheitia* Adams & Adams in lowland to mountain forests of the western Pacific, are terrestrial. These terrestrial species generally have restricted distributions. Clench and Turner (1948) catalogued the family, but there have been few studies of the anatomy (e.g. Kosuge, 1966; Rosenberg, 1996b) or ecology (Rosenberg, 1989), and the systematic relationships within the family have yet to be examined critically. None the less, Rosenberg (1996a) found support in anatomical and molecular characters for the placement of *Geomelania* in a subfamily (Geomelaniinae) separate from *Truncatella* and *Taheitia* (Truncatellinae), as earlier suggested by Kobelt and von Möllendorff (1897).

The Hydrobiidae have a poor fossil record, but probably arose in the Jurassic (Knight *et al.*, 1960a). Ponder (1988) suggested that the hydrobiid radiation was initiated in the western-most part of the Tethys Sea in the Upper Jurassic and was widely established prior to the separation of the laurasian and gondwanan supercontinents in the Jurassic. Many modern hydrobiids, including *Hydrobia* Hartmann and *Tatea* Woods, live on estuarine mud, which is the presumptive habitat of the first hydrobiids. However, the hydrobiid radiation, encompassing some 140 genus–group taxa (Ponder, 1988), has involved establishment in brackish water and, more extensively, freshwater. The great diversity of modern hydrobiids is the result of their low vagility, with isolation in confinement to permanent water bodies resulting in genetic differentiation and speciation at small spatial scales (e.g. Ponder *et al.*, 1994, 1995). These processes have led to particularly spectacular radiations in freshwater stream habitats in North America and Mexico, in springs and seepages in the Great Artesian Basin of South Australia and Queensland, and in subterranean waters in Europe, North America and New Zealand (references in Ponder and de Keyzer, 1998). Some hydrobiids have become amphibious, but only *Falniowskia* Bernasconi and *Terrestribythinella* Sitnikova, Starobogatov & Anistratenko have become terrestrial.

The littorinimorphans arose from early sorbeoconchan caenogastropod stock that had achieved a grade of evolution characterized by: (i) the pallial cavity with a single set of pallial organs, including a monopectinate ctenidium, monotocardian heart and an osphradium comprised of a simple ridge but distinctly structured in having three special cell types; (ii) a gonochoristic or protandrous hermaphrodite reproductive system, with the bursa copulatrix located in the visceral cavity and opening to the posterior of the pallial cavity, and the pallial gonoduct comprising an open glandular groove; the females producing gelatinous spawn masses containing non-encapsulated eggs that yield planktotrophic larvae; the males with a penis and producing dimorphic male gametes (euspermatozoa and paraspermatozoa); (iii) the snout-bearing head containing a buccal mass with paired jaws, odontophoral cartilages reduced to a single pair, and a radula reduced to the taenioglossate configuration; (iv) the oesophagus provided with a dorsal food channel and oesophageal glands; (v) the stomach equipped with a gastric shield and a protostyle; (vi) the intestine rather long, with the rectum situated on the right side of the pallial cavity, divorced from the pericardium, to open through the anus located near the mantle collar; (vii) a streptoneurous nervous system, with pedal cords, and the pleural ganglia sited close to the cerebral ganglia (epiathroidy); (viii) bifurcated tentacular nerves; (ix) loss of the epipodial skirt and associated sense organs; and (x) rigiclaudent, multispiral operculum.

Various combinations of these basal littorinimorphan features are shared with some archaeogastropod groups and Architaenioglossa, reflecting similar grades of evolution. Many of these features, that can be considered plesiomorphic to Littorinimorpha, are modified substantially in various littorinimorphan groups.

In common with basal sorbeoconchans (e.g. some cerithioideans), the left nephridium of littorinimorphans is primitively entirely or primarily visceral, a condition unlike the primarily pallial position of the left nephridium in vetigastropods, advanced architaenioglossans and hetero-branchs. The loss of the functional right renal organ in ancestral caenogastropods required the left nephridium to deal with nitrogenous excretion along with its other functions (Andrews, 1985). The right dorsal wall of the nephridium, which receives nearly all the blood from the head–foot and visceral hump on its way to the ctenidium, extracts nitrogenous waste and discharges it to the lumen. The dorsal wall is modified and folded to provide sufficient resorptive area. In many littorinimorphans, the nephridium is enlarged. The auricular wall remains the primary site of filtration (Andrews, 1988), but, in contrast to the vetigastropods and architaenioglossans, the podocytes occur in a chamber-like appendage on the inner wall of the auricle. The blood filtrate from the heart passes to the nephridium via the renopericardial duct, where its composition is altered by ionic control and resorption of solutes in a nephridial gland before being added to the nitrogenous excretory material from the dorsal wall. This modified fluid is released to the pallial cavity.

In terrestrial littorinoideans, namely the Pomatiasidae, the filtration chambers in the auricle are reduced in area, cutting the filtration rate and volume. In *Pomatias* Studer, uric acid accumulates in a so-called concretion gland (Creek, 1951; Kilian, 1951) – perhaps an enlarged area of connective tissue around a blood vessel that is often used for storage (Fretter *et al.*, 1998). In these animals, the nephridium is pallial and there is provision there for long-term storage of nitrogenous waste (Kilian, 1951; Martoja, 1975) because there is neither an exhalant water current nor a ureter to carry away urine. The nephridium apparently secretes a stream of almost pure water that moistens the surface of the pallial cavity.

The pallial cavity in the littoral marine forms of Littorinidae generally is equipped with a well-developed ctenidium. The ctenidium is retained in *Cremnoconchus*, the only Lacuninae genus found outside the marine eulittoral or sublittorial habitat. Numerous Littorininae occur in the littoral fringe. Among these, species of *Tectarius* Valenciennes and *Nodilittorina* von Martens living at supratidal levels of rocky shores, and of *Littoraria* Griffith & Pidgeon in mangrove trees, the gill leaflets are relatively reduced in size, forming mere wrinkles on the surface of the mantle skirt for much of their length (Reid, 1986). This trend for gill size reduction in apparent association with terrestrial existence evidently is expressed in the Pomatiasidae, exemplified by *Pomatias* where the ctenidium is reduced to a few folds of epithelium. Here, the general surface of the mantle skirt assumes the main respiratory surface (Delhaye, 1974a).

The ctenidium is similarly a plesiomorphic trait of the Rissooidea. In many rissooideans, however, the ctenidium is more or less reduced. Different degrees of gill reduction can occur in closely related clades – for example, in the Pomatiopsidae, the ctenidium is well developed in *Pomatiopsis* and *Tomichia*, but greatly reduced or absent in *Blanfordia*. Reduction and loss of the ctenidium often have been assumed to be associated with amphibious and terrestrial life, but such modifications can occur in fully aquatic taxa and, in part, are correlated with small body size. In aquatic forms, the water current normally produced by the ctenidial leaflets is often produced by the enlarged osphradium (Haszprunar, 1985a).

Based on osphradial fine structure, Haszprunar (1985a) showed that the great majority of Caenogastropoda (i.e. those in Sorbeoconcha) share a highly specialized and diagnostic osphradial type and thus form a holophyletic group. The osphradium has a sensory epithelium overlaying the osphradial nerve and thus forming a raised central zone, which is bordered by two strips of epithelium, the lateral zones, each comprising microvillus Si1 cells and ciliated Si2 cells. It is these Si1 and Si2 cells that are particularly diagnostic of the sorbeoconchan Caenogastropoda (Haszprunar, 1985a, 1988a,b). Haszprunar (1985a) found the lateral zones typically to be developed symmetrically about the central zone in the osphradia of sorbeoconchans, including those of the littorinid *Littorina littorea* (Linnaeus). In *Littorina saxatilis* (Olivi), *Littorina neritoides* (Linnaeus) and the hydrobiid *Hydrobia ulvae* (Pennant), Haszprunar (1985a) noted an asymmetry of the lateral zones, and correlated this with a supralittoral habitat. In *Pomatias elegans* (Müller), this trend is further advanced, with the lateral zone (comprising only Si2 cells) present on the right side only. Thiele (1927) earlier had observed a strongly asymmetrical osphradium, similar to that in *Pomatias*, in the assimineid genus *Pseudocyclotus*. In *Assiminea infima* Berry, the left side of the pallial floor is densely ciliated (Hershler, 1987), while in *Assiminea grayana* Fleming two ciliated ridges or tracts are present, one on the mantle floor and one on its roof (Fretter and Graham, 1978). Two opposed ciliary tracts in the pallial cavity have been advanced as a synapomorphic character of the Heterobranchia (Haszprunar, 1985c). These tracts produce the water current in all those heterobranchs in which a water-filled pallial cavity is present. However, it appears that the potential homology of the lateral zone Si2 cells of the caenogastropod osphradium, the ciliated ridges in Assimineidae and the ciliated tracts in Heterobranchia has not been investigated.

The terrestrial rissooideans possess omniphoric grooves (Davis, 1967), which apparently are modifications of the ciliated, often unpigmented, strips that are present in most aquatic rissooideans (e.g. Johansson, 1939; Marcus and Marcus, 1963) in which the left assists with water flow into the pallial cavity and the right carries waste and sediment from this cavity. In the terrestrial groups, the omniphoric grooves are utilized to carry waste from the pallial cavity on both sides. In at least some terrestrial assimineids, this system is modified further, with mucus pumped down the almost tube-like grooves by their mobile edges and streams from the grooves over the side of the foot. As pointed out by Ponder (1988), these modifications appear to be related to the amphibious life, where an inhalant current cannot be created by cilia in the air, but cleansing of the pallial cavity is still necessary. In the case of Assimineidae, the mucus flow required for binding and transporting waste from the pallial cavity is also utilized as a means of keeping the active animal moist.

As indicated above, littorinimorphans are derived from proso-branchiate stock that, while streptoneurous, had achieved epiathroidy in the configuration of the CNS. Indeed, Haszprunar (1988a,b) considered the epiathroid condition and the bifid tentacle nerve as synapomorphies of higher gastropods, of which the sorbeoconch caenogastropods are basal. None the less, the functional reasons for the change from the hypoathroid to the epiathroid condition, and the bifurcation of the tentacle nerve, are not clear (Haszprunar, 1988b). Independently of epiathroidy, the sorbeoconchan visceral nerve ring has become noticeably more compact than in vetigastropods. The pedal cords have shortened to ovoid ganglia in nearly all littorinimorphans, and are connected by a short commissure. In several littorinimorphan families, including Pomatiop-sidae, two pairs of cords, each bearing a single ganglion, emerge from the anterior–ventral and ventral edges of the pedal ganglia: the upper pair are the propodial ganglia and the lower pair the metapodial ganglia. In more advanced littorinimorphans, the propodial and metapodial ganglia are subsumed within the pedal ganglia. In Pomatiasidae, as in most other littorinimorphans, the pedal ganglia possess a second, weaker commissure (in addition to the main commissure), reminiscent of the numerous commissures linking the pedal cords in ancestral stock. This concentration of the pedal nervous system, resulting in true pedal ganglia, is convergent with that in some archaeogastropods. The degree of conden-sation of the visceral neural loop varies greatly among littorinimorphans, but the lack of available data does not allow us to discern with great confidence trends related to phylogeny or ecology. However, among the families of interest from the perspective of terrestriality, the rather long right pleuro-supraoesophageal connective is retained in Littorinidae and some Hydrobiidae, is somewhat shortened in Pomatiopsidae, and is very much shortened in Assimineidae, Truncatellidae and some Hydrobiidae. Condensation may be related to the size of the animals (Davis *et al.*, 1976). A concentrated nervous system generally is assumed to be more efficient in neuromuscular coordination and neurosecretory functions. None the less, neurosecretory cells occur in the majority of ganglia in the littor-inimorphan CNS (e.g. Andrews, 1968), reminiscent of the widespread neurosecretory activity in Architaenioglossa and archaeogastropods. The littorinimorphan gastropods retain the bifid condition of the tentacular nerve, with certain exceptions that may be related to a very small body size (Haszprunar, 1988b; Huber, 1993).

The reproductive tract of littorinimorphans exhibits a high degree of convergence in structure with that in the architaenioglossans. With the exceptions of the littorinid *Mainwaringia rhizophila* Reid, and the vitrinellid *Cyclostremiscus beauii* Fischer, which are protandrous her-maphrodites, all littorinimorphans apparently have remained dioecious. At least one hydrobiid reproduces primarily through parthenogenesis, with males infrequent or absent in many populations.

As in other prosobranchs, the female reproductive tract comprises a proximal ovary and oviduct, a short renal section, and a large distal part formed from the mantle wall. Primitively there are two sperm sacs associated with the renal gonoduct. As in Architaenioglossa, during copulation spermatozoa are deposited in the bursa copulatrix, a sac on the distal renal oviduct and located primarily among the visceral organs but opening via a duct to the posterior of the pallial cavity. Although spermatozoa may be attached temporarily to the lining of the bursa, they are soon transferred to the seminal receptacle, a bulbous sac incorporated in or linked to the renal gonoduct. Considerable variation in the structural relationships of the renal gonoduct, receptaculum seminis, bursa copulatrix and pallial gonoduct has developed in the littorinimorphan female reproductive tract, as illustrated in families with terrestrial representatives. During ontogeny, the pallial gonoduct of littorinids develops as an open groove, which closes to form a laterally compressed, glandular tube (Guyomarc'h-Cousin, 1976), rather similar to that to be observed in architaenioglossans. In all littorinids, the pallial gonoduct takes the form of a complex, glandular tube, comprising a posterior (largely visceral in location) albumen gland and an anterior (entirely pallial) oviductal gland, with the path of the fertilized oocytes within being twisted and spiralled (Reid, 1986, 1988). In all other littorinimorphans, the pallial oviduct is basically a straight tube, so the spiral form in Littorinidae is apomorphic. Reid (1989) recognized the differentiation of the littorinid oviductal gland into three parts, two associated with production of egg capsules, and the anterior-most one with a gelatinous matrix. The degree of development of these different components varies depending on whether reproductive strategy relies on pelagic egg capsules, benthic gelatinous spawn, ovoviviparity or viviparity. In the marine genera of Lacuninae, and *Laevilitorina* Pfeiffer, the sole (marine) genus of Laevilitorininae, the albumen gland is composed of tall, glandular, epithelial cells alternating with ciliated supporting cells (Bedford, 1965; Reid, 1989). This is also the case for both albumen and capsule glands of *Melarhaphe* Menke, the presumptive basal-most genus in Littorininae (Reid, 1989). In all other littorinid genera, including *Cremnoconchus* from Lacuninae, the secretory cells of the albumen and capsule glands are subepithelial, lying in closely packed groups beneath the basement membrane (Linke, 1933; Reid, 1989). The information available from Skeneopsidae (Fretter, 1948), as well as from members of the Cingulopsoidea (Fretter and Patil, 1958; Ponder, 1968), indicates the absence of subepithelial glandular cells as plesiomorphic in Littorinimorpha.

The opening of the female tract is small in Littorinidae, located at the anterior apex of the oviductal gland complex near the mantle collar. The bursa copulatrix has become pallial in these animals, arising as a diverticulum from the pallial gonoduct. Anterior to the point of separation, the bursa stem is continuous with a deep groove in the lumen of the pallial gonoduct, while posteriorly a ventral seminal groove leads to the seminal receptacle. There is a trend for the bursa to be reduced in size in littorinids, and it is apparently absent in the genera *Pellilitorina* Pfeiffer and *Peasiella* Nevill, where reception of spermatozoa during copulation may occur in the pallial gonoduct. Generally, the seminal receptacle is a sacculate structure on the renal gonoduct, but, in the Lacuninae genera *Pellilitorina*, *Lacuna* Turton and *Cremnoconchus*, it is located within the renal gonoduct lumen. Buckland-Nicks and Darling (1993) presented evidence for storage of allospermatozoa in the ovary of *Lacuna variegata* (Carpenter), which may indicate that the entire renal gonoduct in these animals functions as a seminal receptacle. Separated from the lumen of the pallial gonoduct almost to the anterior opening of the oviduct, the seminal receptacle in *Bembicium* and *Risellopsis* is reached by a long duct. Reid (1989) considered this feature a synapomorphy of *Bembicium* and *Risellopsis*, postulating an anterior migration and lengthening of the originally short duct joining the receptacle to the seminal groove. This is the only known case of a diaulic pallial gonoduct in the Littorinidae.

In contrast to the condition of the females, the pallial gonoduct of the males in most littorinid genera is an open groove, communicating with the pallial cavity all along its length. Only in a small number of littorinid genera, including *Cremnoconchus*, has the prostatic gland become a closed, tubular structure. Creek (1951) considered this an adaptation to terrestrial life. Reid (1989) suggested that the closed prostatic gland is plesiomorphic within the Littorinidae, with secondary opening in a number of littorinid genera. However, Reid acknowledged the convergence of closure of the pallial gonoduct in many caenogastropods and could not offer a functional explanation for the postulated reversal of this trend in Littorinidae. Plesiomorphy of the open spermiduct is the most parsimonious explanation. Generally, the glandular tissue of the prostatic gland is epithelial, with some variation in histochemical character along its length in some littorinids. In *Cremnoconchus*, however, the prostatic gland has subepithelial glandular follicles (Linke, 1935; Reid, 1989).

Spermatozoa are transmitted along a ciliated vas deferens to the penis, which is located below and behind the right cephalic tentacle. The vas deferens may be an open groove or a closed duct, the latter condition occurring in *Cremnoconchus*. In most cases, the closure of the spermiduct is superficial, the duct being connected to the penial epidermis by a strip of epithelial cells, probably reflecting its origin by infolding of the surface epithelium (Reid, 1989). Only in the Lacuninae genera *Pellilitorina*, *Lacuna*, *Cremnoconchus* and *Bembicium* is the spermiduct separated from the epidermis and completely surrounded by muscle tissue. The penis in Littorinidae is usually differentiated into a basal region and a more slender, terminal filament, which enters the bursa copulatrix of the female during copulation. Many littorinid genera possess characteristic types of penial mucous glands (Marcus and Marcus, 1963; Reid, 1989; Buckland-Nicks and Worthen, 1992), which represent different solutions to the problem of securing the penis in position during copulation (Bingham, 1972; Reid, 1986). Reid (1989) could find no obvious correlation between the habitats of species and the presence of different penial glands. *Cremnoconchus* is unique among littorinids in that the filament is retracted into the base when the penis is at rest (Linke, 1935; Reid, 1989).

Several authors have postulated that the Pomatiasidae were derived directly from stock that gave rise to the modern Littorinidae. The structure of the female reproductive tract strongly indicates, however, that the pomatiasids evolved from stock more primitive than modern littorinids. In particular, the Pomatiasidae evolved from stock that had not developed closure of the pallial gonoduct. In *P. elegans* and *Maganipha rhecta* Thompson, the best studied pomatiasids (Creek, 1951; Thompson, 1978), there is a widened section of the renal oviduct which functions as a seminal receptacle, with large numbers of spermatozoa oriented towards the epithelium. More anteriorly, there is a sac at the posterior end of the pallial oviduct that functions as a bursa copulatrix. This bursa opens directly and widely into the pallial cavity, so that the mate's penis can be inserted into this sac at copulation. As in Littorinidae, the pallial oviduct of pomatiasids develops as an open groove. However, unlike in littorinids, the pallial gonoduct in these animals does not close over during ontogeny and remains widely open to the pallial cavity as a ventral slit. Creek (1951) suggested that the extensive opening of the pallial gonoduct in the female of *Pomatias* has been acquired secondarily, in connection with the production of the larger eggs necessary for terrestrial development. Thompson (1978) supported this hypothesis on the basis that both *P. elegans* and *M. rhecta* produce relatively large eggs. However, no evidence has been presented for the secondary nature of the open pallial gonoduct, and I favour the idea that the condition in *Pomatias* is homologous to the plesiomorphic condition of Littorinimorpha. The albumen gland remains located in the visceral cavity, immediately behind the posterior wall of the pallial cavity, while the oviductal gland extends along the full length of the pallial cavity. *Pomatias* has only epithelial glandular cells in the pallial gonoduct (Creek, 1951), which also favours the concept of plesiomorphy. The oviductal glands comprise a reduced capsule gland, responsible for a very thin layer around the albumen-enriched embryo in the egg, and a jelly gland responsible for coating the egg with a gelatinous matrix. Each egg as laid has a bright parchment-like surface, but the female covers this with a layer of soil particles as she deposits them.

The male pallial gonoduct in Pomatiasidae is closed to the pallial cavity. The prostatic gland is a broad, spindle-shaped sac, partially embedded in the visceral hump in *Pomatias* but entirely pallial in *Maganipha* Thompson. Based on histochemistry, the prostate can be divided into posterior and anterior sections (Creek, 1951). Throughout, the glandular tissue is epithelial and thus conforms to the basic pattern in Littorinidae. In all pomatiasids for which the anatomy is known, the vas deferens runs forward as a closed duct, immediately beneath the epithelium covering the floor of the pallial cavity, to eventually penetrate the penis. The phallic organ is concealed within the pallial cavity, and thus located a little further posteriad than in Littorinidae. While it is usually differentiated into a basal region and a more slender, terminal filament, it differs from that in Littorinidae in lacking conspicuous mucous glands. During copulation, the penis becomes considerably lengthened and distended with blood so that it may extend through the length of the pallial cavity of the female to the bursa copulatrix. Large variation in penial size occurs amongst populations within pomatiasid species, as evident in Madagascan *Tropidophora* (Emberton, 1995b).

Ponder (1988) pointed out the importance of innervation in determining the homology of the littorinimorphan penis. The penial nerve originates from the right pedal ganglion in littorinids, as is also the case in Skeneopsidae (Ponder, 1988) and *Pomatias* in Pomatiasidae (Garnault, 1887; Creek, 1951). Thompson (1978) indicated that, while the penis of *Pomatias* is of pedal origin, that of *Maganipha* is innervated from the left pleural ganglion via the suboesophageal nerve and is therefore of pallial origin. However, Reid (1989) points out that the base of the penis in *Pomatias* receives nerves from both the right pedal ganglion and the suboesophageal ganglion, suggesting the possibility that the penis is indeed of pedal origin in Pomatiasidae and homologous to that in Littorinidae. A penis innervated from the pedal ganglia and thus of presumed pedal origin also occurs in all rissooidean families, with the apparent exceptions of the Anabathridae and Emblandidae, where penial innervation is from the cerebral ganglion (Ponder, 1988). Ponder (1988) considered Anabathridae and Emblandidae as the most primitive of the Rissooidea and, because of the apparent non-homology of their phallic organs to that of other Rissooidea, argued that the pedal penis had arisen independently among Rissooidea and Littorinoidea from an aphallic condition. In support of this conclusion, Ponder cites small differences in the origin of the penial nerve – from the anterior part of the pedal ganglion in *Littorina*, and from the base of the pleuropedal connective or further up this connective in the Rissooidea. Reid (1989) was able to confirm the origin of the penial nerve in the pedal ganglion for most littorinid genera. However, as the penial nerve of *Cremnoconchus* and *Pomatias* arises at the base of the pleuropedal connective (Linke, 1935; Reid, 1989), Reid questioned whether the penial innervation was informative about homology of the phallic structure in Littorinoidea and Rissooidea. None the less, he concluded that the pedal penis of the Littorinoidea is a synapomorphy of the superfamily, derived independently from that of the Rissooidea.

In females of the rissooidean families Rissoidae, Caecidae, Barleeidae, Anabathridae and Emblandidae, the glandular tissue of the capsule gland is simple, comprising elongate secretory cells whose histochemical properties indicate two or three distinct zones. There may be some infolding of the glandular walls, reminiscent of that in Littorinidae. In all other rissooideans, including those with representatives in the terrestrial environments and thus of interest here (i.e., Pomatiopsidae, Truncatellidae, Assimineidae and Hydrobiidae), the glandular tissue of the capsule gland has been modified by folding into vase-like or tubular structures, the secretory cells are shortened and the histochemical zonation of the capsule gland is lost. A similar modification has occurred in the glandular tissue of the prostatic gland in males of the same families. This change in the cytology of the capsule gland is a major departure from the primitive state in the group and, because of the uniformity in cytological change in affected families, is thought to have occurred once (Ponder, 1988). Closure of the pallial gonoduct clearly occurred after the divergence in the cytology of the capsule glands, as both open and closed conditions of the pallial gonoducts are seen among members of both groups.

In rissooideans, the bursa copulatrix is located primarily in the visceral cavity, as a diverticulum on the renal oviduct that passes forward into the albumen gland and thence to the capsule gland. Amongst the Pomatiopsidae and Truncatellidae, and in a number of other rissooidean families, several clades retain the primitive feature of utilization of the renal opening for reception of spermatozoa during copulation. In Pomatiopsidae (Triculinae), a renopericardial duct takes spermatozoa from the nephridium via the pericardium (Ponder, 1988), or the seminal duct opens at the posterior of the pallial cavity after passing through the renal wall (Davis *et al.*, 1976; Davis, 1979), or opens into the pericardium and thence into the pallial cavity (Davis, 1968; Davis *et al.*, 1983, 1984). In Truncatellidae (Truncatellinae), the bursa lies within the nephridium with which it communicates by way of a narrow opening, either directly through the bursal wall or by a short duct from the bursal duct (Fretter and Graham, 1962; Kosuge, 1966; Ponder, 1988). A second connection, via a well-developed gonopericardial canal, apparently allows spermatozoa to enter the renal gonoduct from the pericardium via the renopericardial opening. Pomatiopsidae (Pomatiopsinae) differ from Pomatiopsidae (Triculinae), and Truncatellidae (Geomelaniinae) differ from Truncatellidae (Truncatellinae), in that the bursal connection to the renal system is lost, and an elongated seminal duct extends toward the anterior end of the pallial cavity. The seminal duct of the Pomatiopsidae forms during ontogeny as a bud from the bursa (Davis *et al.*, 1976).

Some rissooideans retain the gonopericardial duct, but acquisition of spermatozoa in these animals no longer involves the renal system. Rather, in these rissooideans, allospermatozoa are received primitively through a seminal duct from the bursa copulatrix that is open to the posterior of the pallial cavity. From this primitive ground plan, two different courses of evolutionary development have taken place with respect to the seminal duct. Firstly, the duct is often lengthened towards the anterior of the pallial cavity. In some cases, the anterior section of this seminal duct has fused with the pallial gonoduct and the bursa itself has been displaced anteriorly to lie within the pallial cavity. Secondly, in other rissooideans, the bursa has lost its opening to the pallial cavity and an allospermatozoa-conducting groove has developed along the ventral longitudinal axis of the pallial gonoduct. In the more advanced stages of this latter evolutionary trend, evident in ontogeny in hydrobiid *Spurwinkia* Davis, Mazurkie

& Mandracchia (Davis *et al.*, 1982) and in transitional series amongst species in several families (Ponder, 1988), a seminal duct forms as the ventral channel closes off and separates from the pallial gonoduct. It is evident that both types of pallial seminal ducts have arisen several times, independently, amongst Rissooidea, and indeed both may be present within one family or even subfamily. Thus, in Hydrobiidae and Assimineidae, we see an array of seminal duct forms, but in the absence of comparative ontogenetic information it is often impossible to determine the origin of the seminal duct. This has led to ambiguous scoring of character states and uncertainty about the robustness of phylogenetic reconstructions. My interpretation of the polarity of bursa and seminal duct changes in Rissooidea differs substantially from that of Ponder (1988). Davis (1981) regarded the elongated, pallial seminal duct condition, as seen in Pomatiopsinae, as a pre-adaptation to an amphibious–terrestrial existence, enabling successful copulation and sperm transfer out of water.

Rissooidean males are invariably phallate, with the living animal carrying the penis tightly coiled over the neck. The penis is often simple, with an epithelium populated by glandular cells but lacking specialized glands. However, in most rissooidean families, the males possess complex glandular penial structures, comprising glandular ridges in which rows of small glands discharge through a central slit, apocrine glands, glandular papillae or mammiform glands. Often there is a ciliated anterior epithelium. The prostatic gland in male rissooideans shows all gradations from a channel widely open to the pallial cavity to being entirely closed and sacculate. Similarly, the vas deferens varies from an external groove to a muscular duct that runs internal to the penis. However, both the prostatic gland and the vas deferens invariably are closed in the Pomatiopsidae, Assimineidae, Hydrobiidae and Truncatellidae. The vas deferens penetrating the penis often ends as a papilla, which may be eversible through the action of retractor muscles.

The Littorinimorpha primarily produce spawn in which several to numerous eggs are embedded in a gelatinous mass within a rind. These spawn differ from those of archaeogastropods in that each egg has its own supply of albumen within a thin membrane. Furthermore, the trochophore larval stage invariably passes within the egg, thus the veliger is the only free larval form in Littorinimorpha. Primarily the veliger has a long planktonic life and feeds actively (planktotrophy) before metamorphosis, with the larval shell produced successively by the mantle margin prior to metamorphosis and settlement.

Direct development, where larval life and metamorphosis occur within the confines of the egg, has evolved repeatedly in Littorinimorpha. A further trend common in Littorinimorpha and other caenogastropods is the encapsulation of the spawn within a multilayered, species-diagnostic capsule, which represents elaboration of the simple rind that encases the gelatinous spawn. As indicated by species of *Bembicium*, in Lacuninae, the littorinids primitively conform to the basic littorinimorphan plan of benthic gelatinous spawn and planktotrophic veligers. However, most species in the subfamilies Laevilitorininae and Lacuninae complete larval development within the spawn mass. The eggs and mode of larval development in *Cremnoconchus* presently are unknown. In the more highly derived subfamily Littorininae, the eggs are mostly pelagic, enclosed in capsules. Most pelagic capsules yield veliger larvae that undergo planktotrophic development. In some *Littorina* species, the capsules are benthic and development is direct. Yet further *Littorina* species are ovoviviparous, brooding the embryos either in the pallial cavity for a period before releasing planktotrophic veligers or in the modified pallial gonoduct until completion of larval development. Pomatiasidae produce simplified, spherical capsules, in each of which a single embryo undergoes direct development, although details are available only for *Pomatias* (Creek, 1951; Kilian, 1951). In the case of *P. elegans*, each spherical egg capsule is about 2 mm in diameter and, upon oviposition, is manipulated by the female so as to receive a coating of soil before being buried beneath the soil surface. The egg contains little yolk but a large supply of albumen.

Rissooideans typically deposit the eggs in hemispherical capsules that are attached to hard substrates. In the marine families, planktotrophic development of veligers is plesiomorphic, with this form of development probably having evolved in caenogastropods independently of that in Neritopsina (Haszprunar, 1995). However, direct development has evolved repeatedly in rissooideans and was pre-adaptive to colonization of freshwater and terrestrial environments. Hydrobiidae retain the hemispherical egg capsules. The estuarine species often produce planktotrophic larvae, but the freshwater and terrestrial hydrobiids have direct development. The egg capsules of Assimineidae, Pomatiopsidae and Truncatellidae are spherical or globular, contain a single zygote and have a sticky mucous coating that causes soil and detritus to become attached. With the exception of some estuarine species of Assimineidae, the embryos of these latter three families undergo direct development. Information on the reproductive biology of the terrestrial rissooideans generally is confined to *Pomatiopsis lapidaria* (Say) (e.g. Abbott, 1948; Dundee, 1957).

Littorinimorphans generally are microphagous, feeding on finely particulate organic debris, algal and bacterial films and, in the terrestrial environment, on algal and fungal floras associated with decaying vegetation and on the surface of live leaves. In addition to their microphagous diet, some marine Littorinidae are known to be herbivorous on macroalgae, and members of the terrestrial sister family Pomatiasidae feed on leaf litter. The digestive system is only slightly modified from that seen in Vetigastropoda, with a general trend towards simplification. The modern Littorinimorpha clearly have been derived from stock in which the dentition had been reduced to the taenioglossate form. Only in some members of the rissooidean Emblandidae has the dentition been reduced further. The radular ribbon generally remains long in Littorinoidea, but generally is shortened in Rissooidea. The number of cartilages in the buccal mass is reduced to one pair. The single pair of jaws is generally well developed in Littorinimorpha, but has been lost in a number of lineages, including Pomatiasidae and all Littorinidae except *Pellilitorina* amongst Littorinoidea, and some Assimineidae amongst Rissooidea. The oesophageal gland is a plesiomorphic feature, but in most Littorinimorpha the glandular tissue comprises a raised ventral fold, sometimes interrupted along the midventral line by a zone that marks where the ventral ciliated tract once stood. Lateral pouches of the oesophageal gland persist in reduced form in a number of Littorinimorpha, including Littorinidae and some Assimineidae and Hydrobiidae. The crystalline style, which represents a specialization of the protostyle, is developed, perhaps independently, in several sorbeoconch lineages where more or less continuous microphagous or filter-feeding occurs and free proteolytic enzymes are absent (Robson, 1922; Yonge, 1930, 1932; Graham, 1939). The crystalline style is known from members of Hydrobiidae, Assimineidae, Pomatiopsidae, Truncatellidae and Pomatiasidae, but its occurrence in terrestrial species has been confirmed only for some species of Pomatiasidae, in part because of the lack of detailed anatomical study devoted to land-dwelling littorinimorphans. The oesophageal glands become vestigial in the presence of a crystalline style (Graham, 1939). Thus, in Hydrobiidae, Assimineidae, Pomatiopsidae, Truncatellidae and Pomatiasidae, the oesophageal gland is very much reduced or absent, and the entire oesophagus takes on the appearance of the posterior oesophagus of more basal gastropods in being essentially circular in cross-section and lined with a ciliated epithelium thrown into numerous longitudinal folds. Irrespective of the presence or absence of the style, the style sac is a prominent feature of the stomach. The digestive gland often is reduced to a single opening to the stomach, although occasionally there are multiple openings. The intestinal length is reduced, and typically the intestine runs posteriorly along the style sac and then makes an abrupt turn to run anteriorly; the proximal part of the intestine contains a pronounced typhlosole. The rectum is no longer associated with the pericardium and runs directly to the anus located at the anterior of the pallial cavity; its course is usually straight, but in some rissooideans the rectum forms one or two loops on the pallial roof. These animals produce faecal pellets.

Long cephalic tentacles are plesiomorphic in Littorinimorpha. However, the cephalic tentacles are rather short in *Cremnoconchus* and *Blanfordia*, and often are vestigial or entirely absent in Assimineinae. The eyes are of advanced closed-vesicle lens type, located centrally at the base of the tentacles, on basal bulges of the tentacles or on short peduncles adjacent to the tentacles. In some deep marine and phreatic freshwater rissooideans, the eyes are without pigment or are entirely absent.

The foot sole has a distinct propodium and the suprapedal gland opens in the groove along the anterior edge of the foot between this and the mesopodium. Amongst Littorinidae *Cremnoconchus* is an exception, lacking a transverse propodial groove and exhibiting a much reduced suprapedal gland. The Pomatiasidae do not have a discernible propodium nor transverse groove, and the suprapedal gland opens by a pore at the front end of the longitudinal groove in the sole.

A second pedal gland, the metapodial mucous gland, is found beneath the sole in the centre of the foot of basal sorbeoconch caenogastropods. The gland opens to a longitudinal slit that may reach back to the posterior edge of the foot. The posterior pedal gland is common in marine forms of Littorinimorpha, but rarely persists in species occupying estuarine environments and invariably is absent from those in freshwater and terrestrial environments. A tubular gland occurs in the foot of Pomatiasidae, consisting of a mass of tubules in the haemocoel and opening by a pore halfway along the longitudinal groove dividing the sole. It is not a mucus gland, but is concerned with osmoregulation (Delahaye, 1974a) and is therefore not homologous to the posterior pedal gland.

In most Littorinidae, the sole of the foot is divided longitudinally by a deep groove extending from the posterior edge almost to the anterior transverse groove. This is a sign of the functional division of the foot into two halves, responsible for the retrograde ditaxic style of locomotion (Miller, 1974). However, in a number of littorinid genera, including *Cremnoconchus*, no such division of the foot occurs, although locomotor waves apparently are still retrograde and ditaxic (Miller, 1974; Reid, 1989). The Pomatiasidae do have a similar longitudinal division of the sole, but here the ditaxic waves are direct (Miller, 1974). Although in prosobranchs as a whole, retrograde and ditaxic locomotion is the most common type, in sorbeoconch caenogastropods it is uncommon (Miller, 1974), and Ponder (1988) has suggested that ditaxic locomotion may be a synapomorphy of the Pomatiasidae and Littorinidae. However, the different direction of the pedal waves indicates probable independent acquisition of the bipartite foot in these two families. The rissooideans generally travel by ciliary gliding. The truncatellids have modified their essentially ciliary method of locomotion by involving the snout in a stepping motion (Fretter and Graham, 1962; Kosuge, 1966). Some pomatiopsids and assimineids also involve the snout for locomotion, albeit intermittently (Davis, 1967; Ponder, 1988).

Ellobioid pulmonates

The family Ellobiidae has a worldwide distribution. Five subfamilies generally are recognized (Morton, 1955b; Martins, 1996a,b), although Harbeck (1996) treats Carychiinae as a distinct family and Starobogatov (1976) recognizes Melampodidae, Pythiidae, Pedipedidae, Cassidulidae, Leucophytiidae, Ellobiidae and Carychioidae. The Ellobiinae, Pedipedinae, Melampinae and Pythiinae (*sensu* Morton and Martins) collectively constitute a characteristic component of the mollusc fauna of the upper and supralittoral zones of the mangrove forests and muddy shores of the tropical regions, and salt marshes and upper littoral rocky areas of temperate regions. Among these, *Ovatella aequalis* (Lowe) in Pythiinae and some *Melampus* de Montfort in Melampinae live just above the high tide. Some estuarine species, particularly in the genus *Melampus*, spend considerable amounts of time out of water, often on vegetation. Among these subfamilies only the western Pacific *Pythia* Röding in Pythiinae has achieved true terrestriality; its members always frequent moist places, generally near the shore out of reach of the highest tides, but occasionally up to 1 km inland and at elevations up to 100 m. Species such as *Pythia scarabaeus* Linnaeus are arboreal in mangrove trees. The subfamily Carychiinae is exclusively terrestrial, with the Holarctic *Carychium* Müller living inland in very humid environments, frequently under forest leaf litter or in wetlands, and one species, *Carychium stygium* Call, a troglophile of North America. The European *Zospeum* Bourguignat generally is associated with Karst caves. Thus, the Ellobiidae exhibit rather modest diversity and rather little adaptive radiation on land. Contrary to the perception of Morton and Miller (1973), Starobogatov (1976) and others that the Ellobiidae represent a terrestrial lineage that has returned to the marine habitat ('. . . returning between tides from an atmospheric habit.' Morton and Miller, 1973, p. 237), it is clear that terrestriality in ellobiids has been derived repeatedly from stock living in the marine littoral zone.

The fossil record of the Ellobiidae, reviewed by Martins (1996b), is relatively poor and provides only skeletal information on their evolutionary history. The oldest known fossil ellobiids are the European *Carychiopsis* Sandberger from the Palaeozoic of France. This genus resembles *Carychium*, which is known from the Jurassic of Asia, Europe, America and the West Indies (Zilch, 1959–60). Presumptive halophilic Pedipedinae or Pythiinae are known from the Palaeozoic of Europe and Melampinae from the Cretaceous of North America. It appears, then, that the ellobiids had already invaded the terrestrial habitat through the Carychiinae during the Palaeozoic, long antedating the arrival of the higher caenogastropods such as Littorinimorpha.

The ellobiid shell, oval–conic in shape and with apertural dentition, ranges from 100 mm in *Ellobium* Röding to barely 1 mm in *Leuconopsis* Hutton and some *Pythia*. *Pythia* is unique in having a laterally flattened shell. The shell is heterostrophic in that the larval shell or protoconch is hyperstrophically sinistral and, due to reorientation during ontogeny, sits atop of the dextral, orthostrophic adult shell. Hyperstrophy is suppressed in ellobiid species with direct development. Hyperstrophy is apparently rare in gastropods and has been considered by some malacologists to have arisen only once, thus constituting a synapomorphic character of the Heterobranchia (Robertson, 1985; Haszprunar, 1985c). However, the occurrence of heterostrophy in trochoidean vetigastropods, including cases of hyperstrophy (e.g. Hadfield and Strathmann, 1990), raises the possibility that this geometric expression of shell development had its origins in the archaeogastropods or has been derived independently in several gastropod lineages (Hickman, 1992). All ellobiids, except *Pedipes* Scopoli and *Creedonia* Martins in Pedipedinae, are characterized by partial to complete resorption of internal shell walls. The animal is completely retractable into the shell. In gastropods, the columellar retractor muscle is primarily adnate to the body wall throughout its length, and this pattern is retained in basal Heterobranchia and indeed persists into lower pulmonates. In Ellobiidae, however, there is a trend towards the main stem of the columellar retractor muscle becoming free from the body wall from its origin on the columellar. The result is inversion of the cephalopedal mass on retraction of the ellobiid animal into its shell, rather than simple retraction. The operculum is retained in the veliger larval stage (and in juveniles in *Blauneria heteroclita* (Montagu)), but is absent in the adult.

The Ellobiidae represent the most primitive of the heterobranch clades, and indeed of the pulmonate clades, that have colonized the land. They therefore provide a useful framework for considering the adaptive morphology and ecology of the pulmonate terrestrial radiation. Like that in the Caenogastropoda, the pulmonate radiation is usually assumed to have occurred after the attainment of the monotocardian condition (heart with a single auricle), with concomitant loss of the right nephridium and development of elaborate pallial gonoducts. As their name suggests, the pulmonate gastropods breathe with a 'lung' or pulmonary cavity; this consists of the pallial cavity richly supplied with blood vessels in the form of a capillary network and which is closed except for a small opening, the pneumostome, to the exterior. Contrary to some earlier reports, but consistent with interpretations of Hubendick (1945), Yonge (1952), Brace (1983) and Morton (1988), Ruthensteiner (1997) has demonstrated through studies of ontogeny that the pulmonary cavity of ellobiids and other pulmonates is homologous to the pallial cavity of prosobranchs. The pulmonary cavity thus does not represent a synapomorphic character of pulmonates. A morphocline from a widely open pallial cavity to one closed but for the pneumostome is evident in both pulmonates and opisthobranchs, indicating that the pneumostome too is not a synapomorphic character of pulmonates. The pneumostome is contractile in ellobiids and thus more advanced than the simple, passive orifice of the more basal pulmonates such as the marine Amphibolidae and the freshwater Basommatophora. However, both simple and contractile pneumostome conditions are represented in the marine Siphonariidae, suggesting that the contractile pneumostome may have arisen independently in several pulmonate clades. The pulmonary vein and its pallial branches extend across the dorsal wall of the pallial cavity, permitting respiratory exchange in the absence of a ctenidium, but generally vascularization of the pallial cavity is poorly developed in ellobiids and other basal pulmonates. Having the pallial cavity enclosed reduces water loss by evaporation and enables the carrying of a reserve supply of water. The marginal gain provided by such a modified pallial cavity permits pulmonates to be active not only out of the aqueous environment, but also under relatively low humidity conditions. The development of the enclosed, vascularized pallial cavity was clearly pre-adaptive to terrestriality and was, at least partly, responsible for the more successful terrestrial radiation of pulmonates than that of neritopsins and caenogastropods.

Consistent with a heterobranch level of evolution, the nephridium has both secretory and resorption functions (Delhaye and Bouillon, 1972a), and has assumed the role of ultrafiltration of the blood, a function that was undertaken by the heart in lower gastropods. The nephridium is located entirely within the pallial cavity and, in contrast to the visceral supply in Architaenioglossa and Littorinimorpha, the blood supply to the nephridium in Ellobiidae and other Heterobranchia is pallial. The nephridium passes blood directly to the heart. The ellobiid nephridium is elongated, rather like that in Pyramidellidae with the greater axis orientated anterior–posteriorly with respect to the animal's body, is sacculate, and is subdivided internally by transverse lamellae. The orifice is located at the anterior apex, a little short of the anterior margin of the pallial cavity. In some ellobiids, the anterior-most section of the nephridium, immediately behind the orifice, is free of lamella and thus essentially forms a distal ureteric pouch (termed an orthureter). A nephridial gland is absent. The ventricle is located posterior to the auricle in the pericardium, and gives rise to an aorta that runs forward through the first loop of the intestine before dividing into anterior and posterior branches.

The pallial cavity in ellobiids variously combines retention of the hypobranchial gland located between the nephridium and rectum, two conspicuous glandular pads associated with the pneumostome (pneumostomal glands), and a darkly pigmented excretory organ (pallial organ) to the left of the pericardium and linked by a vessel to the pulmonary vein. All three elements are combined in *Melampus* (Marcus and Marcus, 1965a). In form, the hypobranchial gland of ellobiids is unremarkable. The larvae of Heterobranchia often possess an unpaired (usually) pigmented mantle organ located close to the rectum that probably had/has an excretory function (i.e. black larval kidneys). The lower heterobranchs (e.g. Omalogyridae, Rissoellidae, Pyramidellidae, Architectonicidae and Orbitestellidae) generally retain their 'larval kidneys' in the adult, whereas it is only a larval organ in the higher groups (Haszprunar, 1988b, and references therein). Haszprunar (1988b) indicated that this might be due to paedomorphosis or may reflect that status of the lower Heterobranchia as ancestral with respect to the pulmonate/opisthobranch groups. As in other pulmonates, the larval kidneys are reduced during metamorphosis and do not persist in post-larval stages in most Ellobiidae. However, in a number of ellobiid genera, including *Melampus*, *Carychium* and *Pythia*, a pigmented organ is found in the roof of the pallial cavity of the adult, in a position corresponding to that in adults of lower heterobranchs. In addition, two conspicuous glandular pads are often associated with the pneumostome (pneumostomal glands), described by Marcus and Marcus (1965a) as occupying the roof and floor of the cavity, respectively, in *Melampus*. Similar structures are seen in other basal pulmonates such as *Otina* Gray of Otinidae (Morton, 1955c).

The ellobiid CNS is hypoathroid in as much as the pleural and pedal ganglia are adjacent. Haszprunar (1985b,c, 1988b) indicated that all archaeogastropods and lower caenogastropods with a hypoathroid CNS possess a simple tentacular nerve, while gastropods that have an epiathroid CNS have a bifurcated tentacle nerve (with certain exceptions in very small species). The true epiathroid type has probably evolved only once, being present in Caenogastropoda and in primitive Heterobranchia. Haszprunar (1988b) regarded the hypoathroid nervous system in Ellobiidae, shared among pulmonates only with Succineoidea, Stylommatophora and some Trimusculidae, as a secondary phenomenon. This contention is supported by the occurrence of the bifid tentacle nerve in Ellobiidae (Huber, 1993), although the tentacular nerve is (? secondarily) unbranched in some ellobiids (Marcus and Marcus, 1965b).

An essential autapomorphy of the pulmonate CNS is the development of the special neurosecretory system comprising procerebrum and cerebral gland. According to Pelseneer (1901) and van Mol (1967, 1974), the procerebrum may contain either large cells, small cells (i.e. globineurons) or both. The presence of large cells alone is correlated with the existence of an osphradium, and a capacity to breath when submerged. These cells occur in Siphonariidae, Amphibolidae, Chilinidae, Latiidae and higher freshwater groups such as Lymnaeidae and Planorbidae. Contrary to the view of van Mol (1967, 1974), the presence of large cells alone is certainly primitive for the pulmonates (Haszprunar and Huber, 1990). The presence of globineurons is correlated with the existence of a contractile pneumostome and exclusive air-breathing (Harry, 1964), and is to be regarded as a synapomorphy of the higher pulmonates Ellobiidae, Succineidae, Athoracophoridae, Trimusculidae, Otinidae, Smeagolidae, Onchidiidae, Vaginulidae, Rathouisiidae and the numerous families in Stylommatophora. Among these, the Trimusculidae alone have retained the original large cells in addition to the acquisition of globineurons. As outlined by van Mol (1967, 1974), the cerebral gland exhibits a trend from a tube external to the cerebral ganglion in Ellobiidae, Trimusculidae, Siphonariidae and Succineidae, to a gland internalized within the cerebral ganglion, without an external opening, in the remainder of studied Pulmonata; Haszprunar and Huber (1990) suggest that this process has occurred in various pulmonate clades. In adult Ellobiidae, the tube of the cerebral gland can be traced from the exterior of the animal to the cerebral ganglion (Ruthensteiner, 1999), reflecting its origin as an ectodermal invagination, and seen in histological preparations to be filled in its proximal part by a glandular follicle (Lever et al., 1959). In contrast to lower gastropods but in common with pyramidelloidean Amathinidae and opisthobranchs (Bullock, 1965; Weiss and Kupfermann, 1976; Ponder, 1987), the neurons of the central ganglia in pulmonates increase in size as the animals grow, leading to neuronal giantism (Gillette, 1991).

During ontogeny, the procerebrum arises independently of the cerebral ganglion to which it is linked by two connectives. The homology of the pulmonate procerebrum to the rhinophoral ganglion of the

Pyramidelloidea and Opisthobranchia has not been resolved satisfactorily (for discussion, see van Mol, 1967; Haszprunar, 1988b; Haszprunar and Huber, 1990; Huber, 1993), although Ruthensteiner (1999) provides evidence that these two structures are the same and probably homologous to the tentacular ganglion of caenogastropods. The cerebral gland also arises independently of the cerebral ganglion, and may be homologous to the subtentacular ganglion of the Opisthobranchia.

A thick cerebral and a thin subcerebral commissure link the cerebral ganglia. The subcerebral commissure is present in most Pentaganglionata, and these taxa therefore share a character with the Vetigastropoda and Neritopsina. This character is not found in the basal Heterobranchia. The pedal ganglia in genera such as *Melampus* and *Pythia* are often united by two commissures, the posterior one being weaker. In other ellobiids, this posterior commissure is lacking. The Ellobiidae thus developed from gastropods with two or more pedal commissures.

The step from the triganglionate to the pentaganglionate level of organization of the visceral neural loop has been suggested as one of the most important synapomorphic characters uniting the Opisthobranchia and Pulmonata as a monophyletic group within the Heterobranchia (Haszprunar, 1985c). The pentaganglionate condition arises through the addition of a pair of parietal ganglia to the visceral loop that plesiomorphically comprises visceral, suboesophageal and supraoesophageal ganglia. According to Hoffmann (1939), Bullock (1965) and Regondaud *et al.* (1974), the parietal ganglia formed by separation of portions of the original pleural ganglia that supply the lateral body wall and parts of the mantle, in association with the elongation and modification of the head. Although these parietal ganglia often are fused with other ganglia, they are clearly visible in primitive pulmonates, at least during ontogenesis. However, there is evidence that accessory ganglia on the visceral loop, presumed homologous to the parietal ganglia, are present in some lower heterobranchs (e.g. *Gegania valkyrie* Powell, Mathildidae; Mikkelsen, 1996), indicating an origin earlier than that postulated by Haszprunar (1985c).

In Chilinidae and Latiidae, the suboesophageal and visceral ganglia are separated by short connectives. In all other pulmonates, these two ganglia are in contact or, more commonly, fused. In pulmonates, the supraoesophageal ganglion tends to be in contact with or fused with the right parietal ganglion. In Ellobiidae, these ganglia are fused. A moderately long visceral loop, with traces of chiastoneury, can persist in the presence of these trends, as evidenced by the condition in Chilinidae and Ellobiidae. Euthyneury in pulmonates is brought about by both concentration of the visceral loop and detorsion associated with migration of the osphradium and supraoesophageal ganglion from the left to the right side of the pallial cavity (Haszprunar, 1985c). The visceral loop is variably concentrated in Ellobiidae, but invariably without fusion of the pleural and parietal ganglia. Several species of Ellobiidae (e.g. *Ellobium dominicense* (de Férussac), *Myosotella myosotis* (Draparnaud)),

Chilinidae and Latiidae have an accessory ganglion at the left side of the visceral loop and thus exhibit a hexaganglionate condition. In the majority Ellobiidae, this accessory ganglion is incorporated in the left parietal ganglion (Martins, 1996a,b), thus the hexaganglionate condition may be plesiomorphic in higher heterobranchs, or simply a convergent innovation in Chilinidae, Latiidae and Ellobiidae.

Heterobranchs are hermaphrodites, with the primitive state being a gonad (ovotestis) producing both spermatozoa and ova. Reproduction in Ellobiidae varies from protandric to simultaneous hermaphroditism. Heterobranchs share a modified spermatozoon that possesses a distinctive acrosome, spiral nucleus and a complex mitochondrial derivative, reflecting a different mode of spermiogenesis from that in lower gastropods (Healy, 1988a,b, 1993). Martins (1996a,b) recognized five types of organization in the reproductive system of Ellobiidae.

1. The Pythiinian type, monaulic[1], with the anterior oviductal gland and prostatic gland running parallel to each other and covering the pallial gonoduct as far as the vaginal atrium.
2. The Ellobiinian type, diaulic with gonoducts separating immediately below the receptaculum seminis, with the anterior oviductal and prostatic glands covering the entire length of the pallial gonoducts.
3. The Carychiinian type, monaulic, with the pallial gonoduct glandular and the prostatic gland concentrated distally in the gonoduct.
4. The Pedipedian type, which is monaulic or incipient semidiaulic, with the anterior oviductal gland and prostatic gland covering only the proximal half of the pallial gonoduct.
5. The Melampinian type, which is advanced semidiaulic, with a very short spermoviduct and separated long vagina and posterior vas deferens.

The phenomenon of agglomeration of the oviductal glands around the proximal part of the pallial gonoduct, in a somewhat diverticular configuration, is a highly distinctive feature of ellobiids. However, this feature of the pallial gonoducts is present in basal heterobranchs such as Pyramidellidae, Cornirostridae and Valvatidae, and retained in varying degrees in opisthobranchs and in Ellobiidae, Latiidae, Physidae, Siphonariidae, Otinidae, Smeagolidae, Onchidiidae, Athoracophoridae and Succineidae among pulmonates. This is suggestive of a probable plesiomorphy in the Heterobranchia. The albumen gland departs from

---

[1] The terms monauly and diauly are applied here only with respect to those sections of the reproductive tract primitively associated with the pallial cavity, and essentially concern the relationship of the proximal spermiduct and its prostatic tissues to that of the female oviduct gland. Thus monauly refers to the situation in which the male and female pallial gonoducts open to a single lumen, albeit that the lumen is often morphologically or physiologically subdivided by longitudinal folds. Diauly refers to the situation where the male and female pallial gonoducts are separated. This interpretation differs from that expressed by some authors (e.g. Nordsieck, 1985) who confound several different structures by also taking into account vas deferens, free oviduct/vagina and union of the male and female orifices.

that seen in lower gastropods in that the secretory products drain via a duct to the pallial gonoduct; thus the zygotes do not passage through the gland: the spatially separated albumen gland is probably a synapomorphy of pulmonates. *Pythia* primitively retains a spacious lumen in the albumen gland, but in most ellobiids the gland is complexly lobed with narrow collecting ductules.

Unlike the condition in lower gastropods with the pallial gonoduct in the pallial cavity roof, the reproductive system of basal heterobranchs such as Rissoelloidea, Glacidorboidea, Pyramidelloidea, Acteonoidea, Ringiculoidea and Diaphanoidea is characterized by the pallial gonoduct located in the pallial cavity floor. This trend is elaborated further in higher opisthobranchs, and in pulmonates including Ellobiidae, with the pallial gonoduct being displaced from the pallial cavity to be free, along with the gonadial or mesodermal part of the reproductive tract, in the body cavity. Components of the developing reproductive system in *Ovatella* Bivona differentiate in the post-metamorphic animal on the ventral epithelium of the pallial cavity. The bursa copulatrix together with the duct and the anlage of the pallial glandular portion of the gonoduct are first formed as grooves and subsequently constricted from this epithelium (Ruthensteiner, 1997).

The bursa copulatrix of Ellobiidae arises from the pallial gonoduct. The bursal sac retains the primitive association with the region of the visceral cavity above the pericardium, which necessitates a rather long duct when the opening to the pallial gonoduct is displaced anteriad. The bursa functions as a gametolytic organ (spermatheca, bursa resorbiens), being used for the digestion and resorption of excess spermatozoa in addition to reception of spermatozoa during copulation. The allospermatozoa storage organ, the receptaculum seminis, is unremarkable in being a diverticulate sac on the lower renal gonoduct or, perhaps more commonly, within the lumen of this duct. The seminal receptacle opens distally to a small sacculate chamber that serves as the site of fertilization. This fertilization chamber is an essential autapomorphy of the heterobranchs, but exhibits particular elaboration in ellobiids and other pulmonates.

A tubular diverticulum associated with the terminus of the female portion of the pallial gonoduct is apparently plesiomorphic in Ellobiidae, being present in *Pythia*, *Carychium* and *Ovatella*. It has been lost in some other genera. It evidently is a reproductive structure, given that its presence in *Carychium* varies with the phallic states of individuals (Harry, 1952) and has been found to contain spermatozoa in *Pythia* (Berry *et al.*, 1967). Morton (1955a) termed this structure the pallial gland, but showed that it has no opening into the pallial cavity. Further, Berry *et al.* (1967) found it to be lined by a ciliated epithelium within substantial muscle, and it thus does not conform to a gland. This structure may be homologous to, and thus represent a distal displacement of, the diverticulum on the oviduct part of the female pallial gonoduct in basal pulmonates such as Chilinidae, Otinidae, Siphonariidae and Latiidae.

The Ellobiidae are ditrematous. The female aperture is in the body wall below the pneumostome at the mantle margin, a location corresponding to that in caenogastropods when allowance is made for migration of the pallial gonoducts into the body cavity. The male aperture is anteriad, under the right cephalic tentacle, and thus is consistent with the usual gastropod location of the male pore. The Ellobiidae evidently are derived from marine gastropods with an open sperm groove, as this condition is present in *Pythia*. Indeed, *Pythia* alone among the pulmonates has retained this primitive arrangement for conduction of autospermatozoa. In other ellobiids, a closed vas deferens lies embedded in the body wall in the cephalic region; the latter taxa, however, retain the (? non-functional) ciliated groove along the right side of the body. Mikkelsen (1996) noted the simultaneous occurrence of the external groove and internal duct in shelled sacoglossan opisthobranchs, but erroneously discounted the possibility of homology of the external groove to the functional equivalent groove in Caenogastropoda. In all ellobiids, including *Pythia*, a closed vas deferens emerges from the body wall inward near the male aperture to run to its insertion at the apex of the penis. Primitively, this portion of the spermiduct adheres to the penial wall, as evident in *Blauneria* Shuttleworth in Ellobiinae and *Myosotella* Monterosato in Pythiinae, but in most ellobiid genera the vas deferens is freed from the penis. The penis is internal but everts during mating to penetrate the lower arm of the female section of the pallial gonoduct in the mate. Retraction after mating is effected by a retractor muscle, which is anchored variously to the cephalic body wall, columellar muscle or the floor of the pallial cavity. The insertion of the vas deferens in the penis is elaborated as a vergic papilla that is fleshy and conical in genera such as *Pythia* and *Cylindrotis* Möllendorff (Ellobiinae), a thin filament in *Ellobium*, conical and penetrated by a thin filament in *Auricula* de Lamarck, and vestigial in *Carychium*. Haszprunar (1985c) regarded the protrusible penis, retractable primarily by a branch from the columellar muscle, as a synapomorphic character of the higher pentaganglionate gastropod groups, but internalization of the penis probably developed repetitively and independently with sinking of the pallial gonoduct into the haemocoel (Ponder, 1991). None the less, the uniformity in penial structure suggests a single derivation in the stem that led to the terrestrial pulmonates. The penial nerve emerges from the cerebral ganglion, but data are lacking on the ultimate source within the CNS. While spermatophores are produced commonly in heterobranchs, they apparently are absent in Ellobiidae. Just proximal to the female aperture in several ellobiids, a minute duct, the *canalis junctor*, links the male pallial duct with the bursa copulatrix (Marcus and Marcus, 1965b).

Morton's (1955a) study indicated seasonal protandry in *Carychium tridentatum* (Risso), with euphally in summer and aphally for the remainder of the year. Doll (1982) observed copulation in this species. In contrast, Bulman (1990) indicated that copulation was absent in colonies of *C. tridentatum* maintained in the laboratory over four generations, and suggested that this species has some form of uniparental reproduction. Harry (1952) found the ratio of euphallic to aphallic individuals to vary between populations of *Carychium exiguum* (Say).

The Melampinae produce jelly-like masses containing 5–2000 egg capsules, each with an embryo. These eggs yield free-swimming, plankto-trophic veliger larvae. The veliger stage and metamorphosis in the other ellobiid subfamilies occur within the egg, and hatching occurs at the crawling juvenile stage. *Cassidula* de Férussac, *Ovatella* and *Leucophytia* Winckworth produce egg capsules, numbering as many as 200 and each containing a single embryo, embedded in a mucoid mass. *Auricula* produces several hundred egg capsules that are linked in a string by short chalaziform filaments. In laboratory colonies, Bulman (1990) found *C. tridentatum* to deposit eggs singly, 1–8 per day and 1–13 per individual's lifetime. A single embryo is contained in each egg capsule that is spherical, and gelatinous but for a thin, elastic envelope that is free of calcium carbonate crystals. The eggs are each covered in faeces by the parent; Bulman found that egg viability was lost if the faeces were removed experimentally.

Ellobiidae generally have an unspecialized diet of finely divided plant material and debris, and thus closely approximate the microphagy that is probably the primitive feeding mode in the Heterobranchia. *C. tridentatum*, however, is known to ingest larger fragments of leaf litter. Pelseneer (1894) recognized the rhipidoglossate nature of the radula in pulmonate and basal opisthobranchs, and Morton (1955b) recognized that Ellobiidae and other pulmonate lines must have arisen prior to entrenchment of the taenioglossate radula among monotocardian gastro-pods. This interpretation is accepted here. In contrast, Ponder (1991) considered the taenioglossate radula, or a radula readily derived from it, found in Rissoellidae, Mathildidae and Architectonicidae, Valvatidae, Cornirostridae and Orbitestellidae, to be the primitive one for the Heterobranchia. From this perspective, the broad radulae comprised of numerous similar teeth seen in pulmonates and opisthobranchs is a derived condition, analogous to that seen in Triphoroidea and some other Caenogastropoda. Pulmonates differ fundamentally from prosobranchs in that the number of teeth per transverse radular row is not constant, but in the course of growth new teeth are added on either side, and the shape of the teeth is determined by the profile of the odontoblasts rather than the odontoblastic cushions (Mischor and Märkel, 1984). Some ellobiid taxa exhibit striking ontogenetic change in radular morphology (Martin 1996b), as do many gastropod groups, and the great diversity of radular dentition seen in Ellobiidae encompasses many of the dental forms that have been afforded systematic significance in other pulmonate groups. According to Haszprunar (1988b), the buccal cartilages that support the radula in archaeogastropods and taenioglossate caenogastropods are lacking in all Heterobranchia. Of the presence of cartilage-like structures in some Heterobranchia, Haszprunar (1988b, p. 392) stated that these 'differ entirely in their structure from the original cartilages and obviously are secondary structures'. However, it has long been recognized that in gastropods a continuum exists from vesicular chondroid, matrix-embedded vertebrate-like cartilages in prosobranchs to a matrix-free myocellular complex in pulmonates and opisthobranchs (e.g. Schaffer, 1913, 1930; Person and Philpott, 1963, 1969; Curtis and Cowden, 1977). Therefore, it is likely that the radular supporting structures in hetero-branchs are modified buccal cartilages, rather that new structures. The buccal mass in basal pulmonates is equipped plesiomorphically with a jaw comprised of three plates, namely paired lateral and a single dorsal element. In ellobiids, the lateral plates of the jaws often are reduced or absent.

The oesophagus is simplified, with longitudinal folds but entirely lacking pouches, ventral folds and glands that occur in lower gastropods. Often the oesophagus is dilated to form a crop. The oesophagus generally opens into an anterior atrium of the stomach, into which the anterior lobe of the digestive gland opens. The stomach of ellobiids such as *Ophicardelus* Beck and *Pythia* retains some archaic features such as the ciliary currents, protostyle with typhlosole and food string, and a posterior caecum behind the entrance of the posterior digestive gland opening. The sorting area has become vestigial. A new development, relative to that in caenogastropods, is the strong coat of muscle surrounding the stomach to form a contractile gizzard, accompanied by loss of the gastric shield and spread of cuticle over the whole lumen. The gizzard is elaborated variously in ellobiids, with investment in stronger musculature and a trend towards structural isolation of the gizzard as a triturating organ from the rest of the stomach; these features are shared with other basal pulmonates, such as Amphibolidae. Often the structural isolation of the gizzard involves transformation or loss of the gastric caecum and/or the posterior lobe of the digestive gland. Despite these modifications, the paired typhlosoles still link the digestive gland openings to the intestine.

The gizzard is a synapomorphy of the pulmonates, but often is reduced secondarily. Several groups of primitive opisthobranchs have developed a gizzard anterior to the openings of the digestive gland, which accordingly has been considered an oesophageal gizzard (e.g. Ponder, 1991). However, the variable placement of the gizzard with respect to the digestive gland openings among the various groups of lower pulmonates (e.g. compare Chilinidae and Lymnaeidae with Ellobiidae) indicates a strong possibility of homology throughout the higher Heterobranchia.

The heterobranch stomach is equipped primarily with a style. When retained, the style generally is crystalline, as in Orbitestellidae, Planorbidae and Acroloxidae, and this may be plesiomorphic. In a number of pulmonates (Siphonariidae, Otinidae and Ellobiidae), the style is similar in structure to the protostyle of lower gastropods. The fact that both crystalline and non-crystalline styles occur in Ellobiidae suggests that the latter is a secondary modification of the former. Often in Heterobranchia the style is absent.

The ellobiids generally retain the rather long intestine that encompasses an anteriorly directed loop. The rectum runs along the upper margin of the body whorl to terminate in the anus located within the roof of the pallial cavity, just short of the pneumostome. A small sacculate anal gland opens on either side of the anus. A folded, ciliate groove leads the faeces to the exterior. The posterior position of the anus appears to be a primitive feature of heterobranchs, being the case in all of the lower groups except valvatids and cornirostrids. The anterior position in the pulmonates has been derived because the absence of water current through the pallial cavity necessitates the discharge of faecal material at the pulmonary opening (Ponder, 1991). Ellobiids produce faecal strings.

The cephalic tentacles in Ellobiidae are hollow, and the two bundles of the tentacular nerve run in the centre. More peripheral are the muscular retractors of the tentacle, either as longitudinal fibres in the integument (e.g. *B. heteroclita*) or free within the tentacle cavity (e.g. *Ellobium pellucens* (Menke)). The tentacle retractors originate on either side of the head and insert proximal to the sensory apex, which often is thickened as a sensory knob. The cephalic tentacles generally are only partially contractile or subretractile. In *E. pellucens*, among others, these tentacles can be retracted almost completely into the head: as part of the outer wall of the tentacle is turned inwards when it is retracted, this retraction is, at least in part, an invagination. The sessile eyes, when present, generally are located medial to the base of the tentacles. In some *Melampus*, however, the eyes may be as much as 0.3 times the length of the tentacle from the base. From the information presented by Kowslowsky (1933), Hubendick (1945) interpreted the rudimentary muscles of the ellobiid tentacles as foreshadowing the retractor system and tentacle invagination in Stylommatophora (see below). The anterior margin of the snout, which functions as labial palps, often is elaborated, albeit weakly, into one or more pairs of non-ocular tentacles that are contractile. Based on current information on ontogeny and adult morphology it remains uncertain whether these accessory tentacles represent the rudiments or remnants of more definitive structures.

The head is separated from the sole by a transverse groove, into which the suprapedal mucous gland opens. The foot is entire in Pythiinae, *Microtralia* Dall in Pedipedinae, and *Ellobium* in Ellobiinae. In all other ellobiids, the sole is divided into anterior propodium and posterior metapodium parts. Irrespective of the presence or absence of the transverse groove, most ellobiids progress by direct pedal waves. In *Blauneria*, the wave of contraction, which originates at the hind end of the foot, stops for a moment when it reaches the transverse groove, but then continues to the anterior pedal border. However, *Melampus bidentatus* Say progresses in a crawl-step motion in which the metapodium slides by direct pedal waves while the anterior is lifted and placed (Moffett, 1979). The pedal waves are initiated in the centre of the metapodium. The transverse pedal groove is a feature shared among pulmonates only in Otinidae and Smeagolidae. However, the mechanics of locomotion in these latter taxa are different (Vlés, 1913; G.M. Barker, personal observation), probably indicating convergence in morphology with that in ellobiids. Further, locomotion in *Melampus* appears to be a secondary phenomenon, being derived from that exhibited in the less specialized members of the family that possess more pronounced direct pedal waves. Morton (1955b) postulated that the sole morphology in Ellobiidae is adaptive to substrate conditions, but Fukuda (1994) shows that it has a strong phylogenetic basis. The posterior extremity of the foot is bifid in Melampinae, but all ellobiids lack a metapodial mucous gland.

In summary, if we are to accept Ellobiidae as indicative, then the Eupulmonata arose after loss of the ctenidium, buccal cartilages, oesophageal pouches and glands, metapodial mucous gland and paraspermatozoa, and after the closure of the pallial gonoducts. In the Eupulmonata archetype, the nephridium had become pallial in location and blood supply; the pallial gonoduct had sunk into the body cavity but with the bursa maintaining a close proximity to the pericardium; both the albumen gland and more posterior sections of the oviductal glands had become diverticular to the gonoduct; the female orifice had become extrapallial; the penis had become internal and extrovertible; the spermatozoa had attained spiral nuclei and tail pieces; and the nervous system had attained a hypoathroid configuration, with a procerebrum equipped with globineurons. The lineage from which types are derived had, however, not developed an introvert proboscis; had not developed union of the male and female portions of the pallial gonoduct; had not developed closure of the sperm tract in the cephalic region; and had not reduced the radular dentition to the taenioglossan type.

Onchidioidean pulmonates

The Onchidioidea comprise a single family, the Onchidiidae, which in turn comprises only slug-like species (Starobogatov (1976) overemphasizes the diversity with recognition of several superfamilies). There is no fossil record. Extant Onchidiidae are mostly marine, and live submerged in the subtidal zone, amphibiously in the littoral zone or terrestrially in the supralittoral zone. They occur throughout the world, except in the Polar regions, with over 100 species. The highest species richness occurs in littoral–supralittoral habitats of South-East Asia, with as many as 20 species sympatric in mangrove systems (Britton, 1984). A few are reported to inhabit brackish and freshwater. One genus of the family, *Platevindex* Baker, is represented in both littoral marine and montane rainforest terrestrial habitats of Borneo and the Philippines (Plate, 1893; Collinge, 1901; Tillier, 1983). *Platevindex apoikistes* Tillier of Mindoro, the Philippines, found at 400–600 m altitudes in dipterocarp rainforest, associated with rocks covered in moss and algae, is the best studied representative of this small terrestrial radiation. Tillier (1984a) concluded that no obvious adaptive morphological change in *Platevindex* relates to colonization of land from the intertidal environment of presumptive ancestral species.

Some 40 generic names have been applied to onchidiids, but Britton (1984) considered only 12 genera as well characterized. These animals are usually oval with a leathery notum that is smooth, or more usually covered with papillae or tubercles that may contain accessory eyes. Siliceous spicules occur in the notal tissues of some onchidiids, prompting Labbé (1934) to suggest affinities with the Opisthobranchia. At the periphery of the notum there are often repugnatory glands, in structure somewhat reminiscent of the repugnatory glands at the mantle edge of caenogastropods such as *Calyptraea* de Lamarck and *Crepidula* de Lamarck and in various opisthobranchs. The notum extends ventrally, as the hyponotum, to flank the pedal sole, from which it is separated by a parapodial groove. Their digestive, female reproductive and respiratory orifices open into the hyponotum or peripedal groove. In front, below the notal edge and above the mouth, is a strong frontal shield. These external morphological characters constitute gymnomorphy, which it is shared with rathouisioideans and is approached in some opisthobranchs, Smeagolidae, some Succineoidea and some advanced stylommatophorans.

Onchidiids possess one pair of cephalic tentacles. They exhibit further elaboration of the trend in Ellobiidae, where the hollow tentacles are retractile due to provision of internal bands of retractor muscles, and the eyes are displaced towards the tentacle apex. In onchidiids, the eyes occur at the apex, lying just beneath the sensory epithelium. The tentacles vary from short and conical to long and cylindrical, but in all cases are fully retractable into the head. Retraction is effected by means of a pair of rather short, stout muscles, which arise separately from the body wall on each side of the head next to the cerebral ganglia, and are inserted in the extremities of the tentacles without marked subdivision. The tentacular nerves are separate from the retractor muscles until they reach the ends of the tentacles, where each nerve divides into short branches, one of which is the optic nerve innervating the eye. In some onchidiids, such as *Onchidium nigrans* (Quoy & Gaimard), retraction is brought about by invagination of the sensory, eye-bearing apex of the tentacle into the conical base and then contraction of that basal region (G.M. Barker, personal observation). In others, such as *Onchidella celtica* (Cuvier) and *Onchidella evelinae* Marcus & Burch, retraction is brought about by complete invagination of the tentacle (Fretter, 1943; Burch, 1968). Onchidiids are characterized by broad labial palps, which comprise sensory lappets on either side of the mouth. On the anterior face, above these labial palps, there are often rudiments of sensory knobs, similar to those seen in some ellobiids.

Progression is accomplished by means of pedal waves originating at the posterior end of the foot and coursing anteriorly (direct waves). Ordinarily, only one of these waves, extending across the whole width of the foot (monotaxic), is present on the foot at a time.

Onchidiidae retain the epiathroid configuration of the CNS, which is the presumptive plesiomorphic condition in Heterobranchia. However, affinities with the Pulmonata are clearly indicated by the possession of the complex neurosecretory system consisting of procerebrum, cerebral gland and dorsal bodies. The cerebral gland is internalized within the procerebrum and thus, among globineuron-equipped pulmonates, the Onchidiidae are rather advanced and certainly more so than the Ellobiidae. The more derived condition, relative to ellobiids, is also indicated by the cerebral commissure and by all connectives being short. However, the onchidiids share with ellobiids the fusion of the suboesophageal ganglion with the visceral ganglion and of the supraoesophageal ganglion with the right parietal ganglion. Interpretation made through the origin of the left pallial nerve indicates that the pleural and parietal ganglia, in contrast to the ganglion separation in Ellobiidae, form a single mass at each extremity of the visceral chain in Onchidiidae: thus the onchidiids are secondarily triganglionate in configuration of the visceral loop. Generally, the connectives which join the suboesophageal–visceral ganglion complex with the pleuro-parietal ganglia on each side are of unequal length, the right one being twice as long as the left, which is very short. In *P. apoikistes*, however, further concentration of the CNS has led to the visceral and parietal ganglia being adjacent, but not fused. The pedal ganglia are united by two commissures. Usually these comprise a broad, very short commissure towards the anterior, and a narrow, longer commissure that connects their posterior extremities. In a number of onchidiids, including *P. apoikistes*, these commissures are more equal in development.

The larval pallial cavity of *Onchidium* Buchanan is formed during the last quarter of the planktotrophic larval phase and has a configuration remarkably similar to that in the veligers of Ellobiidae (Ruthensteiner, 1997). The pallial cavity is voluminous and widely open on the right side. The anus lies within the pallial cavity next to its opening. The nephridium opens within the pallial cavity, on its right side. An osphradium is present in the most dorso-anterior region. A ciliary band leads from the posterior left of the pallial cavity towards the opening but, unlike that in Ellobiidae, ends well short of the opening to the exterior.

The shell is cast off shortly after larval metamorphosis. At this time, detorsion results in rotation of the pallial complex and its displacement towards the posterior of the animal, and gives rise to a pallial cavity and cloaca (Fretter, 1943; Ruthensteiner, 1997). Tillier (1984b) and Tillier and Ponder (1992) maintain that the onchidiid nephridium is located in the middle of the pallial cavity with no common wall with the visceral cavity, and thus approaches the condition seen in lower pulmonates such as Amphibolidae and Siphonariidae. This condition was suggested as a synapomorphy of onchidiids because the nephridium sharing at least one wall with the visceral cavity is plesiomorphic in gastropods. However, during ontogeny, the nephridium of onchidiids is isolated from the pallial cavity anlage (Ruthensteiner, 1997) and in adult animals is often not completely surrounded by the pallial roof but by a wall abutting the visceral cavity (Fretter, 1943; Marcus and Marcus, 1956). The relative position of the nephridium in onchidiids is thus not greatly different from that in other gastropods, including lower opisthobranchs and other lower pulmonates such as Ellobiidae. In *Platevindex* and *Oncidina* Semper, the pallial cavity lies along the body wall at the right posterior of the animal, with the pneumostome at the very posterior right. The pericardium is located at the anterior of the pallial cavity, indicating an approximately 90° clockwise rotation from the ancestral condition indicated in lower gastropods. In *Onchidiella* Gray, the rotation has progressed slightly further, resulting in the pallial cavity lying transversely to the body axis at the very posterior, and with a pneumostome opening at the posterior apex of the body (Hoffmann, 1929; Tillier, 1984b). In these taxa, the nephridium assumes a bilobed configuration, with the ureter arising from the left side of the bridge between the respective lobes. In *Hoffmannola* Strand, this rotation is continued further, so that the left lobe is considerably larger than the right. The nephridium is sacculate, generally lacking lamellae but often with a few narrow irregular folds. The ureter originates from the posterior extremity of the nephridium to open into the rectum just before the anus. Tillier (1983) suggests that the ureter opening to the anus is probably a paedomorphic character resulting from the retention of the larval disposition of the nephridium.

Branching diverticula of the pallial cavity penetrate between the lobes of the nephridium and push into the tissues of the notum, so that they come in intimate contact with the blood spaces. In small species, the pallial cavity is internally smooth, but in the larger species the internal surface is richly provisioned with a venous network, thus increasing the respiratory surface. In some onchidiids, the pallial cavity is entirely filled with shallow alveoli. In amphibious species, the pneumostome is closed when the animals are submerged, but widely open when out of the water. On the basis of her ontogenetic studies on *O. celtica*, Fretter (1943) claimed that the 'lung' of onchidiids is fundamentally different from the pulmonate lung, and is a secondary structure. However, recent investigations on organogenesis revealed that the 'lung' of onchidiids is homologous to that in other pulmonate groups and to the pallial cavity of other gastropods (Haszprunar, 1985c; Ruthensteiner, 1997). *Peronia* Fleming, *Labbella* Starobogatov, *Quoyella* Starobogatov and *Lessonia* Starobogatov possess branching papillae or gills as accessory respiratory structures.

Onchidiids are hermaphrodites, with a level of reproductive tract organization similar to that in Ellobiidae. The male and female pallial gonoducts are fused (monauly) and the openings of the pallial gonoduct and penis are separated (ditrematous condition). The male gonopore (exceptionally more than one pore) is located at the right side of the head, either inside or outside the right cephalic tentacle, although in *Hoffmannola* it occurs in the centre of the head. The development of the genital tract from the epithelium of the pallial cavity, after metamorphosis and detorsion during the complex ontogeny of the pallial complex

(Ruthensteiner, 1997), means that the female opening is necessarily retained in the plesiomorphic position near the pneumostome (contrary to Nordsieck, 1992). In the adult animal, the female opening generally lies at the posterior end of body, not greatly distant from the pneumostome. In some species of *Paraoncidium* Labbé and *Platevindex*, the female orifice is not closely associated with the pneumostome but has come to be located in the right lateral to posterio-lateral part of the hyponotum. As noted by Tillier (1984b), the position of the female orifice in onchidiids is thus a secondary phenomenon associated with limacization and detorsion, but the derivation of variation in location of the female opening with respect to the pneumostome seen in onchidiids can only be determined by further comparative ontogenetic studies. As in ellobiids, the albumen gland is diverticulate to the lumen of the pallial gonoduct and not passaged by the zygotes. In most onchidiids, there are two albumen glands, reminiscent of the condition seen in some ellobiids. One side of the spermoviduct has a thin wall lined by a ciliated epithelium. The other side has a thick glandular wall, which is expanded to form hollow projecting folds. Most genera, including *Peronina* Plate, *Onchidella*, *Paraoncidium* and *Hoffmannola*, have two spiralled, agglomerate oviductal mucous glands, as also seen in some ellobiids. However, *Onchidium* has only one, and *Oncidina* several. Eggs of *Onchidella* passage through the pouches formed by the agglomerate oviductal mucous glands in order that the various layers of the egg capsule are laid down (Fretter, 1943).

Generally, two allosperm sacks are present. The bursa copulatrix is in the derived distal location, on a duct originating from the lower reaches of the female pallial gonoduct. The sacculate reservoir retains its primary position near the pericardium. Mating in onchidiids generally is reciprocal, with spermatozoa deposited directly into the bursa copulatrix. However, Stringer (1963) noted that for *O. nigricans*, copulation may involve several individuals in the form of a ring, and McFarlane (1979) observed copulation chains of *Onchidium verruculatum* Cuvier, suggestive of non-reciprocate gamete exchange. Like the same structure in ellobiids, the onchidiid bursa has allosperm reception, and gametolytic and resorptive functions. *Onchidella* possess an elongate diverticulum at the distal extremity of the female pallial gonoduct, which is possibly homologous to a similar diverticulate structure in the same position in Ellobiidae, but its function has not been studied in either ellobiids or onchidiids. The receptaculum seminis remains located in the primary proximal position in most onchidiids. However, as in many heterobranchs, there is trend in onchidiids for this storage structure for allospermatozoa to be displaced distally, to open below the ducts from the albumen gland(s). This is exemplified by *Onchidium simrothi* (Plate).

The prostatic gland exhibits considerable variation in form among onchidiids. In *Platevindex*, the prostatic gland is a longitudinal glandular fold of the spermoviduct, distal to the oviductal glands. In *Onchidella*, the prostatic gland comprises a sacculate lateral fold or diverticulum in the wall of the spermoviduct, often with the narrowed opening to the latter.

In *Onchidina* Semper and *Onchidium*, the prostatic gland is a large sacculate structure on a short duct from the spermoviduct. The polarity of this variation within Onchidiidae remains uncertain, given the equally variable configuration of the prostatic gland in other heterobranchs. As in semi-diaulic Melampinae of Ellobiidae, the spermoviduct in Onchidiidae separates into the vaginal duct and the vas deferens a short distance below the prostatic gland. The vas deferens separates as a closed duct and takes a sinuous course to the floor of the body cavity, where, just anterior to the female aperture, it passes deeply into the thickness of the body wall to run in an anterior direction on the right side adjacent to the perinotal groove. At the body anterior, the vas deferens returns to the haemocoel and runs as a highly convoluted duct to insert in the penis. This route for the vas deferens through the body wall, notwithstanding the presence of an external ciliated groove between the genital pores, is remarkably similar to that in ellobiids (with the exception of *Pythia*) and several other basal pulmonate groups. The importance of this character in phylogenetic reconstruction is generally reduced by the multiple transitions from open external groove to closed internal duct in Gastropoda.

The penial complex of most onchidiids includes an eversible phallic structure, that closely resembles the penis in Ellobiidae, and an auxiliary copulatory structure that has generally been termed penial gland (e.g. Britton, 1984; Tillier, 1984a). In *Quoyella* and *Peronina*, the penis and penial gland have separate, albeit closely approximated apertures, while in *Platevindex* and *Onchidium* the penis and penial gland open independently into a common male atrium and thus share a single opening to the exterior. The penis is usually equipped with a slender papilla, sometimes with a cuticular ring or shield at its base. The papilla is often reduced secondarily. The lining of the papilla and/or the anterior penial cavity is often equipped with cartilaginous hooks that function to anchor the male organ within the vagina of the mate during copulation. Such armature of the penial surface is to be found widely in heterobranchs and may be plesiomorphic in these gastropods. A retractor muscle effects retraction of the penis, as in ellobiids. Plate (1893) recognized three configurations in the penial retractor: type I arising in the body wall at the level of the CNS, type II at the level of the pericardium and type III in the posterior of the body cavity.

Aphally is known from onchidiids (e.g. Labbé, 1934), but its frequency or importance in their reproductive biology are presently unknown. The penial gland, when present, usually opens through a prominent papilla into the male atrium. In some cases, this penial gland papilla is elaborated into a cartilaginous stylet (Plate, 1893; Awati and Karandikar, 1948). There have been few studies of the copulatory behaviour of onchidiids, and the relative role of the penis and penial gland as stimulatory organs and in transfer of semen remains unknown (Tillier, 1983, 1984a). In *Paraoncidium*, *Hoffmannola*, *Onchidina*, *Lessonia* and *Onchidella*, the penial gland is absent, which is likely to be the derived condition.

There is considerable variation among onchidiids in reproductive strategy. For example, the Australian *Onchidium damelii* Semper produces irregular ovoid egg masses containing 100,000–200,000 eggs, attached to the buttress roots of mangrove trees in estuarine habitats. These eggs yield planktotrophic veliger larvae. Among the littoral species of New Zealand, *O. nigricans* lays approximately 500–11,000 eggs in a band-like mass. This species too has indirect development, with a free-swimming veliger. *Onchidella flavescens* (Wissel) deposits 22 to about 90 eggs in an irregular hemispherical mass. The egg masses of *Onchidella campbelli* (Filhol) were similar to those of *O. flavescens*, but larger, and comprise 28–120 eggs. In these latter two species, the veliger stage is passed within the egg capsule, and hatching occurs as crawling juveniles. The presence of planktonic veliger larvae strongly suggests a primary marine association for the family, rather than a reinvasion of the marine habitat from the terrestrial environment suggested by Arey and Crozier (1921), Starobogatov (1976) and Climo (1980). The truly terrestrial species, such as *P. apoikistes*, can be expected to produce eggs in which development is direct, but information is not available presently on the numbers of eggs produced or the degree to which the intracapsular veliger stage is suppressed. In all onchidiids studied to date, the operculum and shell are cast off, and the visceral hump gradually recessed, in the post-veliger (or equivalent) stage.

Littoral forms generally feed on algae. However, Marcus (1979) found numerous small bivalves in the digestive tract of *O. celtica*, suggesting carnivory. The radula is the modified rhipidoglossate type. Primarily the jaw comprises numerous, fused plates, but this structure is often reduced in *Onchidella*. Many onchidiids are characterized by large labial palps. Further, their digestive tract exhibits the basic 90° rotation of pulmonates, as seen in ellobiids, resulting in the opening of the oesophagus into the stomach on the right side, and in the opening of the stomach into the intestine on the left. The oesophagus is sometimes dilated as a crop, with low internal folds that run in a longitudinal direction. The stomach generally comprises an anterior chamber, a medial gizzard and a posterior caecum.

The anterior lobe of the digestive gland opens into the anterior chamber of the stomach through two ducts, and the caecum is reduced to a posterior chamber that receives the duct of the posterior lobe of the digestive gland. The muscular fibres of the gizzard form two muscular plates. In *Onchidina australis* Semper, the gizzard, caecum and duct to the posterior lobe of the digestive gland are diverticulate to the main body of the stomach. That part of the stomach leading into the intestine often retains the character of the style sac, being sacculate with confluence of the major and minor typhlosoles and the presence of the protostyle (Marcus and Marcus, 1956; G.M. Barker, personal observation). As noted by Tillier (1984b), although not combined in a single suprageneric group, all the characters of the morphology of the onchidiid stomach are to be found in

Ellobiidae, suggestive of a relatively close relationship between the two families. Many of these features are also shared with Amphibolidae.

There is considerable variation in intestinal length and coiling in onchidiids. Labbé (1934) recognized five configurations. A moderately long intestine with two basic loops is retained in *Platevindex*, *Paraoncidium*, *Lessonia* and some *Peronia* and *Onchidium* species. In one direction of modification, seen in *Labbella* and some *Onchidium*, these two loops are secondarily coiled in a spiral. In the other direction of modification, in *Hoffmannola*, *Onchidella*, *Quoyella* Starobogatov and some *Peronia*, the intestine is considerably shortened, with a reduced anterior loop and a more or less direct passage posteriorly to the anus. In all cases, the intestine exhibits only one inflexion point corresponding to the limit between the basic periaortic and prerectal bends. The major typhlosole is well developed and in many onchidiids runs the entire length of the intestine. The intestine is usually characterized by low folds that run diagonally from the typhlosole (Watson, 1925; Fretter, 1943; Marcus, 1971; Tillier, 1983). *Labbella*, *Lessonia*, *Quoyella*, *Onchidina*, *Semperoncis* Starobogatov and several species of *Onchidium* and *Platevindex* possess a tubular rectal gland, of unstudied function.

Rathouisioidean pulmonates

The Rathouisioidea comprise two families of strictly terrestrial pulmonates, the Rathouisiidae and Vaginulidae. They comprise solely slug forms and thus lack a fossil record. The body form is similar to that in onchidiids, with the dorsal surface covered by the mantle (notum) that is separated from its ventral extension, the hyponotum, by a sharp keel, the perinotum. The ventral surface has a central sole, separated by a parapodial groove from the lateral hyponota. The notum of rathouisiids is often produced into a mid-dorsal keel and, recalling the onchidiid condition, is supplied with small tubercle-like processes. Vaginulidae too exhibit similarity to the onchidiids in possessing multicellular repugnatory glands along the perinotum. Rathouisiids are known from south-east China, Tonkin, Thailand, Burma, Malaya, Indonesia, the Philippines, New Guinea, the Bismarcks and northern Queensland, but the family is in need of revision in order to determine the true extent of specific and supraspecific diversity. The Vaginulidae generally are described as pantropical. However, while many species do occur in warm, humid tropical environments, the family is well represented both at high southern latitudes and at high altitude of low latitude, in environments that are best described as cool temperate. Vaginulids are most diverse in the Americas (Baker, 1925a; Thomé, 1975, 1976, 1989) and Africa (Forcart, 1953). These gastropods are generally ground-dwellers, living under stones, decaying wood and leaf litter, or subterranean in the soil, but some species are also arboreal.

The rathouisioidean cephalic tentacles carry eyes at their apex, as in onchidiids. However, unlike onchidiids, the cephalic tentacles of rathouisioideans are only contractile, and lack the ability to invaginate, meaning that the eyes cannot be retracted fully into the body cavity. The differences in mode of tentacle withdrawal between onchidiids and rathouisioideans relate to the arrangement of the retractor muscles, being free in the lumen in the former and bound to the tentacle wall in the latter. Rathouisioideans also differ from onchidiids in possessing a second pair of fully developed tentacles, which are characterized by bifid apices and partial fusion with the lower fringe of the snout that represents modified labial palps. In location and innervation, these rathouisioidean inferior tentacles recall the sensory knobs of the anterior face of the labial palps in ellobiids and some onchidiids.

A single tentacular nerve runs from each cephalic and inferior tentacle to the cerebral ganglia, which are similar to those of onchidiids in that the cerebral gland is internalized and located amongst the globineuron neurosecretory cells of the procerebrum. In the case of vaginulids, the similarity of the CNS to that of onchidiids also extends to the short cerebral commissure and the short connectives. However, contrary to Tillier (1984a), the vaginulid nervous system is more primitive than that of onchidiids and ellobiids as the suboesophageal ganglion has not fused with the visceral ganglion and the supraoesophageal ganglion has not fused with the right parietal ganglion (Oberzeller, 1969, 1970; G.M. Barker, personal observation). Shortening of the visceral loop in vaginulids has proceeded only to the extent that the adjacent ganglia are separated by such short connectives that they abut. The CNS of rathouisiids is more advanced than that of onchidiids and vaginulids, with the ganglia of the visceral loop concentrated into left (pleural, parietal, suboesophageal and visceral) and right (pleural, parietal and supraoesophageal) masses through ganglion fusion. As pointed out by Tillier (1984a), this rathouisiid condition is easily derived from the vaginulid-like CNS. As in onchidiids, the epiathroid configuration of the CNS evidently is derived secondarily, as all other features of the nervous system point to affinities with the exclusively air-breathing pulmonates that are primarily hypoathroid.

The pedal ganglia are linked by two short commissures. In Rathouisiidae, the pedal nerves issuing from the posterior of the pedal ganglia run parallel to each other towards the posterior of the body cavity. However, the course of the pedal nerves is varied in vaginulids, being parallel throughout, diverging from their point of origin, or initially parallel but subsequently diverging; the functional significance of this variation presently is not known. The sole is permanently transversely ridged. The locomotor waves of contraction are superimposed on the ridge structure, and each wave takes in several ridges. The epidermis on the leading face of each ridge is considerably thicker than that on the trailing face. Mucous cells are ducted through this thickened face and thus contribute to the mucus on the underside of the sole. The ventral and trailing faces of each ridge are ciliated (Cook, 1987).

The vaginulids are herbivorous, feeding on live and dead plant material. Many species are subterranean herbivores (e.g. Kulkarni and Nagabhushanam, 1973). The buccal mass is equipped with an arcuate jaw comprised of fused narrow vertical plates. The anterior digestive tract is elaborated into an oesophageal crop, whose walls are thin except for several weak, longitudinal folds. The stomach is basically like that of *Ellobium* in Ellobiidae. It consists of an anterior chamber where the oesophagus, intestine and anterior duct of the digestive gland enter, and a posterior chamber that is surrounded by muscular bands and receives the posterior duct of the digestive gland at its posterior extremity. In some vaginulids, the posterior chamber is more developed, and in *Vaginula* de Férussac a distinct vestige of the gastric caecum occurs.

In vaginulids the proximal intestine is stout, and internally the confluence of the major and minor typhlosoles ends abruptly with a sphincter, rather similar to the style sac condition seen in some onchidiids (Marcus and Marcus, 1956). Unlike onchidiids, however, the typhlosoles are confined to this style sac and do not advance into the intestine. The greater part of the intestine is characterized by low folds that run diagonally from the intestinal groove that represents the continuation of the channel established between the typhlosoles in the style sac. This intestinal condition with diagonal folds recalls that seen in onchidiids. In the majority of vaginulids, the intestine retains the one anteriorly directed loop, without coiling around the oesophageal crop. In *Spirocaulis* Simroth, however, the intestine is secondarily lengthened and hypertorted (Simroth, 1913). The intestine abruptly transforms into the rectum just before entering the body wall on the right side to run posteriorly, to emerge as the anus in the hyponotum near or at the body posterior; the proximal rectum is characterized by an epithelium of longitudinal folds, but the remainder of its length within the body wall often is smooth.

The Rathouisiidae are thought to be primarily carnivores, feeding on other gastropods. They none the less are known also to include fungi and vegetable matter in their diet (see Barker and Efford, 2001). The buccal mass is elaborated so that the radula forms a protrusible proboscis. As a consequence, the jaw is absent in Rathouisiidae. The oesophagus is slender, but its walls are thick in that the greatly folded epithelium is supported by connective tissue and surrounded by muscle. The stomach is not well developed, being a simple sacculated merging of oesophageal, digestive gland and intestinal openings. Rathouisiids have lost the anterior lobe of the digestive gland, and the posterior lobe has a large central cavity. The intestine takes a more or less direct course diagonally across the body cavity to the anus in the parapodial groove at the anterior right. The intestine lacks the diagonal folds that characterize the vaginulids.

The rathouisioidean reproductive system is hermaphroditic, ditrematous and diaulic (contrary to monauly reported by early authors, diauly in rathouisiids was demonstrated by de Wilde, 1984). While the reproductive biology is largely unstudied, hermaphroditism in these animals is thought to be strongly protandric and in many species strongly seasonal (Baker, 1925a; Nagabhushanam and Kulkarni, 1971). As is usual for pulmonates, the ovotestis in vaginulids is posterior, embedded in the digestive gland. In rathouisiids, however, the ovotestis is in an anterior position, at the anterior face of the digestive gland. The albumen gland is diverticulate to the female pallial gonoduct but, unlike the primary bilobed condition in Ellobiidae and Onchidiidae, it comprises a single lobe. The female pallial gonoduct comprises a thin-walled, ciliated ventral channel open to the folds of the thick-walled oviductal glands. This gonoduct is elongated and twisted to varying degrees, often producing a spiral of ascending and descending lengths of the gonoduct. The ciliated ventral channel leads to the proximal seminal receptacle that is either embedded in or diverticulate to the renal gonoduct. The bursa copulatrix is medial in rathouisiids and invariably distal in location on the female part of the vaginulid reproductive tract, and its sac retains the primitive association with the pericardium. In the few studied species, copulation has been shown to be reciprocal, and the bursa duct often is elaborately constructed to accommodate insertion of the mate's penial papilla. In vaginulids, the bursa copulatrix is joined by a *canalis junctor*, a branch of the vas deferens that enters the bursa sac or enters the duct with a confluence to the sac. Tillier (1984b) suggested that diauly is probably derived from primitive monauly in vaginulids, because of the presence of the *canalis junctor* joining the distal portions of the pallial oviduct and bursa copulatrix. However, as noted above, the *canalis junctor* also occurs in Ellobiidae, suggesting that the *canalis junctor* itself is a primitive character in pulmonates. Indeed, the occurrence of the *canalis junctor* in Cornirostridae suggests that it may be plesiomorphic in Heterobranchia.

As in ellobiids and onchidiids, a number of vaginulids, such as those in the genera *Leidyula* Baker and *Imerinia* Cockerell, possess an elongate diverticulum at the distal extremity of the female pallial gonoduct, as in some Ellobiidae and Onchidiidae. The female gonopore is about midway down the right side of the animal, opening into the parapodial groove in Rathouisiidae and into the hyponotum (sometimes very close to the parapodial groove) in Vaginulidae.

Soon after its origin, the rathouisioidean male pallial gonoduct sends a short branch to the sacculate prostatic gland, and thus closely approximates the prostatic condition in some Ellobiidae and Onchidiidae. This similarity with ellobiids and onchidiids is also evident in the route taken by the vas deferens, which enters the body wall adjacent to the vagina to re-emerge at the right anterior of the body cavity and run to the apex of the penis. The male pore is in the parapodial groove adjacent to the mouth. As in ellobiids and onchidiids, the rathouisioidean penis is eversible to yield an erect papilla as an intromittent organ.

In Vaginulidae, the papilla sometimes has a pronounced glans, so well developed in *Semperula* Grimpe & Hoffmann that the aperture of the vas deferens is basal rather than apical on the papilla. In *Filicaulis* Simroth and *Imerinia*, the papilla is equipped with spines or hooks, not unlike onchidiids and other heterobranchs. The penial papilla of rathouisiids is simply conical. Retraction of the penis is effected by a retractor muscle that arises from the body wall at the anterior of the pericardium. Vaginulids possess a ganglion on the penial nerve that arises from the right cerebral ganglion (Hoffmann, 1925; Petrellis and Dundee, 1969; Oberzeller, 1970). According to both Ghosh (1915) and Odhner (1917), the penial nerve in rathouisiids has its origin in the right pleural–parietal–supraoesophageal ganglion complex.

In addition to the penis, the rathouisioidean male genitalia are equipped with a penial gland. That of rathouisiids is remarkably like that in onchidiids in comprising a single, elongated secretory tubule terminating in a papilla. In the case of Rathouisiidae, the papilla is rather inconspicuous. The penial gland in rathouisiids represents the left gland of a pair of glands, the Simroth glands that open into the anterior extremity of the parapodial groove on either side of the mouth. In some rathouisiids, the right gland has been reduced. The fact that these glands are primarily paired and the left gland is not connected with the reproductive system suggests that the copulatory function of the penial gland is derived secondarily (Laidlaw, 1940).

In Vaginulidae, the penial gland comprises a large papilla and a cluster of digitiform secretory tubules. These tubules vary in length with the degree of sexual maturity of the animal (Baker, 1925a) and produce a liquid secretion (Forcart, 1953) that is expelled through a single aperture at the apex of the papilla. Bani and Cecchi (1982) working with *Vaginula borelliana* (Colosi), and Peterellis and Dundee (1969) working with *Angustipes ameghini* (Gambetta) demonstrated that the tubules in the vaginulid penial gland comprise two types, which differ in the nature of their secretory products. The penial gland is reduced secondarily in some vaginulid genera. When present, the penial gland of both Rathouisiidae and Vaginulidae is provided with a retractor muscle, usually arising from the body wall adjacent to, or with a common origin to, the penial retractor near the anterior pericardial wall, but occasionally the retractor arises independently from the floor of the body cavity. The occurrence of a penial gland was considered by Hoffmann (1925) to have evolved independently in onchidiids, rathouisiids and vaginulids. Rightly, this was not accepted by Tillier (1984b). In fact, we can postulate that the paired Simroth glands in rathouisiids approximate the primary condition from which the diversity of penial gland configurations (and atrial and vaginal derivatives) in pulmonates was derived.

Being terrestrial, Rathouisioidea exhibit an entirely intracapsular embryonic development. Nothing is known presently about the reproductive biology of rathouisiids. In oviparous vaginulids, clutches vary greatly among species, for example 3–25 eggs in *A. ameghini* and 46–70 eggs in *Laevicaulis alte* (de Férussac) (Nagabhushanam and Kulkarni, 1971; Tompa, 1980). Individual eggs are embedded in and connected by a chalaziform mucus rope. Tompa (1980) showed that the egg capsules of *A. ameghini* are invested with calcium carbonate spherules that are utilized by the developing embryo as a calcium source. At the time of oviposition, the adult slug applies faeces to the surface of each egg, in a manner similar to that seen in Pomatiasidae. The embryonic development of these oviparous species is characterized by the reduction in the veliger stage. In some vaginulid genera, such as *Pseudoveronicella* Germain, *Leidyula* and *Veronicella* de Blainville, both ovoviviparity and viviparity are known. Baker (1925a) records 35 young in the pallial gonoduct of *Leidyula moreleti* (Crosse & Fischer).

In adult Rathouisiidae, the pallial complex is located mid-dorsally towards the anterior of the animal. The nephridium is generally semi-lunar in shape, its two extremities enclosing the pericardium from behind, and yielding a ureter that arises from its anterior margin but courses posteriorly before turning forward and downward to the excretory orifice. The pallial cavity is located on the right lateral aspect and opens ventrally through the juncture of the notum and pedal musculature, into the parapodial groove as the pneumostome (Laidlaw, 1940). The internal surface of the pallial cavity is rather smooth. The ureter and rectum descend to the parapodial groove alongside the pallial cavity, the former having a common opening with the pneumostome. Tillier (1984b) suggested that the pallial configuration in rathouisiids is not the result of the detorsional rotation of the pallial complex, but rather the result of translation and compression to the right side of the body cavity. However, this conclusion disregards the fact that the pallial complex in rathouisiids retains its primary mid-dorsal location and exhibits little evidence of lateral compression. Indeed, the rathouisiid condition is readily attained by an approximately 180° clockwise (i.e. detorsion) rotation of the pallial complex seen in pulmonate snails if we accept that the rathouisiid nephridium was primarily transverse in relation to the body's longitudinal axis, or had attained such a transverse deposition during the early phases of limacization (both conditions evident in Stylommatophora), and with the ureter arising from the nephridial orifice adjacent to the pericardium (i.e. in close proximity to renopericardial and nephridial pores) and running across the anterior face of the nephridium before passage to the pneumostome (a condition also evident in Stylommatophora). The female opening has maintained its position relative to that of the pneumostome, suggesting that the terminal female genitalia were involved in the detorsional process. A gland lies to the right of the pallial cavity, against the body wall, and opens into the pneumostomal cavity independently of the ureter and rectum (Sarasin and Sarasin, 1899).

Sarasin and Sarasin (1899) showed that, in vaginulids, the pallial cavity, ureter and rectum differentiate from a single invagination of the embryonic ectoderm. The nephridium in adult vaginulids is lanceolate, lying against the body wall on the right, with the pericardium anterior–lateral. It is equipped with a folded primary ureter that for a large part is embedded in the adjacent body wall. The secondary ureter runs posteriad within the body wall parallel to the rectum. The ureter and rectum are interconnected by one or more fine tubules (Degner, 1934). In *Vaginina* Simroth, the ureter, while connected to the rectum by multiple tubules, opens separately from the anus at the body posterior. In *Vaginula*, *Leidyula* and *Pseudoveronicella*, the secondary ureter has one or several proximal connections with the rectum and has a common opening with the rectum behind the sole. In *Laevicaulis* Simroth, the secondary ureter is connected to the middle part of the rectum, but the ureter and rectum open separately to the exterior. The pallial cavity comprises a narrow, richly vascularized region of the right posterior body wall surrounding the rectum and secondary ureter, and opening via the pneumostome in the parapodial groove behind the sole; the vaginulid pallial cavity is not absent, as frequently suggested in the literature, it is just greatly reduced in extent. While the 'translation and compression to the right side of the foot cavity' noted by Tillier (1984a, p. 356) does not apply to rathouisiids, it is clearly evident in the pallial configuration of vaginulids. Indeed, the maintenance of the relative positions of the various elements, including the close proximity of renopericardial and nephridial pores (Pelseneer, 1901; Tillier, 1984a) and right anterior orientation of the pericardium, indicates derivation of the vaginulid configuration from a detorted, rathouisiid-like one. From the inclusion of the rectum and ureter in the body wall from the nephridial region, Tillier (1984a) recognized the secondary migration of the anus and pneumostome to the caudal extremity of the foot in vaginulids. In these animals, the female opening was included in the initial detorsional process but not in the secondary phase that resulted in posteriad migration of the anal and pallial cavity openings. The rectum and genitalia generally enter the body wall in close juxtaposition, but in some vaginulid species there is separation.

Succineoidean pulmonates

The extant Succineoidea comprise two families, Succineidae and Athoracophoridae. The earliest known Succineidae are from the Palaeocene of Europe, and fossils occurring in Tertiary deposits in Africa, Europe and North America. The family is now cosmopolitan, but with greatest diversity in the islands of the Pacific, in the Indian subcontinent and the Americas. Succineidae generally are small animals with a thin, high-spired shell whose dimensions and shape are generally of little diagnostic value below the family level except for several notable genera. In several parts of the world, a trend to shell reduction and slug form has occurred. In *Succinea* (*Hyalimax*) Adams & Adams and *Omalonyx* (*Neohyalimax*) Simroth, the shell is reduced to the extent that it is internalized beneath the mantle shield. All succineids have a weakly developed lateral groove pattern and ill-defined tubercles on the dorsal integument. Further, they lack the hyponotum, but possess two shallow parapodial grooves that run parallel to each other posteriorly from the labial palp on either side, to separate the sole from the lateral body wall. The life strategy of Succineidae is often generalized incorrectly as amphibious (e.g. Duncan, 1960; Rigby, 1965). It is true that many species of Succineidae occur in marshes, swamps and at lake margins, and are to be found on moist soil or on emergent vegetation in freshwater systems. However, these species are strictly air-breathers and do not enter the water actively for feeding or reproduction, although several species are known to disperse passively attached to the surface film of water (e.g. Rao, 1925). Only species associated with continuously wet rock surfaces near waterfalls, such as *Lithotis rupicola* Blanford of western Ghats, India, and *Succinea bernardii* Récluz of Tahiti (Patterson, 1973), can truly be regarded as amphibious. The family is also well represented on the ground and arboreally both in moist environments such as rainforests and marshes and in dry environments such as sand dunes, savannah woodlands and seasonally dry stream banks. Succineid snails are able to persist through seasonally dry periods in an aestivating state, with the animals retracted into the shell whose aperture is sealed and attached to the substrate by a sheet of dried mucus, the epiphragm. In the semi-slug and slug forms, this type of aestivation is not possible and these taxa are confined to permanently moist sites. No comprehensive monograph of the Succineidae is available, although the foundations for such have been established by numerous anatomical studies published over the past 75 years. Patterson (1971a,b) recognized two subfamilies and 12 genera.

Athoracophoridae are confined to the south-west Pacific region, with the subfamily Aneitinae occurring from New Guinea through eastern Australia to New Caledonia and Vanuatu, and the Athoracophorinae in New Zealand and the Subantarctic. Their dorsal integument has grooves, with conical tubercles in the intervening panels, both most strongly developed in Athoracophorinae. The Aneitinae are characterized by a triangular cephalic shield, carrying the cephalic tentacles, at the anterior end of the body and extending back to abut the triangular mantle shield located mid-dorsally in the anterior third of the body; the cephalic and mantle shields are bounded by grooves. The anus, pneumostome and excretory orifice occur at the lateral, right apex of the mantle shield. The shell is vestigial, comprising one or several calcareous granules under the mantle shield. Where mantle shield is most strongly developed, in some members of the New Caledonian *Aneitea* Gray, the body form in dorsal aspect closely approaches that in *Omalonyx* (*Neohyalimax*) in Succineidae. The sole is broad, without a hyponotum, although in locomotion the lateral zones of the foot are lifted from the substrate. In Athoracophorinae, the cephalic and mantle shields are reduced and only defined, somewhat weakly in some genera, by grooves in the dorsal integument. Further, the anal orifice has migrated towards the perinotum, the excretory orifice towards the anterior apex of the vestigial mantle, and the respiratory orifice towards the centre of the mantle. The internalized shell comprises numerous small granules, often scribing a broad arc. In common with the Onchidiidae and Rathouisioidea, the Athoracophorinae possess a hyponotum, separated from the dorsum by a pronounced perinotum. Various subgroups of Athoracophoridae have been the subjects of systematic and anatomical treatments by, for example, Plate (1897, 1898), Pfeiffer (1900), Glamann (1903), Suter (1913), Grimpe and Hoffmann (1925), Solem (1959), Burton (1963, 1977, 1980, 1981, 1982, 1983), Climo (1973) and Barker (1978), but these do not document adequately the evolutionary diversity within the family. There are extensive radiations, particularly in New Caledonia and New Zealand, which previously have been overlooked. The family presently is the subject of a monographic revision by the present author. The family has no fossil record.

The foot sole of Succineidae and Athoracophoridae is tripartite, with locomotion effected by multiple, direct pedal waves in the central zone. As in some ellobiids, the pedal waves are initiated at the posterior extremity of the sole. The trend apparent in Ellobiidae for the columellar muscle and its branches to run in the haemocoel, free of the body wall, to insert on the anterior organs and cephalic structures, is developed further in Succineidae. As in ellobiids, this provides for inversion of the cephalopedal mass on retraction of the animal into its shell. In Athoracophoridae, the free muscle system is reduced to the extent that the columellar muscle is absent. In Succineidae, cephalic tentacle withdrawal is accomplished by invagination of the apical part of the tentacle and contraction of the broad basal part (Burch and Patterson, 1969). In the Athoracophoridae, the apical part of the tentacles is a solid rod, which is neither contractile nor invaginable, but can be withdrawn into the body cavity by telescoping into the sleeve-like basal part that is both contractile and invaginable (Burch, 1968). Solem (1978) presents a convincing argument for the tentacle retractor conditions in Succineidae and Athoracophoridae being related to space limitations in the cephalic region and thus secondarily derived from ancestral conditions with fully invaginable tentacles (as in Stylommatophora, see below). However, this does not accord well with the fact that the succineid tentacle structure and retraction modes provide transitional states between the clearly plesiomorphic, contractile tentacles in Rathouisioidea and the apomorphic, invaginable tentacles in Stylommatophora, despite the succineid free muscle system being well developed. One has to acknowledge the reduction in the free muscle system in Athoracophoridae, as eluded to by Solem (1978), but their retracted, solid tentacles occupy as much cephalic space as fully invaginated ones in typical stylommatophorans and thus run counter to Solem's arguments based on space-saving adaptations. Solem's interpretations apparently were coloured by his treatment of Succineidae and Athoracophoridae as advanced stylommatophorans, rather than as sister taxa to the Stylommatophora.

Succineidae generally possess a second pair of short tentacles at the anterior face, below the cephalic, eye-bearing tentacles. In this regard, they are similar to the rathouisioideans, except for the singular rather than bifid apex of these inferior tentacles, and that they are invaginable and thus can be retracted into the body cavity. These inferior tentacles are vestigial in at least *Lithotis* Blanford and *Omalonyx* (*Neohyalimax*), with the retention of the retractor muscle and nerve bundles indicating a secondary loss. These tentacles and their retractors and nerve supply are absent in all Athoracophoridae.

The hypoathroid CNS of Succineidae and Athoracophoridae approximates that in Ellobiidae in possessing a prominent globineuron-containing procerebral lobe and a tubular cerebral gland that extends from the procerebrum to the skin near the base of the cephalic tentacle (Cook, 1966; van Mol, 1967; G.M. Barker, personal observation). While the cerebral gland maintains a tube open to the body exterior in adult Ellobiidae, the orifice to the body exterior has disappeared by the time Succineidae and Athoracophoridae hatch from the egg (Cook, 1966). As in Ellobiidae and Onchidiidae, the Succineoidea possess the subcerebral commissure. With respect to the length of the connectives, the succineoidean CNS is highly concentrated relative to that in ellobiids, but of similar concentration to that in onchidiids and vaginulids. The general pattern of ganglion configuration in the pentaganglionate visceral loop of succineoideans is fusion of the right pleural, parietal and visceral ganglia. The left pleural ganglion generally abuts the left parietal ganglion, which in turn is either separated from the visceral ganglion by a short but distinct connective, or pressed against it. Several Succineidae and Athoracophoridae exhibit even more concentration, with the visceral ganglion seemingly fused to both parietal ganglia and consequently tending towards a medial location on the visceral chain.

Succineoidea are simultaneous hermaphrodites. The reproductive system in Succineidae is semi-diaulic in that the spermiduct initially comprises a ciliated groove on the ventral floor of the female pallial gonoduct, forming a spermoviduct, and separates from the female pallial gonoduct before giving rise to the prostatic gland and continuing as the vas deferens to the penis. Generally, the length of the spermoviduct is short, being associated with the oviductal gland only in its most proximal part. However, in some succineids, such as *Catinella vermeta* (Say), the spermiduct is maintained as a ventral groove for almost the entire length of the glandular oviductal part of the female pallial gonoduct, but none the less separates before giving rise to the vas deferens and associated prostatic gland (G.M. Barker, personal observation). Generally, the oviductal gland is comprised of voluminous folds whose lumina are tangential to the long axis of the gonoduct, recalling the condition in *Ellobium* of the Ellobiidae. During ontogeny, the oviductal gland folds of the female pallial gonoduct develop as outgrowths of the wall of an initially undulating tube (Hoagland and Davis, 1987).

The Athoracophoridae are predominantly diaulic. In *Aneitea* and *Triboniophorus* Humbert in Aneitinae, and *Amphikonophora* Suter in Athoracophorinae, the male and female pallial gonoducts separate immediately below the fertilization chamber. The female pallial gonoducts in these athoracophorid genera generally lack the oviductal folds that characterize Succineidae, but possess a more tubular structure, albeit with often thick, glandular walls. Often the female pallial gonoduct is of considerable length. These features recall the female gonoduct of onchidiids, rathouisiids and several ellobiid genera. In the athoracophorine *Pseudaneitea* Cockerell, the oviductal gland is produced as a single but often massive glomerate outgrowth to the main axis of the otherwise tubular female pallial gonoduct (Burton, 1980). The prostatic gland in all these athoracophorids is basically of the same structure as that in Succineidae in being a compacted mass of follicles opening to the vas deferens.

Remarkably, in *Athoracophorus* Gould and two undescribed genera of the Athoracophorinae, a monaulic condition is present (Barker, 1978; Burton, 1978; G.M. Barker, unpublished), which combines the female oviductal gland configuration of Succineidae with an ellobiid-like diffuse ribbon of prostatic gland follicles opening into the spermiduct that constitutes a ventral groove throughout the length of the pallial gonoduct.

Unlike that in ellobioid, onchidioid and rathouisioid pulmonates, the vas deferens of succineoideans is a closed duct throughout and does not enter the body wall. That part which descends from the pallial gonoduct to reach the anterior of the body evidently is derived from the spermiduct that is embedded in the body wall of the onchidiids, vaginulids, rathouisiids and most ellobiids, but has become free in the haemocoel. The anterior extremity none the less maintains a close association with the connective tissue of the body wall. That part which runs to insert in the penis is homologous to the closed vas deferens in the lower pulmonates. In many succineids and athoracophorids, this latter part of the vas deferens is differentially constructed, with the musculature of the walls considerably thickened. While spermatophores are not produced, these particular succineoideans tend to transfer spermatozoa in distinct mucous packages (para-spermatophores; Webb, 1977a,b,c). Other Succineidae and Athoracophoridae evidently transfer spermatozoa simply suspended in seminal fluid.

The capacity for self-fertilization is widespread in Succineidae (Patterson, 1970), but the extent to which this reproductive strategy is employed in natural populations presently is not known. Rao (1925) raised the possibility of a form of self-copulation in *L. rupicola* as one specimen was found with the penis everted and deeply inserted into its own vagina. Where mating occurs, it apparently invariably involves reciprocal gamete exchange in both Succineidae and Athoracophoridae. In the majority of species, spermatozoa are deposited in the free oviduct of the mate and subsequently transferred to the receptaculum seminalis, with the excess directed to the bursa copulatrix (Webb, 1953; Hecker, 1965; Rigby, 1965; Jackiewicz, 1980; G.M. Barker, personal observation). In some succineids, however, mating occurs without intromission, with gamete exchange effected by deposition of spermatozoa on the mate's everted penis (Webb, 1977a,b,c; Emberton, 1994). Asami *et al.* (1998) categorized mating in succineids as non-reciprocal, with one animal functioning as the 'male' that achieves copulation by mounting the 'female's' shell. Villalobos *et al.* (1995) observed that reciprocity in gamete exchange was achieved by *Succinea cosaricana* von Martens through male–female role reversal in a subsequent copulation.

The seminal receptacle is present primarily as a diverticulate sac opening adjacent to the renal gonoduct in a distinct fecundation pouch. In Succineidae and some Athoracophorinae, the seminal receptacle comprises a number of chambers and thus is lobate in appearance. The diverticulum is absent in Aneitinae and the seminal receptacle is represented as a distended terminal section of the renal gonoduct. In Succineidae, the bursa copulatrix is primarily on a long duct opening to the distal part of the female pallial gonoduct. However, the duct is often reduced and may be located secondarily on the genital atrium or the male genitalia. While its duct generally is reduced, the bursa invariably is located on the female pallial gonoduct in Athoracophoridae.

A diverticulum is present on the oviductal gland region of the female pallial gonoduct in Aneitinae. The occurrence of similar diverticula in Ellobiidae, Otinidae, Chilinidae, Siphonariidae and Latiidae indicates a possibly plesiomorphy in basal pulmonates, but homology has yet to be demonstrated.

The male genitalia of Succineidae and Athoracophoridae exhibit remarkable similarity, and comprise a tubular, thick-walled penis that in some species bears one or more caecal flagella; the penis and its flagella become everted during copulation. The entrance of the vas deferens into the penis is through a simple pore that often is located at the apex of a small conical or, rarely, coiled papilla. The epithelium lining the penis lumen generally is thrown into low folds or papillae that in Succineidae may be covered in a cuticle and/or bear crystalline rosettes (Quick, 1933, 1936; Tillier, 1981; G.M. Barker, personal observation). In Athoracophoridae, the epithelium comprises numerous scale- or hook-like cuticular papillae, of species-specific form (G.M. Barker, personal observation). In both families, the penis and terminal part of the vas deferens may be enclosed entirely or partially within a thin, muscular sheath.

The penial retractor in Athoracophoridae arises from the dorsal body wall, usually in the vicinity of the pallial complex, and runs directly forward to insert on the penial complex. This is essentially the same configuration as in onchidiids, rathouisiids and vaginulids, and is easily derived from the retractor system in ellobiids that is essentially columellar. As recognized by Tillier (1984c), the condition is fundamentally different in Succineidae. Here the retractor arises from the mantle adjacent to the origin of the aorta, inside the periaortic, anterior-most intestinal loop, and in running to the penis the muscle passes under that intestinal loop. This implies some modification of the ontogenesis of the male genitalia, or the retractor muscle, in succineids. Interestingly, A. Dutra-Clarke (personal communication) observed both under- and over-intestinal routes for the penial retractor within a single population of *Omalonyx matheroni* (Potiez & Michaud). Furthermore, the penial retractor is absent in *Indosuccinea* Rao and *Lithotis.*

The penial gland generally is absent in Succineidae, but a thick-walled appendix opening through the lateral wall of the penis in *C. vermeta*, supplied with a branch of the penial retractor muscle, may represent a modified penial gland; in other members of the genus *Catinella* Pease, the penial gland is reduced further. In *Neosuccinea* Matekin, an elongate appendix, with a strongly muscular base and more sacculate apex, occurs on the vagina. That the bursa duct varies in its site of insertion suggests that there is considerable evolutionary plasticity in the configuration of the terminal genitalia, and the penial gland that initially was adjacent to the genital atrium, and the bursa initially opening to the female pallial gonoduct, may have become variously incorporated into the terminal genitalia. The current literature suggests absence of the penial gland in Athoracophoridae, but recently discovered species in New Caledonia possess elaborate male genitalia that include a massive, stylet-bearing stimulator dominating the penis in a position equivalent to the penial gland in *Catinella* (G.M. Barker, unpublished).

While primarily ditrematous, the Pulmonata have achieved mono-trematry in several lineages. Among succineids, the male and female genitalia open separately, but closely juxtaposed, to the exterior in *Succinea* Draparnaud and *Lithotis*. In other Succineidae, and in Athoracophoridae, the male and female ducts open to a common genital atrium immediately inside the body wall and thence through a single orifice to the exterior. The genital opening is located behind the right cephalic tentacle, somewhat posteriad in most Succineidae but immediately adjacent to the tentacle in *Omalonyx* (*Neohyalimax*). The latter position is characteristic of all Athoracophoridae.

There has been much debate on the origin of hermaphroditism in pulmonates. Pelseneer (1896) thought that a monaulic, monotrematic prosobranch female system became hermaphrodite, and cainogenetically developed a male organ with functional connection to the pallial gonoduct opening through an open groove, to give rise to the pulmonate reproductive system. Based on variation in the configuration of the terminal genitalia, Visser (1977, 1981) interpreted hermaphroditism as being derived in pulmonates from a monaulic, prosobranch condition, but with development of female features in a male system leading to the basommatophoran condition, and male features in a female system lead to the stylommatophoran condition. Monauly as the plesiomorphic condition in pulmonates is also supported by ontogeny of the pallial gonoducts in stylommatophorans (e.g. Hochpöchler, 1979) and the wide-spread occurrence of the monaulic (spermoviduct) condition in lower Heterobranchia (e.g. Pyramidelloidea). Diauly has thus been treated as a derived condition (e.g. Duncan, 1960; Berry *et al.*, 1967; Visser, 1981; Martins, 1996b).

However, the widespread occurrence of hermaphroditism, including the presence of both dioecious and hermaphroditic species within some families, and the changes from one sex to the other in response to environmental stress or natural protandry, point to hermaphroditism as plesiomorphic in Gastropoda. It seems that hermaphroditism is often suppressed for periods in the evolutionary history of various gastropod groups, only to re-emerge under certain evolutionary or ecological conditions. In hermaphrodites, there may be separate ovary and testis, or various forms of unification of these into a functional ovotestis. None the less, in all extant gastropods, there is a single renal gonoduct. Contrary to the view of Visser (1981), paired gonads and mesodermal gonoducts are not necessarily a prerequisite for diauly in the pallial gonoducts. It is clear that the ditrematous condition precedes the monotrematic condition in the evolution of the Pulmonata, but the consolidation of the male and female genital openings that has thus taken place during the phylogenetic development of the various pulmonate groups does not necessarily imply unification of the pallial gonoducts. Further, that the pallial gonoduct has an apparent single ectodermal ontogeny in monaulic, higher pulmonates is not necessarily very informative about the origins of hermaphroditism in ancestral pulmonates – developmental studies are needed in diaulic, lower pulmonates. It is clear that diauly is widespread in the lower pulmonates and may be plesiomorphic for this clade; consistent with the interpretation of Solem (1972b, 1976) that fusion of the male and female pallial gonoducts has occurred repeatedly amongst Stylommatophora. Phylogenetic analyses of both morphological and 16S rDNA sequence character data (G.M. Barker, unpublished) strongly indicate that diauly is plesiomorphic in Athoracophoridae and that monauly has been derived independently four times in this family. The variation in the condition of the pallial gonoducts, even within Ellobiidae and Athoracophoridae, points to remarkable evolutionary instability in the degree to which the male and female components are united, and it is probable that within gastropods there is much homoplasy and possibly reversals.

The Succineoidea are oviparous, producing cleidoic eggs. Many suc-cineids produce firm egg capsules, each with a single embryo, embedded in a mucous jelly (Fischer, 1938; Hecker, 1965; Rigby, 1965). *Succinea putris* (Linnaeus) produces such clutches, comprising up to 50 or 60 eggs. Other Succineidae produce separate eggs. *Austrosuccinea archeyi* (Powell), for example, produces clutches of up to 15 spherical egg capsules that are not embedded in jelly (Powell, 1950). The eggs of *Omalonyx felina* Guppy are not embedded in mucous jelly, but are often linked as the capsule material is drawn out in a string, resembling those in Vaginulidae (Patterson, 1971b). Athoracophoridae generally produced clutches of 10–30 separate, spherical to oval egg capsules, each with a single embryo (e.g. Suter, 1913; Barker, 1978). Recent discoveries (G.M. Barker, unpublished) have shown that the Athoracophoridae produce the spectrum from hard, calcareous shelled egg capsules to firm, multilayer, jellied capsules. The Succineoidea exhibit intracapsular development, with no veliger stage but a gradual assumption of adult features. Post-gastrula stages are characterized by differentiation of rudiments of the foot and viscera, and development of the anterior sac (hepatic mass). The Athoracophoridae lack a shell plate, and the vestiges of the shell in the form of one or more granules seen in the adult animal are laid down at the early post-gastrula stage.

Great evolutionary instability in chromosomal richness is apparent in Succineoidea. While some Succineinae retain a haploid chromosome complement approaching that in basal pulmonates (15–19 haploid chromosomes; Patterson and Burch, 1978), others exhibit apparent chromosomal deletion or addition, resulting in variation from 11 to 25 haploid chromosomes within the subfamily. As an apparent result of considerable chromosome deletion, Catinellinae exhibit extremely low chromosomal richness (five or six haploid chromosomes) (Natarajan *et al.*, 1966; Butot and Kiauta, 1967; Patterson, 1968a,b; Patterson and Burch, 1978). Solem (1978) relates aneuploid reduction in Catinellinae to ecological fitness in ephemeral habitats, but failed to take into account the wide variation of chromosome richness in Succineinae occupying the same types of habitats. Athoracophoridae have a haploid complement of 44 (Burch and Patterson, 1971), suggesting a possible polyploid mutational origin from a succineid ancestor.

The pallial organs of Succineoidea have drawn considerable comment in the literature. None the less, their basic configuration in succineid snails is not unlike that in Ellobiidae, except that the nephridium is wider than it is long, the lamellae-free, distal ureteric pouch is absent, and a tubular ureter is formed. In discussion on the Succineidae, Pilsbry (1948, p. 771) indicated, 'the short, wide shape of the kidney may be related to the shortening of the whole pallial cavity in this group'. Indeed, the arrangement of lamellae strongly points to the derivation of succineid nephridium from elongation subperpendicular to the rectum on the visceral side of a relatively short ancestral nephridium (Tillier, 1989). In their possession of an elongate, tubular ureter, the Succineidae are like Onchidiidae, Vaginulidae and Rathouisiidae; the respective ureters have the same cellular structure (Delhaye and Bouillon, 1972b,c), suggesting homology. In succineid snails, the ureter traverses across the face of the nephridium to reach the posterior extremity of the pallial cavity before running forwards parallel to the rectum to the ureteric orifice at the pneumostome. In many Succineidae, the terminal part of the ureter, near the pneumostome, is continued along the front of the pallial cavity as a blind caecum (Quick, 1933, 1936). The short pallial cavity is highly vascularized. Succineid snails living in the splash zone of waterfalls, and thus with water continually collecting on the shell, frequently have evolved a pneumostomal ridge on the shell to channel water away from the pallial cavity opening (Patterson, 1973). The principal pallial modifications associated with evolution of the slug form in succineids are the further compaction of the respiratory surface, secondary shortening and slight detorsional rotation of the nephridium, and the freeing of the ureter from the anterior face of the nephridium, resulting in some variation in its route to the excretory orifice. The vascularization has become concentrated into pouch-like lobes that hang pendulously from the remaining pallial cavity roof in succineid slugs such as *Omalonyx* d'Orbigny *s.str.*

The general relative disposition of the pallial organs in athoracophorid Aneitinae is similar to what is found in succineid slugs:
nephridium basically in the posterior left part of the pallial cavity, heart
in front of the nephridium, and pneumostome opening together with
rectum and ureter on the right side of the pallial cavity. The respiratory
part of the pallial cavity is completely filled by a net of connective tissue,
including the venous system and delimiting a system of air lacunae; this
represents a more advanced stage in the evolutionary compaction of
the respiratory surface evident in succineid slugs. Early workers drew
analogies between this respiratory configuration and the tracheal system
of insects (e.g. Plate, 1898). Simroth (1918) recognized the highly
compacted nature of the pallial cavity in Athoracophoridae.

The nephridium in Aneitinae is compact, with a lumen strongly
lamellate. Adjacent to the opening to the pericardium, the nephridial pore
with a sphincter gives rise to a tubular ureter that in *Aneitea* is rather short
and runs directly to the renal orifice adjacent to the pneumostome. In
Triboniophorus, the ureter is elongated and scribes several loose loops
before running forward to the renal orifice (Burton, 1981; D.W. Burton,
personal communication; G.M. Barker, unpublished). In Aneitinae, the
ureter opens adjacent to the pneumostome, but disjunct from the anus
(Plate, 1897, 1898; Glamann, 1903; Burton, 1981). The nephridium in
Athoracophorinae is elongated and bilobed, with its long axis across the
full width of the body perpendicular to the body axis. The nephridial orifice is on the right anterior face, adjacent to the renopericardial pore, to
give rise to a long, folded ureter similar to that in *Triboniophorus*, to terminate through an orifice in the anterior of the vestigial mantle shield. A
feature common to all athoracophorids is the presence of one or several
tubular, secretory diverticula opening through multiple but closely grouped
pores under the ureter orifice (Burton, 1981). Early authors described and
depicted the ureter in Athoracophoridae as ramifying into diverticula;
this error has been perpetuated in the literature (e.g. Tillier, 1989). As
demonstrated by Burton (1981), the ureter is a tube which, when long,
becomes folded like that in Vaginulidae. The considerable variation in the
nephridial shape and urethral length in Succineoidea indicates that neither
is particularly informative about higher level systematics in Pulmonata.

In Athoracophorinae, a slender duct arises from near the distal
extremity of the ureter and runs along the rectum to open through a
minute pore into the distal intestine (Burton, 1981). Burton (1981)
postulated that the tubule linking the ureter with the intestine is a
synapomorphy of the Athoracophorinae. He assumed that the absence in
Aneitinae is plesiomorphic, but there is a possibility that the structure in
Athoracophorinae is homologous to the tubule(s) linking the ureter with
the rectum in Vaginulidae.

Succineoidea feed primarily on minute cryptogams (e.g. fungi,
filamentous algae and diatoms) associated with rock and plant surfaces.

Some Succineidae, such as *Succinea putris* and *Omalonyx*, are herbivorous, feeding on living plant tissues. As with Ellobiidae, the great diversity of radular dentition in Succineoidea serves to emphasize the low value afforded by the radulae in supraspecific systematics in pulmonates. An essential synapomorphy of the Succineoidea is the elasmognathous jaw, characterized by the development of an accessory plate to which the buccal musculature attaches. The oesophagus is generally short, with low longitudinal ridges. In some succineid species, these folds delimit two quite pronounced (even pigmented) channels on each of the dorsal and ventral aspects of the oesophageal epithelium. The Succineidae and Athoracophoridae invariably possess an oesophageal crop, which narrows posteriorly to the stomach. The stomach is simplified relative to that in Ellobiidae, Onchidiidae and Vaginulidae – it is a simple curvature with oesophageal and intestinal openings at the respective extremities and thus widely separated, the ducts to the two digestive gland lobes open in its concave arc more or less contiguous in the angle formed by the oesophageal crop and intestine, and all vestiges of the gizzard are absent. These features are secondary and, given that similar configurations have been achieved in other pulmonates, do not accord much systematic significance. The absence of the gizzard is obviously correlated with the more enzymatic breakdown of digested food materials relative to that in lower pulmonates. A small caecum is associated with the duct of the posterior lobe of the digestive gland in both Athoracophoridae and Succineidae. In drawing on the work of Rigby (1965), Solem (1978, p. 85) concluded that the caecum in succineids 'is convergent with similar structures in Basommatophora, other Stylommatophora, and browsing opisthobranchs'. Tillier (1989) too considered this a new structure, developed in compensation for the simplified stomach. However, I consider the caecum in Succineoidea to be homologous to that structure in Ellobiidae, and both homologous reductions of the gastric caecum in lower gastropods. Internally, the stomach wall has low folds, the stronger two of which may be recognized as typhlosoles that converge to form a weak intestinal groove between them in the proximal intestine.

The intestine of Succineidae and Athoracophoridae has a single forward-directed loop, turning posterior after crossing over the aorta. From the aorta, the intestine has one loop backwards before going forward to the anus. In Succineidae, the intestine is rather short and lacks any coiling around the crop. In contrast, the intestine in Athoracophoridae is very long, and often (but no invariably) hypertorted to coil around the oesophageal crop (Tillier, 1984c; G.M. Barker, unpublished), exactly as in the vaginulid *Spirocaulis*; this hypertorsion is achieved during ontogeny after torsion of the embryo. In both families, the intestine generally is without epithelial folds, but the rectum frequently has pronounced longitudinal folds.

Stylommatophoran pulmonates

The extant Stylommatophora are highly diverse, with some 71–92 families recognized (Emberton *et al.*, 1990). As pointed out by Solem (1985), the northern hemisphere regions are rather depleted in stylommatophorans as these regions have been colonized in relatively recent post-glacial times from a few refugia. The greatest diversity today occurs in isolated island and montane refugia systems of the southern hemisphere, and some tropical forest systems. Solem (1974, 1979a,b, 1981), Solem and Yochelson (1979), Nordsieck (1986), Tillier *et al.* (1996), Tracey *et al.* (1993) and others have discussed the stylommatophoran fossil record. Solem and Yochelson (1979) attribute four Upper Mississipian taxa, approximately 320 million years old, to the Stylommatophora, which indicates a stylommatophoran radiation in the Carboniferous shortly after the emergence of the pentaganglionate Heterobranchia.

Nordsieck (1986) advocated a similar timing for the Stylommatophora. With consideration of fossils (in particular, the confirmed diversity of stylommatophorans in the Cretaceous) and molecular phylogeny, Tillier *et al.* (1996) concluded that the emergence of the opisthobranch and pulmonate clades probably occurred earlier than the Upper Carboniferous (though possibly as late as the Triassic), but the emergence of the Stylommatophora is relatively recent, being late Cretaceous–Palaeocene (65–55 million years ago). This implies that the Carboniferous terrestrial taxa are not stylommatophoran. However, the late Cretaceous–Palaeocene origin of the Stylommatophora suggested by Tillier *et al.* (1996) is not consistent with the wide dispersion in pre-Cretaceous times indicated by the intercontinental disjunctions evident in extant members of several stylommatophoran families.

A suite of morphological, physiological and behavioural adaptations that enable regulation of body hydration has been critical to the success of the Stylommatophora (Solem, 1974; Riddle, 1983; Cook, Chapter 13, this volume). Undoubtedly the contractile pneumostome, which reduces the extent to which the pallial cavity is exposed to the evaporative demands of the ambient environment, has been central to this success. Stylommatophorans often take up water through the pneumostome: a rectal pump rapidly conveys this water through the anus into the digestive tract. Under dry environmental conditions, this extrasomal reserve cannot only be utilized by absorption into the blood, but a large portion can be expelled from the anus and directly conveyed to the external body surfaces where water losses due to evaporation and locomotion occur (Neuckel, 1985). Furthermore, these animals frequently store urine in the pallial cavity, which is thought to function as a reservoir of water (Blinn, 1964; Smith, 1981) for subsequent resorption by the renal system. Unlike caenogastropods, ultrafiltration to produce urine does not occur in the pericardial cavity but rather in the nephridium (Vorwohl, 1961; Martin *et al.*, 1965); podocytes are absent in the heart, and ultrafiltration occurs across the lamellae of the nephridium (Potts, 1975). These animals are purinotelic in that the nephridium excretes much solid matter in the form of uric acid, guanine and xanthine concretions, thus enabling nitrogen removal with reduced water loss (Riddle, 1983). Many species have been shown to have remarkable physiological tolerance of the considerable variation in body water content that can occur during the course of daily activity cycles and in the course of aestivation. In addition to the uptake through the pneumostome mentioned above, stylommatophorans are able to rehydrate rapidly by uptake of water through the integument, by drinking and through feeding. As with Succineidae, most stylommatophoran snails are able to aestivate over periods of unfavourable conditions, with the animal retracted into the shell and the shell aperture sealed with one or more epiphragms or cemented to the substrate (Riddle, 1983; Solem, 1985).

The basic configuration of the pallial organs in Stylommatophora resembles that in Ellobiidae. In (dextral) stylommatophoran snails, the nephridium, on the left side of the pallial cavity, generally is triangular in shape and its base shares a common wall with the visceral cavity, where it runs along the periaortic intestinal bend. The pericardium is applied to its left side and is prolonged towards the pallial border by the pulmonary venous system and towards the visceral cavity, outside the pallial cavity, by the aorta. The aorta crosses the intestinal bend before dividing into anterior and posterior branches. The pulmonary vein usually produces extensive vascularization over much of the pallial surface between the nephridium and pneumostome. This is the primary site of respiratory gas exchange. The contractile pneumostome opens and the diaphragm (the floor of the pallial cavity) contracts to dilate the pallial cavity and draw air in. Closure of the pneumostome and relaxation of the diaphragm produce a positive pressure inside the cavity, facilitating gaseous exchange across the venous network on the pallial cavity roof. The pneumostome again opens to initiate repetition of the cycle.

I follow Delhaye and Bouillon (1972a,b,c), Nordsieck (1985) and Tillier (1989) in recognizing the renal organ as comprising three morphologically and functionally distinct parts: (i) a broad proximal, internally lamellate nephridial sac, the nephridium proper; (ii) a distal ureteric pouch (orthureter); and (iii) a ureter that may be represented by an open ciliated groove or a closed tube, of lengths varying among taxa. The nephridium opens to the pericardium by a renopericardial pore and expels the excreta through a nephropore into the orthureter or directly into the ureter. The development of ureters in these terrestrial pulmonates, characterized by internal transverse lamellae (Delhaye and Bouillon, 1972b), may be correlated with the need for increased resorption of ions and water.

The higher systematics of the Stylommatophora generally used today originates from Pilsbry (1900), who founded it on the structure of the excretory system. While being rejected by Simroth and Hoffmann (1908–28), Thiele (1929–35) and others on the grounds that it did not reflect natural relationships, Pilsbry's scheme was generally accepted after further development by Baker (1955, 1956), Solem (1959) and Zilch (1959–60). Pilsbry (1900) proposed the ordinal group Orthurethra for stylommatophorans possessing both nephridium and orthureter, with an anterior nephropore, but usually with only the proximal part of the ureter developed as a groove along the rectal face of the nephridium. The taxon Mesurethra was defined as possessing neither a orthureter on the nephridium nor a closed ureter, while Sigmurethra was defined as being without a orthureter but possessing a ureter running along the anterior of the nephridium to the top of the pallial cavity (primary ureter) and then, adjacent to the rectum, to the pneumostome (secondary ureter).

The possession of an orthureter on the distal part of the nephridium generally has been assumed to be plesiomorphic in Stylommatophora because this type of configuration occurs in Ellobiidae and various aquatic, basal pulmonates (Delhaye and Bouillon, 1972a,c; Nordsieck, 1985; Tillier, 1989). Delhaye and Bouillon (1972b) considered the orthureter to be a secondary specialization of the anterior part of the nephridium. While there has been much incompatibility in the various suprafamilial schemes proposed, even in the past few decades, there have been explicit or implicit inferred sister-group relationships of the orthurethran and non-orthurethran Stylommatophora (Solem, 1985; Tillier, 1989; Emberton and Tillier, 1995; Stanisic, 1998). The retention of the orthureter in the excretory system among stylommatophorans generally has been recognized only in those taxa traditionally grouped as Orthurethra, namely Pupilloidea, Chondrinoidea and Partuloidea. Tillier (1989) has demonstrated, however, that a pouch-like differentiation in the nephridium is present in a number of non-orthurethran families. Despite the homology of these nephridial pouch and orthureter structures not being fully resolved by histological and functional studies, the ordinal status of Orthurethra is called into question. The validity of this ordinal category apparently is weakened further by several orthurethrans possessing a retrograde closed ureter (Watson, 1920; Solem, 1964; Tillier, 1989), by many non-orthurethran taxa exhibiting transitional states between the ureter morphologies used originally to define the Mesurethra and Sigmurethra (Nordsieck, 1985; Tillier, 1989), and the absence of clear-cut sister-group relationships between Orthurethra and other Stylommatophora in recent phylogenetic analyses of molecular (Tillier *et al.*, 1996) and morphological data (Barker, 1999). None the less, support for the monophyly of the Orthurethra is found in 28S rDNA sequences (Wade *et al.*, 2001). If the mesurethran condition is part of the same morphocline as the Sigmurethra, as suggested by Delhaye and Bouillon (1972b) and accepted here, then the Sigmurethra were derived from gastropods without a closed ureter, but after the loss of the orthureter. According to Schileyko (1979), sigmurethry is a feature in embryogenesis, even in orthurethran forms without a ureter in the adult animal.

Tillier (1989) proposed a new higher classification of non-orthurethran Stylommatophora. Using the differentiation of the renal morphology, Tillier recognized two phylogenetic lines within the

Sigmurethra, one in which at first a dedifferentiation of the nephridium took place which was partly followed by a closure of the ureter (Brachynephra *sensu* Tillier), and another in which at first a closure of the ureter took place which was partly followed by a dedifferentiation of the nephridium (Dolichonephra *sensu* Tillier). I concur with Nordsieck (1992) in considering Tillier's hypothesis as an overvaluation of the differentiation of the nephridium.

Pilsbry created the ordinal category Heterurethra for the Succineidae (placed in Stylommatophora, contrary to relationships suggested here), whose excretory system was said to differ from that of the Sigmuethra in that the nephridium extends transversely, and the primary and secondary arms of the ureter are at right angles to each other. Baker (1955) applied the term Heterurethra to all stylommatophorans (inclusive of Succineidae) possessing a nephridium transverse with respect to the body axis but, as pointed out by Nordsieck (1985), this configuration in Stylommatophora is really a form of sigmurethry. If we accept that the Succineoidea are not part of the Stylommatophora clade, then the trend to heterurethrary in the stylommatophoran Sigmurethra will be recognized as convergent on that in Succineidae.

Space is limited here, so the great diversity within the Stylommatophora is necessarily treated collectively. They are primarily snails, whose shells vary from very elongate to globose, depressed and discoidal, with the coils rounded, angular, shouldered or flattened, with shallow or impressed sutures, and with many to few whorls. The outer lip of the shell may be either thin in species with indeterminate growth, and variously thickened and sometimes turned back or reflected in species with determinate growth. The interior walls of the shell often possess folds of varying types: those that occur on the shell axial walls or columella are referred to as columellar lamellae, those on outer shell walls are referred to as palatal plicae. The palatal plicae result from folding of the secretory surface of the mantle and are produced throughout the animal's life, but most frequently at the termination of post-embryonic development. In contrast, the columellar lamellae appear even in embryogenesis and continue to lengthen as the shell grows. Based on patterns seen in ontogeny, Schileyko (1979) regarded the presence of lamellae as a plesiomorphic character, while the plicae were considered to have evolved independently in several lineages. Nordsieck (1986) and Pokryszko (1997) emphasized, however, that the presence of lamellae and plicae, irrespective of the ontogeny in extant taxa, must be evaluated as plesiomorphic because these characters are present in taxa (Ellobiidea, shelled opisthobranchs) which they considered indicative of the clades ancestral to the Stylommatophora. At the very lip of the shell, folds, wrinkles and nodules often complement these plicae and lamellae. There has been much debate about the function of these apertural shell structures, with roles in shell carrying, respiration, epiphragm building and protection against predation or environmental adversities (e.g. Schileyko, 1984; Solem, 1972a, 1976; Suvorov and Schileyko, 1991; Suvorov, 1993;

Gittenberger, 1996; Pokryszko, 1997). Independently developed in a number of stylommatophoran families, including Vertiginidae, Chondrinidae, Streptaxidae, Polygyridae and Bulimulidae, the last part of the body whorl may be twisted down or upwards, and lengthened, in such a way that the body whorl leading to the shell aperture is narrowed locally. Internal thickening often narrows the apertural lip in many families. These apertural modifications have been assumed to have an anti-predator function (Gittenberger, 1996). Often the periostracum of the protoconch and teleoconch is produced into an elaborate microsculpture of spiral striae and transverse striae and ribs on the shell external surface. Periostracal hair-like or scale-like processes are not uncommon in a number of stylommatophoran families. A number of hypotheses have been advanced for the functional role of these periostracal processes, including: (i) 'to repel moist particles' (Solem, 1974) and prevent wet leaves from adhering to the shell; (ii) to defend against predators (Webb, 1950); and (iii) to camouflage the shell by trapping soil and debris (Pilsbry, 1940). These hypotheses are yet to be tested. Most characters of the shell vary considerably within the higher taxon categories and so have only little importance for the higher systematics of the Stylommatophora. While their value at lower taxonomic levels is complicated by convergences and divergences, there is increasing realization that shells can yield morphological data useful for resolving evolutionary patterns (e.g. Emberton, 1996).

The operculum is absent in Stylommatophora, even in the embryonic stage. However, a small number of stylommatophorans have secondarily developed structures that function in a manner similar to the operculum in prosobranch snails. Some Charopidae (Solem *et al.*, 1984) are characterized by the presence of a discoidal, callus-like tissue on the dorsal aspect of the posterior foot that functions to seal the shell aperture when the animal retracts into the shell. The camaenid *Thyrophorella thomensis* Greeff produces a lobe-shaped outgrowth from the parietal apertural wall that hinges, along a non-calcified band of periostracum, to close the aperture (Girard, 1895; Gittenberger, 1996). An autapomorphy of the Clausiliidae is the clausilial apparatus, which comprises the clausilium, a spoon-shaped calcareous plate, attached to the shell columella by a long, flexible stalk. It automatically closes the lumen of the shell when the animal has retracted, and is pushed out of the way when the animal extends its head–foot from the shell (Nordsieck, 1982; Gittenberger and Schilthuizen, 1996).

The slug form has evolved many times in Stylommatophora, and many species of widely divergent origins have, by parallel evolution, assumed a remarkable morphological similarity. They have 'sacrificed' the protection offered by possession of a shell for the mobility, fast body movements and ability to occupy very small spaces afforded by the reduction or elimination of the shell. It seems that these selective forces were only able to operate effectively in those environments where moisture was plentiful (Solem, 1974). The presence of a long, closed ureter has been considered a prerequisite for evolution of the slug form (Solem, 1974; Tillier, 1989), but this paradigm is challenged by the absence of a substantive secondary ureter in testacellid (de Lacaze-Duthiers, 1887; Barker, 1999) and some athoracophorid slugs (see Succineoidea above), and the possession of a long ureter in some aquatic pulmonates. In the course of evolution of the slug form, in a process termed 'limacization' by Solem (1974), the number of whorls described by the shell and visceral mass is reduced, and the viscera are incorporated into the foot (van Mol, 1970; Solem, 1974; Tillier, 1984c). In semi-slugs, which represent the intermediate stage in the evolution towards the slug form, the shell is reduced to the extent that the animal can no longer withdraw its head–foot under the protective shell, and the organs associated with the distal part of the visceral mass are incorporated into the anterior head–foot. In these animals, the stomach and associated digestive gland remain in the visceral hump, above the pallial border. In full slugs, the visceral mass is still more reduced or absent, and the stomach is included in the pedal cavity of the foot. The pallial border or mantle collar extends over the surface of the reduced shell and may cover it totally, forming a dorsal mantle shield or 'clypeus'; the shell is absent in some full slugs. In the most limacized slugs, the shield itself can hardly be distinguished from the dorsal surface of the foot. While on a global scale their species number is greatly exceeded by fully shelled taxa, slugs or semi-slugs are highly diverse, and in many regions make up a significant part of the stylommatophoran fauna.

Reduction in shell size and incorporation of the visceral hump into the head–foot, associated with limacization, necessarily involve alteration of the layout and size of the pallial organs described above for fully shelled stylommatophorans. Several patterns of pallial organ modification are evident among lineages that evolved towards the slug form. Surface area for gaseous exchange is maintained by the venous vessels being highly developed in the vestigial pallial cavity or invading the mantle edge. Most semi-slugs and slugs show extensive development of mantle lobes that partially or wholly cover the shell and/or extend forwards along the neck, providing increased surface area for gaseous exchange (Tillier, 1983). In the most highly evolved slugs (e.g. Bulimulidae, van Mol, 1971; Parmacellidae, Tillier, 1982, 1983), the vascularization convergently approaches that in Athoracophoridae, and a large part of the respiration takes place across the integument (e.g. Duval, 1982; Prior et al., 1983). Nephridial size retention often involves its rotation and change in shape, including the development of lateral lobes.

The integument of the head–foot in Stylommatophora typically is thick and with a reticulated, rugose surface pattern. In some stylommatophorans, such as members of the Arionidae and Helicarionidae, the tegument is produced into conical or hemispherical tubercles, recalling the condition in Onchidiidae and Succineoidea. The dorsal aspect of the tail in snails is generally without important features, but species in some families (e.g. Lysinoe Adams & Adams in Helminthoglyptidae, Oxychona

Mörch in Bulimulidae) possess distinct keels. Such keels are even more prominent in slugs of the families Milacidae, Limacidae, Agriolimacidae, Parmacellidae, Trigonochlamydidae and the arionid subfamily Ariopeltinae.

The sole is primarily uniform, although commonly the outer edges are of a different colour from that of the central zone. Locomotion is by ciliary gliding in many small species. In others, locomotion is by multiple direct pedal waves. At the onset of locomotor activity, the first waves are initiated in the medial region of the sole, reminiscent of that in Ellobiidae. Often the sole is tripartite, with only the central zone involved in the locomotive pedal waves. In some Stylommatophora, a pair of parapodial grooves occur along the side of the animal, just above the foot margin, so that the ciliated sole extends above the foot margin. This has been termed the aulacopod condition. In others, the holopod condition occurs, where the parapodial grooves may be fused and sited at the margin of the foot, such that the ciliated sole reaches to the lateral margin of the foot but not extending above. On the basis of these differences in parapodial grooves, Pilsbry (1896) divided the stylommatophoran order Sigmurethra into two suborders Aulacopoda and Holopoda. Wächtler (1935), however, showed that the parapodial grooves are not absent from the Holopoda, but are merely indistinct. He showed that these grooves are similarly present but indistinct in various Orthurethra. Webb (1961) independently repeated Wachtler's discovery of the universality of parapodial grooves, and dismissed the division of Sigmurethra according to distinctiveness of the grooves, on the grounds of ecologically induced convergence. Webb (1961) thought the aulacopod condition better adapted to burrowing in soil, while the holopod condition, with less pronounced grooves, was more dry-adapted. Without reference to Wächtler's or Webb's viewpoints, Solem (1978) maintained that the suborders 'seemed coherent assemblages' and were thus retained in his sigmurethran classification. Drawing on the comparative morphology of the parapodial grooves throughout the Stylommatophora, Schileyko (1979) maintained that aulacopody arose repeatedly from the holopod condition.

In an analysis of pallial and foot character evolution in Stylommatophora, Baker (1955) concluded that the ancestors of the Sigmurethra probably had, among other features, an aulacopod foot. Tillier (1989) noted the general correlation between an aulacopod foot and a shell without determinate growth. With the observation of aulacopody in embryos and hatchlings of Orthurethra and holopod Sigmurethra, Tillier surmised that both the aulacopod foot and indeterminate growth result from paedomorphosis. Aulacopody is often associated with a caudal pit, commonly but wrongly referred to as the caudal mucous gland, developed on the tail just behind where parapodial grooves meet and sometimes produced as a projection or horn above this juncture. Muratov (1999) postulated that the caudal pit functions primarily in resorption of pedal mucus produced during locomotor activity.

Stylommatophora primarily possess two pairs of tentacles on the head, the cephalic tentacles with an eye at the apex of each, and inferior tentacles on the anterior face of the snout below the cephalic tentacles. Unlike the condition in Succineoidea, Onchidiidae, Vaginulidae and Rathouisiidae, the rugose integument of the cephalic body wall extends over all but the very apex of the cephalic tentacles. Furthermore, the retraction of the cephalic tentacles into the body by invagination seen in Succineidae is more developed in Stylommatophora (Burch and Patterson, 1969). The inferior tentacles are similar to those of Succineidae in possessing a single apex, and are retractable by invagination. Paralleling the situation in Succineoidea, these inferior tentacles are secondarily absent in some species of Urocoptidae, Vertiginidae and Pupillidae.

Retraction of the body into the shell in stylommatophoran snails is by inversion of the cephalopedal mass, rather like the inversion of the tentacles during their retraction into the anterior body. The columellar retractor muscle effects the retraction. Like that in Ellobiidae, the columellar has become largely free in the haemocoel, anterior of its columellar origin, to run forward to attach on the cephalic organs and anterior body wall. In large animals, the primary form of the columellar muscle is retained, with a broad fan attaching to the pedal musculature and numerous branches attaching to the organs and body wall in the body anterior. In the majority of Stylommatophora, there has been a general reduction in the bulk of this free muscle system, with the pedal fan greatly reduced and the retractors to the various anterior organs separated as slender muscle strands. The buccal retractor inserts under the buccal mass, while the tentacular retractors divide distally into ocular and inferior tentacle branches. In the primary state, the right tentacular retractor passes through the penial–oviductal angle, over the base of the penis to reach the ocular tentacle. The alternative course of the retractor, outside the penial–oviductal angle, has arisen in many lineages. During limacization, the columellar stem to the pedal musculature frequently is lost, since there is no need for withdrawal of the anterior head–foot. In many slugs, the buccal and tentacular retractors have lost their association with the columella (or its region of origin) and instead arise from the body wall.

In the CNS each cerebral ganglion has a prominent globineuron-containing procerebral lobe, as in Succineoidea. However, the cerebral gland is reduced further than that in Succineoidea, being largely incorporated into the procerebrum and lacking all but vestiges of the tube that opened to the body exterior in the ancestral pulmonates. The Stylommatophora often retain the subcerebral commissure. In their possession of commissural nerves arising from the cerebral commissure these animals differ from lower pulmonates, including Succineoidea (Nolte, 1965; Cook, 1966). Van Mol (1967) recognized variation of the microscopic structure of the cerebral ganglion among stylommatophorans, but the small sample size effectively prevents generalization of these observations to diagnoses for higher taxonomic categories.

The stylommatophoran CNS is primarily hypoathroid. The least concentrated of the CNSs found amongst Stylommatophora closely approximate that found in Ellobiidae with respect to the lengths of the commissures and connectives. In these stylommatophorans, as in onchidiids, ellobiids, succineids and athoracophorids, the suboesophageal ganglion apparently has fused with the visceral ganglion, and the supraoesophageal ganglion with the right parietal ganglion, to form a visceral chain that is structurally pentaganglionate but on external morphology has a triganglionate appearance. Beyond these oesophageal ganglion fusions, the degree of shortening of the commissures and connectives can vary substantially within families. Only where the commissures and connectives are highly shortened does the stylommatophoran CNS approach that in Rathouisioidea and Succineoidea. In some cases, the shortening of the connectives associated with ganglionic concentration has led to convergence on the epiathroid configuration, with the pleural ganglia sited closer to the cerebral than to the pedal ganglia. The proximity of the right parietal and visceral ganglia (often fused) seems to be a synapomorphy of all Stylommatophora (Tillier, 1989), indicating that shortening of the visceral chain in these animals initially involves shortening of the visceral–right parietal connective. Within this constraint, a moderately long visceral chain, with more or less separate ganglia, remains evident in some members of the families Valloniidae, Acavidae, Corillidae, Zonitidae and Charopidae. The pattern in shortening of the left pleural–left parietal, left parietal–visceral and right parietal–right pleural connectives, and fusion of contiguous ganglia, varies among stylommatophorans, as documented by Bargmann (1930), Tillier (1989) and others. In all Stylommatophora, the pedal ganglia are interconnected by way of two commissures.

It has been assumed generally that compaction in the CNS is largely irreversible. However, the shortening of the connectives and fusion of ganglia are ontogenetic processes, and therefore secondarily unfused or separated ganglia can arise by neoteny (Henchman, 1890). Indeed, cases of reversal towards a less concentrated CNS have been noted during the ontogeny of invertebrates, including gastropods (e.g. Kriegstein, 1977). This suggests that reversals may occur during evolution. The extreme length of the lateral and visceral connectives in some carnivorous species, as an apparent functional necessity for extending around the voluminous buccal mass, is clearly an example of such secondary elongation. None the less, as Tillier (1989, p. 46) points out, 'It seems that some connectives might lengthen in the course of evolution insofar as they are not very short, but the secondary development of a connective between two appressed ganglia is unlikely'.

Stylommatophora are hermaphrodites. The gonad, typically embedded in the upper lobe of the digestive gland, produces both oocytes and spermatozoa. While male and female gametes may be produced simultaneously, there is often a degree of protandry. The single renal gonoduct conveys both gamete types to the more distal parts of the reproductive system. In Stylommatophora, its medial part is generally dilated, and occasionally structurally elaborated with several lateral pouches, as a storage site for autospermatozoa. The external form of the seminal receptacle varies greatly among stylommatophorans, from a simple U-shaped bend in the terminal section of the renal gonoduct, to either a single or a cluster of minute diverticular sacs whose stem arises from the juncture of the renal gonoduct with the fecundation pouch. It lies embedded in the columellar side of the albumen gland. The fecundation pouch generally is sacculate, but varies greatly in size and thus prominence among taxa.

The pallial gonoducts of Stylommatophorans, as in Succineoidea, are located entirely within the body cavity. There has been considerable debate in the literature as to the plesiomorphic state of the pallial gonoducts in the Stylommatophora. For the reasons elaborated above in the discussion on Succineoidea, diauly of the pallial gonoducts is to be regarded as the primary condition in the Stylommatophora. Diauly is retained in Endodontidae and some members of Euconulidae, Achatinellidae and Camaenidae. In a great many Stylommatophora, including representatives of Achatinellidae, Vertiginidae, Partulidae, Pupillidae, Sagdidae, Bulimulidae, Valloniidae, Streptaxidae, Camaenidae and Charopidae, the male pallial gonoduct, and associated prostatic tissues, is fused to and opens on to a very short, proximal section of the female gonoduct. This often has been described as a semi-diaulic condition, but functionally it is monaulic in that the prostatic secretions are delivered directly into the lumen of the oviductal gland, albeit in a ventral channel or groove. Beyond this section of male and female union, the male duct is continued distally as a closed, tubular duct, the vas deferens. In the remaining Stylommatophora, the section of male and female pallial gonoduct fusion is more extensive, but none the less the spermiduct and its associated prostatic tissues open to a groove in the oviductal gland, before again separating as the vas deferens. Often longitudinal folds provide for division of the spermoviduct lumen into grooves that respectively function for transport of autospermatozoa, allospermatozoa, and eggs. In some stylommatophorans, such as Agriolimacidae, the fusion of male and female pallial gonoducts is so complete that only physiological division is evident.

The walls of the oviductal gland are typically thick, comprising for the most part large secretory cells that shed into the capacious lumen. Often the oviductal gland is differentiated into several regions of contrasting histochemical properties and clearly responsible for producing different mucopolysaccharide layers and calcium crystals in the egg capsule. The seminal channel within the spermoviduct has a ciliated epithelium. The spermiduct typically is lined with ciliated epithelium and large flask-shaped secretory cells. The prostatic gland comprises few to many multicellular tubules that open separately into the sperm duct or groove. Irrespective of the fusion of male and female gonoducts, oviparous species commonly possess an elongate oviductal gland, and the spermiduct and ribbon of prostatic follicles are similarly elongated. In

some oviparous groups, there has been some secondary shortening of
the oviductal gland and of that section of the spermiduct from which
prostatic follicles issue. Both long and short forms of the oviductal gland
are represented in taxa that have acquired ovoviviparity or viviparity. The
phenomenon of agglomeration of the oviductal glands in the proximal
part of the gonoduct, developed to varying degrees in Ellobiidae,
Onchidiidae, Rathouisioidea and Succineoidea, is entirely absent from
Stylommatophora.

Several groups of Stylommatophora exhibit a trend toward separation
of the allospermiduct from the female gonoduct, as a closed tube, to
form a secondary diaulic condition. In some cases, the allospermiduct is
produced as a diverticulate sac on the female gonoduct, but it remains
uncertain whether this represents an early phase of change towards the
secondary diaulic condition or an advanced phase that follows secondary
diauly. Whatever the sequence of change, the diverticulate sac on the
pallial gonoduct may be homologous to the similar structure on the pallial
gonoduct of some Ellobiidae, Onchidiidae and Athoracophoridae. Some
authors have regarded the separation of the allospermiduct as lending to
the evolutionary development of a diverticulum on the bursa copulatrix
duct (e.g. Visser, 1973, 1981), but there is no convincing evidence for this.

In Stylommatophora, the predominant form of the bursa copulatrix is
a sacculate reservoir lying adjacent to the pericardium and bound by
connective tissue and muscle to the diaphragm that constitutes the floor
of the pallial cavity, and a distinct duct running to the anterior part of the
female reproductive tract. The opening of the bursa duct to the female
tract typically occurs in the distal section of the female gonoduct to as far
forward as the common genital orifice. In some taxa, it has even migrated
to the male genitalia, facilitated by syntrematry. The length of the bursa
duct in these stylommatophorans is thus a function of the degree of forward displacement of the union of bursa duct to the female (or even male)
tract, the amount of relative forward extension of the female tract associated with placement of the common genital orifice, and the relative size of
the cephalic part of the body as a whole. In some Achatinellidae, such as
Tornatellides Pilsbry, the duct of the bursa arises from the proximal
female pallial gonoduct, recalling the condition seen in some Ellobiidae.

The diverticulum on the bursa copulatrix duct is widespread in the
Stylommatophora and apparently is plesiomorphic. The diverticulum
is specifically constructed and positioned, relative to the bursa duct
opening, to function during mating as the site of spermatophore receipt
from the animal's copulatory partner. In many taxa, the diverticulum
is reduced or absent and the bursa copulatrix duct itself assumes the
role of spermatophore receiver. Furthermore, in many Stylommatophora,
spermatophores, or seminal mass where spermatophores are not
produced, are not received directly into the bursa copulatrix duct or its
diverticulum, but are reposited in the oviduct. As is typical in Pulmonata,
the bursa reservoir has gametolytic and resorption functions rather than
a storage function (Tompa, 1984). The receptaculum is the site for storage

of allospermatozoa and it is only the excess allospermatozoa that are conveyed into the bursa copulatrix for breakdown and resorption.

The plesiomorphic reproductive system of Stylommatophora has no simple structure in the terminal genitalia, but rather has a complex set of copulatory organs with structures for producing and receiving spermatophores, and an auxiliary stimulatory organ. As in Succineoidea, the vas deferens does not enter the body wall but runs free in the haemocoel, generally running forward to the penial–vaginal juncture before producing a limb that ascends to insert in the penis. The packaging of spermatozoa for insemination in the form of a spermatophore is of wide occurrence in the Gastropoda, including opisthobranch and various basal pulmonates. Transfer of spermatozoa by spermatophores is neither an adaptation to terrestrial life (Nordsieck, 1985) nor a trait developed independently in several stylommatophoran groups (Solem, 1978; Muratov, 1999), but a plesiomorphic trait in Stylommatophora inherited from their marine ancestors. In Stylommatophora, the spermatophore is produced in the distal part of the vas deferens, which has a special structure and is termed the epiphallus. In the plesiomorphic state, the vas deferens does not insert terminally, and therefore the epiphallus has an appendage termed the flagellum; this flagellum has a role in spermatophore formation. The shape of the epiphallus and its internal folds mould the spermatophores into species-specific structures. In many stylommatophoran groups, there is an evident trend towards insemination without the need for spermatophores. In its early stages, this evolutionary trend is manifest by simplified epiphallic structures, without a flagellum. Further epiphallus simplification has resulted in spermatophores with greatly simplified shape and ornamentation and with reduced thickness of the matrix encasing the spermatozoa. In many Stylommatophora, a structure recognizable as an epiphallus is absent, and insemination is achieved by spermatozoa suspended in a viscous fluid. In the plesiomorphic state of the Stylommatophoran genitalia, the epiphallus opens through the perforated papilla into the tubular phallus. Only rarely in the Stylommatophora is the papilla covered in a hardened cuticle and never does it function as a stylet as seen in some Basommatophora and Opisthobranchia. In many Stylommatophora, the penial papilla is reduced in size or entirely absent. This is a secondary phenomenon.

The eversible and thus protrusible penis of Stylommatophora is of the same general form as that in Succineoidea. The scale- or hook-like cuticular papillae lining the penial lumen, and functioning as hold-fast structures during copulation, are retained in many Stylommatophora and indeed can be lengthened secondarily (e.g. Streptaxidae and Zonitidae) and elaborated (e.g. Trigonochlamydidae). In addition, the penial wall often is produced into pilasters that, on eversion of the penis, function as stimulatory surfaces during courtship. The plesiomorphic condition of the penial retractor muscle is considered a branch of the right tentacular retractor division of the columellar muscle. With the trend to division of the columellar muscle into specialist retractor bands, the phallus retractor apparently became isolated from the remaining muscle bands at all but their origin at the columella. Further reductions firstly involved a shifting forward of the penial retractor root to the pallial cavity floor, and then to the body wall outside the pallial region. In cases of extremes of reduction, the penial retractor is represented by rather weak muscle strands to the pallial gonoduct, or is entirely absent.

The penial nerve originates from the pedal ganglion (de Nabias, 1894). In many stylommatophorans, the penial nerve runs, however, through the cerebral ganglion en route to the male genitalia. Changes in the route of the penial nerve have occurred repeatedly during the evolution of the Stylommatophora and this is not very useful as a character in phylogenetic analyses.

The reproductive system is syntrematic, with the common genital orifice opening on the right side. As in some Succineidae, the male and female openings in some Stylommatophora are merely closely apposed (e.g. *Tryonigens* Pilsbry in Helminthoglyptidae). In the great majority of Stylommatophora, the male and female orifices open to a common chamber, termed the atrium, rather than directly to the body exterior. Earlier, I (Barker, 1999) suggested that the plesiomorphic state within the Stylommatophora was the common genital opening located in the main body of the visceral stalk, not too distant from the pneumostome. However, given that the penis is almost invariably cephalic in Heterobranchia, and indeed usually placed close to the right cephalic tentacle, it seems probable that this represents the plesiomorphic phallic condition in Stylommatophora, and that syntrematry was achieved by forward extension of the female tract to open with the male orifice (not by development of a new female structure as suggested by Sirgel (1990), because the free oviduct was already present in ditrematous ancestors as exemplified by Ellobiidae). In many lineages, there apparently has been a subsequent shift posteriad of the common male–female orifice, often as far back as the visceral stalk to lie below the pneumostome and therefore to be sited close to the probable plesiomorphic location for the female orifice in pulmonates. An alternative hypothesis is that both forward migration of the female opening and posteriad migration of the male opening occurred, resulting in the varied location of the common orifice seen in extant Stylommatophora. The latter is most strongly supported in parsimony analysis.

Stimulatory organs can be found in the terminal genitalia of many Stylommatophora. There are different opinions concerning the evolution of these organs; some authors such as Thiele (1929–35), Solem (1978), Tompa (1984) and Muratov (1999) thought that the various types of stimulatory organs had evolved independently, while others, such as von Ihering (1892) and Schileyko (1979), suggested all or only part of them to be homologous. The stimulatory organ of stylommatophorans traditionally classified as Orthurethra (*sensu* Pilsbry) opens to the atrium adjacent to the penis or is developed as a penial appendage, and consists of a perforated papilla in a sheath, an adjoining gland, and a retractor muscle that is a branch of the penial retractor. In many cases, the bifid retractor is retained despite the stimulatory organ being somewhat reduced. In other taxa, the retractor to the stimulatory organ is retained but has become separated from the penial retractor. Some support for the primitiveness of joined retractors is given in the ontogeny of *Achatinella* Swainson, where the retractors originate as a single muscle and become progressively more separated as the animal matures (Pilsbry *et al.*, 1928).

In at least one group of non-orthurethran Stylommatophora, the Gastrodontidae, a stimulatory organ of very similar structure to that in Orthurethra, and supplied with a retractor muscle, is present as an appendix to the penis. It contains, however, a well-developed cartilaginous dart. In Sagdidae too, the stimulatory organ on the phallus is remarkably like that seen in so-called Orthurethra, except for the absence of a retractor muscle and the presence of a vestigial dart. In a number of stylommatophoran families, not traditionally included in Orthurethra, this dart-bearing stimulatory organ has been retained on the atrium or displaced to the vagina, and the retractor muscle is still evident (e.g. Ariophantidae, Urocyclidae and Vitrinidae). There is little doubt as to the homology of the unarmed stimulatory organ characteristic of the Orthurethra and the dart-equipped organs of these latter Stylommatophora. The difficulty lies in deciding which is the more plesiomorphic. The occurrence of a dart in the stimulatory organ across many superfamilies of non-orthurethran Stylommatophora (the occurrence in Orthurethra has yet to be confirmed, see Tompa, 1984), including the vestigial dart of Sagdidae in the stimulatory organ built like the orthurethran stimulator, suggests that the dart-bearing structure is the primitive feature of Stylommatophora. Nordsieck (1985, 1992) and Hausdorf (1998) reached the same conclusion. This interpretation of plesiomorphy is supported further by the occurrence of auxiliary copulatory organs associated with the penis in many other pulmonates, including Siphonariidae, Amphibolidae, Ancylidae, Onchidiidae, Vaginulidae, Rathouisiidae and Athoracophoridae. Indeed, the stimulatory organ of Onchidiidae and Rathouisiidae is remarkably similar in morphology to that in Orthurethra and Sagdidae and would serve as a model for the plesiomorphic condition in Stylommatophora. Further, the presence of a cartilaginous stylet in a stimulatory organ of some Onchidiidae suggests that the dart arose early in the evolution of Pulmonata and certainly was present in the ancestors of the Stylommatophora.

Loss of the stimulatory organ is widespread in the Stylommatophora. During evolutionary degeneration of the stimulatory organ, dart loss is a stage that precedes full organ loss; this pathway is evident, for example, in helicoid snails. In many taxa, dart loss is accompanied by modification of the previously dart-bearing papilla into a fleshy or cuticle-coated sarcobelum. The branch of the penial retractor muscle inserting on a small lateral penial caecum in Clausiliidae, Zonitidae, Euconulidae and Ferussaciidae, as examples, is testament to the former wide occurrence of a more fully developed stimulatory organ.

Simultaneous, reciprocal exchange of sperm generally has been considered the norm in Stylommatophora. However, Asami *et al.* (1998) have demonstrated that the shell-shape bimodality evident in stylommatophoran snails, where snails either carry a high-spired (height: diameter > 1 mm) or low-spired (height: diameter ≤1 mm) shell (Cain, 1977), is associated with discrete mating behaviours. In general, flat-shelled species mate reciprocally, face-to-face, while tall-shelled species mate non-reciprocally: the 'male' copulates by mounting the 'female's' shell. External sperm exchange, by which spermatozoa are deposited on the mate's everted penis without intromission, has evolved among some polygyrid clades (Webb, 1948, 1961, 1974, 1983; Emberton, 1994), paralleling the dichotomous situation in Succineidae. The Agriolimacidae, Limacidae and discine Helicodiscidae (Webb, 1968; Barker, 1999) have abandoned intromission entirely and mate by external exchange of spermatozoa in a manner similar to that in some Polygyridae.

Self-fertilization is a common but not a universal phenomenon in Stylommatophora, with the frequency of self-fertilization varying greatly among species. Heller (1993, Chapter 12, this volume) provides reviews of self-fertilization in these animals. While the hermaphroditic reproductive system of stylommatophorans typically possesses both male and female genitalia (euphally), aphallic individuals that lack the male copulatory organs are known from Pyramidulidae, Vertiginidae, Pupillidae, Valloniidae, Chondrinidae, Arionidae, Gastrodontidae, Agriolimacidae and Helminthoglyptidae (Watson, 1923; Riedel, 1953; Els, 1978; Barker, 1985; Pokryszko, 1987, 1990; Baur and Chen, 1993; Baur *et al.*, 1993; Heller, 1993). The frequency of expression of genital dimorphism among populations can be high in some species and has been shown to have genetic and environmental components (Baur *et al.*, 1993). Aphallic individuals can self-fertilize, or can outcross as female but not as male.

The majority of Stylommatophora produce cleidoic eggs. Often the eggs are produced with a calcareous shell (Standen, 1917; Tompa, 1976, 1980), although the shell is often itself surrounded by a jelly-like, mucopolysaccharide layer of the egg capsule (Bayne, 1968; Tompa, 1984). Tompa (1984) considered the provision of the embryo with calcium, by ionic mobilization of body shell and/or digestive gland calcium stores and deposition as a calcite egg shell by a specialized gonoduct epithelium, to be a synapomorphy of Stylommatophora. A hard, calcite egg shell was thus considered by Tompa (1980, 1984) to be plesiomorphic in Stylommatophora, and the production of eggs with diminished calcium provision as apomorphic. However, the fact that the mechanism for calcium provision of the eggs differs among Neritopsina, Architaenioglossa, Sorbeoconcha and Pulmonata (Tompa, 1980) indicates independent acquisition of calcareous egg capsules amongst gastropod groups and calls into question the calcareous nature of the plesiomorphic egg capsule in Stylommatophora. The absence of a calcareous egg shell in basal pulmonates, including terrestrial taxa such as Ellobiidae, Onchidiidae and Vaginulidae, suggests that eggs with a calcareous shell developed in Stylommatophora, possibly repeatedly. Ground-dwelling species generally deposited their eggs amongst litter and other debris on the soil surface, or in the soil. Arboreal species generally deposit their eggs in humus suspended in epiphytes or the axils of branches, although many return to the ground to oviposit among the leaf litter. Several arboreal species are known to lay their eggs in the trees, in brood chambers made from leaves (Peake, 1968). As a modification of simple oviparity, some stylommatophorans have adopted embryo brooding, whereby the eggs are retained in the oviduct until environmental conditions become favourable for oviposition, or are retained for longer periods and deposited when embryo development is at an advanced stage. In other cases, that of ovoviviparity, the eggs are retained in the oviduct until hatching (Tompa, 1979a,b, 1984; Baur, 1994; Heller, Chapter 12, this volume). In some species, reproduction can be oviparous, egg-retaining or ovoviviparous.

The Stylommatophora exhibit intracapsular, direct development. The occurrence of larval organs and the formation of transitory structures in the midgut show that there is none the less a complex metamorphosis (Weiss, 1968). A particular advance over other pulmonates, including Succineoidea, was the development of the podocyst, a thin-walled, pulsatile sack attached to the foot of the embryo (Cather and Tompa, 1972) that has been ascribed circulatory (Kuchenmeister et al., 1996) and albumenotrophic (Cather and Tompa, 1972) functions. The stylommatophoran snails produce an embryonic shell, the protoconch, which is generally of 1–2 whorls consistent with the generally short embryonic development. In some taxa, such as members of the Acavidae, the embryonic growth is more prolonged, and 3–4 shell whorls are produced before hatching. In the most advanced stylommatophoran slugs, the embryonic shell is reduced to a few calcareous granules embedded within the mantle (e.g. Carrick, 1938), which may give rise to a shell plate or granules within the mantle in the adult animal.

The haploid chromosomes in Stylommatophora range only from 20 to 34, indicating a general cytological conservatism despite the substantive radiation.

The greater majority of stylommatophorans are detritivores, feeding on decaying plant material. Some also feed on living plant tissues. Facultative or obligate carnivory has developed in a number of stylommatophoran clades (Barker and Efford, 2001).

There is a large range in radular tooth form in Stylommatophora. Because of the apparent frequency of diet-related convergence in tooth form (e.g. Solem, 1973; Breure and Gittenberger, 1982), the radula generally has been regarded as useful in systematics only at generic and species level (Solem, 1978). None the less, a character-state tree can be constructed which is indicative of several courses of change in radular tooth form. This is illustrated by Barker and Efford (2001) in relation to carnivory. In herbivores and detritivores, the buccal mass is relatively small and spheroidal. Carnivory, however, is usually accompanied by elongation and increased muscularization of the buccal apparatus, associated with the mechanics of operating an odontophore that supports a radular with enlarged teeth, and to allow ingestion of large live prey. In its fullest development, the buccal apparatus of carnivores occupies a substantial part of the body cavity, and other organ systems are modified to accommodate this. The jaw commonly is reduced or lost in these carnivores.

The oesophagus opens dorsally from the anterior buccal mass and, in snails, runs backward along the parietal side of the visceral cavity, and most generally expands into a gastric crop a short distance above the top of the pallial complex. The oesophagus might be partly differentiated into an inflated oesophageal crop but, as pointed out by Tillier (1989), contrary to the statement of most treatises of zoology, many stylommatophoran species do not have an oesophageal crop. Two salivary glands lie adjacent to the oesophagus. The gastric crop is prolonged by the gastric pouch or stomach, from which the intestine opens forward and ventrally. The intestine runs along the columellar side of the visceral mass, turns to the left under the anterior gastric crop or the posterior oesophagus, and turns upward to cross over the aorta clockwise in dorsal view (periaortic bend) before turning forward again (prerectal bend). The rectum runs along the suture from the summit of the pallial complex to the roof of the pneumostome. The anterior duct of the digestive gland opens into the concavity of the stomach, between the openings of the oesophagus and proximal intestine. The posterior duct of the digestive gland generally opens through the parietal wall of the stomach, not greatly distant from the opening of the anterior lobe. When present, the ventral groove of the stomach leads to the opening of the anterior duct, from which one usually short typhlosole emerges into the proximal intestine. A second, longer typhlosole, issuing from the opening of the posterior duct, runs parallel to the first into the proximal intestine and reaches at most the beginning of the periaortic intestinal loop.

Tillier (1989) considered the simple stomach, with contiguity of the openings of the ducts of the digestive gland, as being plesiomorphic in Stylommatophora (in which he included Succineidae and Athoraco-phoridae). However, in some Stylommatophora, such as *Megalobulimus* Miller in Acavidae, the arrangement of the digestive gland openings is similar to that in many lower gastropods and basal pulmonates, with the anterior lobe opening to the posterior section of the oesophageal crop and the posterior lobe opening in the curvature of the stomach near the intestinal origin. That this *Megalobulimus* arrangement is plesiomorphic is indicated by the very muscular, gizzard-like nature of the stomach (Leme, 1973; Simone and Leme, 1998). In the majority of Stylommato-phora, the stomach is poorly invested with muscles.

The intestine of stylommatophoran snails primarily has a single forward-directed loop, turning posterior after crossing over the aorta. From the aorta, the intestine has one loop backwards before going forward to the anus. As illustrated by Tillier (1984c), this represents hypertorsion of the condition seen in the post-torsional embryo. In a case study with

Helicarionidae *sensu lato*, Tillier (1984c) demonstrated that this hypertorsion generally is retained during the initial phases of limacization. However, whether this hypertorsion is retained, lost or exaggerated in the later phases of limacization was shown to depend on how the visceral mass was reduced and incorporated into the foot cavity in the slug. Where the hypertorsion is lost, the intestine in stylommatophoran slugs closely approaches the configuration seen in Succineidae, while that where hypertorsion is most exaggerated approaches the Athoracophoridae and some Vaginulidae. *Megalobulimus* has well-developed epithelial folds in the intestine and rectum, which are oblique to the long axis in the distal intestine (Simone and Leme, 1998), similar to those in Onchidiidae. This contrasts with the majority of Stylommatophora where the intestinal and rectal epithelial folds, where present, are longitudinal (e.g. Leal-Zanchet, 1998). The Agriolimacidae and Limacidae are characterized by a caecum on the distal section of the intestine, which has been lost secondarily in some species.

Acknowledgements

I am deeply indebted to Dr Eduardo Ferrari, Dr Alan Potts, Dr Mary Jefferson, Dr Peter B. Mordan, Dr Sergio E. Miquel and Karen Mahlfeld for valuable discussions on pulmonate phylogeny and cladistics that provided the motivation for the preparation of this synopsis. Many of the anatomical data that constituted the raw material for this project were based on dissection of specimen material collected by the author, or kindly made available from various institutions over many years, particularly by the late Dr Alan Solem (Field Museum of Natural History, Chicago), Dr Peter B. Mordan (The Natural History Museum, London), Bruce A. Marshall (Museum of New Zealand, Wellington), Dr Simon Tillier (Muséum National d'Histoire Naturelle, Paris), the late Jim Goulstone (Auckland Institute and Museum, Auckland), Ian Loch (Australian Museum, Sydney), Sergio E. Miquel (Museo de La Plata, Buenos Aires), Dr José W. Thomé (Pontifícia Universidade Católica do Rio Grande do Sul, Porto Alegre) and Dr William F. Sirgel (University of Stellenbosch, Stellenbosch). Dr Chris Wade kindly provided access to unpublished data on phylogenetic reconstruction of the Stylommatophora based on 28S rDNA sequences. Dr Eduardo Ferrari and two anonymous referees provided constructive comment on the manuscript.

References

Abbott, R.T. (1948) A potential snail host of Oriental Schistosomiasis in North America (*Pomatiopsis lapidaria*). *Proceedings of the United States National Museum* 98, 57–68.

Abbott, R.T. (1949) New syncerid mollusks from the Marianas (sic.) Islands (Gastropoda, Prosobranchiata, Synceridae). *Occasional Papers of the B.P. Bishop Museum* 19, 261–274.

Abbott, R.T. (1958) The gastropod genus *Assiminea* in the Philippines. *Proceedings of the Academy of Natural Sciences of Philadelphia* 110, 213–278.

Adanson, M. (1757) *Histoire naturelle du Sénégal (Coquillages). Avec la Relation Abrégée d'un Voyage Fait en ce Pays Pendant les Années 1749–53.* Claude-Jean-Baptiste Bauche, Paris.

Andrews, B.C. (1981) Osmoregulation and excretion in prosobranch gastropods. Part 2. Structure in relation to function. *Journal of Molluscan Studies* 47, 248–289.

Andrews, E. (1937) Certain reproductive organs in the Neritidae. *Journal of Morphology* 61, 525–561.

Andrews, E.B. (1965) The functional anatomy of the mantle cavity, kidney and blood system of some pilid gastropods (Prosobranchia). *Journal of Zoology* 146, 70–94.

Andrews, E.B. (1968) An anatomical and histological study of the nervous system of *Bithynia tentaculata* (Prosobranchia), with special reference to possible neurosecretory activity. *Proceedings of the Malacological Society of London* 38, 213–232.

Andrews, E.B. (1979) Fine structure in relation to function in the excretory system of two species of *Viviparus. Journal of Molluscan Studies* 45, 186–206.

Andrews, E.B. (1985) Structure and function in the excretory system of archaeogastropods and their significance in the evolution of gastropods. *Philosophical Transactions of the Royal Society of London, Biological Sciences* 310, 383–407.

Andrews, E.B. (1988) Excretory systems of molluscs. In: Trueman, E.R. and Clarke, M.R. (eds) *The Mollusca,* Vol. 2, *Form and Function.* Academic Press, San Diego, pp. 381–448.

Andrews, E.B. and Little, C. (1971) Ultrafiltration in the gastropod heart. *Nature* 234, 411–412.

Andrews, E.B. and Little, C. (1972) Structure and function in the excretory system of some terrestrial prosobranch snails (Cyclophoridae). *Journal of Zoology* 168, 395–422.

Andrews, E.B. and Little, C. (1982) Renal structure and function in relation to habitat in some cyclophorid land snails from Papua New Guinea. *Journal of Molluscan Studies* 48, 124–143.

Arey, L.B. and Crozier, W.J. (1921) On the natural history of *Onchidium. Journal of Experimental Zoology* 32, 443–502.

Asami, T., Cowie, R.H. and Ohbayashi, K. (1998) Evolution of mirror images by sexually asymmetric mating behavior in hermaphroditic snails. *American Naturalist* 152, 225–236.

Awati, P. and Karandikar, K. (1948) *Onchidium verruculatum,* Cuv. (Anatomy, embryology and bionomics). *University of Bombay Memoirs in Zoology* 1, 1–53.

Baker, H.B. (1922) Notes on the radula of the Helicinidae. *Proceedings of the Academy of Natural Sciences of Philadelphia* 75, 29–67.

Baker, H.B. (1923) Proserpinidae. *Nautilus* 36, 84–85.

Baker, H.B. (1924) New land operculates from the Dutch Leeward Islands. *Nautilus* 37, 89–94.

Baker, H.B. (1925a) North American Veronicellidae. *Proceedings of the Academy of Natural Sciences of Philadelphia* 77, 157–184.

Baker, H.B. (1925b) Anatomy of *Hendersonia*: a primitive helicinid mollusk. *Proceedings of the Academy of Natural Sciences of Philadelphia* 77, 273–303.

Baker, H.B. (1928) Mexican mollusks collected for Dr. Bryant Walker in 1926, I. *Occasional Papers of the Museum of Zoology, University of Michigan* 193, 1–64, pls. 1–6.

Baker, H.B. (1955) Heterurethrous and aulacopod. *Nautilus* 68, 109–112.

Baker, H.B. (1956) Family names for land operculates. *Nautilus* 70, 28–31.

Bandel, K. (1982) Morphologie und Bildung der frühontogenetischen Gehäuse bei conchiferen Mollusken. *Facies (Erlangen)* 7, 1–198, i–xxii.

Bandel, K. (1993) Caenogastropoda during Mesozoic times. *Scripta Geologica, Special Issue* 2, 7–56.

Bani, G. and Cecchi, R. (1982) Structure of the finger-like gland of *Vaginula borelliana* (Colosi) (Gastropoda Pulmonata). *Monitore Zoologico Italino* 16, 345–358.

Bargmann, H.E. (1930) The morphology of the central nervous system in the Gastropoda Pulmonata. *Journal of the Linnean Society, Zoology* 37, 1–58.

Barker, G.M. (1978) A reappraisal of *Athoracophorus bitentaculatus*, with comments on the validity of genus *Reflectopallium* (Gastropoda: Athoracophoridae). *New Zealand Journal of Zoology* 5, 281–288.

Barker, G.M. (1985) Aspects of the biology of *Vallonia excentrica* (Mollusca – Valloniidae) in Waikato pastures. In: Chapman, R.B. (ed.) *Proceedings of the 4th Australasian Conference on Grassland Invertebrate Ecology*. Caxton Press, Christchurch, New Zealand, pp. 64–70.

Barker, G.M. (1999) *Naturalised Terrestrial Stylommatophora (Mollusca: Gastropoda)*. Fauna of New Zealand 38. Manaaki Whenua Press, Lincoln, New Zealand.

Barker, G.M. and Efford, M.G. (2002) Predatory gastropods as natural enemies of terrestrial gastropods and other invertebrates. In: Barker, G.M. (ed.) *Natural Enemies of Terrestrial Molluscs*. CAB International, Wallingford, UK.

Bartsch, P. (1946) The operculate land molluscs of the family Annulariidae of the island of Hispaniola and the Bahama archipelago. *Bulletin of the United States National Museum, Smithsonian Institution* 192, i–iv, 1–264.

Bartsch, P. and Morrison, J.P.E. (1942) The cyclophorid operculate land molluscs of America. *Bulletin of the United States National Museum, Smithsonian Institution* 181, 142–282.

Baur, B. (1994) Parental care in terrestrial gastropods. *Experientia* 50, 5–14.

Baur, B. and Chen, X. (1993) Genital dimorphism in the land snail *Chondrina avenacea*: frequency of aphally in natural populations and morph-specific allocation to reproductive organs. *Veliger* 36, 252–258.

Baur, B., Chen, X. and Baur, A. (1993) Genital dimorphism in natural populations of the land snail *Chondrina clienta* and the influence of the environment on its expression. *Journal of Zoology* 231, 275–284.

Bayne, C.J. (1968) Survival of the embryos of the grey field slug *Agriolimax reticulatus*, following desiccation of the egg. *Malacologia* 9, 391–401.

Bedford, L. (1965) The histology and anatomy of the reproductive system of the littorinid gastropod *Bembicium nanum* (Lamarck) (family Littorinidae). *Proceedings of the Linnaean Society of New South Wales* 90, 95–105.

Berry, A.J. (1961) The habitats of some minute cyclophorids, hydrocenids and vertiginids on a Malayan limestone hill. *Bulletin of the National Museum, Singapore* 30, 101–105.

Berry, A.J. (1962) The growth of *Opisthostoma* (*Plectostoma*) *retrovertens* Tomlin, a minute cyclophorid from a Malayan limestone hill. *Proceedings of the Malacological Society of London* 35, 46–49.

Berry, A.J. (1964) The reproduction of the minute cyclophorid snail *Opisthostoma* (*Plectostoma*) *retrovertens* from a Malayan limestone hill. *Proceedings of the Zoological Society of London* 142, 655–663.

Berry, A.J. (1965) Reproduction and breeding fluctuations in *Hydrocena monterosatiana* a Malayan limestone archaeogastropod. *Proceedings of the Zoological Society of London* 144, 219–227.

Berry, A.J. (1966) Population structure and fluctuations in the snail fauna of a Malayan limestone hill. *Journal of Zoology* 150, 11–27.

Berry, A.J. (1975) Patterns of breeding activity in West Malaysian gastropod molluscs. *Malaysian Journal of Science (Series A)* 3, 49–59.

Berry, A.J., Loong, S.C. and Thum, H.H. (1967) Genital systems of *Pythia*, *Cassidula* and *Auricula* (Ellobiidae, Pulmonata) from Malayan mangrove swamps. *Proceedings of the Malacological Society of London* 37, 325–337.

Berthold, T. (1989) Comparative conchology and functional morphology of the copulatory organ of the Ampullariidae (Gastropoda, Monotocardia) and their bearing upon phylogeny and palaeontology. *Abhandlungen, Naturwissenschaftlichen Verein zu Hamburg* 28, 141–164.

Berthold, T. (1991) Vergleichende Anatomie, Phylogenie und historische Biogeographie der Ampullariidae (Mollusca, Gastropoda). *Abhandlungen, Naturwissenschaftlichen Verein zu Hamburg* 29, 1–256.

Bieler, R. (1990) Haszprunar's 'clado-evolutionary' classification of the Gastropoda – a critique. *Malacologia* 31, 371–380.

Bieler, R. (1993) Ampullariid phylogeny – book review and cladistic re-analysis. *Veliger* 36, 291–299.

Bingham, F.O. (1972) Several aspects of the reproductive biology of *Littiorina irrorata* (Gastropoda). *Nautilus* 86, 8–10.

Bishop, M.J. (1980) Helicinid land snails with apertural barriers. *Journal of Molluscan Studies* 46, 241–246.

Blanford, W.T. (1863) Descriptions of *Cremnobates syhadrensis* and *Lithotis rupicola*, two new generic forms of Mollusca inhabiting cliffs in the Western Ghats of India. *Annals and Magazine of Natural History (Series 3)* 12, 184–187.

Blanford, W.T. (1870) Contributions to Indian malacology, No. XI. Descriptions of new species of *Paludomus*, *Cremnoconchus*, *Cyclostoma* and of Helicidae from various parts of India. *Journal of the Asiatic Society, Bengal* 39, 9–25.

Blinn, W.C. (1964) Water in the mantle cavity of land snails. *Physiological Zoology* 37, 329–337.

Boer, H.H., Algera, N.H. and Lommerse, A.W. (1973) Ultrastructure of possible sites of ultrafiltration in some Gastropoda, with particular reference to the auricle of the freshwater prosobranch *Viviparus viviparus* L. *Zeitschrift für Zellforschung und Mikroskopische Anatomie* 143, 329–341.

Boss, K.J. and Jacobson, M.K. (1975) Catalogue of the taxa of the subfamily Proserpinidae (Helicinidae; Prosobranchia). *Occasional Papers on Mollusks, Museum of Comparative Zoology, Harvard University* 4, 93–104.

Bourne, G.C. (1911) Contributions to the morphology of the group Neritacea of the aspidobranch gastropods – Part II. The Helicinidae. *Proceedings of the Zoological Society of London* 60, 759–809.

Brace, R.C. (1983) Observations on the morphology and behaviour of *Chilina fluctuosa* Gray (Chilinidae), with a discussion on the early evolution of pulmonate gastropods. *Philosophical Transactions of the Royal Society of London, Biological Sciences* 300, 463–491.

Breure, A.S.H. and Gittenberger, E. (1982) The rock-scaping radula, a striking case of convergence (Mollusca). *Netherlands Journal of Zoology* 32, 307–312.

Britton, K.M. (1984) The Onchidiacea (Gastropoda, Pulmonata) of Hong Kong with a worldwide view of the genera. *Journal of Molluscan Studies* 50, 179–191.

Buckland-Nicks, J.A. and Darling, P. (1993) Sperm are stored in the ovary of *Lacuna* (*Epheria*) *variegata* (Carpenter, 1864) (Gastropoda: Littorinidae). *Journal of Experimental Zoology* 267, 624–627.

Buckland-Nicks, J.A. and Worthen, G.T. (1992) Functional morphology of the mammiliform penial glands of *Littorina saxatilis* (Gastropoda). *Zoomorphology* 112, 217–225.

Bullock, T.H. (1965) The Mollusca: Gastropoda. In: Bullock, T.H. and Horridge, G.A. (eds) *Structure and Function in the Nervous Systems of Invertebrates*, Vol. 2. W.H. Freeman, San Francisco, pp. 1273–1515.

Bulman, K. (1990) Life history of *Carychium tridentatum* (Risso, 1826) (Gastropoda: Pulmonata: Ellobiidae) in the laboratory. *Journal of Conchology* 33, 321–333.

Burch, J.B. (1968) Tentacle retraction in Tracheopulmonata. *Journal of the Malacological Society of Australia* 11, 6–7.

Burch, J.B. and Patterson, C.M. (1969) The systematic position of the Athoracophoridae (Gastropoda: Euthyneura). *Malacologia* 9, 259–260.

Burch, J.B. and Patterson, C.M. (1971) The chromosome number of the athoracophorid slug, *Triboniophorus graeffei*, from Australia. *Malacological Review* 4, 25–26.

Burton, D.W. (1963) A revision of the New Zealand and Subantarctic Athoracophoridae. *Transactions of the Royal Society of New Zealand, Zoology* 3, 47–75.

Burton, D.W. (1977) Two new species of *Pseudaneitea* Cockerell (Athoracophoridae: Gastropoda). *Journal of the Royal Society of New Zealand* 7, 93–98.

Burton, D.W. (1978) Anatomy, histology, and function of the reproductive system of the tracheopulmonate slug *Athoracophorus bitentaculatus* (Quoy & Gaimard). *Zoology Publications from Victoria University of Wellington* 68, 1–16.

Burton, D.W. (1980) Anatomical studies on Australian, New Zealand, and subantarctic Athoracophoridae (Gastropoda: Pulmonata). *New Zealand Journal of Zoology* 7, 173–198.

Burton, D.W. (1981) Pallial systems in the Athoracophoridae (Gastropoda: Pulmonata). *New Zealand Journal of Zoology* 8, 391–402.

Burton, D.W. (1982) The status of the genus *Reflectopallium* Burton (Athoracophoridae). *Journal of the Royal Society of New Zealand* 12, 59–64.

Burton, D.W. (1983) A new athoracophorid (Pulmonata: Gastropoda) from South Canterbury. *Journal of the Royal Society of New Zealand* 13, 49–52.

Butot, L.J.M. and Kiauta, B. (1967) The chromosomes of *Catinella arenaria* (Bouchard-Chantereaux, 1837) with a review of the cytological conditions within the genus *Catinella* and considerations of the phylogenetic position of the Succineoidea ord. nov. (Gastropoda: Euthyneura). *Beaufortia* 14, 157–164.

Cain, A.J. (1977) Variation in the spire index of some coiled gastropod shells, and its evolutionary significance. *Philosophical Transactions of the Royal Society of London, Biological Sciences* 277, 377–428.

Carmichael, E.B. and Rivers, T.D. (1932) The effect of dehydration on the hatchability of *Limax flavus* eggs. *Ecology* 13, 375–380.

Carrick, R. (1938) The life history and development of *Agriolimax agrestis* L., the grey field slug. *Transactions of the Royal Society of Edinburgh* 59, 563–597.

Cather, J.N. and Tompa, A.S. (1972) The podocyst in pulmonate evolution. *Malacological Review* 5, 1–3.

Checa, A.G. and Jiménez-Jiménez, A.P. (1998) Constructional morphology, origin, and evolution of the gastropod operculum. *Paleobiology* 24, 109–132.

Clench, W.J. and Turner, R.D. (1948) A catalogue of the family Truncatellidae with notes and descriptions of new species. *Occasional Papers on Mollusks, Museum of Comparative Zoology, Harvard University* 1, 157–212.

Climo, F.M. (1970) The systematic position of *Cytora* Kobelt and Moellendorff, 1897 and *Liarea* Pfeiffer, 1853 (Mollusca: Mesogastropoda). *Transactions of the Royal Society of New Zealand, Biological Sciences* 12, 213–216.

Climo, F.M. (1973) The systematics, biology and zoogeography of the land small fauna of Great Island, Three Kings Group, New Zealand. *Journal of the Royal Society of New Zealand* 3, 565–628.

Climo, F.M. (1975) The land snail fauna. In: Kuschel, G. (ed.) *Biogeography and Ecology in New Zealand.* Dr W. Junk bv. Publishers, The Hague, pp. 459–492.

Climo, F.M. (1980) Smeagolida, a new order of gymnomorph molluscs from New Zealand based on a new genus and species. *New Zealand Journal of Zoology* 7, 513–522.

Colgan, D.J., Ponder, W.F. and Eggler, P.E. (2000) Gastropod evolutionary rates and phylogenetic relationships assessed using partial 28S rDNA and histone H3 sequences. *Zoologica Scripta* 29, 29–63.

Collinge, W.E. (1901) On the anatomy of a collection of slugs from N.W. Borneo; with a list of the species recorded from that region. *Transactions of the Royal Society of Edinburgh* 40, 295–312.

Cook, A. (1987) Functional aspects of the mucus-producing glands of the Ststellommatophoran slug, *Veronicella floridana. Journal of Zoology* 211, 291–305.

Cook, H. (1966) Morphology and histology of the central nervous system of *Succinea putris* (L.). *Archives Néerlandaises de Zoologie* 17, 1–72.

Creek, G.A. (1951) The reproductive system and embryology of the snail *Pomatias elegans* (Müller). *Proceedings of the Zoological Society of London* 121, 599–640.

Creek, G.A. (1953) The morphology of *Acme fusca* (Montagu) with special reference to the genital system. *Proceedings of the Malacological Society of London* 29, 228–240.

Cuénot, L. (1899) L'excrétion chez les mollusques. *Archives de Biologie, Paris* 16, 49–96.

Cuezzo, M.G. (1998) Cladistic analysis of the Xanthonychidae (= Helminthoglyptidae) (Gastropoda: Stylommatophora: Helicoidea). *Malacologia* 39, 93–111.

Curtis, S.K. and Cowden, R.R. (1977) Ultrastructure and histochemistry of the supportive structures associated with the radular of the slug, *Limax maximus. Journal of Morphology* 151, 187–212.

Davis, G.M. (1967) The systematic relationship of *Pomatiopsis lapidaria* and *Oncomelania hupensis formosana* (Prosobranchia: Hydrobiidae). *Malacologia* 6, 1–143.

Davis, G.M. (1968) New *Tricula* from Thailand. *Archiv für Molluskenkunde* 98, 291–317.

Davis, G.M. (1979) The origin and evolution of the gastropod family Pomatiopsidae, with emphasis on the Mekong River Triculinae. *Academy of Natural Sciences of Philadelphia, Monograph* 20.

Davis, G.M. (1981) Different modes of evolution and adaptive radiation in the Pomatiopsidae (Prosobranchia: Mesogastropoda). *Malacologia* 21, 209–262.

Davis, G.M. (1982) Historical and ecological factors in the evolution, adaptive radiation, and biogeography of freshwater molluscs. *American Zoologist* 22, 375–395.

Davis, G.M., Kitikoon, V. and Temcharoen, P. (1976) Monograph on '*Lithoglyphopsis*' *aperta*, the snail host of Mekong River schistosomiasis. *Malacologia* 15, 241–287.

Davis, G.M., Mazurkiewicz, M. and Mandracchia, M. (1982) *Spurwinkia*: morphology, systematics, and ecology of a new genus of North American marshland Hydrobiidae (Mollusca: Gastropoda). *Proceedings of the Academy of Natural Sciences of Philadelphia* 134, 143–177.

Davis, G.M., Kou, Y.-H., Hoagland, K.E., Chen, P.-L., Yang, H.-M. and Chen, D.-J. (1983) Advances in the systematics of the Triculinae (Gastropoda: Prosobranchia): the genus *Fenouilia* of Yunnan, China. *Proceedings of the Academy of Natural Sciences of Philadelphia* 135, 177–199.

Davis, G.M., Kou, Y.-H., Hoagland, K.E., Chen, P.-L., Yang, H.-M. and Chen, D.-J. (1984) *Kunmingia*, a new genus of Triculinae (Gastropoda: Pomatiopsidae) from China: phenetic and cladistic relationships. *Proceedings of the Academy of Natural Sciences of Philadelphia* 136, 165–193.

Degner, E. (1934) Westafrikanische Nacktschnecken I: Streptaxiden, Helicarioniden, Vaginuliden. *Zoologische Jahrbücher Systematik* 65, 209–308.

de Lacaze-Duthiers, H. (1887) Histoire de la Testacelle. *Archives de Zoologie Expérimentale et Générale (Series 2)* 5, 459–596.

de la Torre, C. and Bartsch, P. (1938) The Cuban operculate land shells of the subfamily Chondropominae. *Proceedings of the United States Natural Museum, Smithsonian Institution, Washington DC* 85, 193–403.

de la Torre, C. and Bartsch, P. (1941) The Cuban operculate land molluscs of the family Annulariidae, exclusive of the subfamily Chondropominae. *Proceedings of the United States Natural Museum, Smithsonian Institution, Washington DC* 89, 131–385, i–x.

de la Torre, C., Bartsch, P. and Morrison, J.P.E. (1942) The cyclophorid operculate land molluscs of America. *Bulletin of the United States Natural Museum, Smithsonian Institution, Washington DC* 181, i–iv, 1–306.

Delhaye, W. (1974a) Recherches sur les glandes pédieuses des gastéropodes prosobranches, principalements les formes terrestres et leur rôle possible dans l'osmorégulation chez les Pomatiasidae et les Chondropomidae. *Forma Functio* 7, 181–200.

Delhaye, W. (1974b) Histophysiologie comparée des organs excreteurs chez quelques Neritacea (Mollusca-Prosobranchia). *Archiv de Biologie* 85, 235–262.

Delhaye, W. (1976) Histophysiologie comparée des organes rénaux chez les Archaeogastéropodes (Mollusca-Prosobranchia). *Cahiers de Biologie Marine* 17, 305–322.

Delhaye, W. and Bouillon, J. (1972a) L'evolution et l'adaptation de l'organe excreteur chez les Mollusques Gasteropodes Pulmones. *Bulletin Biologique* 106, 45–77.

Delhaye, W. and Bouillon, J. (1972b) L'evolution et l'adaptation de l'organe excreteur chez les Mollusques Gasteropodes Pulmones II. Histophysiologie comparée du rein chez les Stylommatophora. *Bulletin Biologique* 106, 123–142.

Delhaye, W. and Bouillon, J. (1972c) L'evolution et l'adaptation de l'organe excreteur chez les Mollusques Gasteropodes Pulmones III. Histo-physiologie comparée du rein chez les Soleoliferes et conclusions generales pour tous les Pulmones. *Bulletin Biologique* 106, 295–314.

De Nabias, M.B. (1894) Recherches histologiques et organologiques sur les centres nerveux des Gastéropodes. *Actes de la Société Linnéenne de Bordeaux* 47, 11–202.

Deshpande, R.D. (1957) Observations on the anatomy and biology of British trochids. PhD thesis, University of Reading, UK.

De Wilde, J.A. (1984) The anatomy of an *Atopos*-species from Papua New Guinea with a discussion on its systematic position (Mollusca, Gastropoda, Rathouisiidae). *Indo-Malayan Zoology* 1, 113–126.

Doll, W. (1982) Beobachtungen über Lebensweisse und Fortpflanzung von *Carychium tridentatum* Risso im Oberrheingebiet (Pulmonata: Ellobiidae). *Archiv für Molluskenkunde* 112, 1–8.

Douris, V., Cameron, R.A.D., Rodakis, G.C. and Lecanidou, R. (1998) Mitochondrial phylogeography of the land snail *Albinaria* in Crete: long-term geological and short-term vicariance effects. *Evolution* 52, 116–125.

Duncan, C.J. (1960) The evolution of the pulmonate genital system. *Proceedings of the Zoological Society of London* 135, 339–356.

Dundee, D.S. (1957) Aspects of the biology of *Pomatiopsis lapidaria* (Say) (Mollusca: Gastropoda: Prosobranchia). *Miscellaneous Publications Museum of Zoology, University of Michigan* 100, 1–37.

Dupuis, P. and Putzeys, S. (1901) Diagnoses de quelques espèces de coquilles nouvelles et d'un genre nouveau provenant de l'ètat indépendant du Congo, suivies de quelques observations relatives a des espèces déja connues. *Annales Societe Royale Malacologie de Belgique* 36, 34–42.

Duval, A. (1982) Le système circulatoire des limaces. *Malacologia* 22, 627–630.

Els, W.J. (1978) Histochemical studies on the maturation of the genital system of the slug *Deroceras laeve* (Pulmonata, Limacidae) with special reference to the identification of mucosubstances secreted by the genital tract. *Annale Universiteit van Stellenbosch (Series A) 2, Zoologie* 1, 1–116.

Emberton, K.C. (1988) The genitalic, allozymic, and conchological evolution of the eastern North American Triodopsinae (Gastropoda: Pulmonata: Polygyridae). *Malacologia* 28, 159–273.

Emberton, K.C. (1991a) The genitalic, allozymic, and conchological evolution of the Mesodontini trib. nov. (Gastropoda: Pulmonata: Polygyridae). *Malacologia* 33, 71–178.

Emberton, K.C. (1991b) Polygyrid relations: a phylogenetic analysis of 17 subfamilies of land snails (Mollusca: Gastropoda: Stylommatophora). *Zoological Journal of the Linnean Society* 103, 207–224.

Emberton, K.C. (1994) Polygrid land snail phylogeny: external sperm exchange, early North American biogeography, iterative shell evolution. *Biological Journal of the Linnean Society* 52, 241–271.

Emberton, K.C. (1995a) On the endangered biodiversity of Madagascan land snails. In: van Bruggen, A.C., Wells, S.M. and Kemperman, Th.C.M. (eds) *Biodiversity and Conservation of the Mollusca.* Backhuys Publishers, Oegstgeest-Leiden, pp. 69–89.

Emberton, K.C. (1995b) Cryptic, genetically extremely divergent, polytypic, convergent, and polymorphic taxa in Madagascan *Tropidophora* (Gastropoda: Pomatiasidae). *Biological Journal of the Linnean Society* 55, 183–208.

Emberton, K.C. (1996) Microsculptures of convergent and divergent polygyrid land-snail shells. *Malacologia* 38, 67–85.

Emberton, K.C. and Pearce, T.A. (1999) Land caenogastropods of Mounts Mahermana, Ilapiry, and Vasiha, southeastern Madagascar, with conservation statuses of 17 species of *Boucardicus. Veliger* 42, 338–372.

Emberton, K.C. and Tillier, S. (1995) Clarification and evaluation of Tillier's (1989) stylommatophoran monograph. *Malacologia* 36, 203–208.

Emberton, K.C., Kuncio, G.S., Davis, G.M., Phillips, S.M., Monderewicz, K.M. and Hua Guo, Y. (1990) Comparison of recent classifications of stylommatophoran land-snail families, and evaluation of large-ribosomal-RNA sequences for their phylogenetics. *Malacologia* 31, 327–352.

Estabrooks, W.A., Kay, E.A. and McCarthy, S.A. (1999) Structure of the excretory system of Hawaiian nerites (Gastropoda: Neritoidea). *Journal of Molluscan Studies* 65, 61–72.

Felsenstein, J. (1985) Confidence limits on phylogenies: an approach using the bootstrap. *Evolution* 39, 783–791.

Fischer, P.-H. (1938) Remarques sur la ponte et l'éclosion des succinées; précocité du caractère terrestre. *Journal de Conchyliologie* 82, 395–401.

Fischer-Piette, E., Blanc, Ch., Blanc, F. and Salvat, F. (1993) Gastéropodes terrestres prosobranches. *Faune de Madagascar* 80, 1–281.

Fleischmann, A. (1932) Vergleichende Betrachtungen über das Schalenwachstum der Weichtiere (Mollusca). II. Deckel (Operculum) und Haus (Concha) der Schnecken (Gastropoden). *Zeitschrift für Morphologie und Ökologie der Tiere* 25, 549–622.

Forcart, L. (1953) The Veronicellidae of Africa (Mollusca, Pulmonata). *Annales du Musée Royal du Congo Belge Tervuren, Belgique, Sciences Zoologiques* 23, 6–110.

Fowler, B.H. (1980) Reproductive biology of *Assiminea californica* (Tryon, 1865) (Mesogastropoda: Rissoacea). *Veliger* 23, 163–166.

Fretter, V. (1943) Studies in the functional morphology and embryology of *Onchidella celtica* (Forbes and Hanley) and their bearing on its relationships. *Journal of the Marine Biological Association of the United Kingdom* 25, 685–720.

Fretter, V. (1946) The genital ducts of *Theodoxus, Lamellaria* and *Trivia* and a discussion of their evolution in the prosobranchs. *Journal of the Marine Biological Association of the United Kingdom* 26, 312–351.

Fretter, V. (1948) The structure and life history of some minute prosobranchs of rock pools: *Skeneopsis planorbis* (Fabricius), *Omalogyra atomus* (Philippi), *Rissoella diaphana* (Alder) and *Rissoella opalina* (Jeffreys). *Journal of the Marine Biological Association of the United Kingdom* 27, 597–632.

Fretter, V. (1965) Functional studies of the anatomy of some neritid prosobranchs. *Journal of Zoology* 147, 46–74.

Fretter, V. (1984) The functional anatomy of the neritacean limpet *Phenacolepas omanensis* Biggs and some comparison with *Septaria. Journal of Molluscan Studies* 50, 8–18.

Fretter, V. and Graham, A. (1962) *British Prosobranch Molluscs, Their Functional Anatomy and Ecology.* Ray Society, London.

Fretter, V. and Graham, A. (1978) The prosobranch molluscs of Britain and Denmark. Part 3 – Neritacea, Viviparacea, Valvatacea, terrestrial and freshwater Littorinacea and Rissoacea. *Journal of Molluscan Studies*, Supplement 5, 100–152.

Fretter, V. and Patil, A.M. (1958) A revision of the systematic position of the prosobranch gastropod *Cingulopsis* (= *Cingula*) *fulgida* (J. Adams). *Proceedings of the Malacological Society of London* 33, 114–126.

Fretter, V., Graham, A., Ponder, W.F. and Lindberg, D.R. (1998) Prosobranchs introduction. In: Beesley, P.L., Ross, G.J.B. and Wells, A. (eds) *Mollusca: the Southern Synthesis. Fauna of Australia*, Vol. 5, Part B. CSIRO Publishing, Melbourne, pp. 605–638.

Fukuda, H. (1994) The anatomy of *Cylindrotis quadrasi* from Okinawa Island, Japan and the subfamilial position of the genus *Cylindrotis* Möllendorff, 1895 (Archaeopulmonata: Ellobiidae). *Journal of Molluscan Studies* 60, 69–81.

Fukuda, H. and Mitoki, T. (1995) A revision of the Assimineidae (Mollusca: Gastropoda: Neotaenioglossa) stored in the Yamaguchi Museum. Part 1: subfamily Omphalotropidinae. *Bulletin of the Yamaguchi Museum* 21, 1–20.

Fukuda, H. and Mitoki, T. (1996a) A revision of the Assimineidae (Mollusca: Gastropoda: Neotaenioglossa) stored in the Yamaguchi Museum. Part 2: subfamily Assimineinae. (1) Two species from Taiwan. *Bulletin of the Yamaguchi Museum* 22, 1–11.

Fukuda, H. and Mitoki, T. (1996b) A revision of the Assimineidae (Mollusca: Gastropoda: Neotaenioglossa) stored in the Yamaguchi Museum. Part 3: subfamily Assimineinae. (2) *Angustassiminea* and *Pseudomphala. Yuriyagai* 4, 109–137.

Garnault, P. (1887) Recherches anatomiques et histologiques sur le *Cyclostoma elegans. Actes de la Société Linnéenne de Bordeaux* 41, 11–158.

Ghosh, E. (1915) On the anatomy of a Burmese slug of the genus *Atopos. Records of the Indian Museum* 11, 153–161.

Gillette, R. (1991) On the significance of neuronal giantism in gastropods. *Biological Bulletin* 180, 234–240.

Girard, A.A. (1895) Sur le *Thyrophorella thomensis*, Greeff. Gastéropode terrestre muni d'un faux opercule à charnière. *Jornal de Sciencias Mathematicas, Physicas e Naturales, Lisboa* 4, 28–32.

Girardi, E.-L. (1978) The Samoan land snail genus *Ostodes* (Mollusca: Prosobranchia: Poteriidae). *Veliger* 20, 191–250.

Gittenberger, E. (1996) Adaptations of the aperture in terrestrial gastropod-pulmonate shells. *Netherlands Journal of Zoology* 46, 191–205.

Gittenberger, E. and Schilthuizen, M. (1996) Parallelism in the origin of the G-type clausilial apparatus (Gastropoda, Pulmonata, Clausiliidae). In: Taylor, J. (ed.) *Origin and Evolutionary Radiation of the Mollusca.* Oxford University Press, Oxford, pp. 295–300.

Giusti, F. (1973) Notulae Malacologicae, XVIII. I molluschi terrestri e salmastri delle Isole Eolie. *Lavori della Società Italiana de Biogeografia (Nuova Seria)* 3, 113–306.

Glamann, G. (1903) Anatomisch Systematische Beiträge zur Kenntnis der Tracheopulmonaten. *Zoologische Jahrbücher, Abtheilung für Anatomie und Ontogenie der Thiere* 17, 679–762.

Golikov, A.N. and Starobogatov, Y.I. (1975) Systematics of pros pods. *Malacologia* 15, 185–232.

Golikov, A.N. and Starobogatov, Y.I. (1988) Problems of phylogen the prosobranchiate gastropods. *Proceedings of the Zoologi Leningrad* 187, 4–77 (in Russian).

Gorf, A. (1961) Untersuchungen über Neurosekretion bei der Su necke *Vivipara vivipara* L. *Zoologische Jahrbücher Physiologie* 69, 379–404.

Graham, A. (1939) On the structure of the alimentary canal of style-bearing prosobranchs. *Proceedings of the Zoological Society of London, Series B* 109, 75–112.

Grimpe, G. and Hoffmann, H. (1925) Die Nacktschnecken von Neu-Caledonien, den Loyalty-Inseln und den Neuen Hebriden. In: Sarasin, F. and Roux, J. (eds) *Nova Caledonia, Zoology* 3, 339–476.

Guyomarc'h-Cousin, C. (1976) Organogenèse descriptive de l'appareil génital chez *Littorina saxatilis* (Olivi), gastéropode prosobranche. *Bulletin de la Société Zoologique de France* 101, 465–476.

Hadfield, M.G. and Strathmann, M.F. (1990) Heterostrophic shells and pelagic development in trochoideans: implications for classification, phylogeny and palaeoecology. *Journal of Molluscan Studies* 56, 239–256.

Harasewych, M.G., Adamkewicz, S.L., Blake, J.A., Saudek, D., Spriggs, T. and Bult, C.J. (1997) Phylogeny and relationships of pleurotomariid gastropods (Mollusca: Gastropoda): an assessment based on partial 18S rDNA and cytochrome c oxidase I sequences. *Molecular Marine Biology and Biotechnology* 6, 1–20.

Harbeck, K. (1996) Die Evolution der Archaeopulmonata. *Zoologische Verhandelingen* 305, 1–133.

Harry, H.W. (1952) *Carychium exiguum* (Say) of Lower Michigan; morphology, ecology, variation and life history (Gastropoda, Pulmonata). *Nautilus* 66, 5–7.

Harry, H.W. (1964) The anatomy of *Chilina fluctuosa* Gray re-examined, with prolegomena on the phylogeny of the higher limnic Basommatophora (Gastropoda: Pulmonatsa). *Malacologia* 1, 355–385.

Haszprunar, G. (1985a) The fine morphology of the osphradial sense organs of the Mollusca. I. Gastropoda, Prosobranchia. *Philosphical Transactions of the Royal Society of London, Biological Sciences* 307, 457–496.

Haszprunar, G. (1985b) The fine morphology of the osphradial sense organs of the Mollusca. II. Allogastropoda (Architectonicidae and Pyramidellidae). *Philosphical Transactions of the Royal Society of London, Biological Sciences* 307, 497–505.

Haszprunar, G. (1985c) The Heterobranchia – a new concept of the phylogeny of the higher Gastropoda. *Zeitschrift für Zoologische Systematik und Evolutionsforschung* 23, 15–37.

Haszprunar, G. (1987) The Vetigastropoda and the systematics of streptoneurous Gastropoda (Mollusca). *Journal of Zoology* 211, 747–770.

Haszprunar, G. (1988a) A preliminary phylogenetic analysis of the streptoneurous Gastropod. In: Ponder, W.F. (ed.) *Prosobranch Phylogeny*. Proceedings of the 9th International Malacological Congress, Edinburgh 1986, *Malacological Review Supplement* 4, pp. 7–16.

Haszprunar, G. (1988b) On the origin and evolution of major gastropod groups, with special reference to the Streptoneura. *Journal of Molluscan Studies* 54, 367–441.

Haszprunar, G. (1990) Towards a phylogenetic system of Gastropoda. Part 1: traditional methodology – a reply. *Malacologia* 32, 195–202.

Haszprunar, G. (1993) Sententia. The Archaeogastropoda: a clade, a grade or what else? *American Malacological Union Bulletin* 10, 165–177.

Haszprunar, G. (1995) On the evolution of larval development in the Gastropoda, with special reference to larval plantotrophy. *Notiz* 16, 5–13.

Haszprunar, G. and Huber, G. (1990) On the central nervous system of Smeagolidae and Rhodopidae, two families questionally allied with the Gymnomorpha (Gastropoda: Euthyneura). *Journal of Zoology* 220, 185–199.

Hausdorf, B. (1998) Phylogeny of the Limacoidea *sensu lato* (Gastropoda: Stylommatophora). *Journal of Molluscan Studies* 64, 35–66.

Healy, J.M. (1988a) Sperm morphology and its systematic importance in the Gastropoda. In: Ponder, W.F. (ed.) *Prosobranch Phylogeny*. Proceedings of the 9th International Malacological Congress, Edinburgh, 1986, *Malacological Review, Supplement* 4, pp. 251–266.

Healy, J.M. (1988b) The ultrastructure of spermatozoa and spermiogenesis in pyramidellid gastropods, and its systematic importance. *Helgolander Meeresuntersuchungen* 42, 303–318.

Healy, J.M. (1993) Comparative sperm ultrastructure and spermiogenesis in basal heterobranch gastropods (Valvatoidea, Architectonicoidea, Rissoelloidea, Omalogyroidea, Pyramidelloidea) (Mollusca). *Zoologica Scripta* 22, 263–276.

Hecker, U. (1965) Zur Kenntnis der mitteleuropäischen Bernsteinschnecken (Succineidae). I. *Archiv für Molluskenkunde* 94, 1–45.

Heller, J. (1993) Hermaphroditism in molluscs. *Biological Journal of the Linnean Society* 48, 19–42.

Henchman, A. (1890) The origin and development of the central nervous system of *Limax maximus*. *Bulletin of the Museum of Comparative Zoology* 20, 169–208.

Henderson, J.B. and Bartsch, P. (1920) A classification of the American operculate land mollusks of the Family Annulariidae. *Proceedings of the United States National Museum* 58, 49–82.

Hershler, R. (1987) Redescription of *Assiminea infirma* Berry, 1947, from Death Valley, California. *Veliger* 29, 274–288.

Hickman, C.S. (1988) Archaeogastropod evolution, phylogeny and systematics: a re-evaluation. In: Ponder, W.F. (ed.) *Prosobranch Phylogeny*. Proceedings of the 9th International Malacological Congress, Edinburgh 1986, *Malacological Review, Supplement* 4, pp. 17–34.

Hickman, C.S. (1992) Reproduction and development of trochacean gastropods. *Veliger* 35, 245–272.

Hoagland, K.E. and Davis, G.M. (1987) The succineid snail fauna of Chittenango Falls, New York: taxonomic status with comparisons to other relevant taxa. *Proceedings of the Academy of Natural Sciences of Philadelphia* 139, 465–526.

Hochpöchler, F. (1979) Vergleichende Untersuchungen über die Entwicklung des Geschlechtsapparates der Stylommatophora (Gastropoda). *Zoologischer Anzeiger* 202, 289–306.

Hoffmann, H. (1925) Die Vaginuliden. Ein Beitrag zur Kenntnis ihrer Biologie, Anatomie, Systematik, geographischen Verbreitung und Phylogenie. (Fauna et Anatomia ceylanica, III, Nr. 1). *Jenaische Zeitschrift für Naturwissenschaft* 61, 1–374.

Hoffmann, H. (1929) Zur Kenntnis der Oncidiiden (Gastrop. Pulm.). II. Teil. Phylogenie und Verbreitung. *Zoologische Jahrbücher Systematik* 57, 253–302.

Hoffmann, H. (1939) Opisthobranchia. In: *Bronns Klassen und Ordnungen des Tierreiches.* Akademische Verlagsges, Leipzig, pp. 1–1247.

Houssay, F. (1884) Recherches sur l'opercule et les glandes du pied des gastéropodes. *Archives de Zoologie Expérimentale et Générale* 2, 171–288.

Hrubesch, K. (1965) Die santone Gosau-Landschneckenfauna von Glanegg bei Salzburg, Österreich. *Mitteilungen der Bayerische Staatssammlung für Paläontologie und Historische Geologie* 5, 83–120.

Hubendick, B. (1945) Phylogenie und Tiergeographie der Siphonariidae. Zur Kenntnis der Phylogenie in der Ordnung Basommatophora und des Ursprungs der Pulmonatengruppe. *Zoologiska Bidrag fran Uppsala* 24, 1–216.

Huber, G. (1993) On the cerebral nervous system of marine Heterobranchia (Gastropoda). *Journal of Molluscan Studies* 59, 381–420.

Ivanov, D.L. (1996) Origin of aculifera and problems of monophyly of higher taxa in molluscs. In: Taylor, J. (ed.) *Origin and Evolutionary Radiation of the Mollusca.* Oxford University Press, Oxford, pp. 59–65.

Jackiewicz, M. (1980) Some observations on biology of reproduction of *Succinea* Draparnaud (Gastropoda, Pulmonata). *Annales Zoologici* 35, 65–68.

Johansson, J. (1939) Anatomische Studien über Die Gastropodenfamilien Rissoidae und Littorinidae. *Zoologiska Bidrag fran Uppsala* 18, 239–396.

Jonges, K. (1980) Genital ducts in several species of the cyclophorid genus *Leptopoma* (Mollusca, Mesogastropoda). *Bijdragen tot de Dierkunde* 50, 292–302.

Kasinathan, R. (1975) Some studies of five species of cyclophorid snails from Peninsula India. *Proceedings of the Malacological Society of London* 41, 379–394.

Kershaw, R.C. (1983) The gastropods *Assiminea* and *Hydrococcus* in Tasmania (Mollusca: Gastropoda). *Records of the Queen Victoria Museum* 83, 1–24.

Kessel, E. (1942) Über Bau und Bildung des Prosobranchier-Deckels. *Zeitschrift für Morphologie und Ökologie der Tiere* 38, 197–250.

Kilian, E.F. (1951) Untersuchungen zur Biologie von *Pomatias elegans* (Müller) und ihrer 'Kronkrementdrüse'. *Archiv für Molluskenkunde* 80, 1–16.

Knight, J.B., Cox, L.R., Keen, A.M., Batten, R.L., Yochelson, E.L. and Robertson, R. (1960a) Systematic descriptions. In: Moore, R.C. (ed.) *Treatise on Invertebrate Paleontology. Part I. Mollusca 1.* Geological Society of America and University of Kansas Press, Boulder, Colorado and Lawrence, Kansas, pp. 1169–1310.

Knight, J.B., Batten, R.L., Yochelson, E.L. and Cox, L.R. (1960b) Paleozoic and some Mesozoic Caenogastropoda and Opisthobranchia. In: Moore, R.C. (ed.) *Treatise on Invertebrate Paleontology. Part I. Mollusca 1.* Geological Society of America and University of Kansas Press, Boulder, Colorado and Lawrence, Kansas, pp. 1310–1331.

Kobelt, W. and von Möllendorff, O.F. (1897) Catalog der gegenärtig lebend bekannten Pneumonopomen. *Nachrichtsblatt der Deutschen Malakozoologischen Gesellschaft* 29, 73–88.

Kosuge, S. (1966) Anatomical studies on Japanese Rissoacea (III). On *Truncatella kiusiuensis* Pilsbry. *Science Report of the Yokosuka City Museum* 12, 18–25.

Kowslowsky, F. (1933) Zur Anatomie der Auriculidae *Melampus boholensis* H. & A. Adams. *Jenaische Zeitschrift für Naturwissenschaft* 68, 117–192.

Kriegstein, A.R. (1977) Stages in the post-hatching development of *Aplysia californica*. *Journal of Experimental Zoology* 199, 275–288.

Kuchenmeister, G.M., Prior, D.J. and Welsford, I.G. (1996) Quantification of the development of the cephalic sac and podocyst in the terrestrial gastropod *Limax maximus* L. *Malacologia* 38, 153–160.

Kulkarni, A.B. and Nagabhushanam, R. (1973) Studies on the ecology and control of the slug, *Laevicaulis alte* (Mollusca: Stylommatophora: Vaginulidae) from Marathwada region, Maharashtra. *Journal of the Zoological Society of India* 25, 71–81.

Labbé, A. (1934) Les Silicodermes Labbé du Muséum d'histoire naturelle de Paris. Première partie. Classification, formes nouvelles ou peu connues. *Annales de l'Institut Océanographique, Paris* 14, 173–246.

Laidlaw, F.F. (1940) Notes on some specimens of the genus *Atopos* (Mollusca Pulmonata) with microphotographs illustrating points in the anatomy of the genus. *Bulletin of the Raffles Museum, Singapore* 16, 121–132.

Lauterbach, K.-E. (1984) Das phylogenetische System der Mollusca. *Mitteilungen der Deutschen Malakozoogischen Gesellschaft* 37, 66–81.

Leal-Zanchet, A.M. (1998) Comparative studies on the anatomy and histology of the alimentary canal of the Limacoidea and Milacidae (Pulmonata: Stylommatophora). *Malacologia* 39, 39–57.

Leme, J.L.M. (1973) Anatomy and systematics of the Neotropical Strophocheiloidea (Gastropoda, Pulmonata) with the description of a new family. *Arquivos de Zoologia, Museu de Zoologia, Universidade de Sao Paulo* 23, 295–337.

Lever, J., Boer, H.H., Duiven, R.J.TH., Lammens, J.J. and Wattel, J. (1959) Some observations on follicle glands in pulmonates. *Proceedings Koninklijke Nederlandse Akademie van Wetenschappen, C* 61, 235–242.

Linke, O. (1933) Morphologie und Physiologie des Genitalapparates der Nordseelittorinen. *Wissenschaftliche Meeresuntersuchungen, Abteilung Helgoländer* 19, 1–60.

Linke, O. (1935) Zur Morphologie und Physiologie des Genitalapparatus der Süsswasser-littorinide *Cremnoconchus syhadrensis* Blanford. *Archiv für Naturgeschicte* 4, 72–87.

Little, C. (1972) The evolution of kidney function in the Neritacea (Gastropoda, Prosobranchia). *Journal of Experimental Biology* 56, 249–261.

Little, C. and Andrews, E.B. (1977) Some aspects of excretion and osmoregulation in assimineid snails. *Journal of Molluscan Studies* 43, 263–285.

Madge, E.H. (1939) Further notes on non-marine Mollusca from the Mascarene Islands. A review of the genus *Omphalotropis* Pfeiffer in Mauritius and Rodriguez, with descriptions of two new species. *Mauritius Institute Bulletin* 1, 8–33.

Marcus, Ev. (1971) On some euthyneuran gastropods from the Indian and Pacific Oceans. *Proceedings of the Malacological Society of London* 39, 355–369.

Marcus, Ev. (1979) The Atlantic species of *Onchidella* (Gastropoda Pulmonata) Part 2. *Boletim Zoologia Universidade de Sao Paulo* 4, 1–38.

Marcus, Ev. and Marcus, Er. (1956) Zwei atlantische Onchidellen (Ergebnisse der Reisen) A. Remane's nach Brasilien und den Kanaren. *Kieler Meeresforschungen* 12, 76–84.

Marcus, Ev. and Marcus, Er. (1963) Mesogastropoden von der Küste São Paulos. *Abhandlungen der Mathematisch-Naturwissenschaftlichen Klasse. Akademie der Wissenschaften und der Literatur, Mainz* (1963), 1–105.

Marcus, Ev. and Marcus, Er. (1965a) On Brazilian supratidal and estuarine snails. *Boletim de Faculdade de Filosofia, Ciencias e Letras, Universidade de Sao Paulo, Zoologia* 287, 29–82.

Marcus, Ev. and Marcus, Er. (1965b) On two Ellobiidae from southern Brazil. *Boletim de Faculdade de Filosofia, Ciencias e Letras, Universidade de Sao Paulo, Zoologia* 287, 425–453.

Martin, A.W., Stewart, D.M. and Harrison, F.M. (1965) Urine formation in the pulmonate land snail, *Achatina fulica. Journal of Experimental Biology* 42, 99–124.

Martins, A.M.F. (1996a) Relationships within the Ellobiidae. In: Taylor, J. (ed.) *Origin and Evolutionary Radiation of the Mollusca.* Oxford University Press, New York, pp. 285–294.

Martins, A.M.F. (1996b) Anatomy and systematics of the Western Atlantic Ellobiidae (Gastropoda: Pulmonata). *Malacologia* 37, 163–332.

Martoja, M. (1975) Le rein de *Pomatias* (= *Cyclostoma*) *elegans* (Gastéropode, Prosobranche): données structurales et analytiques. *Annales des Sciences Naturelles, Zoologie* 17, 535–558.

McFarlane, I.D. (1979) Ecology and behaviour of the intertidal pulmonate mollusc *Onchidium peronii* in Kuwait. *Journal of the University of Kuwait* 6, 169–180.

Michelson, E.H. (1961) On the generic limits in the family Pilidae (Prosobranchia: Mollusca). *Breviora* 133, 1–10.

Mikkelsen, P.M. (1996) The evolutionary relationships of Cephalaspidea s.l. (Gastropoda: Opisthobranchia): a phylogenetic analysis. *Malacologia* 37, 375–442.

Miller, S.L. (1974) The classification, taxonomic distribution and evolution of locomotor types among prosobranch gastropods. *Proceedings of the Malacological Society of London* 41, 233–272.

Minichev, Y.S. and Starobogatov, YI. (1979) The subclasses of Gastropoda and their phylogenetic relations. *Zoologischeskii Zhurnal* 58, 293–305.

Mischor, B. and Märkel, K. (1984) Histology and regeneration of the radula of *Pomacea bridgesi* (Gastropoda, Prosobranchia). *Zoomorphology* 104, 42–66.

Moffett, S. (1979) Locomotion in the primitive pulmonate snail *Melampus bidentatus*: foot structure and function. *Biological Bulletin* 157, 306–319.

Morton, J.E. (1952) A preliminary study of the land operculate *Murdochia pallidum* (Cyclophoridae, Mesogastropoda). *Transactions of the Royal Society of New Zealand* 80, 69–79.

Morton, J.E. (1955a) The functional morphology of the British Ellobiidae (Gastropoda, Pulmonata) with special reference to digestive and reproductive systems. *Philosophical Transactions of the Royal Society of London, Biological Sciences* 238, 89–160.

Morton, J.E. (1955b) The evolution of the Ellobiidae with a discussion on the origin of the Pulmonata. *Proceedings of the Zoological Society of London* 125, 127–168.

Morton, J.E. (1955c) The functional morphology of *Otina otis*, a primitive marine pulmonate. *Journal of the Marine Biological Association of the United Kingdom* 34, 113–150.

Morton, J.E. (1963) The molluscan pattern: evolutionary trends in a modern classification. *Proceedings of the Linnaean Society of London* 174, 53–72.

Morton, J.E. (1988) The pallial cavity. In: Trueman, E.R. and Clarke, M.R. (eds) *The Mollusca. 11. Form and Function*. Academic Press, New York, pp. 253–286.

Morton, J.E. and Miller, M. (1973) *The New Zealand Sea Shore*, 2nd edn. Collins, London.

Muratov, I.V. (1999) Analysis of the phylogenetic relationships and their systematic implications in the Limacoinei (= Zonitinia) infraorder (Gastropoda, Pulmonata, Geophila). *Ruthenica* 9, 5–26.

Nagabhushanam, R. and Kulkarni, A.B. (1971) Reproductive biology of the land slug, *Laevicaulis alte*. *Rivista di Biologia* 64, 15–44.

Natarajan, R., Hubricht, L. and Burch, J.B. (1966) Chromosomes of eight species of Succineidae (Gastropoda, Stylommatophora) from the southern United States. *Acta Biologica Academiae Scientiarum Hungaricae* 17, 105–120.

Neuckel, W. (1985) Anal uptake of water in terrestrial pulmonate snails. *Journal of Comparative Physiology* 156B, 291–296.

Nolte, A. (1965) Neurohämal- 'Organe' bei Pulmonaten (Gasteropoda). *Zoologische Jahrbücher Anatomie* 82, 365–380.

Nordsieck, H. (1982) Die Evolution des Verschlussapparates der Schliessmundschnecken (Gastropoda: Clausiliidae). *Archiv für Molluskenkunde* 112, 27–43.

Nordsieck, H. (1985) The system of the Stylommatophora (Gastropoda), with special regard to the systematic position of the Clausiliidae. I. Importance of the excretory and genital systems. *Archiv für Molluskenkunde* 116, 1–24.

Nordsieck, H. (1986) The system of the Stylommatophora (Gastropoda), with special regard to the systematic position of the Clausiliidae. II. Importance of the shell and distribution. *Archiv für Molluskenkunde* 117, 93–116.

Nordsieck, H. (1992) Phylogeny and system of the Pulmonata (Gastropoda). *Archiv für Molluskenkunde* 121, 31–52.

Oberzeller, E. (1969) Die Verwandtschaftsbeziehungen der *Rhodope veranii* Kölliker zu den Oncidiidae, Vaginulidae und Rathouisiidae in bezug auf das Nervensystem. *Malacologia* 9, 282–283.

Oberzeller, E. (1970) Ergebnisse der Österreichischen Neukaledonien-Expedition 1965. Terrestrische Gastropoda, II: Veronicellidae und Athoracophoridae. *Annalen der Naturhistorisches Museum, Wien* 74, 325–341.

Odhner, N.H. (1917) Results of Dr. E. Mjöbergs Swedish scientific expeditions to Australia 1910–1913. 17 Mollusca. *Kungliga Svenska Vetenskapsakademiens Handlingar* 52, 1–115.

Patterson, C.M. (1968a) Taxonomic studies of the Succineidae (Gastropoda, Stylommatophora). In: *Proceedings of the Symposium on Mollusca held at Cochin from January 12 to 16, 1968. Part I*. Marine Biological Association of India Symposium Series 3, pp. 46–50.

Patterson, C.M. (1968b) Chromosomes of molluscs. In: *Proceedings of the Symposium on Mollusca held at Cochin from January 12 to 16, 1968. Part II*. Marine Biological Association of India Symposium Series 3, pp. 636–686.

Patterson, C.M. (1970) Self-fertilization in the land snail family Succineidae. *Journal de Conchyliologie* 108, 61–62.

Patterson, C.M. (1971a) Generic and specific characters in the land snail family Succineidae. *Malacological Review* 6, 54–56.

Patterson, C.M. (1971b) Taxonomic studies of the land snail family Succineidae. *Malacological Review* 4, 131–202.

Patterson, C.M. (1973) Parallel evolution of shell characters in succineids inhabiting waterfalls. *Bulletin of the American Malacological Union* (1973), 28.

Patterson, C.M. and Burch, J.B. (1978) Chromosomes of pulmonate molluscs. In: Fretter, V. and Peake, J. (eds) *Pulmonates, Volume 2A, Systematics, Evolution and Ecology.* Academic Press, London, pp. 171–217.

Peake, J.F. (1968) Habitat distribution of Soloman Island land Mollusca. In: Fretter, V. and Peake, J.F. (eds) *Studies in the Structure, Physiology and Ecology of Molluscs.* Symposium of the Zoological Society of London 22, pp. 319–346.

Pearce, T.A. (1990) Phylogenetic relationships of *Micrarionta* (Gastropoda: Pulmonata) and distinctiveness of the species on San Nicolas Island, California. *Malacological Review* 23, 1–37.

Pelseneer, P. (1894) Recherches sur divers Opisthobranches. *Mémoires Couronnés et Mémoires des Savants Étrangers* 53, 1–157.

Pelseneer, P. (1896) Hermaphroditism in Mollusca. *Quarterly Journal of Microscopical Science* 37, 19–46.

Pelseneer, P. (1901) Études sur des Gasteropodes Pulmonés. *Mémoires de l'Académie Royale des Sciences, des Letters et des Beaux-arts de Belgique* 54, 1–76.

Perrier, R. (1889) Recherches sur l'anatomie et l'histologie du rein des gastéropodes Prosobranchiata. *Annales des Sciences Naturelles, Zoologie (Series 7)* 8, 61–315.

Person, P. and Philpott, D.E. (1963) Invertebrate cartilages. *Annals of the New York Academy of Sciences* 109, 113–126.

Person, P. and Philpott, D.E. (1969) The nature and significance of invertebrate cartilages. *Biological Reviews* 44, 1–16.

Peterellis, L.S. and Dundee, D.S. (1969) *Veronicella ameghini* (Gastropoda): reproductive, digestive and nervous systems. *Transactions of the American Microscopical Society* 88, 547–558.

Pfeiffer, L. (1876) *Monographia Pneumonopomorum Viventium, Accedente Fossilium Enumeratione. Supplementum Tertium.* T. Fischer, Cassel.

Pfeiffer, W. (1900) Die Gattung *Triboniophorus. Zoologische Jahrbücher, Abtheilung für Anatomie und Ontogenie der Tiere* 13, 293–318.

Pilsbry, H.A. (1896) The Aulacopoda: a primary division of the monotremate land Pulmonata. *Nautilus* 9, 109–111.

Pilsbry, H.A. (1900) On the zoological position of *Partula* and *Achatinella. Proceedings of the Academy of Natural Sciences of Philadelphia* (1900), 561–567.

Pilsbry, H.A. (1940) Land Mollusca of North America (North of Mexico), Volume 1, Part 2. *Monograph of the Academy of Natural Sciences of Philadelphia* 3, 575–994.

Pilsbry, H.A. (1948) Land Mollusca of North America (North of Mexico) Volume 2, Part 2. *Monograph of the Academy of Natural Sciences of Philadelphia* 3, i–xvii, 1–1113.

Pilsbry, H.A., Cooke, C.M. and Neal, M.C. (1928) Land snails from Hawaii, Christmas Island and Samoa. *B.P. Bishop Museum Bulletin* 47, 1–49.

Plate, L. (1893) Studien über opisthopneumone Lungenschnecken. II. Die Oncidiiden. *Zoologischen Jahrbüchern, Abteilung für Anatomie und Ontogenie der Tiere* 7, 93–234.

Plate, L. (1897) Uber einen Typus der Lungenathmung, die Niere und ein subcutanes Sinnesorgan bei Nacktschnecken aus der Familie der Janellen.

Sitzungsberichte der Gesellschaft natürforschender Freunde zu Berlin (1897), 141–145.

Plate, L. (1898) Beitage zur Anatomie und Systematik der Janelliden. *Zoologische Jahrbücher, Abteilung für Anatomie und Ontogenie der Tiere* 11, 193–269.

Pokryszko, B.M. (1987) On the aphally in the Vertiginidae (Gastropoda: Pulmonata: Orthurethra). *Journal of Conchology* 32, 365–375.

Pokryszko, B.M. (1990) Life history and population dynamics of *Vertigo pusilla* O. F. Müller, 1774 (Gastropoda: Pulmonata: Orthurethra), with some notes on shell and genital variability. *Annales Zoologici* 43, 407–430.

Pokryszko, B.M. (1997) Land snail apertural barriers – adaption or hindrance? (Gastropoda: Pulmonata). *Malakologische Abhandlungen* 18, 239–248.

Ponder, W.F. (1968) The morphology of some small New Zealand prosobranchs. *Records of the Dominion Museum, Wellington* 6, 61–95.

Ponder, W.F. (1987) The anatomy and relationships of the opisthobranch limpet *Amathina tricarinata* (Mollusca: Gastropoda). *Asian Marine Biology* 4, 1–34.

Ponder, W.F. (1988) The truncatelloidean (= rissoacean) radiation – a preliminary phylogeny. *Malacological Review, Supplement* 4, 129–164.

Ponder, W.F. (1991) Marine valvatoidean gastropods – implications for early heterobranch phylogeny. *Journal of Molluscan Studies* 57, 21–32.

Ponder, W.F. and de Keyzer, R.G. (1998) Superfamily Rissooidea. In: Beesley, P.L., Ross, G.J.B. and Wells, A. (eds) *Mollusca. The Southern Synthesis. Fauna of Australia*, Vol. 5, Part B. Australian Biological Resources Study, CSIRO Publishing, Melbourne, pp. 745–766.

Ponder, W.F. and Lindberg, D.R. (1996) Gastropod phylogeny – challenges for the 90s. In: Taylor, J. (ed.) *Origin and Evolutionary Radiation of the Mollusca*. Oxford University Press, Oxford, pp. 135–154.

Ponder, W.F. and Lindberg, D.R. (1997) Towards a phylogeny of gastropod molluscs: an analysis using morphological characters. *Zoological Journal of the Linnean Society* 119, 83–265.

Ponder, W.F. and Warén, A. (1988) Classification of the Caenogastropoda and Heterostropha – a list of the family-group names and higher taxa. *Malacological Review, Supplement* 4, 288–326.

Ponder, W.F., Colgan, D.J., Clark, G.A., Miller, A.C. and Terzis, T. (1994) Microgeographic genetic and morphological differentiation of freshwater snails – the Hydrobiidae of Wilsons Promontory, Victoria, south-eastern Australia. *Australian Journal of Zoology* 42, 557–678.

Ponder, W.F., Eggler, P. and Colgan, D.J. (1995) Genetic differentiation of aquatic snails (Gastropoda: Hydrobiidae) from artesian springs in arid Australia. *Biological Journal of the Linnean Society* 56, 553–596.

Potts, W.T.W. (1975) Excretion in the gastropods. *Fortschritte der Zoologie* 23, 75–88.

Powell, A.W.B. (1946) Shellfish of New Zealand, 2nd Edition. Auckland, Whitcombe and Tombs.

Powell, A.W.B. (1950) Life history of *Austrosuccinea archeyi*, an annual snail, and its value as a post-glacial climatic indicator. *Records of the Auckland Institute and Museum* 4, 61–72.

Prince, G.A. (1967) Über Lebensweise, Fortpflanzung und Genitalorgane des terrestrischen Prosobranchiers *Cochlostoma septemspirale*. *Archiv für Molluskenkunde* 96, 1–18.

Prior, D.J., Hume, M., Varga, D. and Hess, S.D. (1983) Physiological and behavioural aspects of water balance and respiratory function in the terrestrial slug, *Limax maximus*. *Journal of Experimental Biology* 104, 111–127.

Pruvot-Fol, A. (1954) Le bulbe buccal et la symétrie des mollusques. II. *Archives de Zoologie Expérimentale et Générale* 91, 235–330.

Quick, H.E. (1933) The anatomy of British Succineæ. *Proceedings of the Malacological Society of London* 20, 295–318.

Quick, H.E. (1936) The anatomy of some African Succineæ, and of *Succinea hungarica* Hazay and *S. australis* Férussac for comparison. *Annals of the Natal Museum* 8, 19–45.

Rao, H.S. (1925) On certain succineid molluscs from the Western Ghats, Bombay Presidency. *Records of the Indian Museum* 27, 385–400.

Rees, W.J. (1964) A review of breathing devices in land operculate snails. *Proceedings of the Malacological Society of London* 36, 55–67.

Regondaud, J., Brisson, P. and de Larambergue, M. (1974) Considérations sur la morphogénése et l'évolution de la commissure viscerale chez les Gasteropodes Pulmones. *Haliotis* 4, 49–60.

Reid, D.G. (1986) *The Littorinid Molluscs of Mangrove Forests in the Indo-Pacific Region: the Genus* Littoraria. British Museum (Natural History), London.

Reid, D.G. (1988) The genera *Bembicium* and *Risellopsis* (Gastropoda: Littorinidae) in Australia and New Zealand. *Records of the Australian Museum* 40, 91–150.

Reid, D.G. (1989) The comparative morphology, phylogeny and evolution of the gastropod family Littorinidae. *Philosophical Transactions of the Royal Society of London, Biological Sciences* 324, 1–110.

Reid, D.G. (1998) Superfamily Littorinoidea. In: Beesley, P.L., Ross, G.J.B. and Wells, A. (eds) *Mollusca. The Southern Synthesis. Fauna of Australia*, Vol. 5, Part B. Australian Biological Resources Study, CSIRO Publishing, Melbourne, pp. 737–741.

Riddle, W.A. (1983) Physiological ecology of land snails and slugs. In: Russell-Hunter, W.D. (ed.) *The Mollusca*, Vol. 6, *Ecology*. Academic Press, London, pp. 431–461.

Riedel, A. (1953) Male copulatory organ deficiency in the Stylommatophora with a special reference to *Retinella nitens*. *Annales Musei Zoologici Polonici* 15, 83–100.

Rigby, J.E. (1965) *Succinea putris*: a terrestrial opisthobranch mollusc. *Proceedings of the Zoological Society of London* 144, 445–486.

Robertson, R. (1985) Four characters and the higher category systematics of gastropods. *American Malacological Bulletin Special Edition* 1, 1–22.

Robson, G.C. (1922) On the style-sac and intestine in Gastropoda and Lamellibranchia. *Proceedings of the Malacological Society of London* 15, 41–46.

Rosenberg, G. (1989) Phylogeny and evolution of terrestriality in Atlantic Truncatellidae (Prosobranchia, Gastropoda, Mollusca). PhD dissertation, Harvard University, Cambridge, Massachusetts.

Rosenberg, G. (1996a) Independent evolution of terrestriality in Atlantic truncatellid gastropods. *Evolution* 50, 682–693.

Rosenberg, G. (1996b) Anatomy and morphometric analysis of the truncatellid nervous system (Rissoacea, Gastropoda). *Journal of Molluscan Studies* 62, 507–516.

Rosenberg, G., Kuncio, G.S., Davis, G.M. and Harasewych, M.G. (1994) Preliminary ribosomal RNA phylogeny of gastropod and unionoidean bivalve mollusks. *Nautilus, Supplement* 2, 111–121.

Rosenberg, G., Tillier, S., Tillier, A., Kuncio, G.S., Hanlon, R.T., Masselot, M. and Williams, C.J. (1997) Ribosomal RNA phylogeny of selected major clades in the Mollusca. *Journal of Molluscan Studies* 63, 301–309.

Roth, B. (1996) Homoplastic loss of dart apparatus, phylogeny of the genera and a phylogenetic taxonomy of the Helminthoglyptidae (Gastropoda: Pulmonata). *Veliger* 39, 18–42.

Rudman, W.B. and Willan, R.C. (1998) Opisthobranchia introduction. In: Beesley, P.L., Ross, G.J.B. and Wells, A. (eds) *Mollusca. The Southern Synthesis. Fauna of Australia*, Vol. 5, Part B. Australian Biological Resources Study, CSIRO Publishing, Melbourne, pp. 915–942.

Ruthensteiner, B. (1997) Homology of the pallial and pulmonary cavity of gastropods. *Journal of Molluscan Studies* 63, 353–367.

Ruthensteiner, B. (1999) Nervous system development of a primitive pulmonate (Mollusca: Gastropoda) and its bearing on comparative embryology of the gastropod nervous system. *Bollettino Malacologico* 34, 1–22.

Salvini-Plawén, L.v. (1972) Zur Morphologie und Phylogenie der Mollusken: Die Beziehungen der Cauofoveata und der Solenogastres als Aculifera, als Mollusca und als Spiralia. *Zeitschrift für Wissenschaftlische Zoologie* 184, 205–394.

Salvini-Plawén, L.v. (1980) A reconsideration of systematics in the Mollusca (Phylogeny and higher classification). *Malacologia* 19, 249–278.

Salvini-Plawén, L.v. (1984) Die Cladogenese der Mollusca. *Mitteilungen der Deutschen Malakozoologischen Gesellschaft* 37, 89–118.

Salvini-Plawén, L.v. (1990) Origin, phylogeny and classification of the phylum Mollusca. *Iberus* 9, 1–33.

Salvini-Plawén, L.v. and Steiner, G. (1996) Synapomorphies and plesiomorphies in higher classification of Mollusca. In: Taylor, J. (ed.) *Origin and Evolutionary Radiation of the Mollusca*. Oxford University Press, Oxford, pp. 29–51.

Sarasin, P. and Sarasin, F. (1899) *Materialien zur Naturgeschichte der Insel Celebes. II. Band: Die Land-Mollusken von Celebes*. Kreidel's Verlag, Wiesbaden.

Schaffer, J. (1913) Über den feineren Bau und die Entwicklung des Knorpelgewebes und über verwandte Formen der Stützsubstanz. IV. *Zeitschrift für Wissenschaftliche Zoologie* 105, 280–347.

Schaffer, J. (1930) Die Stützgewebe. In: von Möllendorf, W. (ed.) *Handbuch der Mikroskopischen Anatomie des Menschen*, 2 (Part 2), pp. 1–390.

Schileyko, A.A. (1979) Sistema otryada Geophila (= Helicida) (Gastropoda Pulmonata). *Trudy Zoologicheskogo Instituta Akademii Nauk SSSR* 80, 44–69. [English translation: The system of the order Geophila (= Helicida) (Gastropoda Pulmonata) by A.A. Schileyko (Boss, K.J. and Jacobson, M.K., eds) Department of Mollusks, Harvard University, Special Occasional Publication 6, 1–45, 1985.]

Schileyko, A.A. (1984) Nazemnye mollyuski podotryada Pupillina fauny SSSR (Gastropoda, Pulmonata, Geophila). *Fauna SSSR, Mollyuski* 3(3). Akademija Nauk, Leningrad.

Scott, B. (1996) Phylogenetic relationships of the Camaenidae (Pulmonata: Stylommatophora: Helicoidea). *Journal of Molluscan Studies* 62, 65–73.

Simone, L.R.L. and Leme, J.L.M. (1998) Two new species of Megalobulimidae (Gastropoda, Strophocheiloidea) from North São Paulo, Brazil. *Iheringia, Zoologia* 85, 189–203.

Simroth, H. (1913) Uber die von Herrn Prof. Voeltzkow auf Madagaskar und irt Ostafrika erbeuteten Vaginuliden, nebstverwandtem Zmaterial von ganz Afrika. *Voeltzkow Reise in Ostafrika in der Jahren 1903–1905* 3, 129–216.

Simroth, H. (1918) Über einige Nacktschnecken vom Malayischen Archipel von Lombok an östwarts bis zu Gesellschafts-Inseln. *Abhandlungen der Senckenbergischen Naturforschenden Gesellschaft* 35, 261–302.

Simroth, H. and Hoffmann, H. (1908–28) Pulmonata. In: *H.G. Bronn's Klassen und Ordnungen des Tier-Reichs. III Mollusca, 2 Gastropoda Buch 2.*

Sirgel, W.F. (1990) Phylogeny of the Arionidae and other pulmonates based on the terminal genital ducts (Mollusca). *Annals of the Natal Museum* 31, 233–238.

Smith, B.J. and Stanisic, J. (1998) Pulmonata introduction. In: Beesley, P.L., Ross, G.J.B. and Wells, A. (eds) *Mollusca. The Southern Synthesis. Fauna of Australia*, Vol. 5, Part B. Australian Biological Resources Study, CSIRO Publishing, Melbourne, pp. 1037–1061.

Smith, L.H. (1981) Quantified aspects of pallial fluid and its affects on the duration of locomotor activity in the terrestrial gastropod *Triodopsis albolabris*. *Physiological Zoology* 54, 407–414.

Solem, A. (1956) The helicoid cyclophorid molluscs of Mexico. *Proceedings of the Academy of Natural Sciences of Philadelphia* 108, 41–59.

Solem, A. (1959) Systematics and zoogeography of the land and fresh-water Mollusca of the New Hebrides. *Fieldiana, Zoology* 43, 1–359.

Solem, A. (1961) A preliminary review of the pomatiasid land snails of Central America (Mollusca, Prosobranchia). *Archiv für Molluskenkunde* 90, 191–213.

Solem, A. (1964) *Amimopina*, an Australian enid land snail. *Veliger* 6, 115–120.

Solem, A. (1972a) Microarmature and barriers in the apertures of land snails. *Veliger* 15, 81–87.

Solem, A. (1972b) *Tekoulina*, a new viviparous tornatellinid landsnail from Rarotonga, Cook Islands. *Proceedings of the Malacological Society of London* 40, 93–114.

Solem, A. (1973) Convergence in pulmonate radulae. *Veliger* 15, 165–171.

Solem, A. (1974) *The Shell Makers*. John Wiley & Sons, New York.

Solem, A. (1976) *Endodontid Land Snails from Pacific Islands. Part 1: Family Endodontidae*. Field Museum of Natural History, Chicago.

Solem, A. (1978) Classification of the land Mollusca. In: Fretter, V. and Peake, J. (eds) *Pulmonates*, Vol. 2A, *Systematics, Evolution and Ecology*. Academic Press, London, pp. 49–97.

Solem, A. (1979a) Biogeographic significance of land snails, Paleozoic to Recent. In: Gray, J. and Boucot, A.J. (eds) *Historical Biogeography, Plate Tectonics, and the Changing Environment*. Oregon State University Press, Corvallis, pp. 277–287.

Solem, A. (1979b) A theory of land snail biogeographic patterns through time. In: van der Spoel, S., van Bruggen, A.C. and Lever, J. (eds) *Pathways in Malacology*. Bohn, Scheltema & Holkema, Utrecht, and W. Junk b.v., The Hague, pp. 225–249.

Solem, A. (1981) Land-snail biogeography: a true snail's pace of change. (collective references pp. 539–584) In: Nelson, G. and Rosen, D.E. (eds) *Vicariance Biogeography. A Critique*. Columbia University Press, New York, pp. 197–221.

Solem, A. (1984) A world model of land snail diversity and abundance. In: Solem, A. and van Bruggen, A.C. (eds) *World-wide Snails: Biogeographical Studies on Non-marine Mollusca.* Brill, Leiden, pp. 6–22.

Solem, A. (1985) Origin and diversification of pulmonate land snails. In: Trueman, E.R. and Clarke, M.R. (eds) *The Mollusca*, Vol. 10, *Evolution.* Academic Press, New York, pp. 269–293.

Solem, A. (1988) Non-camaenid land snails of the Kimberley and Northern Territory, Australia. I. Systematics, affinities and ranges. *Invertebrate Taxonomy* 4, 455–604.

Solem, A. and Yochelson, E.L. (1979) North American Paleozoic land snails, with a summary of other Paleozoic non-marine snails. *United States Geological Survey, Professional Paper* 1072, 1–42.

Solem, A., Girardi, E.L., Slack-Smith, S. and Kendrick, G.W. (1982) *Austro-assiminea letha*, gen. nov., sp. nov., a rare and endangered prosobranch snail from south-western Australia (Mollusca: Prosobranchia: Assimineidae). *Journal of the Royal Society of Western Australia* 65, 119–129.

Solem, A., Tillier, S. and Mordan, P. (1984) Pseudo-operculate pulmonate land snails from New Caledonia. *Veliger* 27, 193–199.

Standen, R. (1917) On the calcareous eggs of terrestrial Mollusca. *Journal of Conchology* 15, 154–167.

Stanisic, J. (1998) Suborder Stylommatophora. In: Beesley, P.L., Ross, G.J.B. and Wells, A. (eds) *Mollusca. The Southern Synthesis. Fauna of Australia*, Vol. 5, Part B. Australian Biological Resources Study, CSIRO Publishing, Melbourne, pp. 1079–1080.

Starobogatov, Y.I. (1976) Composition and systematic position of marine pulmonate molluscs. *Soviet Journal of Marine Biology* 4, 206–212.

Stringer, B.L. (1963) Embryology of the New Zealand Onchidiidae and its bearing on the classification of the group. *Nature* 197, 621–622.

Suter, H. (1913) *Manual of the New Zealand Mollusca.* Government Printer, Wellington, (1915, Atlas of 72 plate).

Suvorov, A.N. (1993) Some aspects of functional morphology of the aperture in the Pupillina suborder (Gastropoda Pulmonata). *Ruthenica* 3, 141–152.

Suvorov, A.N. and Shileyko, A.A. (1991) Functional morphology of the apertural armature in the Lauriinae subfamily (Gastropoda, Orculidae) and some problems of the systematics of the group. *Ruthenicia* 1, 67–80.

Swofford, D.L. (1998) PAUP*. *Phylogenetic Analysis using Parsimony (*and Other Methods. Version 4.* Sinauer Associates, Sunderland, Massachusetts.

Thiele, J. (1902) Die systematische Stellung der Solenogastren und die Phylogenie der Mollusken. *Zeitschrift für Wissenschaftliche Zoologie* 72, 249–466.

Thiele, J. (1910) Über die Anatomie von *Hydocena cattoroensdis* Pf. *Abhandlungen der Senckenbergischen Naturforschenden Gesellschaft* 32, 351–358.

Thiele, J. (1927) Über die schneckenfamilie Assimineidae. *Zoologische Jahrbücher, Abteilung für Systematik* 53, 113–146.

Thiele, J. (1929–35) Handbuch der systematischen Weichtierkunde. Gustav Fischer, Jena.

Thomaz, D., Guiller, A. and Clarke, B. (1996) Extreme divergence of mitochondrial DNA within species of pulmonate land snails. *Proceedings of the Royal Society of London, Biological Sciences* 263, 363–368.

Thomé, J.W. (1975) Os gêneros da familia Veronicellidae nas Américas (Mollusca; Gastropoda). *Iheringa, Zoologia* 48, 3–56.

Thomé, J.W. (1976) Revisão do gênero *Phyllocaulis* Colosi, 1922 (Mollusca; Veronicellidae). *Iheringa, Zoologia* 49, 67–90.

Thomé, J.W. (1989) Annotated and illustrated preliminary list of the Veronicellidae (Mollusca: Gastropoda) of the Antilles, and Central and North America. *Journal of Medical and Applied Malacology* 1, 11–28.

Thompson, F.G. (1967) A new cyclophorid land snail from the West Indies (Prosobranchia), and the discussion of a new subfamily. *Proceedings of the Biological Society of Washington* 80, 13–18.

Thompson, F.G. (1969) Some Mexican and Central American land snails of the family Cyclophoridae. *Zoologica* 54, 35–77.

Thompson, F.G. (1978) A new genus of operculate land snails from Hispaniola with comments on the status of family Annulariidae. *Nautilus* 92, 41–54.

Thompson, F.G. (1980) Proserpinoid land snails and their relationships within the Archaeogastropoda. *Malacologia* 20, 1–33.

Thompson, F.G. and Dance, S.P. (1983) Non-marine mollusks of Borneo II Pulmonata: Pupillidae, Clausiliidae. III Prosobranchia: Hydrocenidae, Helicinidae. *Bulletin of the Florida State Museum, Biological Sciences* 29, 101–152.

Tielecke, H. (1940) Anatomie, Phylogenie und Tiergeographie der Cyclophoriden. *Archiv für Naturgeschicte* 9, 317–371.

Tillier, S. (1981) South American and Juan Fernandez succineid slugs (Pulmonata). *Journal of Molluscan Studies* 47, 125–146.

Tillier, S. (1982) Structures respiratoires et excrétrices secondaries des limaces (Gastropoda: Pulmonata: Stylommatophora). *Bulletin de la Société Zoologique de France* 108, 9–19.

Tillier, S. (1983) Structures respiratoires et excrétrices secondaries des limaces (Gastropoda: Pulmonata: Stylommatophora). *Bulletin de la Société Zoologique de France* 108, 9–19.

Tillier, S. (1984a) A new mountain *Platevindex* from Philippine Islands (Pulmonata: Onchidiidae). *Journal of Molluscan Studies, Supplement* 12A, 198–202.

Tillier, S. (1984b) Relationships of gymnomorph gastropods (Mollusca: Gastropoda). *Zoological Journal of the Linnean Society* 82, 345–362.

Tillier, S. (1984c) Patterns of digestive tract morphology in the limacization of helicarionid, succineid and athoracophorid snails and slugs (Pulmonata: Stylommatophora). *Malacologia* 25, 173–192.

Tillier, S. (1989) Comparative morphology, phylogeny and classification of land snails and slugs (Gastropoda: Pulmonata: Stylommatophora). *Malacologia* 30, 1–303.

Tillier, S. and Ponder, W.F. (1992) New species of *Smeagol* from Australia and New Zealand, with a discussion on the affinities of the genus (Gastropoda: Pulmonata). *Journal of Molluscan Studies* 58, 135–155.

Tillier, S., Masselot, M., Philippe, H. and Tillier, A. (1992) Phylogénie moléculaire des Gastropoda (Mollusca) fondée sur le séquençage partiel de l'ARN ribosomique 28 S. *Comptes Rendus de l'Académie des Sciences Paris* 314, 79–85.

Tillier, S., Masselot, M., Guerdoux, J. and Tillier, A. (1994) Monophyly of major gastropod taxa tested from partial 28S rRNA sequences, with emphasis on Euthyneura and hot-vent limpets Peltospiroidea. *Nautilus, Supplement* 2, 122–140.

Tillier, S., Masselot, M. and Tillier, A. (1996) Phylogenetic relationships of the pulmonate gastropods from rRNA sequences, and tempo and age of the stylommatophoran radiation. In: Taylor, J.D. (ed.) *Origin and Evolutionary Radiation of the Mollusca*. Oxford University Press, Oxford, pp. 267–284.

Tompa, A.S. (1976) A comparative study of the ultrastructure and mineralogy of land snail eggs. *Journal of Morphology* 150, 861–888.

Tompa, A.S. (1979a) Studies on the reproductive biology of gastropods: part 1. The systematic distribution of egg retention in the subclass Pulmonata (Gastropoda). *Journal of the Malacological Society of Australia* 4, 113–120.

Tompa, A.S. (1979b) Oviparity, egg retention and ovoviviparity in pulmonates. *Journal of Molluscan Studies* 45, 155–160.

Tompa, A.S. (1980) Studies on the reproductive biology of gastropods: part III. Calcium provision and the evolution of terrestrial eggs among gastropods. *Journal of Conchology* 30, 145–154.

Tompa, A.S. (1984) Land snails (Stylommatophora). In: Wilbur, K.M. (ed.) *The Mollusca. Vol. 7. Reproduction*. Academic Press, New York, pp. 47–140.

Tracey, S., Todd, J.A. and Erwin, D.H. (1993) Mollusca: Gastropoda. In: Benton, M.J. (ed.) *The Fossil Record 2*. Chapman & Hall, London, pp. 131–167.

Turner, R.D. and Clench, W.J. (1972) Land and freshwater snails of Savo Island, Solomons, with anatomical descriptions (Mollusca: Gastropoda). *Steenstrupia* 2, 207–232.

van Benthem Jutting, W.S.S. (1948) Systematic studies on the non-marine Mollusca of the Indo-Australian Archipelago. 1. Critical revision of the Javanese operculate land-shells of the families Hydrocenidae, Helicinidae, Cyclophoridae, Pupinidae and Cochlostomatidae. *Treubia* 19, 539–604.

van Bruggen, A.C. (1995) Biodiversity of the Mollusca: time for a new approach. In: van Bruggen, A.C., Wells, S.M. and Kemperman, Th.C.M. (eds) *Biodiversity and Conservation of the Mollusca*. Backhuys Publishers, Oegstgeest-Leiden, pp. 1–19.

van Mol, J.-J. (1967) Evolution phylogénétique du ganglion cérébroïde des Gastéropodes Pulmonées (Mollusques). *Mémoires de l'Académie Royale de Belgique, Classe des Sciences* 37, 1–168.

van Mol, J.-J. (1970) Révision des Urocyclidae (Mollusca, Gastropoda, Pulmonata). Première partie. *Annales du Musée Royal de l'Afrique Centrale, series in-8⁰, Sciences Zoologiques* 180, 1–234.

van Mol, J.-J. (1971) Notes anatomiques sur les Bulimulidae (Mollusques, Gasteropodes, Pulmones). *Annales Societe Royale Zoologique de Belgique* 101, 183–226.

van Mol, J.-J. (1974) Evolution phylogénétique du ganglion cérébroide chez les gastéropodes pulmones. *Haliotis* 4, 77–86.

Varga, A. (1984) The *Cochlostoma* genus (Gastropoda, Prosobranchiata) in Yugoslavia. I. Anatomical notes. *Miscellanea Zoologica Hungarica* 2, 51–64.

Villalobos, C., Monge-Nájera, J., Barrientos, Z. and Franco, J. (1995) Life cycle and field abundance of the snail *Succinea costaricana* (Stylommatophora: Succineidae), a tropical agricultural pest. *Revista de Biologia Tropical* 43, 181–188.

Visser, M.H.C. (1973) The ontogeny of the reproductive system of *Gonaxis gwandaensis* (Preston) (Pulmonata, Streptaxidae) with special reference to the phylogeny of the spermatic conduits of the Pulmonata. *Annale Universiteit van Stellenbosch* 48, 1–79.

Visser, M.H.C. (1977) The morphology and significance of the spermoviduct and prostate in the evolution of the reproductive system of the Pulmonata. *Zoologica Scripta* 6, 43–54.

Visser, M.H.C. (1981) Monauly versus diauly as the original condition of the reproductive system of Pulmonata and its bearing on the interpretation of the terminal ducts. *Zeitschrift für Zoologie, Systematik und Evolutionsforschung* 19, 59–68.

Vlés, F. (1913) Observations sur la locomotion d'*Otina otis* Turt. Remarques sur la progression des gasteropodes. *Bulletin de la Société Zoologique de France* 38, 242–250.

von Ihering, H. (1892) Morphologie und Systematik des Genitalapparates von *Helix. Zeitschrift für Wissenschaftliche Zoologie* 54, 386–520.

Vorwohl, G. (1961) Zur Funktion der Exkretionsorgane von *Helix pomatia* (L.) und *Archachatina ventriosa* (Gould). *Zeitschrift für Vergleichende Physiologie* 45, 12–49.

Wächtler, W. (1935) Beiträge zur Anatomie der stylommatophoren Lungenschnecken. Über die Gattung *Ferussacia* Risso. *Zoologische Jahrbücher, Abteilung für Systematik, Ökologie und Geographie der Tiere* 67, 117–194.

Wade, C.M., Mordan, P.B. and Clarke, B.C. (1998) Towards a molecular phylogeny for the pulmonate land snails. In: Bieler, R. and Mikkelsen, P.M. (eds) *Abstracts, World Congress of Malacology, Washington, DC, 1998*. Unitas Malacologica, Chicago, p. 346.

Wade, C.M., Mordan, P.B. and Clarke, B.C. (2000) A phylogeny of the land snails (Gastropoda: Pulmonata). *Proceedings of the Royal Society of London, B* 268, 413–422.

Wagner, A. (1907–11) Die Familie der Helicinidae. In: Küster, H.C. (ed.) *Systematisches Conchylien-Cabinet von Martini und Chemnitz 1*, 18. Nurnberg.

Watson, H. (1920) The affinities of *Pyramidula, Patulastra, Acanthinula* and *Vallonia. Proceedings of the Malacological Society of London* 14, 6–30.

Watson, H. (1923) Masculine deficiencies in the British Vertigininae. *Proceedings of the Malacological Society of London* 15, 270–280.

Watson, H. (1925) The South African species of the molluscan genus *Onchidella. Annals of the South African Museum* 20, 237–307.

Webb, G.R. (1948) The mating of *Stenotrema fraternum. Nautilus* 62, 8–12.

Webb, G.R. (1950) Shell-spinules as defense structures. *Nautilus* 63, 107.

Webb, G.R. (1953) Anatomical studies on some Midwestern Succinidae and two new species. *Journal of the Tennessee Academy of Science* 28, 213–220.

Webb, G.R. (1961) The phylogeny of American land snails with emphasis on the Polygyridae, Arionidae and Ammonitellidae. *Gastropodia* 1, 31–44.

Webb, G.R. (1968) Observations on the sexology of the endodontid land-snail, *Anguispira alternata* (Say). *Gastropodia* 1, 66–67.

Webb, G.R. (1974) The sexual evolution of the polygyrid snails. *Gastropodia* 1, 85–90.

Webb, G.R. (1977a) On the sexology of *Catinella* (*Mediappendix*) *avara* (Say) or *C.* (*M.*) *vermeta* (Say). *Gastropodia* 1, 100–102.

Webb, G.R. (1977b) Some sexologic observations on *Oxyloma retusa* Lea. *Gastropodia* 1, 102–104.

Webb, G.R. (1977c) The comparative sexology of several Succineidae. *Gastropodia* 1, 105–107.

Webb, G.R. (1983) On the sexology of *Mesodon kiowaensis* (Simpson) Pulmonata, Polygyridae, Polygyrinae. *Gastropodia* 2, 19–20.

Weber, L. (1924) Die Mantel- und Geschlechtsorgane von *Cyclophorus ceylanicus* (Sowerby). *Jenaische Zeitschrift für Naturwissenschaft* 60, 397–438.

Weiss, K.R. and Kupfermann, I. (1976) Homology of the giant serotonergic neurons (metacerebral cells) in *Aplysia* and pulmonate molluscs. *Brian Research* 117, 33–49.

Weiss, M. (1968) Zur embryonalen und postembryonalen Entwicklung des Mitteldarmes bei Limaciden und Arioniden. *Revue Suisse de Zoologie* 75, 157–226.

Wenz, W. (1939–44) *Gastropoda, 1.* In: Schindewolf, O.H. (ed.) *Handbuch der Paläozoologie, 6 (Lief) 1–7.* Borntraeger, Berlin.

Winnepenninckx, B., Steiner, G., Backeljau, T. and de Wachter, R. (1998) Details of gastropod phylogeny inferred from 18S rRNA sequences. *Molecular Phylogenetics and Evolution* 9, 55–63.

Yonge, C.M. (1930) The crystalline style of the Mollusca and a carnivorous habit cannot normally co-exist. *Nature* 125, 444–445.

Yonge, C.M. (1932) Notes on feeding and digestion in *Pterocera* and *Vermetus* with a discussion of the occurrence of the crystalline style in the Gastropoda. *Scientific Report of the Great Barrier Reef Expedition 1928–29* 1, 259–281.

Yonge, C.M. (1952) The mantle cavity in *Siphonaria alternata* Say. *Proceedings of the Malacological Society of London* 29, 190–199.

Zilch, A. (1959–60) Gastropoda, Euthyneura. In: Wenz, W. *Gastropoda. Handbuch der Paläozoologie 6.* Gebrüder Borntraeger, Berlin.

Appendix 1.1. Morphological Characters (and their States) Used in Analysis of Phylogenetic Relationships in the Gastropoda

Alimentary characters

1. *Radular membrane structure.* 0 – stereoglossate; 1 – flexoglossate; 2 – radular membrane absent.
2. *Odontophoral cartilages.* 0 – two pairs; 1 – three or more pairs, or two pairs plus accessory cartilages; 2 – one pair; 3 – four single cartilages; 4 – cartilages vestigial or absent.
3. *Radular diverticulum.* 0 – present; 1 – absent.
4. *Radular dentition.* 0 – docoglossate; 1 – rhipidoglossate; 2 – taeniglossate; 3 – rachiglossate; 4 – ptenoglossate; 5 – toxoglossate; 6 – teeth absent.
5. *Snout.* 0 – snout well developed, separated from foot by deep cleft, without proboscis; 1 – snout weakly developed, separated from foot by shallow but none the less distinct cleft, without proboscis; 2 – snout greatly reduced, separated from sole by indistinct cleft, without proboscis; 3 – snout well developed, separated from foot by deep cleft, with pleurembolic proboscis intraembolic; 4 – snout well developed, separated from foot by deep cleft, with acrembolic proboscis; 5 – snout well developed, separated from foot by deep cleft, with pseudoacrembolic proboscis; 6 – snout well developed, separated from foot by deep cleft, with intraembolic proboscis.
6. *Jaw.* 0 – paired dorsal, paired lateral plates, fused; 1 – paired lateral plates and single dorsal plate; 2 – paired lateral plates; 3 – single plate derived from fused lateral plates; 4 – single plate derived from fused lateral plates, with accessory dorsal plate; 5 – modified into stylet; 6 – absent.
7. *Oesophageal folds.* 0 – dorsal and ventral folds; 1 – dorsal folds only; 2 – dorsal folds only, partially or completely separated as part of the gland of Leiblein or poison gland; 3 – absent.
8. *Oesophageal glands.* 0 – present, non-papillate; 1 – present, papillate; 2 – present, separated as the gland of Leiblein or poison gland; 3 – present, as two distinct lobes of differing histology; 4 – absent.
9. *Salivary glands.* 0 – present, without ducts; 1 – present, with ducts; 2 – absent.
10. *Gastric caecum.* 0 – present; 1 – absent.
11. *Stomach.* 0 – gastric region with gastric shield and sac containing protostyle; 1 – gastric region with gastric shield and sac containing crystalline style; 2 – gastric region with gastric shield or vestiges of same, style absent; 3 – gastric region with gastric shield elaborated into gizzard with plates (gizzard may be displaced anteriad); 4 – gastric region simple, without gizzard or plates.
12. *Anterior intestinal* loop. 0 – present; 1 – absent.
13. *Rectum.* 0 – passes through or openly widely into heart ventricle within pericardium; 1 – passes through pericardium; 2 – not passing through pericardium.

Circulatory and renal characters

14. *Excretory organs.* 0 – left and right ones present, both extra-pallial; 1 – left and right ones present, left one pallial, right one extra-pallial; 2 – left one only, extra-pallial; 3 – left one only, pallial.
15. *Number of auricles.* 0 – one pair; 1 – one auricle.
16. *Circulation patterns.* 0 – nephridium(a) → ctenidium(a) → auricle(s), with or without blood passing through a nephridial gland (ctenidia may be absent); 1 – mantle → nephridium(a) → heart (secondary gill may be present).
17. *Ctenidium.* 0 – bipectinate, without skeletal rods and bursicles; 1 – bipectinate, with skeletal rods but without bursicles; 2 – bipectinate, with skeletal rods and bursicles; 3 – monopectinate, without skeletal rods and bursicles; 4 – monopectinate, with skeletal rods but without bursicles; 5 – monopectinate, with skeletal rods and bursicles; 6 – greatly modified or secondary, without skeletal rods and bursicles, or absent.
18. *Pallial cavity.* 0 – widely open; 1 – opening narrowed to pneumostome; 2 – opening narrowed to contractile pneumostome; 3 – reduced, without opening to exterior.
19. *Pallial ciliary tracts.* 0 – absent; 1 – present; 2 – secondarily lost.

Cytological characters

20. *Haploid chromosome number.* 0 – <11; 1 – 11 to 15; 2 – 16 to 20; 3 – 21 to 25; 4- 26 to 30; 5 – 31 to 35; 6 – 36 to 40; 7 – 41 to 45; 8 – 46 to 50; 9 – >50; ? – no data.
21. *Sex chromosomes.* 0 – absent; 1 – present.

Development and larval characters

22. *Larval development.* 0 – trochophore and lecithotrophic veliger larva free-living; 1 – trochophore development in the egg, lecithotrophic veliger larva free; 2 – trochophore in the egg, planktotrophic veliger larva free; 3 – no trochophore, veliger stages completed in egg; 4 – no trochophore or veliger stages, direct development in the egg.

Appendix 1.1. *Continued*

23. *Nurse egg cells.* 0 – absent; 1 – present.
24. *Echinospira larvae.* 0 – absent; 1 – present.
25. *Larval operculum.* 0 – present; 1 – absent.

Muscle and pedal characters

26. *Adult shell muscles.* 0 – left and right muscles present, divided into several bundles; 1 – left and right muscles present; 2 – left muscle only; 3 – left muscle only, at most vestigial strands to tentacles.
27. *Suprapedial gland.* 0 – present; 1 – absent.
28. *Metapodial mucous gland.* 0 – absent; 1 – present.

Nervous system and sensory characters

29. *Inferior tentacles.* 0 – absent; 1 – present.
30. *Eye type* (when present). 0 – open vesicle or pit; 1 – vesicle or pit, filled with vitreous mass; 2 – closed vesicle with lens.
31. *Eye position.* 0 – embedded centrally or adjacent to base of cephalic tentacle, often closely associated with cerebral ganglion; 1 – on or in bulbous swelling on outer base of cephalic tentacle; 2 – on short accessory pedicle, adjacent to cephalic tentacle when latter are present; 3 – at tip of more or less elongate, retractile but non-invaginable cephalic tentacle; 4 – at tip of more or less elongate, retractile and invaginable cephalic tentacle; 5 – at tip of solid tentacle that can be retracted (but not invaginated).
32. *Cephalic tentacles.* 0 – present; 1 – absent.
33. *Tentacular nerve.* 0 – simple; 1 – bifurcated.
34. *Visceral loop of central nervous system.* 0 – strephoneurous; 1 – euthyneurous, but rather long; 2 – euthyneurous, highly concentrated.
35. *Cerebral gland.* 0 – absent; 1 – present.
36. *Cerebral–pleural ganglion association of central nervous system.* 0 – hypathroid, with pleural ganglia close to pedal ganglia; 1 – dystenoid with pleural ganglia equally distant from cerebral and pedal ganglia; 2 – epiathroid, with pleural ganglia close to cerebral ganglia.
37. *Pedal element of central nervous system.* 0 – not concentrated, in form of cords with numerous commissures; 1 – concentrated into ganglia, with 1–2 connectives; 2 – concentrated into ganglia, with 1–2 connectives and lateral nerves.
38. *Parietal ganglia of central nervous system visceral loop.* 0 – absent; 1 – present.
39. *Procerebrum of cerebral ganglia.* 0 – absent (anlage comprising tentacular ganglion); 1 – present, containing large cells; 2 – present, containing numerous small (globineurons) and some large cells; 3 – present, containing only small cells; 4 – modified to form a rhinophoral ganglion.
40. *Statocyst position.* 0 – lateral to pedal nerve cords or ganglia; 1 – dorsal or just posterior to pedal ganglia; 2 – anterior to pedal ganglia.
41. *Statocysts.* 0 – with several statoliths; 1 – with one statolith; 2 – absent.
42. *Cephalic lappets.* 0 – absent; 1 – present.
43. *Tentacle neck lobes.* 0 – absent; 1 – present.
44. *Epipodial skirt.* 0 – absent; 1 – present.
45. *Epipodial tentacles.* 0 – absent in adults; 1 – multiple pairs in adults; 2 – one pair in adults.
46. *Epipodial sense organs.* 0 – absent; 1 – present.

Reproductive characters

47. *Hermaphroditism of gonad.* 0 – sexes separate (dioecious); 1 – protandic hermaphrodite; 2 – simultaneous hermaphrodite.
48. *Fertilization.* 0 – external; 1- internal.
49. *Gonopericardial duct.* 0 – present; 1 – absent.
50. *Pallial gonoduct.* 0 – absent, gametes released into posterior of mantle cavity; 1 – present, an open groove in mantle cavity; 2 – present, a closed tube in mantle cavity; 3 – present, a closed tube in body cavity.
51. *Seminal duct and bursa copulatrix.* 0 – absent; 1 – bursa copulatrix a sac separate from genital duct, orifice open to posterior of mantle cavity, seminal tract confined to renal gonoduct; 2 – bursa copulatrix a diverticular sac associated with renal gonoduct, orifice at posterior of mantle cavity,

seminal tract confined to renal gonoduct; 3 – bursa copulatrix a sac separate from genital duct, orifice open to posterior of mantle cavity, seminal tract fused with pallial gonoduct; 4 – bursa copulatrix a sac associated with renal gonoduct, widely open to posterior of mantle cavity, seminal tract fused with pallial gonoduct; 5 – bursa copulatrix a diverticular sac associated with renal gonoduct, seminal duct separate from pallial gonoduct and having its orifice at anterior of mantle cavity (orifice may be contiguous with female orifice); 6 – bursa copulatrix a diverticular sac associated with renal gonoduct, seminal tract fused with pallial gonoduct; 7 – bursa copulatrix a diverticular sac associated with anterior of pallial gonoduct, seminal duct separate from pallial gonoduct and having its orifice at anterior of mantle cavity contiguous with female orifice; 8 – bursa copulatrix a diverticular sac associated with anterior of pallial gonoduct, seminal tract fused with pallial gonoduct; 9 – bursa copulatrix, as a diverticular sac, secondarily absent, its function assumed by pallial gonoduct lumen.

52. *Albumen gland.* 0 – absent, ovum supplied with albumen from ovaries; 1 – present, traversed by fertilized zygote; 2 – present, separated from gonoduct, not traversed by fertilized zygote.

53. *Genital orifice.* 0 – male groove or duct terminus in cephalic region, female orifice associated with pallial cavity or pallial opening; 1 – male terminal genitalia absent, common orifice associated with pallial cavity or pallial opening; 2 – male groove or duct terminus in cephalic region, female orifice extra-pallial but separate from male orifice and opening in lateral body wall; 3 – male groove or duct terminus in cephalic region, female orifice extra-pallial, through body wall but separate from male orifice and displaced to posterior with detorsion of viscera; 4 – male and female orifices fused or nearly so in cephalic region.

54. *Chalazae.* 0 – absent; 1 – present.

55. *Penis.* 0 – when present, external; 1 – when present, internal.

56. *Extrapallial sperm duct.* 0 – open groove or absent; 1 – closed duct.

57. *Spermatophores.* 0 – absent; 1 – present.

58. *Eusperm nucleus.* 0 – without basal invagination, or with basal invagination but not penetrated by axonome/centriolar; 1 – with basal invagination housing axonome/centriolar; 2 – tubular, not penetrated at all by axonome; 3 – tubular, penetrated in basal part by axonome; 4 – tubular, penetrated fully by axonome; 5 – tubular, penetrated fully by axonome and partially by mitochondia; 6 – coiled round mitochondria; 7 – partially enclosed by mitochondria.

59. *Eusperm nucleus coiling.* 0 – absent; 1 – present.

60. *Eusperm mid-piece coarse fibres.* 0 – absent; 1 – present.

61. *Euperm mid-piece glycogen helices.* 0 – absent; 1 – one present; 2 – two present; 3 – three present; 4 – four present.

62. *Eusperm mid-piece mitochondrial form.* 0 – spherical, clustered around centrioles; 1 – spherical, clustered around centrioles and axonome; 2 – elongate strands around axonome, unmodified cristae; 3 – elongate strands around axonome, cristae in form of parallel plates; 4 – cylindrical around centriolar derivative, cristae unmodified; 5 – cylindrical around centriolar derivative, cristae as lamellae; 6 – cylindrical around centriolar derivative, cristae as parallel plates; 7 – cylindrical around axonome, cristae unmodified; 8 – cylindrical around axonome, cristae as lamellae; 9 – cylindrical around axonome, cristae as parallel plates.

63. *Eusperm mid-piece coiling.* 0 – absent; 1 – present.

64. *Eusperm mid-piece paracrystalline material.* 0 – absent; 1 – present, between mitochondrial strands; 2 – present, between mitochondrial strands and around perimeter of mid-piece; 3 – present, around perimeter of mid-piece; 4 – present, around axonome and around perimeter of mid-piece.

65. *Eusperm glycogen piece.* 0 – absent; 1 – present as tracts of glycogen granules around axonome; 2 – present as tracts of glycogen but no axonome.

66. *Eusperm end-piece.* 0 – axonome surrounded by plasma membrane; 1 – plasma membrane only; 2 – absent.

67. *Eusperm acrosome.* 0 – thick walled, with axial rod; 1 – thin walled, with axial rod; 2 – thin walled, with axial rod and apical bleb; 3 – thin walled, with axial rod and apical vesicle; 4 – thin walled, with rudimentary pedestal but no apical vesicle; 5 – thin walled, with rudimentary pedestal and apical vesicle; 6 – thin walled, with well-developed pedestal and apical vesicle; 7 – absent.

68. *Parasperm.* 0 – absent; 1 – present, bipolar tailed neritopsine type; 2 – present, with head and tail tuft; 3 – present, with attached euspermatozoa; 4 – present, vermiform; 5 – present, as 'nurse cells'.

Appendix 1.1. *Continued*

Shell characters

69. *Shell structure.* 0 – shell present, with nacreous layer; 1 – shell present, nacreous layer absent; 2 – shell greatly reduced or absent.
70. *Intersected, cross-platy shell structure.* 0 – absent; 1 – present.
71. *Cross-cone lamellar shell structure.* 0 – absent; 1 – present.
72. *Adult operculum.* 0 – flexiclaudent, multispiral; 1 – rigiclaudent, multispiral; 2 – rigiclaudent, paucispiral; 3 – rigiclaudent, concentric; 4 – absent.

Appendix 1.2. Character States for the Taxa Included in the Parsimony Analysis of Phylogenetic Relationships in Gastropoda

	1	2	3	4	5	6	7	8	9	10	11	12	13	14	15	16	17
Patellogastropoda	0	0,1	0	0	0	0	0	0	1	1	4	0	*0,2	0,1	1	0	0,6
Cocculiniforma	1	0,2	0	*0,2	0,1	*2,6	0,1	0	0,2	0,1	2,4	0,1	*0,2	*1,3	1	0	*3,6
Neritopsina	1	0,1	0	1	0,1	6	0	3	1,2	0,1	2	0	0,2	1	0,1	0	0,6
Vetigastropda	1,2	0,1	0	1,6	0	2,6	0	0,1	0,1	0,1	0,2	0	0	*0,3	0	0	*2,6
Neomphaloidea	1	0,2	0	1	0,1	2	0	0	0	0	2	0	0,2	3	1	0	1
Sequenzioidea	1	0	0	1	0,1	2	0	0,1	0	0	2	0	2	1	1	0	5
Ampullarioidea	1	2	1	2	0	2,3	1	0	1	1	1,2	0,1	2	2	1	0	3
Cyclophoroidea	1	2	1	2	0	2,6	0	0	1	0,1	0	0,1	2	2,3	1	0	6
Stromboidea	1	2	1	2	0	2	1	0	1	0	1	1	2	2	1	0	4
Calyptraeoidea	1	2	1	2	3	2,6	1	0	1	0,1	1,2	1	2	2	1	0	4
Vetutionoidea	1	2	1	2,3	4	2,3	1	0	1	1	2	1	2	2	1	0	4
Cypraeoidea	1	2	1	2	0	5	1	0	1,2	0,1	1,2	1	2	2	1	0	4
Tonnoidea	1	2	1	2	3	2,6	1	0	1	1	2,4	1	2	2	1	0	4
Cerithioidea	1	2	1	2	4	2	1	0	1	0,1	*0,2	0,1	2	2,3	1	0	4
Janthinoidea	1	3	1	6	4	2,3	1	0	1	0	2	1	2	2	1	0	4,6
Eulimoidea	1,2	2	1	*3,6	4	2,6	1	0	1	1	4	1	2	2	1	0	4,6
Triphoroidea	1	2	1	*1,4	4	2,3	1	0	1	0	4	1	2	2	1	0	4
Littorinoidea	1	2	1	2	0	2,6	1	0	1	0,1	1,2	0,1	2	2,3	1	0	4,6
Rissooidea	1	2	1	2	0	2,6	1	0	1	0,1	1,2	0,1	2	2,3	1	0	4,6
Muricoidea	1,2	2	1	3,6	3	6	2	2	1	0,1	*1,4	1	2	2	1	0	4
Conoidea	1	2,4	1	*2,5	6	6	2	2	1	1	4	1	2	2	1	0	4
Rissoelloidea	1	4	1	2,3	0	2	3	4	1	1	2	1	2	2,3	1	1	6
Omalogyroidea	1	4	1	3	0	6	3	4	1	1	1,4	1	2	3	1	1	6
Valvatoidea	1	4	1	*1,3	0	*1,6	1	4	1	0,1	*0,2	1	2	3	1	1	6
Architectonicoidea	1,2	4	1	*1,6	4	2	3	4	1	1	4	1	2	3	1	1	6
Pyramidelloidea	2	4	1	6	4	3,5	3	4	1	1	4	1	2	3	1	1	6
Anaspidea	1	4	1	1,3	2	2	3	4	1	0,1	3,4	0,1	2	3	1	1	6
Tylodinoidea	1	4	1	1	2	2,3	3	4	1	1	3,4	1	2	3	1	1	6
Pleurobranchoidea	1	4	1	1	2	2	3	4	1	1	4	1	2	3	1	1	6
Doridina	1,2	4	1	*1,6	2	2,6	3	4	1	0,1	3,4	1	2	3	1	1	6
Aeolidina	1	4	1	3	2	2	3	4	1	1	3,4	1	2	3	1	1	6
Arminina	1	4	1	1,3	2	2	3	4	1	1	4	1	2	3	1	1	6
Dendronotina	1,2	4	1	*1,6	2	2,6	3	4	1	1	3,4	1	2	3	1	1	6
Sacoglossa	1,2	4	1	3,6	*0,2	6	3	4	1	1	4	1	2	3	1	1	6
Cephalaspidea	1,2	4	1	*1,6	1,2	2,6	3	4	1	0,1	3,4	1	2	3	1	1	6
Smeagolidae	1	4	1	1	1	6	3	4	1	1	2,4	0	2	3	1	1	6
Siphonariidae	1	4	1	1	1	3,6	3	4	1	0	0,4	0	2	3	1	1	6
Trimusculidae	1	4	1	1	2	6	3	4	1	0	3	0	2	3	1	1	6
Amphibolidae	1	4	1	1	0	6	3	4	1	0	3,4	0	2	3	1	1	6
Ellobiidae	1	4	1	1	1	2,3	3	4	1	0,1	*0,3	0	2	3	1	1	6
Chilinidae	1	4	1	1	2	3	3	4	1	0	3,4	0	2	3	1	1	6
Otinidae	1	4	1	1	0	3	3	4	1	1	0	0	2	2	1	1	6
Onchidiidae	1	4	1	1	2	3	3	4	1	0	3	0	2	3	1	1	6
Rathouisiidae	1	4	1	1	2	6	3	4	1	1	4	1	2	3	1	1	6
Vaginulidae	1	4	1	1	2	3	3	4	1	0,1	3,4	0	2	3	1	1	6
Latiidae	1	4	1	1	2	3	3	4	1	0	3	0	2	3	1	1	6
Acroloxidae	1	4	1	1	2	3	3	4	1	0,1	1,4	0,1	2	3	1	1	6
Physidae	1	4	1	1	2	3	3	4	1	1	3,4	0	2	3	1	1	6
Planorbidae	1	4	1	1	2	1,3	3	4	1	0	1,3	0,1	2	3	1	1	6
Lymnaeidae	1	4	1	1	2	1,3	3	4	1	0	2,3	0	2	3	1	1	6
Glacidorboidea	1	4	1	3	1	1	3	4	1	1	4	1	2	3	1	1	6
Succineidae	1	4	1	1	2	4	3	4	1	0	4	0	2	3	1	1	6
Athoracophoridae	1	4	1	1	2	4	3	4	1	0	4	0	2	3	1	1	6
Stylommatophora	1	4	1	1,6	2	3,6	3	4	1	0,1	4	0,1	2	3	1	1	6

Appendix 1.2. *Continued*

	18	19	20	21	22	23	24	25	26	27	28	29	30	31	32	33	34
Patellogastropoda	0	0	0	0	*0,3	0	0	0	0	1	0	0	0	0,2	0	0,1	0,1
Cocculiniforma	0	0	?	0	1,3	0	0	0,1	*0,2	0,1	0	0	1,2	0,2	0	0	0
Neritopsina	0	0	*0,2	0,1	*1,3	0,1	0	0	*0,2	0	0	0	2	*0,2	0,1	0	0,1
Vetigastropda	0	0	*1,3	0	*0,3	0	0	0	*0,2	0,1	0,1	0,1	1,2	0,2	0	0	0
Neomphaloidea	0	0	?	0	1	0	0	0	0,2	0	0	0	?	?	0	0	0
Sequenzioidea	0	0	?	0	1,3	0	0	0	2	0	0	0	?	0	0	0	2
Ampullarioidea	0	0	0,1	0,1	3	0	0	0	2	0	0	0	2	1,2	0	0,1	0
Cyclophoroidea	0	0	1	0	3	0	0	0	2	0	0	0	2	1,2	0	0	0
Stromboidea	0	0	?	0	1,2	0	0	0	2	0,1	0	0	2	1,2	0	1	0
Calyptraeoidea	0	0	5	0	*1,3	0,1	0,1	0	2	0	1	0	2	1	0	1	0,2
Vetutionoidea	0	0	?	0	2	0,1	1	0	1,2	0	1	0	2	1	0	1	1
Cypraeoidea	0	0	?	0	2,3	0	0,1	0	1,2	0	1	0	2	1	0	1	0
Tonnoidea	0	0	?	0	*1,3	0,1	0	0	2	0	1	0	2	1,2	0	0	0
Cerithioidea	0	0	*0,9	0,1	*1,3	0,1	0	0	2	0	0,1	0,1	2	0,2	0	0	0,1
Janthinoidea	0	0	?	0	2,3	0,1	0	0	2	0	0	0	2	0,1	0	1	0
Eulimoidea	0	0	?	0	1,2	0	0	0	2	0	1	0	2	0	0,1	1	1
Triphoroidea	0	0	?	0	*1,3	0	0	0	2	0	1	0	2	0,1	0	1	1
Littorinoidea	0	0	1,2	0	*1,3	0,1	0	0	2	0	0,1	0	2	0,1	0	0,1	0,1
Rissooidea	0	0,1	*1,3	0,1	2,3	0,1	0,1	0	2	0	0,1	0,1	2	0,2	0,1	0,1	0,1
Muricoidea	0	0	*1,6	0	2,3	0,1	0	0	2	0	0,1	0	2	*0,2	0,1	0	0
Conoidea	0	0	?	0	*1,3	0,1	0	0	2	0	0,1	0	2	*1,3	0,1	0	0
Rissoelloidea	0	1	?	0	1,3	0	0	0	1,2	0	1	0,1	2	0,1	0	0	0
Omalogyroidea	0	1	?	0	*1,3	0	0	0	2	0	1	0	2	0	0	0	0
Valvatoidea	0	1,2	0	0	*1,3	0	0	0	2,3	0	0,1	0,1	2	0,1	0	1	*0,2
Architectonicoidea	0	1	?	0	*1,4	0	0	0	2,3	0	1	0	2	0,1	0	1	0
Pyramidelloidea	0	1	2	0	*1,3	0	0	0	2,3	0	0,1	0	2	0,1	0	0,1	*0,2
Anaspidea	1,3	2	2	0	2	0	0	0	3	0	0	0,1	2	0	0,1	0	*0,2
Tylodinoidea	3	2	1	0	2	0	0	1	3	0	0	1	2	0	0	0	2
Pleurobranchoidea	3	2	1	0	1,2	0	0	1	3	0	0,1	0	2	0	0	0	2
Doridina	3	2	1	0	*1,4	0	0	0,1	3	0	1	0,1	2	0	0	0	2
Aeolidina	3	1	1	0	*1,3	0	0	0,1	3	0	0	0,1	2	0	0	0	2
Arminina	3	2	1	0	2	0	0	0	3	0	0,1	0,1	2	0	0	0	2
Dendronotina	3	2	1,2	0	1,2	0	0	0	3	0	0	0	2	0	0	0	2
Sacoglossa	0,3	2	0,2	0	*1,4	0	0	0,1	3	0	0	0,1	2	0,1	0	0	*0,2
Cephalaspidea	*0,3	1	1,2	0	*1,4	0	0	0,1	2,3	0	0,1	0,1	2	0	0,1	0,1	0,1
Smeagolidae	2	2	?	0	3	0	0	?	2	0	0	0	2	0	1	0	2
Siphonariidae	1,2	1	2	0	2,3	0	0	0,1	2	0	0	0	2	0	1	0,1	1,2
Trimusculidae	2	2	?	0	3	0	0	0	2	0	0	0	2	0	0	0	1,2
Amphibolidae	1	1	2	0	2	0	0	0,1	2	1	0	0	2	0	0	1	1
Ellobiidae	2	2	2	0	2,3	0	0	0	2	0	0	0,1	2	0,1	0	0,1	0,2
Chilinidae	1	1	2	0	3	0	0	0	2	1	0	0	2	0,1	0	1	0,1
Otinidae	2	2	?	0	3	0	0	0	2	0	0	0	2	0	0	0	1
Onchidiidae	2	2	2	0	2,3	0	0	0,1	3	0	0	0,1	2	3,4	0	0	2
Rathouisiidae	2	2	?	0	3	0	0	1	3	0	0	1	2	3	0	0	2
Vaginulidae	2	2	2	0	3	0	0	1	3	0	0	0	2	3	0	0	2
Latiidae	1	1	2	0	3	0	0	1	2	1	0	0	2	1	0	1	1
Acroloxidae	1	1	2	0	3	0	0	1	2	1	0	0	2	0	0	0	2
Physidae	1	2	2	0	3	0	0	1	2	1	0	0	2	0	0	0	2
Planorbidae	1	1,2	*1,9	0	3	0	0	1	2	1	0	0	2	0	0	0	1,2
Lymnaeidae	1	2	1,2	0	3	0	0	1	2	1	0	0	2	0	0	1	2
Glacidorboidea	0	2	?	0	*1,4	0	0	0	2	0	0	0	2	0	0	1	2
Succineidae	2	2	*0,3	0	4	0	0	1	2	0	0	0,1	2	4	0	0	2
Athoracophoridae	2	2	7	0	4	0	0	1	3	0	0	0	2	5	0	0	2
Stylommatophora	2	2	*2,5	0	4	0	0	1	2,3	0	0	0,1	2	4	0	0	2

35	36	37	38	39	40	41	42	43	44	45	46	47	48	49	50	51	52	53	54	55
0	0	0	0	0	0	0	0	0	0,1	0	0	*0,2	0,1	0	0	0	0	0	0	0
0	0,1	*0,2	0	0	1,2	0,1	0	0	0	*0,2	0	0,2	1	0	0	0,1	0	0	0	0
0	0	0,1	0	0	1	0	0,1	0	0,1	0,1	0	0,1	1	0,1	2	5,8	1	0	0	0
0	0,1	0,1	0	0	1,2	0	0,1	0,1	0,1	0,1	0,1	0,1	0,1	0	*0,2	1,2	0	0	0	0
0	0,1	0,1	0	0	1	0,1	0	0	1	1	0	0	1	0	2	1	1	0	0	0
0	0	0	0	0	1	0	0	1	0	1	1	0	1	0	2	1	0	0	0	0
0	0,1	0,1	0	0	1	0	0	0	0	0	0	0,1	1	0,1	1	6	1,2	0	0	0
0	0	0	0	0	1	0	0	0	0	0	0	0	1	0,1	1,2	*3,9	1	0	0	0
0	2	1,2	0	0	1	1	0	0	0	0	0	0	1	1	1,2	8	1	0	0	0
0	2	1	0	0	1	1	0	0	0	0	0	0,1	1	0,1	1,2	6,9	1	0	0	0
0	2	1	0	0	1	1	0	0	0	0	0	*0,2	1	0,1	1,3	6	1	0	0	0
0	2	0,1	0	0	1	1	0	0	0	0	0	0,2	1	1	2	8	1	0	0	0
0	2	1	0	0	1	1	0	0	0	0	0	0,1	1	1	2	8	1	0	0	0
0	2	0,1	0	0	1	0,1	0	0	0	0	0	*0,2	1	0	1,2	*3,9	1	0	0,1	0,1
0	2	1	0	0	1	0	0	0	0	0	0	*0,2	1	0	1,2	9	2	0	0,1	0,1
0	2	1	0	0	1	0	0	0	0	0	0	*0,2	1	0,1	1,2	6,8	1	0	0	0,1
0	2	1	0	0	1	0	0	0	1	0,1	0	0	1	1	1,2	6	1	0	0	1
0	2	0,1	0	0	1	1	0	0	0	0	0	0,1	1	0,1	1,2	*4,9	1	0	0	0
0	2	1	0	0	1	0	0	0	0	0	0	0,1	1	0,1	1,2	*5,9	1,2	0	0	0
0	2	1	0	0	1	1	0	0	0	0	0	*0,2	1	0,1	1,2	*5,8	1,2	0	0	0
0	2	1	0	0	1	1	0	0	0	0	0	0,1	1	0	2	8	1,2	0	0	0
0	2	1	0	0	1	0	0	0	0	0	0	2	1	0	2	5,6	1	0	0	0
0	2	1	0	0	1	0	0	0	0	0	0	1,2	1	1	1	8	1	1	0	1
0	2	1	0	0	1	0,1	0	0	0	0	0	1,2	1	1	1	*5,9	1,2	0	0,1	0
0	1,2	1	0,1	0	1	0,1	0	0	0	0	0	*0,2	1	1	1,2	9	1	0	1	1
0	0,2	2	0	4	1	0,1	0	0	0	0	0	0,2	1	1	1,3	9	2	0	0,1	0,1
0	0,1	2	1	4	1	0,1	0	0	0	0	0	2	1	1	2,3	*5,8	2	0	1	1
0	1,2	2	1	4	1	1	0	0	0	0	0	2	1	1	3	8	1,2	0,3	0	0,1
0	2	2	1	4	1	1	0	0	0	0	0	2	1	1	3	*1,8	2	2	0	1
0	2	2	1	4	1	1	0	0	0	0	0	2	1	1	3	*5,8	2	2	0	1
0	2	2	1	4	1	0,1	0	0	0	0	0	2	1	1	3	8	2	2	0	1
0	2	2	1	4	1	1	0	0	0	0	0	1	1	1	3	8	2	2	0	1
0	2	2	1	4	1	0,1	0	0	0	0	0	1	1	1	3	*5,8	2	*0,3	0	1
0	2	2	1	4	1	0,1	0	0	0	0	0	2	1	1	2,3	*5,8	1,2	0,1	1	0,1
1	1,2	2	1	3	1	1	0	0	0	0	0	2	1	1	3	8	2	4	1	1
1	2	2	1	1	1	1	0	0	0	0	0	2	1	1	3	8	2	4	1	1
1	0,1	2	1	2	1	1	0	0	0	0	0	2	1	1	3	8	2	2	0	1
1	2	2	1	1	1	0	0	0	0	0	0	2	1	1	3	8	2	4	1	1
1	0	2	1	3	1	1	0	0	0	0	0	1,2	1	1	3	8	2	2	0,1	1
1	1,2	2	1	1	1	0	0	0	0	0	0	2	1	1	3	8	2	2	0,1	1
1	2	2	1	3	1	1	0	0	0	0	0	2	1	1	3	7	2	4	0	1
1	2	2	1	3	1	1	0	0	0	0	0	2	1	1	3	8	2	3	0,1	1
1	2	2	1	3	1	1	0	0	0	0	0	2	1	1	3	8	2	3	0	1
1	2	2	1	3	1	1	0	0	0	0	0	2	1	1	3	8	2	3	0,1	1
1	2	2	1	1	1	1	0	0	0	0	0	2	1	1	3	8	2	2	0	1
1	2	2	1	1	1	1	0	0	0	0	0	2	1	1	3	8	2	2	0	1
1	2	2	1	1	1	1	0	0	0	0	0	1,2	1	1	3	8	2	2	0	1
1	2	2	1	1	1	1	0	0	0	0	0	2	1	1	3	8	2	2	0	1
0	2	0	0	0	1	0	0	0	0	0	0	1	1	1	2	8	1	0	0	1
1	0	2	1	3	1	1	0	0	0	0	0	2	1	1	3	8	2	4	0	1
1	0	2	1	3	1	1	0	0	0	0	0	2	1	1	3	8	2	4	0	1
1	0	2	1	3	1	1	0	0	0	0	0	2	1	1	3	8,9	2	4	0	1

Appendix 1.2. *Continued*

	56	57	58	59	60	61	62	63	64	65	66	67	68	69	70	71	72
Patellogastropoda	0	0	0	0	0	0	0	?	0	0	0	1,3	0	1	0	1	4
Cocculiniforma	0,1	0	?	?	?	?	?	?	?	?	?	?	0	1	0	1	1,4
Neritopsina	0	0,1	3	0	0	0	2	0,1	0	0	0	1	0,1	1	0	1	*2,4
Vetigastropda	0	0,1	*0,2	0,1	0	0	*0,5	0	0	0	0	0,1	0,2	0,1	1	0,1	*0,4
Neomphaloidea	0,1	0	0	0	0	0	0	0	0	0	0	0	0	0	0	0	4
Sequenzioidea	0,1	0	?	?	?	?	?	?	?	?	?	?	0	0	1	0	2
Ampullarioidea	0	0	1	1	0	0	3	1	0	1	0	1	0,2	1	0	0	2,3
Cyclophoroidea	0,1	0,1	1	?	?	?	3	0	2	1	0	1	2	1	0	0	1
Stromboidea	0	0	?	0	0	0	2	1	0	1	0	2	1	1	0	0	3,4
Calyptraeoidea	0	0	?	0	0	0	2	1	0	1	0	2	0	1	0	0	3,4
Vetutionoidea	0,1	0	?	?	?	?	?	?	?	?	?	?	0	1	0	0	4
Cypraeoidea	0,1	0	?	0	0	0	2	1	1	1	0	2	4	1	0	0	4
Tonnoidea	0,1	0	4	0	0	0	1,2	1	0	1	0	2	4	1	0	0	*2,4
Cerithioidea	0,1	0,1	*1,5	0	0	0	2,3	0	0	1	0	1,2	2	1	0	0	*0,3
Janthinoidea	0,1	1	?	0	0	0	2	1	0	0,1	0	2	1	1	0	0	*2,4
Eulimoidea	0	0	?	0	0	0	2	1	0	1	0	2	0,1	1	0	0	2,4
Triphoroidea	0,1	0,1	?	?	?	?	?	?	?	?	?	?	3	1	0	0	1,2
Littorinoidea	0,1	0	4	0	0	0	2	1	0	1	0,2	2	0,5	1	0	0	1,2
Rissooidea	0,1	0	1,4	0,1	0	0	2,6	1	0,1	1	0	1,2	0	1	0	0	*1,3
Muricoidea	0,1	0	1,4	0	0	0	2	1	0	1	0	2	4	1	0	0	3,4
Conoidea	0,1	0	4	0	0	0	2	1	0	1	0	2	4	1	0	0	3,4
Rissoelloidea	1	0	4	0	1	0,1	2	1	4	?	?	?	0	1	0	0	2,3
Omalogyroidea	1	0	1	1	1	0	9	1	0	?	?	4	0	1	0	0	1,2
Valvatoidea	0,1	0	4	0,1	0	0	8	0	0,3	1	?	7	0	1	0	0	1,4
Architectonicoidea	1	1	*0,7	0,1	0	0	*6,9	1	0	1	0	5	0	1	0	0	1,2
Pyramidelloidea	0,1	0,1	1,4	0,1	1	1	2	1	4	0,1	?	6	0	1	0	0	2,4
Anaspidea	0	0	6	1	0,1	1	2	1	4	1	?	6,7	0	1	0	0	4
Tylodinoidea	0,1	0	6	1	1	1	2	1	4	1	?	6	0	1	0	0	4
Pleurobranchoidea	1	0	1,6	1	1	1	2	1	4	1	0	6	0	1	0	0	4
Doridina	1	0,1	0,1	0,1	1	1	2,7	1	4	*0,2	2	6	0	2	0	0	4
Aeolidina	1	0	1	1	1	1	2,7	1	4	2	?	6	0	1,2	0	0	4
Arminina	1	0	1	1	1	1	2,7	1	4	1	?	?	0	2	0	0	4
Dendronotina	1	0	1	1	1	1	2,7	1	4	1	?	?	0	2	0	0	4
Sacoglossa	1	0	4	1	1	1	2	1	4	1	?	?	0	1,2	0	0	4
Cephalaspidea	0,1	0,1	1	1	0,1	*1,4	2,7	1	4	1	0	6	0	1,2	0	0	*2,4
Smeagolidae	1	0	?	1	?	?	?	?	?	?	?	?	0	2	0	0	4
Siphonariidae	1	1	1	1	1	1	2,7	1	4	1	0	6	0	1	0	0	4
Trimusculidae	1	0	?	?	?	?	?	?	?	?	?	?	0	1	0	0	4
Amphibolidae	1	0	1	1	1	1	2,7	1	4	2	2	6	0	1	0	0	2
Ellobiidae	0,1	0	1	1	1	3	2,7	1	4	0	2	6	0	1	0	0	4
Chilinidae	0,1	1	?	?	?	?	?	?	?	?	?	?	0	1	0	0	?
Otinidae	1	0	?	?	?	?	?	?	?	?	?	?	0	1	0	0	4
Onchidiidae	0,1	0	1	1	1	1	2,7	1	4	1	2	6	0	2	0	0	4
Rathouisiidae	1	0	?	?	?	?	?	?	?	?	?	?	0	2	0	0	4
Vaginulidae	1	0	?	?	?	?	?	?	?	?	?	?	0	2	0	0	4
Latiidae	1	0	?	?	?	?	?	?	?	?	?	?	0	1	0	0	4
Acroloxidae	1	0	?	1	?	1	?	?	?	1	?	6	0	1	0	0	4
Physidae	1	0	1	1	1	3,4	2	1	4	1	0	6	0	1	0	0	4
Planorbidae	1	0	1	1	1	2,4	2,7	1	4	1	0	6	0	1	0	0	4
Lymnaeidae	1	0	1	1	1	2,3	2,7	1	4	1	2	6	0	1	0	0	4
Glacidorboidea	1	0	?	0	?	1	?	?	?	1	?	6	0	1	0	0	2
Succineidae	1	0	1	1	1	1	2,7	1	4	0	2	6	0	1	0	0	4
Athoracophoridae	1	0	1	1	1	1	2,7	1	4	0	2	6	0	1,2	0	0	4
Stylommatophora	1	0,1	1	0,1	1	1	2,7	1	4	0	0,2	6	0	1,2	0	0	4

See Appendix 1.1 for description of the characters. Characters with more than one state indicate polymorphism within the taxon. *signifies that three or more character states were present in the original data but are reduced to two (extremes in arithmetic series) in this tabulation due to space constraints.

Appendix 1.3. The List of the Apomorphies for the Principal Nodes in the Phylogram of Relationships in Gastropoda, shown in Fig. 1.6, based on parsimony analysis as detailed in the legend of Fig. 1.3

Character	Change	Character	Change	Character	Change
root → node 1		12	0 ---> 1	64	0 ==> 2
11	0 ==> 2	36	0 ---> 2	72	2 ==> 1
15	0 ---> 1	46	1 ==> 0	node 6 → node 10	
17	0 ---> 5	51	1 ==> 6	3	0 ==> 1
22	0 ---> 1	52	0 ==> 1	5	0 ==> 4
44	0 ==> 1	node 5 → node 6		7	0 ==> 1
67	0 ==> 1	17	5 ---> 4	13	0 ==> 2
72	0 ==> 4	31	0 ---> 1	14	1 ==> 2
node 1 → Patellogastropoda		node 6 → node 7		22	1 ---> 2
9	0 ==> 1	4	2 ---> 1	28	0 ==> 1
10	0 ==> 1	12	1 ---> 0	33	0 ==> 1
11	2 ==> 4	17	4 ---> 3	69	0 ==> 1
27	0 ==> 1	36	2 ---> 0	node 10 → node 11	
69	0 ==> 1	58	4 ---> 3	44	1 ==> 0
71	0 ==> 1	63	1 ---> 0	45	1 ==> 0
node 1 → node 2		node 7 → Neritopsina		51	6 ---> 8
1	0 ==> 1	6	2 ==> 6	node 10 → Triphoroidea	
4	0 ==> 1	8	0 ==> 3	2	0 ==> 2
6	0 ==> 2	50	1 ==> 2	11	2 ==> 4
20	0 ---> 1	69	0 ==> 1	26	1 ==> 2
26	0 ---> 1	71	0 ==> 1	34	0 ==> 1
30	0 ==> 1	node 7 → node 8		37	0 ==> 1
40	0 ==> 1	14	1 ==> 2	49	0 ==> 1
45	0 ==> 1	58	3 ==> 1	55	0 ==> 1
46	0 ==> 1	node 8 → Neomphaloidea		68	0 ==> 3
51	0 ==> 1	9	1 ==> 0	node 11 → node 12	
58	0 ---> 2	14	2 ==> 3	2	0 ==> 2
62	0 ---> 2	17	3 ==> 1	24	0 ==> 1
node 2 → Vetigastropoda		50	1 ==> 2	65	0 ==> 1
15	1 ---> 0	51	6 ==> 1	node 11 → node 13	
70	0 ==> 1	58	1 ==> 0	4	2 ==> 6
node 2 → node 3		62	2 ==> 0	26	1 ==> 2
14	0 ==> 1	67	1 ==> 0	31	1 ---> 0
48	0 ==> 1	72	2 ==> 4	37	0 ==> 1
58	2 ---> 4	node 8 → node 9		67	1 ==> 2
63	0 ---> 1	2	0 ==> 2	68	0 ---> 1
node 3 → Cocculiniforma		3	0 ==> 1	node 12 → node 14	
44	1 ==> 0	4	1 ---> 2	41	0 ==> 1
46	1 ==> 0	11	2 ---> 1	67	1 ==> 2
69	0 ==> 1	13	0 ==> 2	72	2 ==> 4
71	0 ==> 1	22	1 ==> 3	node 12 → node 15	
node 3 → node 4		26	1 ==> 2	26	1 ==> 2
30	1 ---> 2	44	1 ==> 0	33	1 ---> 0
50	0 ==> 1	45	1 ==> 0	node 13 → Janthinoidea	
72	4 ==> 2	59	0 ---> 1	2	0 ==> 3
node 4 → Sequenzioidea		62	2 ==> 3	28	1 ==> 0
13	0 ==> 2	65	0 ==> 1	51	8 ==> 9
26	1 ==> 2	68	0 ---> 2	52	1 ==> 2
34	0 ==> 2	69	0 ==> 1	57	0 ==> 1
43	0 ==> 1	node 9 → Ampullarioidea		node 13 → Eulimoidea	
44	1 ==> 0	7	0 ==> 1	2	0 ==> 2
50	1 ==> 2	10	0 ==> 1	10	0 ==> 1
70	0 ==> 1	63	0 ---> 1	11	2 ==> 4
node 4 → node 5		node 9 → Cyclophoroidea		34	0 ==> 1
4	1 ---> 2	11	1 ==> 0	65	0 ==> 1
9	0 ==> 1	17	3 ==> 6		

Appendix 1.3. *Continued*

Character	Change	Character	Change	Character	Change
node 14 → node 16		node 5 → node 22		54	0 ==> 1
5	4 ==> 0	2	0 ==> 4	57	0 ==> 1
68	0 ---> 4	3	0 ==> 1	node 25 → Glacidorboidea	
node 14 → node 17		4	2 ---> 3	5	0 ==> 1
20	1 ---> 5	7	0 ---> 1	6	2 ==> 1
37	0 ==> 1	8	0 ==> 4	19	1 ==> 2
51	8 ---> 6	13	0 ==> 2	34	0 ==> 2
node 15 → Cerithioidea		14	1 ==> 2	38	1 ==> 0
24	1 ==> 0	16	0 ==> 1	50	1 ==> 2
63	1 ==> 0	17	5 ==> 6	61	0 ==> 1
68	0 ==> 2	19	0 ==> 1	67	5 ==> 6
node 15 → Rissooidea		20	1 ---> 0	node 24 → Valvatoidea	
5	4 ==> 0	38	0 ==> 1	37	0 ==> 1
37	0 ==> 1	44	1 ==> 0	38	1 ==> 0
node 16 → Cypraeoidea		45	1 ==> 0	63	1 ---> 0
6	2 ==> 5	47	0 ==> 1	67	5 ==> 7
49	0 ==> 1	51	6 ---> 8	72	2 ==> 1,4
50	1 ==> 2	67	1 ==> 5	node 22 → node 23	
64	0 ==> 1	69	0 ==> 1	7	1 ==> 3
node 16 → node 18		node 23 → Rissoelloidea		20	0 ---> 2
24	1 ==> 0	10	0 ==> 1	37	0 ==> 1
26	1 ==> 2	28	0 ==> 1	47	1 ==> 2
node 17 → Calyptraeoidea		38	1 ==> 0	61	0 ---> 1
5	4 ==> 3	50	1 ==> 2	64	0 ==> 4
26	1 ==> 2	51	8 ---> 5,6	67	5 ---> 6
node 17 → Vetutinoidea		56	0 ==> 1	node 23 → node 27	
10	0 ==> 1	60	0 ==> 1	5	0 ==> 2
34	0 ==> 1	node 22 → node 24		11	2 ==> 3
node 18 → Littorinoidea		14	2 ==> 3	22	1 ==> 2
68	4 ---> 0,5	26	1 ==> 2	26	1 ==> 2
72	4 ==> 1,2	33	0 ==> 1	37	1 ==> 2
node 18 → node 19		49	0 ==> 1	39	0 ==> 1
37	0 ==> 1	62	2 ==> 8	41	0 ==> 1
node 19 → Stromboidea		65	0 ==> 1	49	0 ==> 1
11	2 ==> 1	node 24 → node 25		59	0 ==> 1
28	1 ==> 0	7	1 ==> 3	72	2 ==> 4
49	0 ==> 1	10	0 ==> 1	node 27 → node 28	
68	4 ---> 1	11	2 ==> 4	11	3 ==> 4
node 19 → node 20		55	0 ==> 1	14	2 ==> 3
5	0 ==> 3	56	0 ==> 1	39	1 ==> 4
6	2 ==> 6	58	4 ---> 1	60	0 ==> 1
8	0 ---> 2	62	8 ---> 9	node 28 → Pyramidelloidea	
11	2 ---> 4	node 25 → node 26		1	1 ==> 2
33	1 ==> 0	28	0 ==> 1	4	3 ==> 6
node 20 → Muricoidea		67	0 ==> 1	5	2 ==> 4
4	2 ==> 3,4,5,6	59	0 ---> 1	6	2 ==> 3,5
7	1 ==> 2	node 26 → Omalogyroidea		10	0 ==> 1
node 20 → node 21		6	2 ==> 6	38	1 ==> 0
10	0 ==> 1	33	1 ==> 0	51	8 ==> 9
50	1 ==> 2	38	1 ==> 0	52	1 ==> 2
node 21 → Tonnoidea		53	0 ==> 1	node 28 → node 29	
8	2 ---> 0	60	0 ==> 1	26	2 ==> 3
49	0 ==> 1	67	5 ==> 4	50	1 ==> 3
node 21 → Conoidea		node 26 → Architectonicoidea		65	0 ==> 1
5	3 ==> 6	5	0 ==> 4	node 29 → Sacoglossa	
7	1 ==> 2	51	8 ==> 9	6	2 ==> 6

Character	Change	Character	Change	Character	Change
10	0 ==> 1	node 34 → node 35		57	0 ==> 1
19	1 ==> 2	5	2 ==> 0	65	0 ==> 1
47	2 ==> 1	6	2 ==> 6	node 38 → node 39	
52	1 ==> 2	34	0 ==> 1	54	1 ==> 0node 39
55	0 ==> 1	35	0 ==> 1	→ node 40	
56	0 ==> 1	50	2 ==> 3	18	1 ==> 2
node 29 → node 30		53	0 ==> 4	19	1 ==> 2
4	3 ---> 1	56	0 ==> 1	36	2 ==> 0
18	0 ==> 3	60	0 ---> 1	39	1 ==> 2
20	2 ==> 1	65	0 ---> 2	node 40 → Trimusculidae	
34	0 ==> 2	66	0 ---> 2	6	3 ==> 6
node 30 → Tylodinoidea		node 35 → Amphibolidae		34	0 ==> 1,2
10	0 ==> 1	14	2 ==> 3	56	0 ==> 1
19	1 ==> 2	27	0 ==> 1	node 40 → node 41	
25	0 ==> 1	33	0 ==> 1	31	0 ---> 1
29	0 ==> 1	41	1 ==> 0	39	2 ==> 3
58	4 ==> 6	72	4 ==> 2	node 41 → Ellobiidae	
node 30 → node 31		node 35 → node 36		5	2 ==> 1
52	1 ==> 2	10	0 ==> 1	61	1 ==> 3
53	0 ==> 2	11	3 ---> 2	66	0 ==> 2
55	0 ==> 1	18	1 ==> 2	node 41 → node 42	
56	0 ==> 1	19	1 ==> 2	11	3 ---> 4
58	4 ==> 1	22	2 ==> 3	31	1 ==> 3
node 31 → Pleurobranchoidea	39	1 ==> 3	34	0 ==> 2
10	0 ==> 1	node 34 → node 37		53	2 ==> 3
19	1 ==> 2	14	2 ==> 3	node 42 → node 43	
25	0 ==> 1	node 36 → Smeagolidae		26	2 ==> 3
node 31 → Aeolidina		5	0 ==> 1	36	0 ==> 2
4	1 ---> 3	14	2 ==> 3	65	0 ---> 1
10	0 ==> 1	32	0 ==> 1	66	0 ---> 2
65	1 ==> 2	34	1 ==> 2	69	1 ==> 2
node 31 → node 32		69	1 ==> 2	node 43 → Onchidiidae	
19	1 ==> 2	node 36 → Otinidae		11	4 ---> 3
66	0 ---> 2	6	6 ==> 3	node 43 → node 44	
69	1 ==> 2	11	2 ==> 0	25	0 ==> 1
node 32 → Doridina		51	8 ==> 7	56	0 ==> 1
28	0 ==> 1	51	1 ==> 0	node 44 → Rathouisiidae	
node 32 → Dendronotina		node 37 → Anaspidea		6	3 ==> 6
10	0 ==> 1	19	1 ==> 2	10	0 ==> 1
node 32 → Arminina		26	2 ==> 3	12	0 ==> 1
10	0 ==> 1	36	2 ==> 0,1	29	0 ==> 1
47	2 ==> 1	39	1 ==> 4	node 42 → node 45	
65	1 ==> 2	58	1 ==> 6	(Stylommatophora)	
node 27 → node 33		65	0 ==> 1	22	3 ==> 4
4	3 ---> 1	node 37 → node 38		25	0 ==> 1
50	1 ==> 2	6	2 ==> 3	31	3 ==> 4
54	0 ==> 1	22	2 ---> 3	53	3 ==> 4
58	4 ==> 1	35	0 ==> 1	56	0 ==> 1
node 33 → Cephalaspidea		50	2 ==> 3	node 45 → node 46	
14	2 ==> 3	53	0 ==> 2	(Succineidae)	
39	1 ==> 4	60	0 ==> 1	6	3 ==> 4
65	0 ==> 1	node 38 → Siphonariidae		66	0 ==> 2
node 33 → node 34		5	2 ==> 1	node 46 → Athoracophoridae	
12	1 ==> 0	32	0 ==> 1	20	2 ==> 7
18	0 ==> 1	34	0 ==> 1,2	26	2 ==> 3
52	1 ==> 2	53	2 ==> 4	31	4 ==> 5
55	0 ==> 1	56	0 ==> 1		

Appendix 1.3   *Continued*

Character	Change	Character	Change	Character	Change
node 39 → node 47		56	0 ==> 1	node 48 → Lymnaeidae	
27	0 ==> 1	node 48 → Acroloxidae		19	1 ==> 2
65	0 ---> 1	34	1 ==> 2	33	0 ==> 1
node 47 → Chilinidae		node 48 → Planorbidae		34	1 ==> 2
33	0 ==> 1	61	1 ---> 3	61	1 ---> 3
41	1 ==> 0	node 48 → Physidae		66	0 ==> 2
57	0 ==> 1	10	0 ==> 1	node 48 → Latiidae	
node 47 → node 48		19	1 ==> 2	31	0 ==> 1
25	0 ==> 1	34	1 ==> 2	33	0 ==> 1
34	0 ==> 1	61	1 ---> 3	61	1 ---> 3

Key: ==> unambiguous change; ---> occurs under some reconstructions, but not others.

Appendix 1.4. Morphological Characters (and their states) Used in Analysis of Phylogenetic Relationships in the Terrestrial Pulmonata

Alimentary characters

1. *Buccal mass.* 0 – spheroidal, small; 1 – cylindrical, enlarged.
2. *Gastric caecum.* 0 – present; 1 – absent.
3. *Rectal caecum.* 0 – absent; 1 – present.
4. *Jaw structure.* 0 – ribbed; 1 – smooth; 2 – absent.
5. *Jaw accessory plate.* 0 – absent; 1 – present.
6. *Radular teeth.* 0 – all teeth tricuspid, on quadrate or rectangular basal plates; 1 – endocones and ectocones retained in marginal teeth but these serrated, on quadrate or rectangular basal plates; 2 – whole radular becoming modified pectinate marginals, teeth on quadrate or broadly rectangular to elongate basal plates; 3 – endocones lost in lateral and marginal teeth; ectocones sometimes lost in central and lateral teeth but generally retained and serrated in marginals; teeth on quadrate or broadly rectangular basal plates; 4 – central and lateral teeth lacking endocones and ectocones but with broad mesocone; marginals tricuspid or pectinate; teeth on quadrate or broadly rectangular basal plates; 5 – mesocones and endocones of lateral and marginal teeth tending to elongate and fuse; teeth on upright, narrowly rectangular basal plates; 6 – lateral and marginal teeth elongated, with endocones lost by complete fusion with mesocones; teeth on upright, narrowly rectangular basal plates; 7 – lateral and marginal teeth elongated, with endocones lost by complete fusion with mesocones, and ectocones becoming serrated; teeth on upright, narrowly rectangular basal plates; 8 – lateral and marginal teeth elongated, with endocones and ectocones lost or nearly so by fusion with the mesocones; teeth on upright, narrowly rectangular basal plates; 9 – whole radular becoming modified, elongate, unicuspid teeth on upright, narrowly rectangular basal plates.
   Prolonged cuspid head on radular teeth 0 – *present; 1 – absent.*
7. *Digestive gland.* 0 – openings of digestive gland lobes disjunct, the anterior opening oesophageal,
8. the posterior opening intestinal; 1 – openings of digestive gland lobes more or less adjacent, openings intestinal.
   Stomach. 0 – strongly muscular, with definitive gizzard; 1 – somewhat simplified, with strong
9. musculature but without definitive gizzard; 2 – greatly simplified, with very poorly developed musculature.
   Diagonal intestinal folds. 0 – present; 1 – absent (folds longitudinal when present).
10. *Intestinal valve.* 0 – proximal intestine with a sphincter or valve; 1 – pre-rectal part of intestine with a
11. sphincter or valve; 2 – absent.

Renal characters

12. *Nephridium internal structure.* 0 – homogeneous, with lamellae reaching the distal region and the level of the nephridial pore; 1 – two distinct morphological regions, the distal one usually lacking lamellae; 2 – two distinct morphological regions, the distal lacking lamellae and greatly extended towards the pneumostome.
13. *Closed retrograde ureter.* 0 – absent; 1 – reaching at most to the top of the pallial cavity; 2 – reaching a point between top of pallial cavity and pneumostome; 3 – reaching the pneumostome.

Cytology characters

14. *Chromosome number.* 0 – <10; 1 – 10 to 15; 2 – 16 to 20; 3 – 21 to 25; 4 – 26 to 30; 5 – 31 to 35; 6 – 36 to 45.

Developmental and larval characters

15. *Embryo brooding.* 0 – absent, oviparous; 1 – present, oviviparous or viviparous.
16. *Podocyst.* 0 – absent; 1 – present.
17. *Egg mass.* 0 – eggs embedded in a jelloid/mucoid mass; 1 – eggs single, not embedded in jelloid/mucoid mass.
18. *Egg capsule.* 0 – partially calcified, with calcite crystals embedded in jelly layers but not forming a distinct shell; 1 – calcified, forming distinct shell.
19. *Larval development.* 0 – trochophore in the egg, planktotrophic veliger larva free; 1 – no trochophore, veliger stages completed in egg; 2 – no trochophore or veliger stages, direct development in the egg.
20. *Larval operculum.* 0 – present; 1 – absent.

Appendix 1.4. *Continued*

Muscle and pedal characters

21. *Right ocular retractor.* 0 – passing between penis and vagina (vas deferens in ellobiids, onchidiids and vaginulids); 1 – not passing between penis and vagina.
22. *Buccal retractors.* 0 – separate from remainder of anterior free muscle system, arising from stem of columellar muscle; 1 – arising from right ocular retractor branch; 2 – arising from left ocular retractor branch; 3 – reduced, not associated with anterior free muscle system that originates from columellar, but arising from body wall, or absent.
23. *Caudal mucous pit or horn.* 0 – absent; 1 – present.
24. *Sole.* 0 – uniform; 1 – tripartite.
25. *Margins of the foot.* 0 – parapodial groove(s) sited at margin of foot (or absent), the ciliated sole reaching to lateral margin of the foot but not extending above; 1 – parapodial groove(s) sited above margin of foot, so that the ciliated sole extends above the margin of the foot.
26. *Cephalic shield.* 0 – present as rigid plate bounded by distinct grooves; 1 – present as not so rigid plate bounded by distinct grooves; 2 – reduced, defined only by vestigial grooves.
27. *Hyponotum.* 0 – present; 1 – absent.
28. *Ocular cephalic tentacle structure and withdrawal.* 0 – contractile but not retractile, may involve partially invagination of apex; 1 – withdrawal comprising three movement elements (a) contraction of the basal rugose part, (b) inversion of the apical chemoreceptor part and (c) invagination of the basal rugose part; 2 – withdrawal comprising three movement elements (a) contraction of the basal rugose part, (b) inversion of the chemoreceptor part at the apex and (c) invagination from the apex; 3 – withdrawal comprising two movement elements (a) contraction and (b) invagination of the basal rugose part.

Nervous system and sensory characters

29. *Cerebro-pedal connectives.* 0 – length more than twice right cerebral ganglion width; 1 – length once or twice right cerebral ganglion width; 2 – length shorter than right cerebral ganglion width.
30. *Cerebral commissure.* 0 – length distinctly greater than right cerebral ganglion width; 1 – length about same as right cerebral ganglion width; 2 – length distinctly shorter than right cerebral ganglion width.
31. *Left parietal ganglion.* 0 – closer to left pleural ganglion than to the visceral ganglion; 1 – closer to the visceral ganglion than to left pleural ganglion; 2 – in contact with the left pleural ganglion only; 3 – in contact with the visceral ganglion only.
32. *Right parietal ganglion.* 0 – closer to right pleural ganglion than to the visceral ganglion; 1 – closer to the visceral ganglion than to right pleural ganglion; 2 – in contact with the right pleural ganglion only; 3 – in contact with the visceral ganglion only; 4 – in contact with both the visceral ganglion and the right pleural ganglion.
33. *Visceral loop of central nervous system.* 0 – strephoneurous; 1 – euthyneurous, but rather long; 2 – euthyneurous, highly concentrated.
34. *Visceral ganglion.* 0 – on right side of the median plane; 1 – median; 2 – on the left side of the median plane.
35. *Pedal ganglia.* 0 – concentrated into ganglia, with lateral nerves and two commissures; 1 – concentrated into ganglia, with lateral nerves and one commissure.
36. *Inferior tentacles.* 0 – present; 1 – absent.
37. *Eye position.* 0 – embedded centrally or adjacent to base of cephalic tentacle; 1 – at tip of more or less elongate cephalic tentacle.
38. *Tentacular nerve.* 0 – bifurcated; 1 – simple.
39. *Cerebral–pleural ganglion association.* 0 – hypathroid, with pleural ganglia close to pedal ganglia; 1 – epiathroid, with pleural ganglia close to cerebral ganglia.
40. *Supraoesophageal ganglion.* 0 – distinct; 1 – fused with the right parietal ganglion.
41. *Suboesophageal ganglion.* 0 – distinct; 1 – fused with the visceral ganglion.

Reproductive characters

42. *Genital orifice.* 0 – male orifice in cephalic region, female orifice extra-pallial but separate from male orifice and opening in lateral body wall; 1 – male orifice in cephalic region, female orifice extra-pallial, through body wall but separate from male orifice and displaced to posterior with detorsion of viscera; 2 – male and female orifices fused or nearly so in cephalic region, near right ocular tentacle; 3 – male and female orifices fused or nearly so in cephalic region, midway between pneumostome and the right ocular tentacle; 4 – male and female orifices fused or nearly so in cephalic region, in the main body of the visceral stalk, near the pneumostome.

Appendix 1.4.   *Continued*

43. *Ovotestis.* 0 – embedded in lobes of the digestive gland or above in early teleconch whorls; 1 – not embedded in digestive gland, but located anteriorly in the body cavity.
44. *Albumen gland.* 0 – comprising two lobes; 1 – comprising a single lobe.
45. *Pallial gonoducts.* 0 – male and female ducts separate below carrefour, with prostate follicles not opening to oviduct; 1 – male and female ducts fused below carrefour to form spermoviduct, prostatic follicles opening to oviductal gland.
46. *Oviducal glands.* 0 – agglomerated a proximal end of pallial gonoduct, somewhat diverticulate to proximal–distal axis of duct; 1 – oviductal gland elongate tube, at proximal end of the pallial gonoduct folded with ascending and descending limbs spiralled around one another; 2 – oviductal gland essentially tubular, with more or less infolded lumenal walls.
47. *Extrapallial sperm duct.* 0 – open groove in foot; 1 – closed duct, embedded in foot; 2 – closed duct, free in body cavity.
48. *Auxillary genital organs.* 0 – stimulatory organ opening to the atrium, bearing a dart; 1 – stimulatory organ opening to the penis, bearing a dart; 2 – stimulatory organ opening to the vagina, bearing a dart; 3 – stimulatory organ opening to the penis, lacking a dart; 4 – stimulatory organ opening to the atrium, lacking a dart; 5 – stimulatory organ opening to the vagina, lacking a dart; 6 – stimulatory organ absent.
49. *Epiphallus.* 0 – present, without flagellum; 1 – present, with flagellum; 2 – absent.
50. *Epiphallic papilla in penis.* 0 – present; 1 – absent.
51. *Bursa copulatrix duct.* 0 – long, >0.75 pallial gonoduct length; 1 – medium in length, 0.5–0.75 pallial gonoduct length; 2 – short, <0.5 pallial gonoduct length.
52. *Bursa copulatrix origin.* 0 – from free oviduct; 1 – from pallial gonoduct.
53. *Bursa copulatrix duct diverticulum.* 0 – absent; 1 – present.
54. *Penial sheath.* 0 – present, partially or completely enclosing or attaching to distal part of vas deferens/epiphallus; 1 – present, not enclosing or attaching to vas deferens/epiphallus; 2 – absent.
55. *Penial retractor muscle.* 0 – a branch of the columellar muscle or arising from columellar independently of the body of the free muscle stem; 1 – arising from the diaphragm, independently of columellar muscle system; 2 – arising from body wall, independently of columellar muscle system; 3 – reduced, arising from lower reproductive tract, or absent.
56. *Allospermiduct of pallial gonoduct.* 0 – groove open to the lumen of the female pallial gonoduct, with a blind, tubular diverticulum variously sited on the female pallial gonoduct; 1 – a groove or channel open to the lumen of the female pallial gonoduct; 2 – a closed tube, separated for its greater part from the female pallial gonoduct, the diverticulum lost.
57. *Penial spines.* 0 – lumen of penis with few to many spines; 1 – lumen of penis lacking spines.

Appendix 1.5. Character States for the Taxa Included in the Parsimony Analysis of Phylogenetic Relationships in Terrestrial Pulmonata

	1	2	3	4	5	6	7	8	9	10	11	12	13	14	15	16	17	18
Achatinellidae	0	1	0	0,2	0	2	1	1	2	1	2	2	0	2,3	0,1	1	1	1
Partulidae	0	0,1	0	0	0	5	0,1	1	2	1	2	2	0	4	1	1	1	1
Amastridae	0	1	0	0	0	3	1	1	2	1	2	2	0	-	0,1	1	1	1
Cochlicopidae	0	1	0	0	0	3	1	1	2	1	2	2	0,1	4	0,1	1	1	0
Pyramidulidae	0	1	0	0	0	3	1	1	2	1	2	2	0	4	1	1	1	1
Vertiginidae	0	1	0	0	0	*1,3	1	1	2	1	2	2	0,1	-	0,1	1	1	0
Orculidae	0	1	0	0	0	3	1	1	2	1	2	2	0	-	0	1	1	-
Chondrinidae	0	1	0	0,1	0	3	1	1	2	1	2	2	0	4	0	1	1	-
Pupillidae	0	1	0	0	0	3	1	1	2	1	2	2	0	4	0,1	1	1	0
Valloniidae	0	0,1	0	0	0	3	1	1	2	1	2	2	0,1	4	0,1	1	1	1
Pleurodiscidae	0	1	0	0	0	3	1	1	2	1	2	2	0	-	1	1	1	-
Enidae	0	1	0	0,1	0	3	1	1	2	1	2	2	0,1	3	0,1	1	1	-
Succineidae	0	0	0	0,1	1	*1,3	0,1	1	2	1	2	0	2	*0,3	0	0	0,1	0
Ariopeltinae	0	1	0	1	0	3	1	1	2	1	2	0	2	-	0	1	1	-
Aneitinae	0	0	0	1	1	1,2	0,1	1	2	1	2	0	2	6	0	0	1	0,1
Athoracophorinae	0	0	0	1	1	2	1	1	2	1	2	0	2	6	0	0	1	0
Charopidae	0	1	0	0	0	*1,3	1	1	2	1	2	0,1	2	-	0	1	1	0
Punctidae	0	1	0	0	0	*1,3	1	1	2	1	2	1	2	-	0	1	1	0
Binneyinae	0	0	0	0	0	3	1	1	2	1	2	0	2	-	0	1	1	-
Ariolimacinae	0	1	0	0	0	3	1	1	2	1	2	0	2	-	0	1	1	-
Anadeninae	0	1	0	0	0	3	1	1	2	1	2	0	2	-	0	1	1	-
Arioninae	0	1	0	0	0	3	1	1	2	1	2	0	2	3,4	0	1	1	0,1
Oopeltinae	0	1	0	1	0	3	1	1	2	1	2	0	2	-	0	1	1	-
Philomycidae	0	1	0	0,1	0	3	1	1	2	1	2	0	2	3,4	0	1	1	0
Vitrinidae	0	1	0	1	0	*5,9	1	1	2	1	2	0	2	4,5	0	1	1	0
Parmacellidae	0,1	1	0	1	0	8	1	1	2	1	2	0	2	-	0	1	1	1
Boettgerillidae	0	1	0	1	0	8	1	1	2	1	2	0	2	-	0	1	1	-
Rhytididae	0	1	0	2	0	9	1	1	2	1	2	0	1,2	4,5	0,1	1	1	1
Urocoptidae	1	1	0	0,1	0	*2,4	1	1	2	1	2	0	*0,2	-	0,1	1	1	1
Trochomorphidae	0	1	0	1	0	*5,8	1	1	2	1	2	1	2	4	0	1	1	1
Endodontidae	0	1	0	0	0	1,3	1	1	2	1	2	0,1	1,2	4,5	0,1	1	1	1
Helminthoglyptidae	0	1	0	0,1	0	*1,4	0,1	1	2	1	2	0,1	2	4	0,1	1	1	0,1
Bradybaenidae	0	1	0	0	0	3	1	1	2	1	2	1	2	4	0,1	1	1	1
Polygyridae	0	1	0	0	0	1,3	1	1	2	1	2	1	2	4,5	0	1	1	0,1
Camaenidae	0	1	0	0,1	0	1,3	0,1	1	2	1	2	0,1	2	4	0,1	1	1	0,1
Sagdidae	0	1	0	0	0	1,3	1	1	2	1	2	0,1	1,2	-	0,1	1	1	1
Haplotrematidae	0	1	0	1,2	0	*7,9	1	1	2	1	2	1	1,2	4	0,1	1	1	1
Ferussaciidae	0,1	1	0	0	0	1,3	1	1	2	1	2	0	2	4	0,1	1	1	1
Subulinidae	0	1	0	0	0	*0,3	1	1	2	1	2	0,1	2	4,5	0,1	1	1	1
Achatinidae	0	1	0	0	0	1,3	1	1	2	1	2	0	2	4	0,1	1	1	1
Sphincterochilidae	0	1	0	0,1	0	1,3	1	1	2	1	2	0	2	-	0	1	1	-
Hygromiidae	0	1	0	0	0	1,3	1	1	2	1	2	0	2	4	0	1	1	1
Helicidae	0	1	0	0,1	0	1,3	1	1	2	1	2	0	2	3,4	0,1	1	1	0,1
Helicodiscidae	0	1	0	0	0	1	1	1	2	1	2	0	2	-	1	1	1	0
Papillodermidae	0	1	0	2	0	9	1	1	2	1	2	0	2	-	0	1	1	-
Milacidae	0	1	0	1	0	8	1	1	2	1	2	0	2	5	0	1	1	0,1
Limacidae	0	1	0,1	1	0	6,8	1	1	2	1	2	0	2	3,5	0	1	1	0
Agriolimacidae	0	1	0,1	1	0	6,8	1	1	2	1	2	0	2	4	0	1	1	0
Euconulidae	0	1	0	1	0	6,7	1	1	2	1	2	1	2	4	0,1	1	1	1
Bulimulidae	0	0,1	0	0,1	0	*1,5	0,1	1	2	1	2	0,1	2	4	0,1	1	1	1
Acavidae	0	0,1	0	0,1	0	3,4	1	0,1	1,2	0,1	1,2	0,1	*0,2	4,5	0,1	1	1	1
Zonitidae	0	1	0	0,1	0	*6,8	1	1	2	1	2	0,1	2	*3,5	0	1	1	1
Vitreidae	0	1	0	0,1	0	8	1	1	2	1	2	1	2	2	0	1	1	1
Daudebardiidae	1	1	0	1	0	9	1	1	2	1	2	0	2	-	0	1	1	-
Gastrodontidae	0	1	0	1	0	6,9	1	0,1	2	1	2	1	2	4	0	1	1	1
Urocylidae	0	1	0	1	0	*5,7	0,1	1	2	1	2	0,1	2	-	0,1	1	1	1
Helicarionidae	0	1	0	1	0	6,8	1	1	2	1	2	1	2	3,4	0	1	1	0
Ariophantidae	0	1	0	1	0	*5,9	1	1	2	1	2	0,1	2	*3,5	1	1	1	0,1
Streptaxidae	0,1	1	0	2	0	9	1	1	2	1	2	0	2	-	1	1	1	1
Oleacinidae	0,1	0,1	0	1,2	0	*3,9	1	1	2	1	2	0	1,2	4	0,1	1	1	1
Testacellidae	1	1	0	2	0	9	1	1	2	1	2	0	2	5	0	1	1	1
Systrophiidae	0	1	0	*0,2	0	9	1	0,1	2	1	2	1	1,2	-	0,1	1	1	-
Clausiliidae	0	1	0	0	0	*1,4	1	1	2	1	2	0	2	3,4	0,1	1	1	0
Oreohelicidae	0	1	0	0	0	3	1	1	2	1	2	0	1	5	0,1	1	1	1
Corillidae	0	1	0	0,1	0	3	1	1	2	1	2	0	0,2	-	1	1	1	-
Trigonochlamydidae	0,1	1	0	*0,2	0	9	1	1	2	1	2	0	2	-	0	1	1	-
Dyakiidae	0	1	0	1	0	8	0,1	1	2	1	2	0	2	4	0	1	1	1
Chlamydephoridae	0,1	1	0	2	0	9	1	1	2	1	2	0	1	-	0	1	1	-
Ellobiidae	0	0,1	0	0,1	0	3,4	0,1	0	0	1	2	0,1	2	2	0	0	0,1	0
Onchidiidae	0	0	0	0	0	3,4	0,1	0	0	0	2	0	2	2	0	0	0,1	0
Vaginulidae	0	0,1	0	0	0	4	0,1	0	1	0	0	0	2	2	0,1	0	0,1	0

19	20	21	22	23	24	25	26	27	28	29	30	31	32	33	34	35	36	37	38	39	40
2	1	0,1	0	0	0,1	0	2	1	2	*0,2	1,2	0	3,4	2	0	0	0	1	1	0	1
2	1	0	2	0	0	0	2	1	2	1,2	1,2	2,3	3,4	2	1	0	0	1	1	0	1
2	1	0	2	0	–	0	2	1	2	1	1	2,3	3,4	2	0,2	0	0	1	1	0	1
2	1	0	2	0	0,1	0	2	1	2	0	0	0	3	2	0	0	0	1	1	0	1
2	1	0	0,2	0	0	0	2	1	2	0,1	1,2	0	3,4	2	0	0	0	1	1	0	1
2	1	0	0	0	0	0	2	1	2	1	2	0	3	2	0	0	0,1	1	1	0	1
2	1	0	0	0	0	0	2	1	2	1	2	2	3	2	1	0	0	1	1	0	1
2	1	0	2	0	0	0	2	1	2	1	0,1	0	3	2	0,1	0	0	1	1	0	1
2	1	0	2	0	0	0	2	1	2	0,1	1,2	0	3	2	0,1	0	0,1	1	1	0	1
2	1	0	2	0	0	0	2	1	2	1,2	1,2	*0,3	3,4	1,2	0	0	0	1	1	0	1
2	1	0	2	0	–	0	2	1	2	1	1	3	4	2	0	0	0	1	1	0	1
2	1	0	2	0	0	0	2	1	2	1,2	1,2	3	3,4	2	0	0	0	1	1	0	1
2	1	0,1	0,2	0	1	1	2	1	1	2	1	3	4	2	1	0	0,1	1	1	0	1
2	1	0	0,2	0,1	0	1	2	1	2	1,2	1	3	4	2	0,1	0	0	1	1	0	1
2	1	0,1	3	0	1	0	0	1	3	2	1,2	3	4	2	0	0	1	1	1	0	1
2	1	0	3	0	1	1	1	0	3	2	2	3	4	2	0	0	1	1	1	0	1
2	1	0,1	0,2	0,1	0	1	2	1	2	1,2	1,2	0,2	3,4	1,2	0	0	0	1	1	0	1
2	1	0	0,2	0	0	1	2	1	2	1	2	0	4	2	0	0	0	1	1	0	1
2	1	0	0	0,1	0,1	1	2	1	2	2	1	2	4	2	0	0	0	1	1	0	1
2	1	0,1	0	0,1	0,1	1	2	1	2	2	1	2	4	2	0	0	0	1	1	0	1
2	1	0	0	0,1	0,1	1	2	1	2	2	1	2	4	2	0	0	0	1	1	0	1
2	1	1	3	1	1	1	2	1	2	2	1,2	3	4	2	0	0	0	1	1	0	1
2	1	0	3	0	0	1	2	1	2	1	0	2	4	2	0	0	0	1	1	0	1
2	1	0	3	0	0,1	1	2	1	2	2	1,2	3	4	2	0	0	0	1	1	0	1
2	1	0,1	0,1	0	1	1	2	1	2	1,2	1,2	2,3	4	2	1,2	0	0	1	1	0	1
2	1	0	0,2	0	1	1	2	1	2	0,1	0,1	3	4	2	0,1	0	0	1	1	0	1
2	1	0	2	0	1	1	2	1	2	2	2	3	4	2	0	0	0	1	1	0	1
2	1	0,1	1	0	0,1	0	2	1	2	*0,2	0,2	*0,3	3,4	2	*0,2	0	0	1	1	0	1
2	1	0,1	0,2	0	0	0	2	1	2	0	2	2	4	2	0,1	0	0,1	1	1	0	1
2	1	0	0	0,1	0	1	2	1	2	2	2	0,3	3,4	1,2	*0,2	0	0	1	1	0	1
2	1	0	0	1	0	1	2	1	2	0,1	1,2	0	3,4	2	0	0	0	1	1	0	1
2	1	0,1	2	0,1	0,1	0,1	2	1	2	*0,2	2	3	4	2	1,2	0	0	1	1	0	1
2	1	0	–	0	1	0	2	1	2	0,1	2	3	4	2	1,2	0	0	1	1	0	1
2	1	0	0	1	0,1	0	2	1	2	*0,2	2	3	4	2	1,2	0	0	1	1	0	1
2	1	0,1	0,1	0	0,1	0	2	1	2	0,1	2	3	4	2	1,2	0	0	1	1	0	1
2	1	0,1	0,2	0	0,1	0	2	1	2	1	2	3	4	2	1,2	0	0	1	1	0	1
2	1	0,1	0	0	0	0	2	1	2	1	2	2,3	4	2	1	0	0	1	1	0	1
2	1	0	2	0,1	0,1	0,1	2	1	2	1	0,2	0,2	3,4	2	0	0	0	1	1	0	1
2	1	0	2	0,1	0	0,1	2	1	2	0,1	2	2,3	3,4	2	0,1	0	0	1	1	0	1
2	1	0	–	0	0	0	2	1	2	1	2	3	4	2	1	0	0	1	1	0	1
2	1	0,1	–	0	1	0	2	1	2	0,1	2	2,3	3,4	2	1	0	0	1	1	0	1
2	1	0,1	0	0	0,1	0	2	1	2	1,2	2	3	4	2	1,2	0	0	1	1	0	1
2	1	0,1	2	0	0,1	0	2	1	2	0,1	0,2	2,3	4	2	*0,2	0	0	1	1	0	1
2	1	0	0	1	0	1	2	1	2	1	2	3	4	2	0	0	0	1	1	0	1
2	1	0	3	0	0	1	2	1	2	1	2	3	3	2	2	0	0	1	1	0	1
2	1	1	0	0	1	1	2	1	2	1	2	3	4	2	0,2	0	0	1	1	0	1
2	1	0,1	0,2	0	1	1	2	1	2	1	2	3	4	2	1	0	0	1	1	0	1
2	1	0,1	0,2	0	1	1	2	1	2	1,2	1,2	3	4	2	1	0	0	1	1	0	1
2	1	0	0,2	0,1	0,1	1	2	1	2	1,2	1,2	0	4	2	0,1	0	0	1	1	0	1
2	1	0	0,2	0	0	0	2	1	2	0,1	*0,2	2,3	4	2	2	0	0,1	1	1	0	1
2	1	0	0,2	0	0,1	0	2	1	2	*0,2	1,2	2,3	3,4	1,2	*0,2	0	0	1	1	0	1
2	1	0,1	*0,2	0,1	0,1	1	2	1	2	0	2	0,3	3,4	1,2	1	0	0	1	1	0	1
2	1	0	0,2	0,1	0	1	2	1	2	0	2	0	4	2	1	0	0	1	1	0	1
2	1	1	–	1	1	1	2	1	2	0	0	0	3	2	1	0	0	1	1	0	1
2	1	0,1	0	0,1	0	0,1	2	1	2	0,1	2	0	3,4	2	0,1	0	0	1	1	0	1
2	1	0,1	1,2	1	1	1	2	1	2	1,2	1,2	2,3	4	2	0,1	0	0	1	1	0	1
2	1	0,1	0	1	1	1	2	1	2	1	2	0,3	3,4	1,2	0	0	0	1	1	0	1
2	1	0	*0,2	1	0,1	1	2	1	2	1,2	0,2	2,3	3,4	2	1	0	0	1	1	0	1
2	1	0,1	0,2	0	0	0,1	2	1	2	*0,2	0,2	2,3	3,4	2	0,2	0	0	1	1	0	1
2	1	0	3	0	1	1	2	1	2	0	2	3	4	2	2	0	0	1	1	0	1
2	1	0	0,2	1	0	0,1	2	1	2	1	2	0	4	2	0	0	0	1	1	0	1
2	1	0,1	1	0	0	0,1	2	1	2	1,2	0	0,2	3,4	2	0,1	0	0	1	1	0	1
2	1	0	*0,2	0	0	0,1	2	1	2	0,2	0,2	2	3,4	2	0	0	0	1	1	0	1
2	1	0	2	0	0	0	2	1	2	0,1	*0,2	0	3	1,2	*0,2	0	0	1	1	0	1
2	1	0,1	0,3	0	1	1	2	1	2	0	0,2	3	4	2	0	0	0	1	1	0	1
2	1	0,1	–	1	1	0,1	2	1	2	1	2	2	4	2	2	0	0	1	1	0	1
2	1	0	3	0	0	1	2	1	2	*0,2	2	3	4	2	2	0	0	1	1	0	1
0,1	0	0	0,2	0	0	0	2	1	0	0,1	*0,2	0,1	*0,2	*0,2	*0,2	0,1	0,1	0	0,1	0	1
0,1	0,1	0	3	0	0	0	2	0	1,2	2	1,2	1,3	2,4	2	0,1	0	0,1	1	1	1	1
2	1	0	3	0	0	1	2	0	0	2	1	0	0	2	1	0	1	1	1	1	0

Appendix 1.5. *Continued*

	41	42	43	44	45	46	47	48	49	50	51	52	53	54	55	56	57
Achatinellidae	1	2	0	1	0,1	2	2	3,6	0,2	0,1	*0,2	0,1	0	2	1	0,1	1
Partulidae	1	3	0	1	1	2	2	6	0,2	0,1	1,2	0	0	2	1	0	0,1
Amastridae	1	3	0	1	1	2	2	3	0	0	*0,2	0	0	2	1	0	1
Cochlicopidae	1	3	0	1	1	2	2	*3,6	0,2	0	1,2	0	0,1	2	1	0,1	0,1
Pyramidulidae	1	4	0	1	1	2	2	3,6	0	0	*0,2	0	0,1	2	1	0	1
Vertiginidae	1	2,3	0	1	0,1	2	2	3,6	0,2	0,1	*0,2	0	0,1	2	1	0,1	1
Orculidae	1	2	0	1	1	2	2	3,6	0,1	0	0,1	0	0,1	2	1	0,1	1
Chondrinidae	1	2	0	1	1	2	2	6	0,1	0	0,1	0	0,1	2	*0,3	0,1	1
Pupillidae	1	2,4	0	1	1	2	2	3,6	0,1	0	*0,2	0	0,1	1,2	1	0,1	0,1
Valloniidae	1	*2,4	0	1	1	2	2	3,6	0,1	0	1,2	0	0,1	2	0,1	0	1
Pleurodiscidae	1	3	0	1	1	2	2	6	0	0	0,1	0	0	0,2	1	0	1
Enidae	1	2,3	0	1	1	2	2	3	1	0	*0,2	0	0,1	1,2	0,1	0	0,1
Succineidae	1	2,3	0	1	0,1	*0,2	2	*3,6	0,2	0,1	*0,2	0	0	1,2	2,3	0	0,1
Ariopeltinae	1	2	0	1	1	2	2	2,5	0	0	1,2	0	0	2	0,2	0	0
Aneitinae	1	2	1	1	0	2	2	*1,6	0	0,1	2	0	0	1	1,2	1	0,1
Athoracophorinae	1	2	1	1	0,1	*0,2	2	3,6	0	0,1	2	0	0	1	2,3	0	0
Charopidae	1	2	0	1	1	2	2	*3,6	*0,2	0,1	0,1	0	0,1	1,2	1	0	1
Punctidae	1	2	0	1	1	2	2	6	0,2	0,1	1	0	0	0,2	1	0	1
Binneyinae	1	2	0	1	1	2	2	3	0	0	1	0	0	2	0,1	0	1
Ariolimacinae	1	2	0,1	1	1	2	2	2,5	0	0,1	2	0	0	1,2	*1,3	0	0,1
Anadeninae	1	2,3	0,1	1	1	2	2	6	0	0	2	0	0	0	1,3	0	0,1
Arioninae	1	2,4	0	1	1	2	2	*4,6	0	0	2	0	0	0,2	1,2	0	1
Oopeltinae	1	2	0	1	1	2	2	4,6	0,2	0	2	0	0	2	1	0	1
Philomycidae	1	2,3	0	1	1	2	2	2,6	2	1	1,2	0	0	0,1	1	0	1
Vitrinidae	1	2	0	1	1	2	2	*0,6	2	0,1	1,2	0	0,1	0,2	1,3	0	0,1
Parmacellidae	1	2	0	1	1	2	2	*0,5	0	0,1	2	0	0	1,2	1	0	0,1
Boettgerillidae	1	2	0	1	1	2	2	6	2	0	2	0	0	2	1	0	1·
Rhytididae	1	*2,4	0	1	1	2	2	6	0,2	0	1,2	0	0	2	2	0	0,1
Urocoptidae	1	2	0	1	1	2	2	3,6	0,2	0,1	0,1	0	0,1	2	*0,3	0	1
Trochomorphidae	0,1	2	0	1	1	2	2	3,6	0	0,1	0,2	0	0	2	*0,3	0	1
Endodontidae	1	2	0	1	0	2	2	6	0,2	0,1	0	0	0	2	0,1	0	1
Helminthoglyptidae	1	2	0	1	1	2	2	*0,6	*0,2	0,1	*0,2	0	0,1	*0,2	1,3	0	1
Bradybaenidae	1	2	0	1	1	2	2	0,2	0,1	0,1	0	0	0	0,1	1	0,1	1
Polygyridae	1	2,3	0	1	1	2	2	3,6	*0,2	0,1	*0,2	0	0,1	*0,2	1	0	0,1
Camaenidae	1	2	0	1	1	2	2	3,6	0,1	0	*0,2	0	0,1	*0,2	1	0	0,1
Sagdidae	1	2	0	1	1	2	2	*1,6	*0,2	0,1	*0,2	0	0,1	*0,2	1,3	0	1
Haplotrematidae	1	2,4	0	1	1	2	2	3,5	0,2	0,1	0,1	0	0	*0,2	1	0	1
Ferussaciidae	1	2,3	0,1	1	0,1	2	2	3,6	0,2	0,1	*0,2	0	0	1,2	0,1	0	1
Subulinidae	1	2,3	0	1	1	2	2	3,6	*0,2	0,1	1,2	0	0,1	*0,2	0,1	0	1
Achatinidae	1	2	0	1	1	2	2	6	0,2	1	1	0	0	0,1	0,1	0	1
Sphincterochilidae	1	2	0	1	1	2	2	4,5	0,1	0,1	*0,2	0	0,1	2	1	0	1
Hygromiidae	1	2,3	0	1	1	2	2	*2,6	0,1	0,1	*0,2	0	0	0,1	1	0	0,1
Helicidae	1	2,3	0	1	1	2	2	*2,6	0,1	0,1	*0,2	0	0,1	1,2	*0,3	0	0,1
Helicodiscidae	1	2	0	1	1	2	2	3,6	0,1	0	0	0	0	2	1	0	1
Papillodermidae	1	2	1	1	1	2	2	6	2	0	0	0	0	2	2	0	1
Milacidae	1	3	0	1	1	2	2	4,5	0	0,1	2	0	0	2	1	0	0,1
Limacidae	1	2,3	0	1	1	2	2	*3,6	2	0,1	2	0	0	2	1	0	1
Agriolimacidae	1	2,3	0	1	1	2	2	3,6	2	1	2	0	0	*0,2	1	0	1
Euconulidae	1	2,3	0	1	0,1	2	2	*1,6	0,1	0,1	2	0	0	*0,2	*0,3	0,1	0,1
Bulimulidae	1	2	0	1	1	2	2	3,6	*0,2	0,1	*0,2	0	0,1	*0,2	1	0	0,1
Acavidae	1	2	0	1	1	1,2	2	*3,6	0,2	0,1	0,2	0	0,1	1,2	1	0,1	0,1
Zonitidae	1	3	0	1	1	2	2	4,6	*0,2	0,1	*0,2	0	0	*0,2	1	0	0,1
Vitreidae	1	2	0	1	1	2	2	*4,6	0,2	0,1	0,2	0	0	*0,2	1	0	1
Daudebardiidae	1	3	0	1	1	2	2	6	0,1	0,1	0,2	0	0	*0,2	1	0	1
Gastrodontidae	1	2,3	0	1	1	2	2	1,3	0,2	0,1	2	0	0	*0,2	1	0,2	0,1
Urocylidae	1	2	0	1	1	2	2	*0,6	0,1	0,1	*0,2	0	0	*0,2	1,3	0	0,1
Helicarionidae	0,1	2	0	1	0,1	2	2	4,6	0,1	0,1	1,2	0	0	0,1	0	0	1
Ariophantidae	1	2	0	1	0,1	2	2	*0,6	0,1	0,1	*0,2	0	0	*0,2	0,1	0	1
Streptaxidae	1	4	0	1	0,1	2	2	3,6	2	0	0	0	0	*0,2	0,1	*0,2	0,1
Oleacinidae	1	*2,4	0	1	0,1	2	2	*3,6	*0,2	0,1	0,1	0	0,1	1,2	0,1	0,2	1
Testacellidae	1	2	0	1	1	2	2	3,6	0,2	1	1,2	0	0,1	0,2	1	0	1
Systrophiidae	1	3	0	1	1	2	2	3,6	0,2	1	0,1	0	0,1	0,1	0	0	1
Clausiliidae	1	2	0	1	1	2	2	3,6	*0,2	1	0,2	0	0,1	1	0	0	1
Oreohelicidae	1	2	0	1	1	2	2	6	0,1	0,1	0,1	0	0	1,2	0,1	0	0,1
Corillidae	1	2	0	1	1	2	2	5,6	0,2	0	0,1	0	0,1	2	1	0,1	1
Trigonochlamydidae	1	2,3	0	1	1	2	2	6	0	1	2	0	0,1	2	1	0	1
Dyakiidae	1	2	0	1	1	2	2	0,2	0	1	2	0	0	2	1	0	1
Chlamydephoridae	1	2	0	1	1	2	2	6	0	1	2	0	0	2	1	0	0
Ellobiidae	0,1	2	0	0	0,1	*0,2	0,1	4,6	2	0	*0,2	0,1	0	0,2	*0,2	0,1	1
Onchidiidae	1	1	0	0,1	0	1	1	*1,6	2	0,1	2	0	0	2	1,2	0,1	0,1
Vaginulidae	0	0	0	0,1	0	1,2	1	3,6	2	0	2	0	0	2	2	0,1	0,1

Appendix 1.6. The List of the Apomorphies for the Principal Nodes in the Phylogram of Relationships in Terrestrial Pulmonata, Shown in Fig. 1.6, Based on Parsimony Analysis as Detailed in the Legend of Fig. 1.3

Character	Change	Character	Change	Character	Change
node 1 ---> node 2		33	1 ==> 2	41	0 ==> 1
9	0 ==> 1	41	0 ==> 1	42	0 ---> 1
11	0 ==> 1	43	0 ---> 1	node 5 ---> Vaginulidae	
19	0 ==> 2	48	0 ==> 1	6	3 ==> 4
20	0 ==> 1	51	0 ==> 1	9	0 ==> 1
32	0 ==> 3	55	0 ==> 1	19	0 ==> 2
33,	0 ==> 1	57	1 ---> 0	20	0 ==> 1
37	0 ==> 1	node 3 ---> node 7		36	0 ==> 1
38	0 ==> 1	9	1 ==> 2	48	1 ==> 3,4,5,6
40	0 ==> 1	10	0 ==> 1	55	1 ==> 2
42	0 ----> 2	11	1 ==> 2	node 7 ---> node 9	
44	0 ==> 1	23	0 ==> 1	6	0 ==> 1
47	0 ==> 2	46	1 ==> 2	13	0 ==> 1
54	0 ----> 1	node 3 ---> node 8		node 7 ---> node 10	
node 1 ---> node 4		6	0 ==> 3	15	0 ---> 1
6	0 ==> 3	7	0 ==> 1	41	0 ==> 1
8	1 ---> 0	25	1 ==> 0	48	0 ==> 3
14	3 ==> 2	31	0 ---> 2	node 9 ---> node 11	
26	0 ==> 2	41	0 ==> 1	31	0 ==> 2
30	0 ---> 1	45	0 ==> 1	33	1 ==> 2
34	0 ---> 1	48	0 ==> 3	41	0 ==> 1
48	0 ==> 1	55	0 ==> 1	node 9 ---> node 12	
49	0 ---> 2	node 4 ---> node 5		2	0 ==> 1
51	0 ---> 1	13	0 ==> 2	6	1 ==> 5
node 2 ---> node 3		22	0 ==> 3	50	0 ==> 1
14	3 ==> 4	29	0 ==> 2	54	1 ---> 0
16	0 ==> 1	33	0 ==> 2	node 10 ---> node 13	
17	0 ==> 1	37	0 ==> 1	6	0 ==> 1
18	0 ==> 1	38	0 ==> 1	23	1 ==> 0
26	0 ==> 2	39	0 ==> 1	node 10 ---> node 14	
27	0 ==> 1	47	0 ==> 1	2	0 ==> 1
28	0 ==> 2	51	1 ==> 2	7	0 ==> 1
53	0 ==> 1	54	0 ==> 2	13	0 ==> 1
node 2 ---> node 6		55	0 ==> 1	49	0 ---> 1
5	0 ==> 1	node 5 ---> Ellobiidae		node 11 ---> node 17	
6	0 ==> 1	10	0 ==> 1	6	1 ==> 3
9	1 ==> 2	11	0 ==> 2	7	0 ==> 1
10	0 ==> 1	25	1 ==> 0	29	0 ---> 1
11	1 ==> 2	27	0 ==> 1	48	0 ==> 2
13	0 ==> 2	40	0 ==> 1	node 11 ---> node 18	
24	0 ==> 1	48	1 ==> 4,5,6	32	3 ==> 4
29	0 ==> 2	node 5 ---> Onchidiidae		34	0 ==> 2
30	0 ==> 1	11	0 ==> 2	45	0 ==> 1
31	0 ==> 3	25	1 ==> 0	node 12 ---> node 19	
32	3 ==> 4	40	0 ==> 1	13	1 ==> 2

Footnote for Appendix 1.5 (opposite)

See Appendix 1.4 for description of the characters. Characters with more than one state indicate polymorphism within the taxon. *signifies that three or more character states were present in the original data but are reduced to two (extremes in arithmetic series) in this tabulation due to space constraints.

Appendix 1.6.   *Continued*

Character	Change	Character	Change	Character	Change
node 12 ---> node 20		30	0 ==> 2	node 26 ---> node 28	
33	1 ==> 2	48	0 ==> 3	6	0 ==> 1
41	0 ==> 1	53	1 ==> 0	15	1 ---> 0
45	0 ==> 1	node 20 ---> node 21		18	1 ==> 0
node 13 ---> node 15		4	0 ==> 1	45	0 ==> 1
12	0 ==> 2	13	1 ==> 2	55	0 ==> 1
25	1 ==> 0	51	0 ==> 1	node 29 ---> node 30	
node 13 ---> node 16		53	1 ==> 0	6	1 ==> 3
2	0 ==> 1	54	0 ---> 1	22	0 ==> 2
7	0 ==> 1	55	0 ==> 1	42	2 ==> 3
45	0 ==> 1	node 20 ---> node 22		45	0 ==> 1
node 14 ---> node 25		6	5 ==> 7	51	0 ==> 1
6	0 ==> 1	7	0 ==> 1	node 29 ---> node 31	
33	1 ==> 2	48	0 ==> 3	2	0 ==> 1
53	1 ==> 0	node 26 ---> node 27		7	0 ==> 1
node 14 ---> node 26		30	1 ==> 2	30	0 ==> 1
13	1 ==> 2	31	0 ==> 2	33	1 ==> 2
30	0 ==> 1	33	1 ==> 2	54	1 ==> 2
node 19 ---> node 23		34	0 ---> 1	55	0 ==> 1
7	0 ==> 1	49	1 ---> 2		

Key: ==> unambiguous change; ---> occurs under some reconstructions, but not others.

2   Body Wall: Form and Function

D.L. LUCHTEL AND I. DEYRUP-OLSEN[1]

Department of Environmental Health, School of Public Health and Community Medicine; [1]Department of Zoology; University of Washington, Seattle, Washington, USA

Introduction

This chapter presents a series of topics that may seem disparate at first glance, but are in fact closely linked in a functional sense. These topics are the structure of the integument, regulation of the water and solute contents of the body, the form and plasticity of the body, the coordinated mechanisms effecting the animal's movements, and self-defence.

The body of the gastropod mollusc is invested with epithelium and, in many cases, is protected further by a shell secreted by specialized epithelial cells. Underlying the epithelial layer of the gastropod's integument, complex arrays of muscle and connective tissue complete its basic structure. The muscles are richly innervated, and movement and changes in body form result from configurations of muscle contraction or relaxation in conjunction with variable distribution of haemolymph. Both stability (adhesion) and movement (locomotion) in space also depend on mucus and other secretions of the gastropod's epithelium and associated glands. Diffusion and active transport processes of the highly permeable integument play major roles in water and ion exchanges. As in all animals, these processes and their exquisite integration are crucial to the gastropod's survival.

The Integument

The term 'integument' is used rather than 'body wall' or 'skin' to designate the tissues that make up the external form of terrestrial gastropods, be it in the area of the head, mantle, sole or the posterior/dorsal integument. The literature on histology of the integument and associated glands was reviewed by Hyman (1967). The ultrastructural literature was reviewed more recently by Luchtel *et al.* (1997).

Epithelium

The integument is composed of an epithelium (with epithelial cells on a basement membrane) and subepithelial muscle and connective tissue. The epithelium is composed primarily of a single layer of two types of columnar- to cuboidal-shaped cells – microvillous cells and ciliated cells. These two cell types are distinguished by the presence of microvilli and cilia on their apical surfaces, respectively (Fig. 2.1). Microvillous cells apparently are the most numerous of the epithelial cells and the apical microvilli have absorptive and transporting functions (Newell, 1974; Machin, 1977; Simkiss, 1988; Prior *et al.*, 1994). The distribution of ciliated cells is variable – they are relatively sparse over much of the dorsal integument, but occur in great abundance in specific areas, such as around the pneumostome, on the mantle and extensively on the pedal sole.

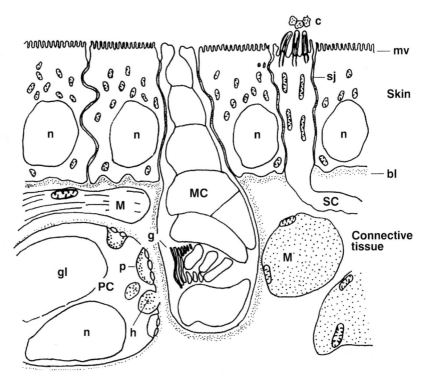

Fig. 2.1. Schematic diagram of the major cell types in the dorsal integument of a generalized stylommatophoran. bl, basal lamina (basement membrane); c, cilia; gl, glycogen; g, Golgi zone; h, haemocyanin; M, muscle; MC, mucus-secreting cell; mv, microvilli (on microvillous cells); n, nucleus; p, pores; PC, pore cell (rhogocyte); SC, primary sensory cell; sj, septate junction. From Newell (1977).

In addition, the epithelium has large secretory cells, which extend into the matrix of muscle and connective tissue that make up the subepithelial part of the integument. These large cells include mucous cells, channel cells and calciferous cells. The older literature often referred to these cells as cellular glands (e.g. calciferous glands) or unicellular glands (Hyman, 1967). Such cells are also sometimes referred to as subepithelial but, structurally, they are a component of the epithelium as they rest on the basement membrane continuous with the basement membrane that underlies the basal surfaces of microvillous and ciliated cells. This is in contrast to nerve endings of sensory receptor cells which innervate the epithelium. The dendritic processes of these cells penetrate through the basement membrane, extend between the epithelial cells and form free nerve endings at the surface of the epithelium (Zylstra, 1972a; Zaitseva, 1984)

The histochemical composition and distribution of the secretory cell types vary greatly from species to species (Simkiss and Wilbur, 1977; South, 1992). Descriptions of these variations have dominated morphological studies of the pulmonate integument (Simkiss, 1988).

Campion (1961) described eight types of secretory cell in the dorsal integument of the helicid *Cantareus aspersus* (Müller) – four types of mucous cell that produced different kinds of mucus, one that secreted protein (probably a channel cell), one with calcium carbonate granules, one that formed a pigmented secretion that had a flavone (considered a waste product) and one with fat globules. Most of the secretory cells on the pedal sole were of a distinct kind that produced mucus combined with protein. Wondrak (1967) identified three types of 'slime-gland-cells' in the dorsal integument of the arionid *Arion ater* Linnaeus – two different types of mucous cells, and a third cell type which was probably a channel cell. Subsequently, Wondrak (1968, 1969b) described the fine structure of epithelial cells in the dorsal integument of this species.

Newell (1974, 1977) compared the epithelial fine structure of the dorsal and pedal integument in *Arion hortensis* de Férussac and the agriolimacid *Deroceras reticulatum* (Müller). The dorsal epithelia in these species are very similar, with narrow intercellular spaces and extensive interdigitation of the lateral plasma membranes between adjacent cells. The sole epithelium is composed of ciliated and microvillous cells, each often arranged in tracts, and with ciliated cells less numerous in *A. hortensis*.

Chétail and Binot (1967) distinguished four types of mucous cells, based on histochemical characteristics in the sole of *Arion rufus* (Linnaeus). Using similar techniques, Cook and Shirbhate (1983) distinguished 14 types of mucous cells in the limacid *Limacus pseudoflavus* (Evans), five of which were distributed over the dorsal surface and nine in the sole. Cook (1987) subsequently described 11 mucous cell types in the vaginulid *Veronicella floridana* (Leidy).

Our own earlier work on the structure of the dorsal integument was on *Ariolimax columbianus* (Gould) (Arionidae) (Luchtel *et al.*, 1984, 1991).

We found that 'rapid freeze' fixation preserves tissue architecture much better than chemical fixation, as the slow rate of chemical fixation allows artefactual distortion of the tissue to take place. We are extending our work to other pulmonate species, and Fig. 2.2A shows a light micrograph

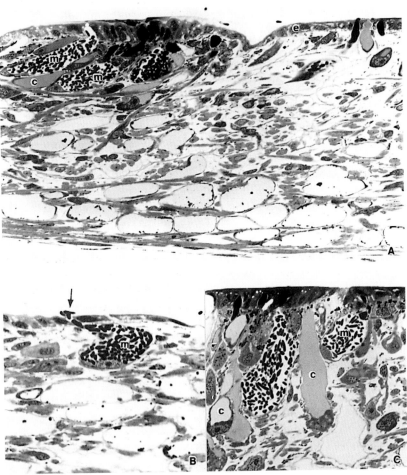

Fig. 2.2. Light micrographs of 1 μm thick epon sections, stained with toluidine blue, of the dorsal integument of *Limax maximus* Linnaeus (Limacidae). (A) The overall histological appearance of the integument of a juvenile animal after ultrarapid freeze fixation. The entire thickness of the integument is shown. The surface epithelium consists of an epithelial cell layer (e), and includes mucous cells (m) and channel (c) cells which extend from the apical surface into the subepithelium. Most of the integument's thickness consists of a subepithelial meshwork of connective tissue and muscle cells. Magnification ×650. (B) A mucous cell (m) sectioned at the location where it opens on to the epithelial surface (arrow) and releases intact mucous granules. The individual mucous granules are pleiotropic in shape as they are released from the mucous cell. Magnification ×710. (C) Mucous cells (m) have a granular appearance because of the darkly staining mucous granules, while the contents of channel cells (c) stain homogeneously. Mucous (m) and channel cells (c) extend for variable depths into the subepithelial tissue. The deepest penetration of these cells is marked by the location of the nucleus at the basal end of the cell. Magnification ×530.

of an epon section of the dorsal integument from a juvenile of the limacid *Limax maximus* Linnaeus, after 'rapid freeze' fixation. Mucous and channel cells extend into the subepithelial connective tissue; the contents of mucous cells have a granular appearance (Fig. 2.2B and C), while those of channel cells stain homogeneously (Fig. 2.2C). Intact mucous granules are secreted on to the surface of the integument (Fig. 2.2B), and then lyse to form a mucous layer (Luchtel *et al.*, 1991).

The channel cell (Fig. 2.2C) is involved in fluid secretion (Martin *et al.*, 1982; Luchtel *et al.*, 1984, 1997). This fluid provides the medium into which the mucous granules are released. The channel cell has a large central channel (hence the name) or reservoir. Fluid and particles apparently are transported to the central channel via cytoplasmic tubules (which resemble tubules of smooth endoplasmic reticulum) that traverse the peripheral cytoplasmic layer of the cell. Previous researchers designated this cell with several names, for example 'protein cell' (Campion, 1961; Wondrak, 1968, 1969b) and 'non-muciparous cell' (Zylstra, 1972b).

Connective tissue

Connective tissue constitutes an important part of the integument, supporting the epithelial cell layer. It comprises an extracellular matrix of more or less densely packed collagen fibrils and various cell types. In the helicid *Helix pomatia* Linnaeus, collagen fibril diameters range from 25 to 70 nm, with a cross-striation pattern of periodicity 53–57 nm (Schmut *et al.*, 1980). The cell types that have been identified in the connective tissue of terrestrial gastropods include rhogocytes (pore cells), calcium cells, granular cells, vesicular connective tissue cells, pigment cells, fibroblasts (fibrocytes), muscle cells, nerve cells and processes, and amoebocytes. However, variations among species in cell structure and variations in terminology applied by researches make comparisons difficult.

Rhogocytes are distinctive cells that have been designated by a variety of names including pore cells, vesicular cells, globular cells, cells of Leydig, Kugelzeller and Blasenzeller (Fig. 2.3). They are described from a number of Stylommatophora (Wondrak, 1969a; Buchholz *et al.*, 1971; Wolburg-Buchholz, 1972; Skelding and Newell, 1975; Beltz and Gelperin, 1977; Jones and Bowen, 1979) and occur throughout the phylum Mollusca (Haszprunar, 1996). Also, cell types at least partially homologous to the rhogocyte are found in other phyla (Haszprunar, 1996). While 'pore cell' has been the term used most commonly to designate this cell type, it is appropriate now to use the name 'rhogocyte' (proposed by Fioroni *et al.*, 1984; supported by Haszprunar, 1996) – from the Greek 'rhogos' which means 'slit' – because the pores are an artefact produced by cutting slits in cross-section.

Rhogocytes are large, usually spherical to elongate in shape, but occasionally branched and somewhat irregular in form. These cells are found freely in the haemocoel as well as in the integument. They possess

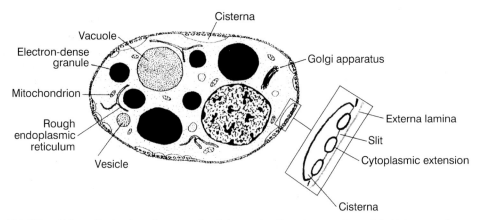

Fig. 2.3.   Schematic drawing of a generalized rhogocyte. Granules usually contain high concentrations of metal ions. In some Stylommatophora, the vacuole greatly expands in volume, is filled with glycogen and displaces other cytoplasmic organelles and the nucleus to the cell periphery. The inset shows an enlarged view of a surface cisterna. The slits are bridged with fine, fibrillar diaphragms. The cistern complex apparently functions as an inward-directed molecular sieve. Modified from Haszprunar (1996).

numerous lysosomes, lipid droplets, granules and abundant rough endoplasmic reticulum. They may have extensive glycogen stores that can displace the spherical nucleus to a peripheral position. The most characteristic feature of this cell type is the cell surface, which has numerous tubular or vesicular invaginations that are partially occluded by cytoplasmic extensions (tongues, bars), forming a system of surface cisternae that is partly enclosed yet communicates with the extracellular space. This system has been termed the Spaltenapparat, or sieve system. An external, finely fibrillar coat (basement membrane or external lamina) surrounds the cell (Skelding and Newell, 1975; Boer and Sminia, 1976).

Haemocyanin molecules have been identified in the endoplasmic reticulum and surface cisternae of rhogocytes (Skelding and Newell, 1975; Sminia and Vlugt-van Daalen, 1977). This implicates rhogocytes in haemocyanin formation, the predominant haemolymph pigment in molluscs. Rhogocytes of the freshwater pulmonate, *Helisoma duryi* Wetherby (Planorbidae) contain ferritin, and these cells may be a source of the ferritin stored in mature oocytes (Miksys and Saleuddin, 1987). On the other hand, because high molecular weight compounds such as ferritin and trypan blue appear in rhogocytes after injection into the haemocoel of the animal, rhogocytes have been ascribed a limited phagocytotic function (Buchholz *et al.*, 1971; Sminia, 1972).

Haszprunar (1996) recently reviewed the structure and function of rhogocytes. He concluded that these cells play an important role in metal ion metabolism, possibly involved in the recycling of respiratory pigments and in metal detoxification. The large amount of glycogen found in rhogocytes of some Stylommatophora suggests that these cells may also be involved in transport or storage of nutrients. Other putative functions have been assigned to rhogocytes, including collagen synthesis (Plummer, 1966) and accumulation of calcium leading to differentiation into calcium cells (Fournié and Chétail, 1984). In *Limax* Linnaeus, interstitial cells identified by Reger (1973) as containing and presumably synthesizing haemocyanin do not correspond structurally to the usual description of rhogocytes.

Calcium cells have been observed in the integument of a wide range of pulmonates (Fournié and Chétail, 1982). These cells typically are packed with vacuoles containing concretions (up to 20 μm in diameter) formed of concentric layers of calcium carbonate and organic materials. Besides their probable roles as storage sites for ions needed in shell growth and repair, reproduction and mucus formation, calcium cells may play a role as a source of bicarbonate ions for pH regulation.

Vesicular connective tissue or glycogen cells (also termed vesicular cells, Leydig cells or chondroid cells) are large cells with a thin rim of cytoplasm surrounding a large nucleus and an inner vacuole presumably packed with glycogen, the contents of which are usually lost during the embedding process. Their role is apparently for nutrient storage and release.

Granular cells are characterized by cytoplasm that contains glycoprotein granules that sometimes pack the cells enough to deform the nuclei. The granules appear to originate in the extensive Golgi apparatus, and are sometimes seen in electron micrographs in fusion with the cell membrane, indicative of exocytosis. Such cells are described in several stylommatophoran species (Wondrak, 1969a; Steinbach, 1977). Their function is not clearly established, beyond the obvious implication that they secrete specific glycoproteins. A possible neurosecretory role was suggested by Steinbach (1977). Glycogen and granular cells may be modified rhogocytes, and Haszprunar (1996) suggested that this may also be the case for calcium cells.

Fibroblasts, also called 'undifferentiated cells', are widely distributed spindle-shaped cells, often with branching processes, of relatively small size. The nucleus is chromatin rich. Some of the cells show intimate connection with the collagen fibrils of the extracellular connective tissue, at invaginations or even discontinuities of the cell membrane, and the presence of fibrils in the cytoplasm. These cells are implicated in producing the collagen fibrils in connective tissue (Wondrak, 1969a).

Pigment cells are found throughout the integument, and frequently are most numerous in the mantle and sole. These cells are irregular in form, often with long processes extending among other connective tissue elements. Characteristically, they contain many pigment granules (Wondrak, 1969b).

Amoebocytes in connective tissue may be a subpopulation of the migratory population of amoebocytes found in pulmonate haemolymph. They are of generally spherical form with irregular pseudopods, and the presence of numerous vacuoles in the peripheral cytoplasm suggests endocytotic activity. Numerous membrane-bounded structures appear to be lysosomes and often contain fibrils, granules and other cell debris. There is a conspicuous Golgi system, and a rough endoplasmic reticulum of varying appearance. Mitochondria are spherical or ovoid. These cells are phagocytotic and capable of ingesting particulate matter (e.g. ink, bacteria and yeast-derived zymosan granules) and then clearing it from the body, either through intracellular digestion or by egress of the cells from the animal's body. Brown (1967) injected *C. aspersus* with thorium dioxide, and observed that amoebocytes phagocytosed the material. Many of the thorium-laden amoebocytes then migrated through the walls of the reproductive organs (common duct, vagina and penis), accumulated in the lumens of the ducts, and were eventually cleared from the animal. Besides their phagocytic roles, amoebocytes are implicated in digestion and transport of energy sources, for example glycogen (Wagge, 1955). Abolins-Krogis (1972), in *H. pomatia*, described the ultrastructural features of amoebocytes involved in shell repair. These cells apparently transport nutritive substances by uptake of materials from granular cells in the digestive gland, and then migrate to the mantle area where repair processes take place.

The functions of the above cell types are far from clearly understood. The rhogocytes are perhaps the most intriguing because of their uniquely complex surface structure and the diversity of functions attributed to them. The granule cells appear to have specific secretory roles that are still unknown, and calcium and vesicular cells have been ascribed a wide variety of functions including mechanical support, mineral and nutrient storage, and detoxification. Nerves provide significant innervation to connective tissue, but their roles are not fully delineated.

Muscle

The structural integrity of the integument is largely dependent on the complex multidimensional array of subepithelial muscle cells and collagenous connective tissue. Muscle cells are organized in the dorsal and pedal integument as sheets, bands or columns, or occur individually in arrays that are longitudinal, transverse, radial or oblique with respect to the major axis of the body (Fig. 2.4). Muscle cells are innervated and, through integrated contraction patterns operating in conjunction with the haemocoelic fluid, serve as the animal's hydrostatic skeleton (Brown and Trueman, 1982; Nicaise and Ansellem, 1983; Kier, 1988).

Ejection of secretions of epithelial secretory cells, on to the surface of the integument, apparently is driven by contraction of fine networks of muscles surrounding them (Campion, 1961). The muscles in the integument may assist in the dispersion of mucus over its surface once it has been ejected, aided in some cases by ciliary action. In most species, the muscles of the sole are involved in locomotion through the generation of pedal waves. In other (mostly small) species, locomotion is based entirely on cilia.

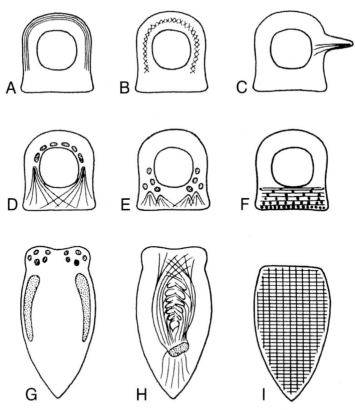

Fig. 2.4. Diagrams of the major muscle systems in the integument of *Lymnaea stagnalis* (Linnaeus) (Lymnaeidae) illustrative of the pattern in pulmanate gastropods. (A–F) Cross-sections of the head–foot region: (G–I) muscle systems of the foot in the floor of the cephalopedal sinus. (A) Circular muscle system. (B) Diagonal muscle system. (C) Ventral tentacle retractor system. (D) Longitudinal muscle system. (E) Columellar muscle system. (F) Horizontal foot muscle system. (G) Longitudinal muscles. (H) Columellar muscle system. (I) Horizontal foot muscle. From Plesch *et al.* (1975).

The muscle cells of the integument are spindle shaped, uninucleate and unstriated. Within the cells, thin (actin) and thick filaments (either paramyosin or myosin, the latter often with a paramyosin core) mingle in precise arrays and interconnect with cross-bridges. Thin filaments attach to or traverse fusiform dense bodies (Z-bodies). These provide the stable basis, in analogy to the Z-discs of vertebrate muscle, for displacement of thick filaments relative to thin filaments during shortening or tension development by the muscle cell.

Muscle cells usually connect directly with the collagenous connective tissue surrounding them, rather than forming muscle-to-muscle connections. Davis and Shivers (1987), in the integument of *L. maximus*, described the structure of electron-dense plaques in the membranes of muscle cells. The plaques are oval, of variable size (~0.2 µm width × 0.6 µm length), and consist of a mesh of electron-dense filaments that is believed to be α-actinin. Delicate filaments arise in the region of these plaques to traverse a thin 'lamina externa' that intimately surrounds the cell, to link with collagen fibrils in the surrounding connective tissue. Presumably this organization ensures that the energized displacement of thick on thin muscle filaments can be transmitted to connective tissue, and integrated to effect tension or contraction in the integument.

The muscle cells of the gastropod integument possess invaginations in the cell membrane, either as tubules lined with basal lamina or as simpler caveolae. These associate with a system of sarcoplasmic vesicles or tubules that connect and penetrate throughout the cell (North, 1963; Nisbet and Plummer, 1968). The properties and organization of muscle proteins in molluscs are reviewed extensively by Chantler (1983). Muscle cells contain mitochondria in varying numbers, and stored glycogen is present reflecting the cell's metabolic demands.

Muscle cells are innervated by nerve fibres originating from the anterior ganglia (circumoesophageal nerve ring, cerebrum) or from numerous peripheral ganglia located throughout the body (reviewed by Bullock and Horridge, 1965). Coordinated contractions of muscles of the body wall play crucial roles in the activities of gastropods, although muscle function is known only in outline.

Nerve axons have been seen to connect with muscle cells in a variety of ways (Nicaise and Ansellem, 1983). Typically nerve endings contain vesicles suggestive of neurotransmitter or neurohormone storage and release (Bogusch, 1972; Heyer *et al.*, 1973). A single neuron can innervate several muscle cells or, conversely, a single muscle cell can be innervated by several neurons.

Mucus and Fluid Secretions of the Integument

The gastropod integument functions in respiratory exchange, water flux and ion regulation. It is a protective barrier as well as a secretory, absorptive and transporting tissue (Machin, 1977; Simkiss, 1988). Perhaps the most conspicuous property of the gastropod integument is its facility to produce copious amounts of mucus and proteineous fluid (Simkiss and Wilbur, 1977; Simkiss, 1988). The secretions are thought to form a continuous layer over the entire body surface, although the ultrastructural characteristics of such a layer are not well demonstrated.

In *L. pseudoflavus*, Cook and Shirbhate (1983) noted that the nature of mucous secretions on the body surface varies, which suggests different functions for mucus on different areas of the epithelium. The dorsal integument is covered by viscous mucus while the pedal mucus is more fluid. The dorsal and pedal sheets of mucus are separated by a weakly acidic mucus secreted by cells in the parapodial groove, situated between the sole and the dorsal body surface.

The viscosity of mucus is highly dependent on its divalent ion content, which can vary between species and with the physiological condition of the animal. In *D. reticulatum*, for example, the mucus in undisturbed animals has a relatively low divalent ion content and is both clear and low in viscosity, but in disturbed animals the mucus is milky white, very thick and high in calcium salt content. Also, some mucus is pigmented, for example the yellow mucus of *Arion subfuscus* (Draparnaud) (South, 1992).

We found that the large mucous cells in the integument of arionid and agriolimacid slugs secreted or liberated intact mucus granules on to the surface of the integument (Deyrup-Olsen *et al.*, 1983). Such intact granules can be observed in droplets of fluid collected from the body surface (Fig. 2.5A and C). Ultrarapid freezing and freeze-substitution preparation techniques are necessary to observe such intact granules at the ultrastructural level as they are not preserved after chemical fixation (Luchtel, 1989; Luchtel *et al.*, 1991). The stability of secreted granules varies with the ionic composition of the isosmotic medium in which they are suspended and they are ruptured rapidly in the presence of adenosine triphosphate (Luchtel *et al.*, 1991; Deyrup-Olsen *et al.*, 1992). In nature, the granules rapidly lyse on the body surface to form a viscous mucous layer of fibro-granular texture (Fig. 2.5B).

Simkiss and Wilbur (1977), using phase contrast and differential interference optical microscopy, and electron microscopy, demonstrated the complexity of the components in mucus in a study on the mucous trails of *C. aspersus*. These authors found the mucous trails to be highly heterogeneous in composition, being comprised of four types of structural elements: (i) fine filaments many hundreds of micrometres long which were oriented parallel to the direction of animal motion; (ii) birefringent granules either free or attached to filaments; (iii) elongate rodlets about

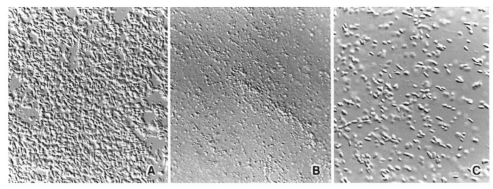

Fig. 2.5. Differential interference contrast micrographs of mucous granules collected on glass slides. (A) Fresh, unfixed mucous granules from *Deroceras reticulatum* (Müller) (Agriolimacidae). Electrical stimulation of the posterior integument produced a fluid secretion that was collected in a buffered medium. A drop was pipetted on to a glass slide and viewed without further treatment. (B) Unfixed mucous layer, from *D. reticulatum*, after the mucous granules lyse and form a fibrogranular-appearing mat. (C) Fresh, unfixed mucus granules from *Prophysaon foliolatum* (Gould) (Arionidae). The fluid secreted by *P. foliolatum* is less concentrated than that of *D. reticulatum* and the individual granules are more dispersed. All at magnification ×350.

4.5 µm long and with an aspect ratio of 30 : 1; and (iv) delicate membranes hundreds of square micrometres in area. The specific nature of these components and how they are formed during and after secretion of mucus granules remain unknown. The elongate rodlets, however, were also observed in samples of mucus taken from within the suprapedal gland, and Simkiss and Wilbur (1977) could not determine whether they represented fragments of cilia or, more probably, normal components of the mucus. Although not mentioned by Simkiss and Wilbur (1977), the membranes might be remnants of those that bound intracellular mucus granules.

Martin and Deyrup-Olsen (1982, 1986) described the secretions that can be obtained after experimental stimulation of the integument of *A. columbianus*. Such secretions are often voluminous, and produced by cells of the epithelium, particularly by the mucous and channel cells (Luchtel *et al.*, 1984, 1991). Our original investigations on the structure of the channel cell were prompted by studies to identify the route by which macromolecules and large volumes of fluid traverse the integument (Luchtel *et al.*, 1984). We found that material injected into the haemocoel, even particles as large as ink particles (estimated at 10–100 nm diameter), entered large (up to 500 µm in length) channel cells and then passed to the exterior via the apical opening of these cells (Fig. 2.6).

As mentioned earlier, the channel cell was designated by several different names in the literature, but earlier investigators did not have the physiological data to determine the unique function of the channel cell as distinct from the roles of other epithelial cell types. Our data indicate that these cells function by ultrafiltration of haemolymph components into their central channels where they are modified prior to

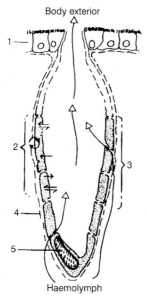

Fig. 2.6. Schematic diagram of a channel cell from the dorsal integument of *Ariolimax columbianus* (Gould) (Arionidae), showing its junction with microvillous cells at the surface of the integument and several of its structural and functional properties (shown in the basal portion of the cell; the apical portion with dashed lines is not drawn to scale). (1) microvillous cells; (2) ion exchange and transport processes modify the composition of fluid in the central vacuole; (3) channel cell cytoplasm with tubules through which haemolymph ultrafiltrate passes into the central vacuole (channel); (4) basement membrane; (5) nucleus. Modified from Deyrup-Olsen and Martin (1987).

expulsion on the body surface (Martin and Deyrup-Olsen, 1982, 1986; Martin *et al.*, 1982, 1990; Luchtel *et al.*, 1984; Deyrup-Olsen *et al.*, 1989). Acetylcholine, 5-hydroxytryptamine (serotonin) and the neurohormone arginine vasotocin stimulated fluid movement through the channel cell, while noradrenaline and atropine inhibited fluid output. These observations led to the conclusion that the production of fluid by channel cells is under neurohormonal control.

Regulation of Water and Solute Exchange

The availability of adequate supplies of water is a crucial requirement for terrestrial animals, and especially so for gastropods with their generally permeable integument and high body content of water (ranging from 78 to 92% of shell-free body weight, Machin, 1975; Lyth, 1983; Arad, 1993a; and others). Their survival depends in large part on their behaviour, but also on their remarkable abilities to tolerate dehydration. Tolerance of water loss of 60–80% of body weight has been demonstrated for several limacid slugs and helicid snails (data summarized by Machin, 1975). Small Pupillid snails (0.5–0.15 mg) tolerated 30–50% loss of body water (Barnhart, 1989). Because of the importance of water in the lives of terrestrial gastropods, this topic has been the subject of extensive research and several reviews (Runham and Hunter, 1970; Machin, 1975; Prior, 1985; South, 1992).

In gastropod snails, the shell serves to reduce the extent of the integument that is exposed to the evaporative demands of air at less than 100% humidity. When terrestrial prosobranch snails withdraw the body into the shell, the shell aperture generally can be closed with a gate, the operculum (among terrestrial prosobranch, the operculum is absent only in some Helicinidae). This structure is absent in stylommatophoran snails, but in some cases is functionally substituted for by the epiphragm, a sheet of dried, and sometimes calcified, mucus (Solem, 1985). Often the epiphragm cements the shell aperture to the substrate. Machin (1967) compared the capacity for conservation of water, hence for survival, of *C. aspersus* with that of the arid-adapted members of the family Helicidae (*Otala lactea* Müller) and Sphincterochilidae (*Sphincterochila boissieri* de Charpentier). The superior performance of the two desert snails correlated with the thickness of the shells and epiphragms, and the relative reductions in the area of the shells' apertures relative to the total area of shell-plus-aperture. In studies on the comparative resistance to dehydration in the caenogastropod *Pomatias olivieri* de Carpentier (Pomatiasidae) and that in *Sphincterochila cariosa* (Olivier) and *Helix engaddensis* Bourguignat, Arad (1993b) and Rankevitch *et al.* (1996) found marked differences in individuals, apparently depending to a large extent on the behaviour of the animals and specific characteristics of the microenvironments to which they were adapted. Arad (1993b) observed that the caenogastropod species was more susceptible to dehydration than the two stylommatophoran species, and questioned the effectiveness of the caenogastropod operculum as a barrier to water loss.

In contrast to snails, gastropods of the slug form have no protective shell, and are limited in their distribution to places that are at least seasonally moist. They rely for survival on their ability to tolerate extreme loss of body water and on behaviour that minimizes water loss.

Terrestrial gastropods have suites of physiological mechanisms to conserve and acquire water. At the cellular level, mantle edge (collar) cells may oppose loss of water from the body under conditions of water stress (Machin, 1966; reviewed by Machin, 1975, 1977). It is not known in detail how this is effected, but in the case of *O. lactea*, Machin and co-workers showed that during aestivation the mantle cells were highly impermeable to water. These cells were observed to have an apical region of very high osmotic pressure that was correlated with high concentrations of potassium and chloride ions. The region was located below the microvillous surfaces of the cells and just above an inner region containing a dense layer of vesicles (Machin, 1974; Newell and Machin, 1976; Appleton *et al.*, 1979; Newell and Appleton, 1979). This unusual structure may explain the cells' resistance to water transfer. Mucus on the body surface of terrestrial gastropods is not a significant barrier to evaporation of water (reviewed by Machin, 1975). Rather, secretion of mucus can be a major route of water loss, and this process is depressed during periods of inactivity or dehydration.

A well-known mechanism for defence against dehydration for terrestrial animals is storage of water within the body; for example, in the urinary bladders of desert amphibians and chelonian reptiles (Bentley, 1966; Minnich, 1979; Petersen, 1996). It is possible that the very large haemolymph volumes of terrestrial gastropods, in combination with their tolerance of dehydration, function in this way. Blinn (1964) observed retention of fluid in the pallial cavity of several stylommatophorans, including *Mesodon thyroidus* (Say), *Allogona profunda* (Say) (Polygyridae) and *H. pomatia*, and suggested that this fluid may serve as a water reservoir. A similar suggestion was made by Deyrup-Olsen and Martin (1987) concerning the retention of hypotonic fluid in the rectum of *A. columbianus*.

The metabolic rate decreases in dehydrated terrestrial gastropods. The mechanism accounting for this is not understood, but it doubtless functions to reduce demands on circulatory, digestive and excretory systems stressed by the inadequacy of water supply. Feeding is deeply depressed or stops altogether in dehydrated animals (Phifer and Prior, 1985). Beyond these generalizations, it is not known how these animals cope with the range of cellular changes that inevitably must accompany drastic dehydration. On the other hand, the rise in osmotic pressure (fall of water concentration) of the body fluids in dehydration reduces the rate of evaporative loss of water from the body. In addition, it facilitates rehydration when water becomes available. On access to water, the animals rehydrate within minutes to hours with fast, parallel reductions in osmotic pressure and ionic composition of the haemolymph.

Although terrestrial gastropods do not absorb water directly from saturated air (Hunter, 1964; and others), they can acquire water by diffusion through the integument of the sole as well as the body as a whole (Burton, 1966). The epithelia of the soles of *D. reticulatum* and *L. maximus* were observed to take up both water and solutes through specialized cell junctions (paracellular fluid transport) (Ryder and Bowen, 1977a,b; Uglem *et al.*, 1985). The sole of *L. maximus* is divided into three longitudinal bands, with the outer pair characterized by microvillous cells thought to subserve fluid uptake, while the central band, invested with ciliated cells, functions chiefly in locomotion (Prior *et al.*, 1994). Prior (1984) described a particularly clear example of this water acquisition mechanism in *L. maximus*. When dehydrated to a certain point (~60–70% of body weight), the animals exhibited a consistent behaviour whereby they moved to a proffered moist surface and applied the sole tightly to the substrate. Water was absorbed through the epithelium until about 93% of the body weight was restored, when the 'foot drinking' was terminated and the animal resumed other activities. Injection of hypertonic fluid, increasing haemolymph osmotic pressure, also activated this 'foot drinking' behaviour though there had been no loss of body water. In strongly dehydrated animals, paracellular pathways could be demonstrated between epithelial cells of the sole, which presumably facilitated osmotic intake of water from the moist substrate (Prior and Uglem, 1984). Dehydrated stylommatophorans may also take water in through the mouth (e.g. Pusswald, 1948; Klein-Rollais and Daguzan, 1988; A.W. Martin and I. Deyrup-Olsen, unpublished observations), but this behaviour may not be universal among these animals (Machin, 1977). Prior (1989) observed that buccal intake of water occurred in *L. maximus*, but only when the available water contained traces of organic compounds (amino acids and carbohydrates), suggesting that this pattern of water intake is a type of feeding, rather than drinking. Prior and co-workers have described other behaviours in *Limax* that are favourable to conservation of water. Under water stress conditions, these animals reduce the duration of the open phase of the pneumostome, decrease motor activity and, if other individuals are present, huddle with them (Prior *et al.*, 1983; Dickinson *et al.*, 1988). Cook (1981) also noted huddling behaviour, conserving water, in caged *L. pseudoflavus* and *Limacus flavus* (Linnaeus).

Terrestrial gastropods are also vulnerable to hyperhydration, for example when animals are in soil crevices flooded by rain. There is relatively little information as to how, and to what extent, excess water can be removed from the body in such circumstances. One obvious route of excretion is through the urine. Aquatic snails, living in water of low salt content, continuously excrete very hypotonic urine. Another major route for removal of excess fluid, documented in *A. columbianus*, is through activation of secretion through epithelial channel cells (Martin and Deyrup-Olsen, 1986; Martin *et al.*, 1990).

One of the most important mechanisms for avoiding excess loss or gain of water is the animal's selection of favourable microhabitats.

Many terrestrial gastropods, for example, seek out the moist, cool crevices among leaf litter and rock rubble, or underground. Desert species may adopt a contrasting strategy, climbing above ground on plant stems and thereby taking advantage of the relatively cool air above the hot desert surface.

Regulation of ion exchange is closely related to water regulation in terrestrial gastropods, and both are central to the organization of body fluids, secretion of mucus and formation of shells (Burton, 1968a,b, 1983; de With and van der Schors, 1983). The univalent ions sodium, potassium and chloride play major roles in determining the osmotic properties of the body fluids, as well as in cellular excitability. These ions are taken in through the epithelia of the gut and integument. Active transport processes mediating sodium and chloride ion uptake were studied extensively in the freshwater pulmonate *Lymnaea* de Lamarck (Lymnaeidae) (Greenaway, 1970; and others). Net loss of ions is partly through the urine, partly at the body surface. Burton (1965) demonstrated that in *H. pomatia*, the mucous secretions of the integument contain univalent ions at concentrations different from those in haemolymph, with increased concentration of potassium and reduction of sodium and chloride levels. In *A. columbianus*, these adjustments were shown to occur through activity of channel cells of the epithelium (Martin and Deyrup-Olsen, 1986).

Physiological and behavioural aspects of water and solute exchange are coordinated by neural and neuroendocrine mechanisms. A strong beginning has been made in elucidating this control, reviewed by Joosse and Geraerts (1983), Joosse (1988) and Boer (1997). Early *in vitro* studies of electrical activity in neurons of the pedal ganglia of *D. reticulatum* demonstrated an increased rate of discharge when the osmotic pressure of the medium was reduced (Hughes and Kerkut, 1956; Kerkut and Taylor, 1956). Thus the pedal ganglia could register and perhaps coordinate information about haemolymph tonicity. More recently, the roles of the central ganglia in water and ion regulation have been investigated in some detail. By far the most data have come from studies of freshwater pulmonates such as *Lymnaea* (reviewed by Joosse, 1988; and others), *Biomphalaria* Preston (e.g. Kahn and Saleuddin, 1979) and *Helisoma* Swainson (e.g. Grimm-Jorgensen and Connelly, 1983). A few investigations have included land-dwelling gastropods, for example, the stylommatophorans *L. maximus* (Dickinson *et al.*, 1988), *A. columbianus* (Sawyer *et al.*, 1984), *Achatina fulica* Bowdich (Achatinidae) (Mizuno and Takeda, 1988) and *Cepaea nemoralis* (Linnaeus) (Helicidae), *C. aspersus* and *D. reticulatum* (Boer and Montagne-Wajer, 1994). Several investigative approaches have been used to identify the central mechanisms of control, namely:

1.   The role of specific neurons in one or more of the ganglia of the central nervous system (CNS) in the synthesis of neurosecretory material or neurohormones, as identified by reactions to cytological stains or, more

recently, by immunocytochemistry and techniques of *in situ* hybridization (Boer and Montagne-Wajer, 1994). de With *et al.* (1993), for example, established the primary structure of the neuropeptide that stimulates influx of sodium ions in *Lymnaea*. This neuropeptide was localized (by immunocytochemical methods) to the visceral, parietal and pleural ganglia, as well as to certain major nerves.

2.   Under differing states of hydration, neurons may show evidence of changing levels of synthesis or discharge of peptides, and removal of these cells may affect water or ion balance differentially. An example is the observation that removal of the dark green cells of the pleural and parietal ganglia in *Lymnaea* results in excessive water retention in the body (reviewed by Joosse, 1988).

3.   Extracts of ganglia of the CNS, injected into living animals, may alter water/ion dynamics. Their effects, or actions of closely related vertebrate neuropeptides, were assessed by administration to gastropods while monitoring parameters such as rates of hydration, sodium and/or chloride exchange, mucous secretion and aspects of behaviour. An example is the inhibition in *L. maximus* of pneumostome opening (conducive to water retention) by injection of the hormones arginine-vasotocin and arginine-vasopressin, as well as by dehydration (Dickinson *et al.*, 1988). Also, arginine-vasotocin initiates or enhances water-seeking behaviour in *L. maximus* (Banta *et al.*, 1990)

The various examples listed above provide evidence for regulation of water and ion exchanges. So far, however, a full and coordinated account of control processes has not been made for any species. The most comprehensive studies to date have focused on freshwater pulmonates, which are in a constant condition of environmental hypotonicity. Information in this area is scanty or totally lacking for most terrestrial gastropods. It would be of great interest to develop models of ionic regulation studies in xeric-adapted terrestrial gastropods.

Control of Body Shape

The body form of land gastropods is highly plastic, with shape varying with degree of contraction in the muscles of the integument and the free muscle system. In snails, the shell encloses and defines the form of the mantle and visceral hump, but the head–foot can be protruded from or be withdrawn into the shell. Dale (1974) gave a detailed account of head–foot retraction into and emergence from the shell in *H. pomatia*. Coordinated contraction of the columellar (free) muscle system (Fig. 2.7), supplemented with contractions of tentacular, buccal retractor and pedal muscles, was suggested to be responsible for withdrawal of the animal's head–foot into the shell. During this manoeuvre, haemolymph was largely displaced from the anterior cephalopedal sinus into the posteriorly located visceral sinus. Retraction or withdrawal is accomplished rapidly,

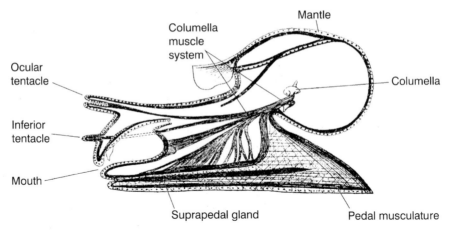

Fig. 2.7. Schematic diagram of *Helix pomatia* Linnaeus (Helicidae) (in longitudinal section, shell removed) to show the columellar muscle system, with its attachments to the head and pedal tissues. Modified from Trappmann (1916).

usually within 2–4 s. Dale described how, in *H. pomatia*, emergence or extrusion of the head–foot was dependent on relaxation of the free muscle system and contraction of muscles in the floor of the mantle cavity. Downward pressure of the mantle floor was described as initiating extrusion of the head–foot. This process was reinforced by hydraulic pressure exerted by movement of haemolymph from the visceral sinus into the cephalopedal sinus, causing distention of the head–foot, emergence from the shell aperture and extension of the tentacles. Dale showed that the process of extrusion is much slower than retraction, requiring about 15 s.

Brown and Trueman (1982) proposed an alternative method for emergence from the shell in the marine caenogastropod, *Bullia digitalis* Meuschen (Nassidae). In this species, the columellar muscle system was shown to provide not only retraction but also extrusion of the head–foot. Contraction of muscle cells oriented in circular and transverse directions within the integument were shown to lengthen the columellar complex, thereby forcing the head–foot out through the shell aperture. These authors suggested that the same processes occur in other gastropods, including the stylommatophoran *Theba pisana* (Müller) (Helicidae).

The bodies of semi-slugs and, more particularly, slugs can be varied in shape more extensively than can the bodies of snails, primarily because form is not constrained by the shell. Those muscles of the integument and foot that are organized in a generally longitudinal direction shorten the body when they contract. Circular muscles, and muscles with direction oblique or tangential to the body surface, cause the body to lengthen. Interaction of these muscles, contracting against the mass of the haemolymph and viscera, effect rapid, diverse alterations in body form. Slugs that are crawling and then are forced to stop can shorten the body to half the active length within 10 s (I. Deyrup-Olsen, unpublished observations on *D. reticulatum*, *L. maximus*, *Arion circumscriptus* Johnston and

Philomycus carolinianus (Bosc) (Philomycidae)). It is not known whether, or how, such compaction of the body affects function of the internal organs.

The ability of slugs to enter narrow crevices such as those in soil undoubtedly reflects the absence of a bulky shell and the great plasticity of their bodies. Snails also use protective sites, but for a given sized animal they have a more limited selection because of their rigid shells. One suggestion to explain the occurrence of shell reduction in diverse stylommatophoran taxa is the advantage conferred by the plasticity of body form resulting from loss of the shell, even though the shell offers significant protection from predation and excess loss or gain of water.

Because the integument in terrestrial gastropods is highly permeable to water, these animals are subject to dehydration and hyperhydration. Changes in body shape that may result from net loss or gain of fluid are not readily demonstrable in snails because these animals retreat into their shells under unfavourable conditions. In slugs, dehydration results in observable shrinkage of the body. Hyperhydration in slugs has not been studied in detail with respect to its effects on body shape, although it is known to occur, as when animals are exposed to excess moisture such as rain-soaked vegetation and soil (Prior, 1985). In favourable circumstances, excess water can be excreted by the nephridium. Another corrective mechanism was observed in *A. columbianus*. In this species, the body has a limited capacity to regulate its volume by both muscular and secretory adjustments (Martin *et al.*, 1990). If the body is distended by excess fluid, muscle contractions occur, raising intrahaemocoelic pressure, and channel cells secrete copiously. This ability for volume autoregulation may be part of the overall physiological response to distention of the body in hyperhydration. Limacids in wet environments have significantly lowered osmotic pressure of the haemolymph (data cited by Prior, 1985). These observations, together with the demonstrated capability of the integument to autoregulate, indicate that in these animals, maintenance of body volume may take precedence over haemolymph concentration in the animal's overall reaction to hyperhydration.

Sudden extrusion of haemolymph through the pneumostome, 'blood venting', in response to aversive stimulation of the head–foot has been noted in both basommatophorans (e.g. *Lymnaea stagnalis* (Linnaeus)) and stylommatophorans (e.g. *Monadenia fidelis* Gray (Helminthoglyptidae), *A. circumscriptus*, *A. ater*, *A. columbianus*, *D. reticulatum*, *L. maximus* and *H. pomatia*) (Bekius, 1972; Martin and Deyrup-Olsen, 1982; and others). The amount of haemolymph lost varies with species and the nature of the stimulus. It can be very large; for example, in *A. ater*, it may be equivalent to the total haemolymph volume, as measured by the isotope dilution method. Haemolymph loss is reflected in corresponding decreases in body size. Fluid volume is quickly restored if the animal has access to water. Blood venting, with resulting body size reduction, may facilitate the slug's access to small crevices, but there is no evidence to confirm or disprove this putative function. Bekius (1972) suggested that blood venting may facilitate rapid withdrawal of the snail's body into its shell.

Locomotion and Adhesion

Most terrestrial gastropods move by contraction of muscles of the sole, or, in the case of very small individuals, by beating of cilia covering the sole. In all cases, mucus secreted over the epithelium provides the linkage between the animal's motors and the substrate. If this coupling is lost, as when mucus becomes hydrated under excessively wet conditions, adhesion fails and the gastropod is immobilized. Beyond these generalizations, terrestrial gastropods show a wide range of locomotor patterns. Many questions remain as to the diversity of these patterns in different taxa, and the degree to which locomotion may differ under varying conditions. Important reviews on this topic are those by Runham and Hunter (1970), Jones (1975), Trueman (1983) and South (1992).

In a dominant pattern of muscular locomotion, waves of contraction sweep over the sole. Usually these pedal muscular waves move in the same direction as the movement of the body – these are termed direct waves. In some circumstances the pedal waves are opposite in direction to the body motion, and are termed retrograde waves. The entire sole may show waves of contraction, or the pedal waves may be restricted to particular areas. For instance, in *Pomatias elegans* Müller the lateral halves of the foot are alternately lifted, moved forward, and affixed to the substrate in a characteristic 'stepping' pattern (cited by Fretter and Graham, 1994). In *L. maximus* only a central longitudinal band of the sole carries pedal waves.

The mechanisms whereby contraction of muscles of the sole effect motion are complex and understood only in outline. Studies in *H. pomatia*, *C. aspersus* and *D. reticulatum* (Lissman, 1945–46, cited by Jones, 1975; Jones, 1973) have been integrated into a hypothesis of locomotory mechanisms in species with pedal muscular waves. Jones (1975) proposed that, at the immediate site of each pedal wave, a series of contractions (i) lift the foot and move it forward, then (ii) draw the overlying muscles and the body as a whole forward. Process (i) is effected by oblique muscles, running from a more dorsally located longitudinal muscle band, to insert on the epithelium posteriorly (anterior oblique muscles). Process (ii) is then the activation of oblique muscles that run from the longitudinal array to insert on the epithelium anteriorly (posterior oblique muscles). This pattern is repeated continuously as the pedal wave sweeps along the length of the sole. Denny (1981) proposed an alternative mechanism for locomotion in *A. columbianus*. He measured forces under an animal crawling across a force plate, and concluded that contractions of the foot musculature during the pedal waves do not lift the sole from the substrate. Rather, they raise the pressure in the haemocoel anterior to the wavefront, pushing the body forward.

In a contrasting pattern of movement, an animal may show 'loping' or 'galloping' locomotion in which the head–neck is lifted from the substrate, the body is extended and arched, and the head then lowered again so the sole makes intermittent contact with the substrate. The resulting slime trail is a series of spots, rather than a continuous path. Jones (1975) and Pearce (1989) cited examples of this type of locomotion in several stylommatophoran species. Pearce noted that 'loping' locomotion was not faster than typical 'gliding' (the sole in continuous contact with the substrate). Various workers have suggested that loping may aid fluid conservation, decrease contact with unfavourable substrates or inhibit trail following (e.g. by predators). None of these hypotheses are validated or disproved as yet.

Neural mechanisms mediating muscular activity of the foot appear to be both peripheral and central. For example, in several terrestrial gastropod species, the sole of the foot, severed from the anterior part of the body, shows coordinated contractions mediated by the rich local plexus of nerves connected with the foot musculature (Bullock and Horridge, 1965). Prior and Gelperin (1973) found that surgical transection of both main nerves from the pedal ganglia to the sole (pedal nerves) of living *L. maximus* abolished pedal waves in denervated sections of the sole. If one nerve was left intact, however, pedal activity was normal. They concluded that central nervous function initiates and modulates the peripheral generation of the pedal wave patterns. In partial confirmation of these results, Broyles and Sokolove (1978) observed that, when the pedal nerves of *L. maximus* were cut, pedal waves initially were absent. After 3–21 days, however, pedal waves reappeared in the denervated sole although reinnervation from the cerebral ganglia had not occurred, as judged by histological methods and tests with electrical stimulation. The newly established pedal waves showed lower frequency and velocity than pedal waves of control animals.

Locomotion and Mucus

Mucus is essential for locomotion and is supplied by a specialized gland, the suprapedal, located in the body cavity and opening to the ventral side of the animal at the anterior margin of the sole. Individual mucous cells are also present in the epithelium of the foot sole as well as over the body (see above). Their potential contributions to locomotion have not been assessed as yet. Subepidermal gland cells also deliver secretory products to the sole and sides of the foot. Together these various cells provide a complex mixture of neutral and acidic mucoproteins and sulphated and neutral mucopolysaccharides, but the specific functions of these components are unknown (Shirbhate and Cook, 1987). The secretion of mucus and contractions of the pedal musculature must be closely coordinated, but details of this control have not been investigated adequately.

Composed primarily of water (~89–98%), mucous secretions obtain their special properties from the presence of huge macromolecules – glycoproteins (mucins) and complex carbohydrates (glycosamino- glycans). They are secreted from epithelial cells encased in membrane- bound granules that are ruptured rapidly and release their contents at the epithelial surface (Deyrup-Olsen *et al.*, 1992; Deyrup-Olsen, 1996; Deyrup-Olsen and Jindrova, 1996; Kapeleta *et al.*, 1996). These linear and flexible molecules are studded with ionized groups (e.g. sulphate and phosphate) and, by virtue of their protean shapes and multiple charges, can react with one another to form films, threads and meshworks. They combine with ions such as Na^+ and Ca^{2+} and, more generally, with vast numbers of components of the extracellular environment. These characteristics account for the adhesive properties of mucus. Since mucus is also extremely hydrophilic, it absorbs and retains copious amounts of water, with resulting lubricant characteristics. In undertaking their locomotor activities, animals use these contrasting properties which can change on a moment by moment basis as the mucous system undergoes stress, is allowed to relax or is exposed to reactants such as H^+ or Ca^{2+}. Denny (1983) summarized studies of the biomechanics of the mucus of terrestrial gastropods, including *H. pomatia* and *A. columbianus*. When unstressed, mucus behaves like a solid, or gel, but when stressed it changes rapidly and reversibly, assuming fluid properties. Such rapid alterations must occur as the animal crawls, alternately adhering to the substrate and slipping over it.

The mucus trail yields evidence of the organization of mucous components during an animal's movements. Simkiss and Wilbur (1977) described granules and fine fibrils in the trails of *C. aspersus*, and Denny and Gosline (1980) also noted threads in the mucus of *A. columbianus* adhering continuously to a vertical surface. The structural changes that occur in mucus secreted by the dorsal integument of terrestrial gastropods can be observed *in vitro* simply by trapping the material between a microscope slide and coverslip. Treated in this way, the mucus forms visible threads when tension is applied, for example by moving the coverslip with respect to the slide. Similarly, mucus *in vitro* can be drawn out into thin, elastic threads, simulating the structures sometimes used in locomotion (Richter, 1980; Deyrup-Olsen *et al.*, 1983).

Escape and Self-defence

Snails contract into their shells when attacked. In prosobranch snails, the operculum closes off the aperture as the foot is drawn into the shell. Lacking the operculum, stylommatophoran snails pull into the shell as deeply as their contracted bodies allow. Lacking the protection of shells, animals of the slug form have to rely on escape and self-defence reactions. Some of these are described by Rollo and Wellington (1979), Richter

(1980), Pakarinen (1992, 1994a,b) and Brown *et al.* (1994). Attacks during territorial disputes with other gastropods, or by predators such as carabid beetles, may be deterred in part by copious secretion of mucus. Pakarinen observed that mucus produced by *D. reticulatum* and *Arion fasciatus* (Nilsson) provided significant protection against generalist carabid beetles, but not when attacks were made by specialist malacophagous species. In many stylommatophorans, the mucus produced by disturbed animals has the golden colour frequently associated with warning and defence reactions in a wide range of animal groups, but the defence role of these mucus secretions has not been demonstrated adequately. In response to disturbance, slug forms such as *D. reticulatum* and *L. maximus* may lift the tail and oscillate it rapidly back and forth in an apparent attempt to repel their attacker (Rollo and Wellington, 1979; Pakarinen, 1994b). Similarly, some slugs (e.g. *D. reticulatum*, *Deroceras laeve* (Müller), *L. maximus* and *A. columbianus*) exhibit a behaviour of lifting and extension of the mantle in response to aversive stimulation (Rollo and Wellington, 1979; and others). This flaring of the mantle is often followed by clamping of the mantle edge to the substrate, thus protecting the underlying head. In contrast, arionids such as *A. fasciatus* and *A. ater* draw the body into a compact hemispheroid shape, clamp the sole tightly to the substrate and, in the case of *A. ater*, rock the whole body from side to side. According to Rollo and Wellington (1979), many slugs abandon these defence displays and take to flight if aggression continues.

Some arionids, agriolimacids and limacids can respond to attack by autotomy – programmed self-section of the posterior part of the foot. This provides a distracting reward to a predator and may allow escape of its prey (Hand and Ingram, 1950; Deyrup-Olsen *et al.*, 1986; Pakarinen, 1994a). Autotomy in *Prophysaon foliolatum* (Gould) (Arionidae) is due to fast contraction of a ring of muscles just anterior to a specialized terminal section of the body ('tail'), and appears to be elicited by peripheral stimulation of nerves since it can occur in preparations in which the head is removed. Following autotomy, the tail is regenerated within a few weeks, restoring the original form of the body.

Although they must be central to the individual's survival, the neural mechanisms associated with escape and self-defence are almost unknown. Terrestrial gastropods follow and perhaps glean information from mucous trails (Wells and Buckley, 1972; Cook, 1977, 1992; and others). Landauer and Chapnick (1981) found that *Lehmannia valentiana* de Férussac (Limacidae) moved more rapidly, as compared with controls, from areas bearing mucus deposited by stressed conspecifics. Pakarinen (1992) tested whether animals would avoid areas where stressed and non-stressed individuals had left mucous trails. *A. fasciatus* avoided mucous trails of stressed conspecifics, but not those of *Limax cinereoniger* Wolf. In contrast, *D. reticulatum* showed no pattern of avoidance of mucus deposited by stressed conspecifics.

Conclusion

Significant progress has been made in elucidating the structure and function of the integument of terrestrial gastropods. Modern methods are needed because of the complexity of the integument and the variety and sensitivity of the cells that make up the integument. Early histological and physiological investigations are now supplemented with the results from better methods of tissue preparation and techniques such as electron microscopy. Precise biophysical methods have been applied to measure forces exerted by the body musculature and internal fluid dynamics. There have been important findings in the area of neuroendocrine control. Review of the field leads to the realization, however, that the results so far are limited to a relatively small number of species. Of these, most are characteristic of temperate environments. Yet caenogastropod and pulmonate gastropods present a great range of biological processes and adaptations. They are distributed from the tropics to cold polar regions of the earth. They inhabit very moist environments, and, at the opposite extreme, live on dry land and in deserts. More investigations of such animals, widely ranging in their ecological adaptations, would be of great intrinsic interest and valuable for elucidating evolutionary patterns of terrestrial molluscs.

References

Abolins-Krogis, A. (1972) The tubular endoplasmic reticulum in the amoebocytes of the shell-regenerating snail, *Helix pomatia* L. *Zeitschrift für Zellforschung und Mikroskopische Anatomie* 128, 58–68.

Appleton, T.C., Newell, P.F. and Machin, J. (1979) Ionic gradients within mantle–collar epithelial cells of the land snail *Otala lactea. Cell and Tissue Research* 199, 83–97.

Arad, Z. (1993a) Water relations and resistance to desiccation in three Israeli desert snails, *Eremina desertorum, Euchondrus desertorum* and *Euchondrus albulus. Journal of Arid Environments* 24, 387–395.

Arad, Z. (1993b) Effect of desiccation on the water economy of terrestrial gastropods of different phylogenetic origins: a prosobranch (*Pomatias glaucus*) and two pulmonates (*Sphinterochila cariosa* and *Helix engaddensis*). *Israel Journal of Zoology* 39, 95–104.

Banta, P.A., Welsford, I.G. and Prior, D.J. (1990) Water-orientation behaviour in the terrestrial gastropod, *Limax maximus*: the effect of dehydration and arginine vasotocin. *Physiological Zoology* 63, 683–696.

Barnhart, M.C. (1989) Evaporative water loss from minute terrestrial snails (Pulmonata: Pupillidae). *Veliger* 32, 16–20.

Bekius, R. (1972) The circulatory system of *Lymnaea stagnalis. Netherlands Journal of Zoology* 22, 1–58.

Beltz, B. and Gelperin, A. (1977) An ultrastructural analysis of the salivary system of the terrestrial mollusc, *Limax maximus. Tissue and Cell* 11, 31–50.

Bentley, P.J. (1966) The physiology of the urinary bladder of amphibia. *Biological Reviews* 41, 275–316.

Blinn, W.C. (1964) Water in the mantle cavity of land snails. *Physiological Zoology* 37, 329–337.

Boer, H.H. (1997) Central nervous system. In: Harrison, F.W. and Kohn, A.J. (eds) *Microscopic Anatomy of Invertebrates*, Vol. 6B. *Gastropoda: Pulmonata.* Wiley-Liss, New York, pp. 661–701.

Boer, H.H. and Montagne-Wajer, C. (1994) Functional morphology of the neuropeptidergic light-yellow-cell system in pulmonate snails. *Cell and Tissue Research* 277, 531–538.

Boer, H.H. and Sminia, T. (1976) Sieve structure of slit diaphragms of podocytes and pore cells of gastropod molluscs. *Cell and Tissue Research* 170, 221–229.

Bogusch, G. (1972) Zur innervation des glatten Penisretraktormuskels von Helix pomatia: Allgemeine Histologie und Histochimie des monoaminergen Nervensystems. *Z. Zellforsch.* 126; 383–401.

Brown, A.C. (1967) Elimination of foreign particles by the snail, *Helix aspersa. Nature* 213, 1154–1155.

Brown, A.C. and Trueman, E.R. (1982) Muscles that push snails out of their shells. *Journal of Molluscan Studies* 48, 97–98.

Brown, G.E., Davenport, D.A. and Howe, A.R. (1994) Escape deficits induced by a biologically relevant stressor in the slug (*Limax maximus*). *Psychological Reports* 75, 1187–1192.

Broyles, J.L. and Sokolove, P.G. (1978) Pedal wave recovery following transection of pedal nerves in the slug, *Limax maximus. Journal of Experimental Zoology* 206, 371–380.

Buchholz, K., Kuhlmann, D. and Nolte, A. (1971) Aufnahme von Trypanblau und Ferritin in die Blasenzellen des Bindegewebes von *Helix pomatia* und *Cepaea nemoralis* (Stylommatophora, Pulmonata). *Zeitschrift für Zellforschung und Mikroskopische Anatomie* 113, 203–215.

Bullock, T.H. and Horridge, G.A. (1965) *Structure and Function of the Nervous System of Invertebrates.* Vol. 2. Freeman and Sons, San Francisco, pp. 1321–1334.

Burton, R.F. (1965) Relationship between the cation contents of slime and blood in the snail *Helix pomatia* L. *Comparative Biochemistry and Physiology* 15, 339–345.

Burton, R.F. (1966) Aspects of ionic regulation in certain terrestrial pulmonates. *Comparative Biochemistry and Physiology* 17, 1007–1018.

Burton, R.F. (1968a) Ionic regulation in the snail *Helix aperta. Comparative Biochemistry and Physiology* 25, 501–508.

Burton, R.F. (1968b) Ionic balances in the blood of Pulmonata. *Comparative Biochemistry and Physiology* 25, 509–518.

Burton, R.F. (1983) Ionic regulation and water balance. In: Saleuddin, A.S.M. and Wilbur, K. (eds) *The Mollusca*, Vol. 5, *Physiology*, Part 2. Academic Press, New York, pp. 291–352.

Campion, M. (1961) The structure and function of the cutaneous glands in *Helix aspersa. Quarterly Journal of Microscopical Science* 102, 195–216.

Chantler, P.D. (1983) Biochemical and structural aspects of molluscan muscle. In: Wilbur, K. (ed.) *The Mollusca*, Vol. 4. Academic Press, New York, pp. 77–154.

Chétail, M. and Binot, D. (1967) Particularites histochimiques de la glande et de la sole pedieuses d' *Arion rufus* (Stylommatophora: Arionidae). *Malacologia* 5, 269–284.

Cook, A. (1977) Mucus trail following by the slug *Limax grossui* Lupu. *Animal Behaviour* 25, 774–781.

Cook, A. (1981) Huddling and the control of water loss by the slug *Limax pseudoflavus*. *Animal Behaviour* 29, 289–298.

Cook, A. (1987) Functional aspects of the mucus-producing glands of the systellommatophoran slug, *Veronicella floridana*. *Journal of Zoology* 211, 291–305.

Cook, A. (1992) The function of trail following in the pulmonate slug, *Limax pseudoflavus*. *Animal Behaviour* 43, 813–821.

- Cook, A. and Shirbhate, R. (1983) The mucus producing glands and distribution of the cilia of the pulmonate slug *Limax pseudoflavus*. *Journal of Zoology* 201, 97–116.

Dale, B. (1974) Extrusion, retraction and respiratory movements in *Helix pomatia* in relation to distribution and circulation of the blood. *Journal of Zoology (London)* 173, 427–439.

Davis, E.C. and Shivers, R.R. (1987) Membrane-associated dense plaques in smooth muscle cells of the common slug, *Limax maximus*: possible sites of transmembrane interaction of filaments. *Journal of Submicroscopic Cytology* 19, 537–544.

Denny, M.W. (1981) A quantitative model for the adhesive locomotion of the terrestrial slug, *Ariolimax columbianus*. *Journal of Experimental Biology* 91, 195–217.

Denny, M.W. (1983) Molecular biomechanics of molluscan mucous secretions. In: Hochachka, P.W. and Wilbur, K.M. (eds) *The Mollusca*, Vol. 1. Academic Press, New York, pp. 431–465.

Denny, M.W. and Gosline, J.M. (1980) The physical properties of the pedal mucus of the terrestrial slug *Ariolimax columbianus*. *Journal of Experimental Biology* 88, 375–393.

Deyrup-Olsen, I. (1996) Product release by mucous granules of land slugs. II. Species diversity in triggering of mucous granule rupture. *Journal of Experimental Zoology* 276, 330–334.

Deyrup-Olsen, I. and Jindrova, H. (1996) Product release by mucous granules of land slugs: *Ariolimax columbianus* as a model species. *Journal of Experimental Zoology* 276, 387–393.

Deyrup-Olsen, I. and Martin, A.W. (1987) Osmolyte processing in the gut and an important role of the rectum in the land slug, *Ariolimax columbianus* (Pulmonata; Arionidae). *Journal of Experimental Zoology* 243, 33–38.

Deyrup-Olsen, I., Luchtel, D.L. and Martin, A.W. (1983) Components of mucus of terrestrial slugs (Gastropoda). *American Journal of Physiology* 245, R448–R452.

Deyrup-Olsen, I., Martin, A.W. and Paine, R.T. (1986) The autotomy escape response of the terrestrial slug *Prophysaon foliolatum* (Pulmonata: Arionidae). *Malacologia* 27, 307–311.

Deyrup-Olsen, I., Luchtel, D.L. and Martin, A. W. (1989) Secretory cells of the body wall of *Ariolimax columbianus*: structure, function and control. In: Henderson, I.F. (ed.) *Slugs and Snails in World Agriculture*. British Crop Protection Council, Monograph No. 41, pp. 407–412.

Deyrup-Olsen, I., Louie, H., Martin, A.W. and Luchtel, D.L. (1992) Triggering by ATP of product release by mucous granules of the land slug *Ariolimax columbianus*. *American Journal of Physiology* 262, C760–C765.

de With, N.D. and van der Schors, R.C. (1983) Neuroendocrine aspects of sodium and chloride metabolism in the pulmonate snail *Lymnaea stagnalis*. In: Lever,

J. and Boer, H.H. (eds) *Molluscan Neuroendocrinology.* North-Holland Publishing Company, Amsterdam, pp. 208–212.

de With, N.D., van der Schors, R.C., Boer, H.H. and Ebberink, R.H.M. (1993) The sodium influx-stimulating peptide of the pulmonate freshwater snail *Lymnaea stagnalis. Peptides* 14, 783–789.

Dickinson, P.S., Prior, D.J. and Avery, C. (1988) The pneumostome rhythm of slugs: a response to dehydration controlled by hemolymph osmolality and peptide hormones. *Comparative Biochemistry and Physiology* 89A, 579–585.

Fioroni, P., Sundermann, G. and Scheidegger, D.P. (1984) Die Ultrastruktur der freien Rhogocyten bei intrakapsulären Veligern von *Nucella lapillus* (Gastropoda, Prosobranchia, Stenoglossa). *Zoologischer Anzeiger* 212, 193–202.

Fournié, J. and Chétail, M. (1982) Accumulation calcique au niveau cellulaire chez les mollusques. *Malacologia* 22, 265–285.

Fournié, J. and Chétail, M. (1984) Calcium dynamics in land gastropods. *American Zoologist* 24, 857–870.

Greenaway, P. (1970) Sodium regulation in the freshwater mollusc *Limnaea stagnalis* (L.) (Gastropoda: Pulmonata). *Journal of Experimental Biology* 53, 147–163.

Grimm-Jorgensen, Y. and Connelly, S. (1983) Effect of thyrotropin releasing hormone (TRH) on the synthesis and secretion of polysaccharides by the integument of gastropods. *General and Comparative Endocrinology* 52, 23–29.

Hand, C. and Ingram, W.M. (1950) Natural history observations on *Prophysaon andersoni* (J.G. Cooper) with special reference to amputation. *Bulletin of the Southern California Academy of Sciences* 49, 15–28.

Haszprunar, G. (1996) The molluscan rhogocyte (pore-cell, blasenzelle, cellule nucale), and its significance for ideas on nephridial evolution. *Journal of Molluscan Studies* 62, 185–211.

Heyer, C.B., Kater, S.B. and Karlsson, U.L. (1973) Neuromuscular systems in molluscs. *American Zoologist* 13, 247–270.

Hughes, G.M. and Kerkut, G.A. (1956) Electrical activity in a slug ganglion in relation to the concentration of Locke solution. *Journal of Experimental Biology* 33, 282–294.

Hunter, W.R. (1964) Physiological aspects of ecology of nonmarine molluscs. In: Wilbur, K.M. and Yonge, C.M. (eds) *Physiology of Mollusca*, Vol. 1. Academic Press, New York, pp. 83–126.

Hyman, L.H. (1967) *The Invertebrates*, Vol. VI. *Mollusca I.* McGraw-Hill, New York, pp. 558–566.

Jones, G.W. and Bowen, I.D. (1979) The fine structural localization of acid phosphatase in pore cells of embryonic and newly hatched *Deroceras reticulatum. Cell and Tissue Research* 204, 253–265.

Jones, H.D. (1973) The mechanism of locomotion of *Agriolimax reticulatus* (Mollusca: Gastropoda). *Journal of Zoology* 171, 489–498.

Jones, H.D. (1975) Locomotion. In: Fretter, V. and Peake, J. (eds) *Pulmonates.* Academic Press, New York, pp. 1–32.

Joosse, J. (1988) The hormones of molluscs. In: Laufer, H. and Downer, R.G.H. (eds) *Endocrinology of Selected Invertebrate Types*, Vol. 2. Alan R. Liss. Inc., New York, pp. 89–140.

Joosse, J. and Geraerts, W.P.M. (1983) Endocrinology. In: Saleuddin, A.S.M. and Wilbur, K.M. (eds) *The Mollusca*, Vol. 4. Academic Press, New York, pp. 317–406.

Kahn, H.R. and Saleuddin, A.S.M. (1979) Osmotic regulation and osmotically induced changes in the neurosecretory cells of the pulmonate *Helisoma*. *Canadian Journal of Zoology* 57, 1256–1383.

Kapeleta, M.V., Jimenez-Mallebrera, C., Carnicer-Rodriguez, M.J., Cook, A. and Shephard, K.L. (1996) Production of mucous granules by the terrestrial slug *Arion ater* L. *Journal of Molluscan Studies* 62, 251–256.

Kerkut, G.A. and Taylor, B.K.R. (1956) The sensitivity of the pedal ganglion of the slug to osmotic pressure changes. *Journal of Experimental Biology* 33, 493–501.

Kier, W.M. (1988) The arrangement and function of molluscan muscle. In: Wilbur, K. (ed.) *The Mollusca*, Vol. 11. Academic Press, New York, pp. 211–251.

Klein-Rollais, D. and Daguzan, J. (1988) Oral water consumption in *Helix aspersa* Müller (gastropod mollusc: Stylommatophora) according to age, reproductive activity and food supply. *Comparative Physiology and Biochemistry* 89A, 351–357.

Landauer, M.R. and Chapnick, S.D. (1981) Responses of terrestrial slugs to secretions of stressed conspecifics. *Psychological Reports* 49, 617–618.

Luchtel, D.L. (1989). Mucous secretion as observed after ultrarapid freezing and freeze substitution. In: Bailey, G.W. (ed.) *Proceedings of the 47th Annual Meeting of the Electron Microscopy Society of America*. San Francisco Press, San Francisco, pp. 740–741.

Luchtel, D.L., Martin, A.W. and Deyrup-Olsen, I. (1984) The channel cell of the terrestrial slug *Ariolimax columbianus*. *Cell and Tissue Research* 235, 143–151.

Luchtel, D.L., Deyrup-Olsen, I. and Martin, A.W. (1991) Ultrastructure and lysis of mucin containing granules in epidermal secretions of the terrestrial slug *Ariolimax columbianus* (Mollusca, Gastropoda, Pulmonata). *Cell and Tissue Research* 266, 375–383.

Luchtel, D.L., Martin, A.W., Deyrup-Olsen, I. and Boer, H.H. (1997) *Gastropoda: Pulmonata*. In: Harrison, F.W. and Kohn, A.J. (eds) *Microscopic Anatomy of Invertebrates*, Vol. 6B. Wiley-Liss, New York, pp. 459–718.

Lyth, M. (1983) Water content of slugs (Gastropoda: Pulmonata) in natural habitats, and the influence of culture conditions on water content stability in *Arion ater* (Linné). *Journal of Molluscan Studies* 49, 179–184.

Machin, J. (1966) The evaporation of water from *Helix aspersa*. IV. Loss from the mantle of the inactive snail. *Journal of Experimental Biology* 45, 269–278.

Machin, J. (1967) Structural adaptation for reducing water-loss in three species of terrestrial snail. *Journal of Physiology* 152, 55–65.

Machin, J. (1974) Osmotic gradients across snail epidermis: evidence for a water barrier. *Science* 183, 759–760.

Machin, J. (1975) Water relationships. In: Fretter, V. and Peake, J. (eds) *Pulmonates*, Vol. 1. Academic Press, London, pp. 105–163.

Machin, J. (1977) Role of integument in molluscs. In: Gupta, B.L., Moreton, R.B., Oschman, J.L. and Wall, B.L. (eds) *Transport of Ions and Water in Animals*. Academic Press, New York, pp. 735–762.

Martin, A.W. and Deyrup-Olsen, I. (1982) Blood venting through the pneumostome in terrestrial slugs. *Comparative Biochemistry and Physiology* 72C, 51–58.

Martin, A.W. and Deyrup-Olsen, I. (1986) Function of the epithelial channel cells of the body wall of a terrestrial slug, *Ariolimax columbianus*. *Journal of Experimental Biology* 121, 301–314.

Martin, A.W., Luchtel, D.L. and Deyrup-Olsen, I. (1982) Mechanism of action of arginine-vasotocin and lysine-vasotocin on surface exudation by a slug body wall *in vitro*. *Physiologist* 25, 298 (abstract).

Martin, A.W., Deyrup-Olsen, I. and Stewart, D.M. (1990) Regulation of body volume by the peripheral nervous system of the terrestrial slug *Ariolimax columbianus*. *Journal of Experimental Zoology* 253, 121–131.

Miksys, S.L. and Saleuddin, A.S.M. (1987) Ferritin in mantle pore cells and its role in reproduction of *Helisoma duryi* (Mollusca: Pulmonata). *Journal of Experimental Zoology* 242, 75–83.

Minnich, J.E. (1979) Reptiles. In: Maloiy, G.M.O. (ed.) *Comparative Physiology of Osmoregulation in Animals*, Vol. 1. Academic Press, London, pp. 391–641.

Mizuno, J. and Takeda, N. (1988) Phylogenetic study of the arginine-vasotocin/ arginine-vasopressin-like immunoreactive system in invertebrates. *Comparative Biochemistry and Physiology* 91A, 739–747.

Newell, P.F. (1974) Étude de l'ultrastructure de l'epithelium dorsal et pedieux des limaces *Arion hortensis* Férussac et *Agriolimax reticulatus* (Müller). *Haliotis* 3, 131–141.

Newell, P.F. (1977) The structure and enzyme histochemistry of slug skin. *Malacologia* 16, 183–195.

Newell, P.F. and Appleton, T.C. (1979) Aestivating snails – the physiology of water regulation in the mantle of the terrestrial pulmonate *Otala lactea*. *Malacologia* 18, 575–581.

Newell, P.F. and Machin, J. (1976) Water regulation in aestivating snails. Ultra-structural and analytical evidence for an unusual cellular phenomenon. *Cell and Tissue Research* 173, 417–421.

Nicaise, G. and Ansellem, J. (1983) Cytology of muscle and neuromuscular junction. In: Wilbur, K. (ed.) *The Mollusca*, Vol. 4. Academic Press, New York, pp. 1–33.

Nisbet, R.H. and Plummer, J.M. (1968) The fine structure of the cardiac and other molluscan muscles. *Symposia of the Zoological Society of London* 22, 193–211.

North, R.J. (1963) The fine structure of the myofibers in the heart of the snail *Helix aspersa*. *Journal of Ultrastructural Research* 8, 206–218.

Pakarinen, E. (1992) The responses of terrestrial slugs *Arion fasciatus* and *Deroceras reticulatum* to the mucus of stressed conspecifics and hetero-specifics. *Animal Behaviour* 43, 1051–1052.

Pakarinen, E. (1994a) Autotomy in arionid and limacid slugs. *Journal of Molluscan Studies* 60, 19–24.

Pakarinen, E. (1994b) The importance of mucus as a defense against caribidae beetles by the slugs *Arion fasciatus* and *Deroceras reticulatum*. *Journal of Molluscan Studies* 60, 149–156.

Pearce, T.A. (1989) Loping locomotion in terrestrial gastropods. *Walkerana* 3, 229–237.

Petersen, C.C. (1996) Seasonal water and solute relations in two populations of the desert tortoise (*Gopherus agassizii*) during chronic drought. *Physiological Zoology* 69, 1324–1358.

Phifer, C.B. and Prior, D.J. (1985) Body hydration and haemolymph osmolality affect feeding and its neural correlate in the terrestrial gastropod *Limax maximus*. *Journal of Experimental Biology* 118, 405–421.

Plesch, B., Janse, C. and Boer, H.H. (1975) Gross morphology and histology of the musculature of the freshwater pulmonate *Lymnaea stagnalis* (L.). *Netherlands Journal of Zoology* 25, 332–352.

Plummer, J.M. (1966) Collagen formation in Achatinidae associated with a specific cell type. *Proceedings of the Malacological Society of London* 37, 189–198.

Prior, D.J. (1984) Analysis of contact-rehydration in terrestrial gastropods: osmotic control of drinking behaviour. *Journal of Experimental Biology* 111, 63–73.

Prior, D.J. (1985) Water-regulatory behaviour in terrestrial gastropods. *Biological Reviews* 60, 403–424

Prior, D.J. (1989) Contact-dehydration in slugs: a water regulatory behaviour. In: Henderson, I.F. (ed.) *Snails and Slugs in World Agriculture.* British Crop Protection Council Monograph No. 41, pp. 217–223.

Prior, D.J. and Gelperin, A. (1973) Behavioral and physiological studies on locomotion in the giant garden slug, *Limax maximus. Malacological Reviews* 7, 50–51.

Prior, D.J. and Uglem, G.L. (1984) Analysis of contact-rehydration in terrestrial gastropods. Absorption of ^{14}C-inulin through the epithelium of the foot. *Journal of Experimental Biology* 11, 75–80.

Prior, D.J., Hume, M., Varga, D. and Hess, S.D. (1983) Physiological and behavioural aspects of water balance and respiratory function in the terrestrial slug, *Limax maximus. Journal of Experimental Biology* 104, 111–127.

Prior, D.J., Maugel, T.K. and Sellers, M. (1994) Morphological correlate of regional partitioning of integumental water absorption in terrestrial slugs. *Tissue and Cell* 26, 421–429.

Pusswald, A.W. (1948) Beiträge zum Wasserhausholt der Pulmonaten. *Zeitschrift für Vergleichende Physiologie* 31, 227–248.

Rankevitch, D., Lavie, B., Nevo, E., Beiles, A. and Arad, Z. (1996) Genetic and physiological adaptations of the prosobranch landsnail *Pomatias olivieri* to microclimatic stresses on Mount Carmel, Israel. *Israel Journal of Zoology* 42, 425–441.

Reger, J.F. (1973) A fine structure study on hemocyanin formation in the slug *Limax* sp. *Journal of Ultrastructural Research* 43, 377–387.

Richter, K.O. (1980) Movement, reproduction, defense, and nutrition as functions of the caudal mucus in *Ariolimax columbianus. Veliger* 23, 43–47.

Rollo, C.D. and Wellington, W.G. (1979) Intra- and inter-specific agonistic behaviour among terrestrial slugs (Pulmonata: Stylommatophora). *Canadian Journal of Zoology* 57, 846–855.

Runham, N.W. and Hunter, P.J. (1970) *Terrestrial Slugs.* Hutchinson, London.

Ryder, T.A. and Bowen, I.D. (1977a) Studies on transmembrane and paracellular phenomena in the foot of the slug *Agriolimax reticulatus* (Mü). *Cell and Tissue Research* 183, 143–152.

Ryder, T.A. and Bowen, I.D. (1977b) The use of X-ray microanalysis to demonstrate the uptake of the molluscicide copper sulphate by slug eggs. *Histochemisty* 52, 55–60.

Sawyer, W.H., Deyrup-Olsen, I. and Martin, A.W. (1984) Immunological and biological characteristics of the vasotocin-like activity in the head ganglia of gastropod molluscs. *General and Comparative Endocrinology* 54, 97–108.

Schmut, O., Roll, P., Reich, M.E. and Palm, W. (1980) Biochemical and electron microscopic investigations on *Helix pomatia* collagen. *Zeitschrift für Naturforschung* 35C, 376–379.

Shirbhate, R. and Cook, A. (1987) Pedal and opercular secretory glands of *Pomatias*, *Bithynia* and *Littorina*. *Journal of Molluscan Studies* 53, 79–96.

Simkiss, K. (1988) Molluscan skin (excluding cephalopods). In: Trueman, E.R. and Clarke, M.R. (eds) *The Mollusca*, Vol. 11. Academic Press, New York, pp. 11–35.

Simkiss, K. and Wilbur, K. M. (1977) The molluscan epidermis and its secretions. *Symposia of the Zoological Society of London* 39, 35–76.

Skelding, J.M. and Newell, P.F. (1975) On the functions of the pore cells in the connective tissue of terrestrial pulmonate molluscs. *Cell and Tissue Research* 156, 381–390.

Sminia, T. (1972) Structure and function of blood and connective tissue cells of the fresh water pulmonate *Lymnaea stagnalis* studied by electron microscopy and enzyme histochemistry. *Zeitschrift für Zellforschung und Mikroskopische Anatomie* 130, 497–526.

Sminia, T. and Vlugt-van Daalen, J.E. (1977) Haemocyanin synthesis in pore cells of the terrestrial snail *Helix aspersa*. *Cell and Tissue Research* 183, 299–301.

Solem, A. (1985) Origin and diversification of pulmonate land snails. In: Trueman, E.R. and Clarke, M.E. (eds) *The Mollusca*, Vol. 10, *Evolution*. Academic Press, Orlando, pp. 277–282.

South, A. (1992) *Terrestrial Slugs*. Chapman and Hall, London.

Steinbach, P. (1977) Granular cells in the connective tissue of *Helix*. *Cell and Tissue Research* 181, 91–103.

Trappmann, W. (1916) Die musculatur von *Helix pomatia*. *Zeitschrift für Wissenschaftlich Zoologie* 115, 489–585.

Trueman, E.R. (1983) Locomotion in molluscs. In: Saleuddin, A.S.M. and Wilbur, K.M. (eds) *The Mollusca*. Vol. 4. Academic Press, New York, pp. 155–198.

Uglem, G.L., Prior, D.J. and Hess, S.D. (1985) Paracellular water uptake and molecular sieving by the foot epithelium of terrestrial slugs. *Journal of Comparative Physiology* 156A, 285–289.

Wagge, L.E. (1955) Amoebocytes. *International Review of Cytology* 4, 31–78.

Wells, M.J. and Buckley, S.K.L. (1972) Snails and trails. *Animal Behaviour* 20, 345–355.

Wohlburg-Buchholz, K. (1972) Blasenzellen im Bindegewebe des Schlundrings von *Cepaea nemoralis* L. (Gastropoda, Stylommatophora). *Zeitschrift für Zellforschung und Mikroskopische Anatomie* 128, 100–114.

Wondrak, G. (1967) Die exoepithelialen Schleimdrüsenzellen von *Arion empiricorum* (Fér). *Zeitschrift für Zellforschung und Mikroskopische Anatomie* 76, 287–294.

Wondrak, G. (1968) Elektronenoptische Untersuchungen der Korperdecke von *Arion rufus* L. (Pulmonata). *Protoplasma* 66, 151–171.

Wondrak, G. (1969a) Die Ultrastruktur der Zellen aus dem interstitillen Bindegewebe von *Arion rufus* (L.), Pulmonata, Gastropoda. *Zeitschrift für Zellforschung und Mikroskopische Anatomie* 95, 249–262.

Wondrak, G. (1969b) Elektronenoptische Untersuchungen der Drüsen- and Pigment-zellen aus der Körperdecke von *Arion rufus* (L) (Pulmonata). *Zeitschrift für Mikroskopische-Anatomische Forschung* 80, 17–40.

Zaitseva, O.V. (1984) Innervation of the integument of pulmonata. *Neuroscience and Behavioral Physiology* 14, 23–29.

Zylstra, U. (1972a) Distribution and ultrastructure of epidermal sensory cells in the freshwater snails *Lymnaea stagnalis* and *Biomphalaria pfeifferi*. *Netherlands Journal of Zoology* 22, 283–298.

Zylstra, U. (1972b) Histochemistry and ultrastructure of the epidermis and the subepidermal cells of the freshwater snails *Lymnaea stagnalis* and *Biomphalaria pfeifferi. Zeitschrift für Zellforschung und Mikroskopische Anatomie* 130, 93–131.

3 Sensory Organs and the Nervous System

R. CHASE

Department of Biology, McGill University, 1205 Av. Docteur Penfield, Montréal, Québec, Canada, H3A 1B1

Introduction

The biology of molluscan neurons has been investigated intensively in recent years. The bulk of this research, including some important studies of learning and memory, has been on marine gastropod molluscs, principally *Aplysia* Linnaeus (Aplysiidae, Opisthobranchia) (reviewed in Kandel, 1979). Although *Aplysia* is not included among the animals considered here, it is now abundantly clear that the cellular properties of neurons in the terrestrial gastropod molluscs are fundamentally similar to those in *Aplysia* and, indeed, similar to those of all animals. The cellular properties of molluscan neurons, including electrophysiology, biochemistry and pharmacology, have been reviewed elsewhere (Koester and Byrne, 1980; S.-Rózsa, 1984; Willows, 1986).

This chapter is concerned mostly with structure and function in terrestrial pulmonates. An invaluable earlier review of this subject was published by Bullock in 1965. Changes in the pattern of research since that time have resulted in a striking reduction in the variety of animals studied. Thus, it must be emphasized that the information in this chapter is based largely on studies from just a few species, principally in the genera *Achatina* de Lamarck (Achatinidae), *Helix* Linnaeus and *Cantareus* Risso (Helicidae), and *Limax* Linnaeus (Limacidae). Further, readers are advised that some of the taxonomic names used here differ from those found in the cited literature. The discrepancy reflects recent taxonomic revisions and the need for consistent usage across chapters in this volume. In particular, it should be noted that the snail referred to as *Helix aspersa* Müller in the neurobiological literature is herein referred to as *Cantareus aspersus* (Müller); for discussion, see Giusti *et al.* (1995).

Sense Organs

Olfactory organs

Terrestrial gastropod molluscs have no acoustic sense and little or no ability to recognize objects by vision. Thus, olfaction is the principal modality for perception at a distance. There are olfactory organs located at the tips of each of the four tentacles. Most experimental observations, and the following description, relate to the posterior cephalic tentacles (Chase, 1986b; Chase and Tolloczko, 1993). An unusual feature of the tentacles is their ability to regenerate subsequent to a total lesion. Not only does the tentacle itself regenerate, but so too does the olfactory organ, the eye and the associated neural components. Further, the regenerated structures are functionally competent (Chase and Kamil, 1983).

The olfactory organ consists of a specialized epithelium located immediately ventral to the eye on the tentacle's terminal knob. The outer surface is covered with a complex microvillar border which may serve to protect sensory structures from desiccation (Emery, 1992). Moving inwards, one finds next the layer of epithelial cells, then muscle and then clusters of bipolar sensory neurons. In *Achatina*, there are approximately 100,000 sensory neurons per tentacle. These cells are morphologically similar to vertebrate olfactory receptors and, like their vertebrate counterparts, the population is regenerated continuously (Chase and Rieling, 1986). The dendrites of these cells extend to the brush border where they terminate in either cilia or microvilli, or a combination of both cilia and microvilli (Emery, 1992). There are also several types of secretory cells within the olfactory organ. These differ in kind and number from those found in the adjacent tentacular skin (Chase and Tolloczko, 1985).

The neural pathways from the olfactory organs to the brain are both direct and indirect (Fig. 3.1; Chase and Tolloczko, 1993). About 10% of the sensory neuron axons travel directly to the procerebrum. Other sensory axons terminate in glomeruli (Fig. 3.2d) located just beneath the sensory epithelium (Chase and Tolloczko, 1986). Still others terminate in the ganglion located at the tip of the tentacles; here they presumably synapse on to the small ganglion cells, which in turn relay the signals to the brain via the olfactory (tentacle) nerve. An unusual feature of the tentacle ganglion concerns the ultrastructure of the synapses. Approximately one-third of the synapses have a symmetrical architecture (Fig. 3.2c), suggestive of bidirectional transmission (McCarragher and Chase, 1985).

The lips are important chemosensory structures that mediate responses to chemicals upon contact (Salánki and van Bay, 1976).

Eyes

Each cephalic tentacle houses an eye. It is located immediately dorsal to the olfactory epithelium, but the eye has its own nerve and it functions

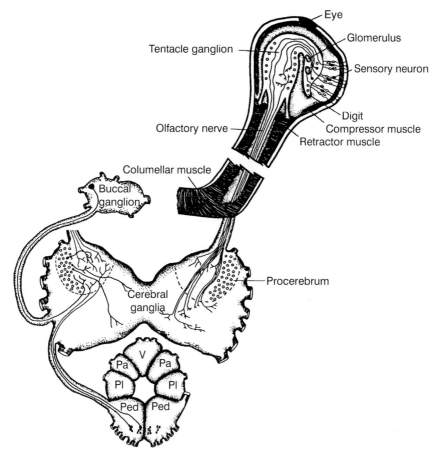

Fig. 3.1. The olfactory system, based on observations in *Achatina fulica* Bowdich (Achatinidae), *Cantareus aspersus* (Müller) (Helicidae) and *Limax maximus* Linnaeus (Limacidae). A single ocular tentacle is drawn on the right side and a single buccal ganglion on the left side; all other structures are shown with bilateral representations, but different aspects of neural circuitry within the procerebrum are shown on the left and right sides. Abbreviations of ganglia: V, visceral; Pa, parietal; Pl, pleural; Ped, pedal.

independently of the nose. Vision is not an important sense in terrestrial gastropods because they are active mostly at night. Behavioural studies suggest that the eye is used for negative phototaxis, to entrain circadian rhythms and to regulate reproductive functions (Sokolove and McCrone, 1978; Bailey, 1981). The eye consists of a cornea, a lens and a retina. There are two morphological types of photoreceptor cells, which together number a few thousand (Brandenburger, 1975). Type I photoreceptors have long microvilli and contain dense aggregations of clear spherical vesicles about 80 nm in diameter, whereas type II photoreceptors have short, irregular villi and no aggregations of vesicles. The spectral sensitivity of the eye of *Limax* has dual peaks at 460 and 480 nm, possibly corresponding to two photoreceptor types (Suzuki *et al.*, 1979). In *Helix*,

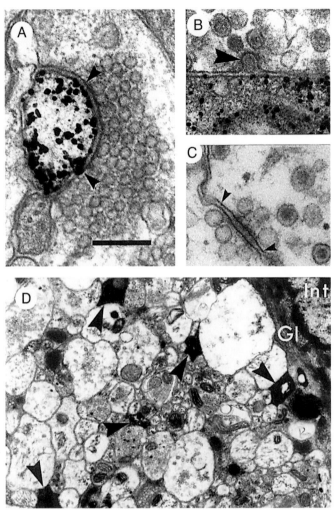

Fig. 3.2. Ultrastructural features of neuropil and synapses. (a) Synapse on to the giant cerebral neuron, here labelled by black grains from silver intensification of a hexamminecobalt injection. Note the aggregation of uniformly clear vesicles on the presynaptic side of the active zone (between the arrowheads). (b) Exocytosis of a dense-cored vesicle (arrowhead) at a non-specialized junction between an unidentified process and a dendrite of the giant cerebral neuron (labelled with silver grains). (c) Example of a symmetrical chemical synapse (between the arrowheads) in the tentacle ganglion. Note the presence of dense-cored vesicles. (d) A typically dense neuropil in a glomerulus underlying the olfactory epithelium. Arrowheads point to processes of sensory neurons labelled with horseradish peroxidase. Gl, a portion of the glial processes that surround the glomerulus; Int, periglomerular interneuron. Micrographs (a) and (b) are from *Rumina decollata* (Linnaeus) (Subulinidae) (Chase and Tolloczko, 1992); (c) is from *Achatina fulica* Bowdich (Achatinidae) (McCarragher and Chase, 1985); (d) is from *A. fulica* Bowdich (Chase and Tolloczko, 1993). Scale bar = (a) and (b) 200 nm; (c) 230 nm; (d) 1 μm.

the composite spectral sensitivity is in the range 390–580 nm, with a single peak at 496 nm, suggesting that the entire sensitivity is determined by rhodopsin (Chernorizov *et al.*, 1994).

Statocysts

The statocysts sense the direction of gravity. They probably play a role in such behaviours as righting, vertical migrations after rainfalls and burrowing. It has also been reported that the statocyst receptor cells respond to low-frequency vibrations, for example of the substrate on which the animals crawl (Kovalev *et al.*, 1981, cited in Zakharov, 1994). These sense organs are oddly placed within the pedal ganglia of the central nervous system (CNS). Each statocyst is connected to the ipsi-lateral cerebral ganglion by a fine sensory nerve that runs alongside the cerebral–pedal connective nerve.

Structurally, the statocyst is a hollow sphere of about 100 μm diame-ter. The centre contains lymph in which float several hundred ellipsoidal calcium particles (statoconia, 3–15 μm in length). Ciliated sensory cells, 10–13 in number, line the inside wall of the cyst (Zaitseva, 1994). When gravity causes the statoconia to fall on to the cilia, a generator potential is produced in the sensory cells which, if large enough, causes spiking in fibres of the statocyst nerve.

The statocyst itself is innervated by fibres that evidently provide for the modulation of statocyst function. Two sources of such modulatory influences are the eyes and olfactory organs (Zaitseva, 1994).

Mechanosensors

The identity and location of the primary mechanosensors have not been clearly established in terrestrial gastropods. There are two likely locations for the mechanosensory cell bodies: either subcutaneous or in the ganglia. Evidence to support the former location is the presence of bipolar receptor cells throughout the integument, even in places that do not appear to be chemosensitive. By matching cell types with their distributions in the skin, it has been proposed that the peripherally located receptors that have large numbers of cilia gathered as bristles are mechanoreceptors (Zaitseva, 1994). However, free nerve endings, presumably mechanosensitive, are also observed in the integument (Zaitseva, 1994), and these are not associated with peripherally located cell bodies. A central location for the mechanoreceptor cell bodies is supported by studies in *Aplysia* in which several clusters of centrally located mechanosensory neurons have been identified (Kandel, 1979). Only one mechanosensory neuron has been identified individually in ter-restrial species, a radular mechanosensor located in the buccal ganglion of *Inciliaria fruhstorferi* Collinge (Philomycidae) (Kawahara *et al.*, 1994).

Peripheral Nervous System

The peripheral nervous system (PNS) is comprised of the nerve cells whose somata lie outside the major ganglia. The neurons of the PNS are found in various tissues, and in various arrangements. One important component of the PNS is the subepithelium, which, as noted above, contains sensory receptor cells. It also contains other intrinsic neurons that may have either bipolar or monopolar morphologies. Still other nerve cells are found scattered among the nerves that form plexuses in the gastric organs and the foot. In addition, there are small ganglia associated with some peripheral organs such as the tentacles and, in some cases, the genitalia. The small size of the peripheral neurons has largely prevented electrophysiological investigations of their properties. Thus, their function is inferred largely from lesion studies as described below.

At least some portions of the PNS contain motoneurons and integrated circuitry because tissues can exhibit reflex movements even after their connections with the CNS have been severed. A good example is the tentacle withdrawal reflex (Prescott and Chase, 1996). This response is mediated by two contractile components: a subepithelial muscle and a retractor muscle that connects with the columellar muscle. The form of the reflex is largely unaffected by the removal of the CNS, and excision of the CNS reduces the amplitude of the response by only about 40%. The relative contribution of the CNS to the total response increases with increasing stimulus strength. The CNS is required for sensitization of the reflex, but not for habituation (Prescott and Chase, 1996).

While it is certain that local reflexes can be mediated entirely by peripheral circuitry, the competence of the PNS to mediate responses at sites remote from the point of stimulation is less clear. Studies performed in *Aplysia* suggest that distance conduction in the periphery is variable depending on stimulus properties (Carew *et al.*, 1979). Early studies in *Limax* suggested that peristaltic waves of muscular contraction in the foot persist even after the tissue is isolated from the CNS. Whether this requires the pedal nerve plexus or is simply a manifestation of muscles responding to a succession of passive forces is uncertain. In contrast, *Helix* is incapable of maintaining the waves after removal of the CNS (Bullock, 1965).

Central Nervous System

Evolution

The absence of segmentation in gastropod molluscs has permitted the CNS to be organized in significantly different ways in different animals. In particular, the ganglia are present in a range of numbers and a variety of positions. A general review of these arrangements is constrained by the nature of the subject, as summarized by Bullock (1965, p. 1288): 'The gross anatomy of the gastropod nervous system embraces some of the most difficult problems in comparative anatomy and has spawned a prodigious literature.'

The early molluscs were symmetrical, and they had a straight-through alimentary tract, that is a mouth at the anterior end, and an anus at the posterior end. The anus was present within the pallial cavity that lay beneath the overhanging shell. Also within the pallial cavity were a pair of ctenidia, or gills. Primitively, the nervous system consisted of five pairs of ganglia. Anteriorly, there were the buccal, cerebral, pleural and pedal ganglia. These were joined, by a long nerve cord, to paired intestinal and visceral ganglia (the latter pair often fused). The succession of ganglia from the left parietal ganglion down to the visceral ganglion, and then back up to the right parietal ganglion, is known as the visceral chain. At some point during the early Cambrian period, animals appeared in which the mantle cavity, with its associated structures, was rotated 180° in the counter-clockwise direction towards the anterior end.

An important result of torsion was the displacement of many interior organs. In the case of the nervous ganglia, the displacements were secondary consequences of adaptations involving other organs, principally the gills. To appreciate what happened, it is useful to understand that torsion is a developmental process as well as a phylogenetic one. Thus, certain ganglia may have their origin on one side of the animal (left or right) but come to reside on the opposite side in the adult. Because the connectives of the visceral chain become twisted in the process, the nervous system acquires a figure of eight appearance. This condition, known as streptoneury, is a plesiomorphic characteristic of the extant gastropod molluscs. However, many gastropod lineages have acquired an untwisted arrangement secondarily, known as euthyneury. Thus, the evolutionary history of the gastropod may be summarized as straight, then twisted, then once again untwisted.

According to one view (Fretter and Graham, 1962; Bullock, 1965; Haszprunar, 1985), euthyneury came about through a combination of detorsion (the secondary posterior migration of the anus and mantle cavity) and condensation of the ganglia. However, other authorities find evidence to exclude condensation as a significant factor (Régondaud, 1964; Page, 1992). Several specific theories to account for the transition from streptoneury to euthyneury have been argued and discussed extensively (see Dorsett, 1986).

In addition to the appearance of euthyneury, the evolution of the higher gastropod clades, the Opisthobranchia and Pulmonata, is marked by two other important developments. First, the pleural ganglia moved from a position next to the pedal ganglia on the ventral side of the gut (hypoathroid condition) to a position next to the cerebral ganglia on the dorsolateral side of the gut (epiathroid condition). This change was first detailed in prosobranchiate gastropods by Fretter and Graham (1962), and it is generally taken as representative of the trend towards concentration of the nervous system in the higher gastropods. However, in the stylommatophoran Pulmonata, the hypoathroid condition once again pre-vails even though the nervous system is highly condensed (see below).

The second major evolutionary event affecting the Pulmonata was the appearance of a new pair of ganglia, variously known as 'pallial' or 'parietal', that evidently derived from the pleural ganglia under adaptive pressures for the innervation of new body surfaces (Brace, 1977). Since the new ganglia comprise part of the visceral loop – together with the suboesophageal, supraoesophageal and visceral ganglia – their presence is said to describe a pentaganglionic condition (Haszprunar, 1985, 1988). In Haszprunar's interpretation, pentagangliony is not only highly correlated with euthyneury but it is actually more diagnostic of the presumed phy-logeny. He has therefore created the taxonomic group Pentaganglionata that, together with the Triganglionata, comprise the Heterobranchia, the most advanced group of gastropods (roughly equivalent to Opisthobranchia plus Pulmonata). While it may be true that the Pentaganglionata are monophyletic, Haszprunar's conclusions are controversial and they have been criticized on methodological grounds (Bieler, 1990).

The CNS of *Achatina fulica* Bowdich is illustrated in Fig. 3.3 to show details of the ganglia, peripheral nerves and connectives. This is a well-studied species, whose CNS organization is often taken to be representative or illustrative of the Stylommatophora.

The variable organization of the nervous system in pulmonate gastro-pods has inspired at least three taxonomic classifications based solely, or predominantly, on the morphology of this system. Bargmann's (1930) classification is based on the extent of concentration along the visceral loop, for which she discerned eight degrees. However, concentration seems to have occurred by various means in several lineages, so the value of this character for phylogenetics is questionable (Haszprunar and Huber, 1990). A novel approach was taken by van Mol (1974) who pointed out that the procerebral lobe of the cerebral ganglion is present in all pulmonates but in no opisthobranch. He also observed that the size of the procerebrum, as well as the size of the neurons in the procerebrum, varies in accordance with habitat and lifestyle, with the greatest development seen in terrestrial families. He therefore divided the Pulmonata into four groups (Fig. 3.4) based primarily on the prominence of the procerebrum, but also incorporating data on the size of the neurons in the procerebrum and the morphology of associated neurosecretory structures. van Mol's conclusions have been criticized on the grounds that he did not examine a sufficiently representative sample of brains (Tillier, 1989). Haszprunar and Huber (1990) re-evaluated van Mol's work in the light of their own view that primitive pulmonates had large neurons in the procerebrum, not small ones as van Mol had suggested. Their phylogeny also incorpo-rates the assumption that the epiathroid condition (the pleural ganglia close to the cerebral ganglia) is primitive, whereas the hypoathroid condition (the pleural ganglia adjacent to the pedal ganglia) is representa-tive of the advanced pulmonates, including the Stylommatophora. The illustration of Haszprunar and Huber's classification scheme (Fig. 3.5)

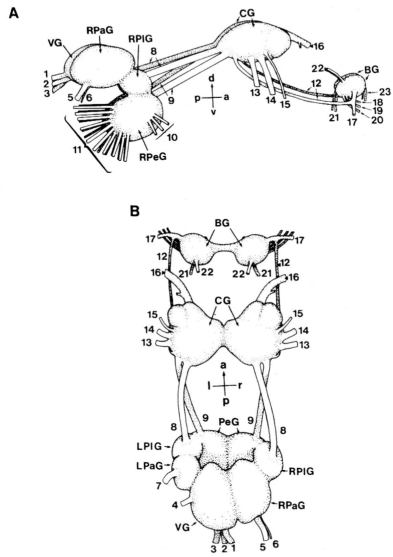

Fig. 3.3. The central nervous system of *Achatina fulica* Bowdich (Achatinidae). (A) Right lateral view. (B) Dorsal view. Abbreviations of ganglia: CG, cerebral ganglia; BG, buccal ganglia; PeG, pedal ganglia; PlG, pleural ganglia; RPaG and LPaG, right and left parietal ganglia; VG, visceral ganglion. Abbreviations of nerves: 1, right posterior pallial; 2, intestinal; 3, anal; 4, left posterior pallial; 5, right parietal; 6, accessory right parietal; 7, left parietal; 8, cerebral–pleural connective; 9, cerebral–pedal connective; 10, cutaneous pedal; 11, inferior pedal; 12, cerebral–buccal connective; 13, interior labial; 14, medial labial; 15, exterior labial; 16, tentacular (olfactory); 17, superficial pharyngeal; 18, anterior lateral buccal; 19, medial lateral buccal; 20, posterior lateral buccal; 21, oesophageal; 22, salivary; 23, medial buccal. From Croll (1988).

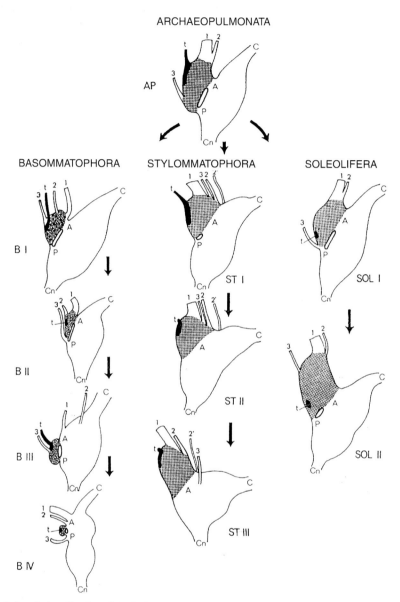

Fig. 3.4.   A classification of the Pulmonata according to the morphology of the procerebrum and related structures. From van Mol (1974). The illustrations are representative of the following families: AP: Ellobiidae; BI: Trimusculidae; BII Amphibolidae, Chilinidae, Latiidae, Ancylidae; BIII: Siphonariidae; BIV: Lymnaeidae, Planorbidae; STI: Succineidae; STII: Orthurèthres; STIII: Sigmurèthres, Achatinidae; SOLI: Onchidiidae; SOLII: Vaginulidae; Rathouisiidae. Other symbols and abbreviations: 1, tentacle nerve; 2, external peritentacular nerve; 2′, internal peritentacular nerve; 3, optic nerve. C, cerebral commissure; Cn, cerebropleural connective; A, anterior procerebral connective; P, posterior procerebral connective; T, cerebral gland.

contains representative examples of various CNS arrangements in the Pulmonata.

A more analytical approach has been taken by Tillier (1989) in his major revision of the Stylommatophora. His phylogeny and classification

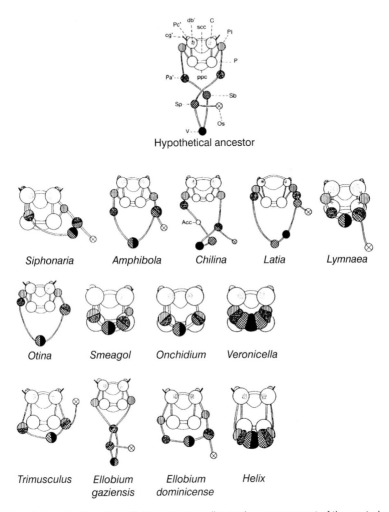

Fig. 3.5. A classification of the Pulmonata according to the arrangement of the central ganglia and the type of procerebrum. Each illustrated genus represents its family. Concentration of the ganglia increases from left to right. (I) Basommatophora; (II) Systellommatophora; (III) Eupulmonata. acc, accessory ganglion; c, cerebral ganglion; db, dorsal body; cg, cerebral gland; os, osphradial ganglion; p, pedal ganglion; pa, parietal ganglion; pc, procerebrum (small stipple, globineurons; large stipple, large neurons); pl, pleural ganglion; ppc, parapedal commissure; sb, suboesophageal ganglion; sp, suproesophageal ganglion; scc, subcerebral commissure; v, visceral ganglion. Modified from Haszprunar and Huber (1990). *Siphonaria* Sowerby: Siphonariidae; *Amphibola* Schumacher: Amphibolidae; *Chilina* Gray: Chilinidae; *Latia* Gray: Latiidae; *Lymnaea* de Lamarck: Lymnaeidae; *Otina* Gray: Otinidae; *Smeagol* Climo: Smeagolidae; *Onchidium* Buchanan: Onchidiidae; *Veronicella* de Blainville; Vaginulidae; *Trimusculus* Schmidt; Trimusculidae; *Ellobium gaziensis* (Preston) and *Ellobium dominicense* (de Férussac): Ellobiidae; *Helix* Linnaeus: Helicidae).

are based on a factor analysis of several morphological systems. For the nervous system, he recognizes four characters and 15 character states, all involving the arrangement of the ganglia in relation to the size and shape of the animal. One of Tillier's major conclusions, in agreement with Haszprunar and Huber (1990), is that proximity of the pleural and cerebral ganglia (epiathroidy) is apomorphic with respect to the proximity of the pleural and pedal ganglia (hypoathroidy). Also, the proximity of the right parietal and visceral ganglia is accepted as a synapomorphy of all Stylommatophora. A critique of Tillier's classification can be found in Emberton and Tillier (1995).

Development

Only the basic features of the embryonic development of the CNS are known for any species of molluscs, and studies of the terrestrial species are especially rare. Neurogenesis occurs locally in the ectoderm, followed by migration of the post-mitotic cells to nearby sites of ganglion formation within the body cavity (Hickmott and Carew, 1991). The commissures and the connective nerves are formed by secondary outgrowth from the ganglia. The order of appearance of the ganglia generally progresses in an anterior to posterior direction.

Not all ganglia develop at the same rate and, even within a single ganglion, there may be regional variations. The mesocerebrum develops relatively late and undergoes a burst of growth just prior to sexual maturation (LaBerge and Chase, 1992). The late growth of the mesocerebrum is attributed to an increase in the size of the neurons already present, rather than to delayed or continuing histogenesis (LaBerge and Chase, 1992). However, in the procerebrum, there is evidence for the proliferation of new neurons as late as 1 month after hatching (Zakharov et al., 1998).

Structure of the ganglia

The ganglia are surrounded by a sheath which functions both as a physical barrier to contain the neurons and as an interface for the exchange of substances between the haemocoel and the nerve cells, perhaps with filtering and transporting properties. The sheath contains a variety of cell types (Fernandez, 1966), for example globular cells containing glycogen and lipofuscin particles, gland cells, pigment cells, and muscle cells whose fibres are partly continuous with extra-ganglionic muscles. Collagen strands form complex lamellae throughout the tissue, but an outer, more loose region (about two-thirds of the total width) can be distinguished from a more dense, inner region. The actual thickness and density of the sheath are variable in different species and, within a species, they vary as a function of ganglion identity, the age of the animal and the physiological state of the animal (e.g. aestivation).

The CNS itself is avascular, but blood vessels are present in the ganglionic sheath. Branches of the anterior aorta enter the sheath overlying each ganglion where they arborize further to form an extensive vascular network. The pattern of vascularization is quite consistent from animal to animal, at least in *Helix* (Hernádi, 1992).

As is the case with other invertebrate ganglia, those of pulmonate gastropods consist of an outer cortex, which contains the nerve cell somata, and an inner neuropil, which contains the dendritic and axonal processes (Fig. 3.6). An exception to this scheme is found in the procerebrum (described below).

Glial cells

Interior to the ganglionic sheath, the CNS contains two cell types, neurons and glia. Glia possess small oval nuclei of 6–8 µm diameter and extended cytoplasmic processes. The glial nuclei can be identified in light

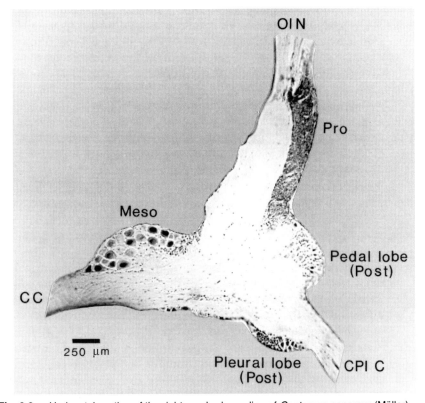

Fig. 3.6. Horizontal section of the right cerebral ganglion of *Cantareus aspersus* (Müller) (Helicidae) stained with toluidine blue. The ganglion comprises three principal areas, or lobes: Pro, procerebrum; Meso, mesocerebrum; Post, postcerebrum (including the pedal and pleural lobes). CC, cerebral commissure; CPl C, cerebro-pleural connective nerve; Ol N, olfactory nerve. Histology prepared by S. Ratté.

microscopy by their very dark appearance after treatment with chromatic stains, while electron microscopy reveals the cytoplasm of glial processes as relatively more dense than the neuronal cytoplasm after osmium staining (Fig. 3.2d). These cells are found principally under the basal lamella of the sheath and in the cortex of the ganglia, but their processes extend well into the neuropil. In general, the glial processes interweave between the neuronal processes in the neuropil (Fernandez, 1966). Multiple functions have been assigned to the glia, consistent with their known or hypothesized roles in vertebrates (i.e. nutrition, mechanical support, regulation of extracellular ion concentrations and transmitter uptake).

The axons of nerve cells generally are surrounded by processes of non-neuronal cells, both in the neuropil and in the nerves. Amoroso *et al.* (1964) identify the supporting cell as a special cell type, which is said to be associated with fibres in nerves, but it is not clear whether these cells are fundamentally different from the glia. In any case, the shape of the supporting cells in the nerves is stellate, and the axons travel in a direction perpendicular to the non-neuronal processes. The larger axons are wrapped individually, whereas the smaller ones are enclosed collectively.

An unusual feature of molluscan axons is the extensive invagination of the membrane, which is especially evident in large-calibre fibres. Such infoldings are occupied by processes of glia cells (or supporting cells). The invaginations can be interpreted as adaptations related to the elastic property of the animals' soft bodies, as they permit the axons to be stretched without damage. Another significant consequence is that the surface area of the membrane for a given axon remains more or less constant regardless of the degree of stretch. This results in a constant conduction velocity for action potentials independent of the animal's state of retraction or extension (Mirolli and Talbott, 1972).

Nerve cells

When the sheath of a ganglion is removed by dissection, the nerve cell somata are readily visible on the surface of the ganglia. They are round or oval in shape and highly variable in size, with diameters ranging from 7 to nearly 200 µm. The prevalence of some very large neurons is characteristic of the opisthobranch and pulmonate molluscs, and contrasts with all other animal groups (Gillette, 1991). All but the smallest neurons are polyploid, and there is evidence for differential DNA endoreplication (Chase and Tolloczko, 1987). Most probably, the size of a neuron, together with its proportional DNA content, reflects the total surface area of the cell, including its processes. As the area increases, so too do the demands for protein production (Gillette, 1991; LaBerge and Chase, 1992). The larger neurons can be identified reliably as unique individuals by their locations as well as by other morphological, physiological and biochemical properties. Some of the individual cells are described later in this chapter.

Neurons in the CNS are generally monopolar (i.e. there is just one process or neurite extending from the soma), and input/output functions are segregated to different branches of the main neurite. A curious feature of many neurons is that they may send multiple, parallel axon branches into the same peripheral or connective nerve (e.g. Fig. 3.7). The most attractive hypothesis to explain this feature is that such an arrangement could increase the efficiency of release of messenger molecules from the neuron into the nerve trunk, and hence into the blood circulation (Pin and Gola, 1984).

There are no accurate counts of the total number of neurons. A consensus estimate for the ten individual ganglia comprising the CNS is 10,000–15,000 neurons (Dorsett, 1986). However, this estimate does not include the procerebrum which alone has approximately 40,000 neurons in *Achatina* and *Helix* (Ratté and Chase, 2000) and 50,000 neurons in *Limax* (Kleinfeld *et al.*, 1994). There are also large numbers of neurons outside the CNS, for example 100,000 sensory cells in the olfactory

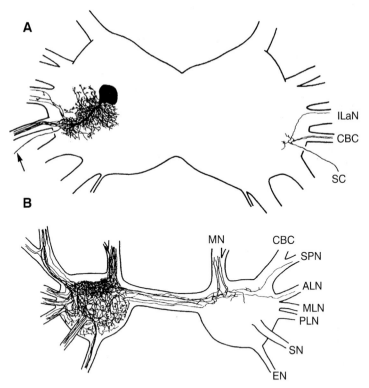

Fig. 3.7. Morphology of the giant cerebral neuron (GCN) in *Achatina fulica* Bowdich (Achatinidae). Drawing from a preparation in which the cell was injected with nickel-lysine to reveal its structure. (A) The cell body and dendrites in the cerebral ganglion, with axonal projections in the interior labial nerve, the cerebral–buccal connective, and the subcerebral commissure (arrow). Note the commissural projection to the contralateral side of the ganglion. (B) Innervation of the buccal ganglion. Nerve abbreviations are given in the legend of Fig. 3.3. Scale bar = 120 μm. From Chiasson *et al.* (1994).

epithelium of each cephalic tentacle, 5000 interneurons in each tentacle ganglion and 3000 neurons in each eye. From these data, it can be calculated that the olfactory system alone accounts for about 250,000 neurons (Chase, 1986b).

Synaptic transmission

Chemical transmission between neurons occurs at synapses that have an ultrastructural appearance similar to that of vertebrate central synapses, except for the absence of a prominent postsynaptic density (Fig. 3.2a). In earlier publications, synapses were reported as rare or uncertain (e.g. Amoroso *et al.*, 1964), but, with the improvement of fixation methods, the observation of well-differentiated synapses in gastropods has become commonplace. Nevertheless, the limited data so far available suggest that the density of synapses in the gastropod brain may be considerably less that than in the CNS of other animals (Chase and Tolloczko, 1992).

Several morphological features of gastropod synapses merit special mention. Some synaptic junctions have an ultrastructural symmetry suggestive of bilateral chemical transmission (Fig. 3.2c). Such synapses have been observed in the tentacle ganglion (McCarragher and Chase, 1985) as well as in the procerebrum (Chase and Tolloczko, 1989). Although not unique to gastropods, nor even to molluscs, bilaterally symmetrical chemical synapses are intriguing structures with an unknown functional significance. Another point to note is that a single nerve fibre profile typically contains a variety of vesicles and granules, ranging from small and clear to large and dense as seen in electron microscopy (Fig. 3.2). It is generally assumed that the small, clear vesicles contain either amino acids or other 'classical' neurotransmitters, while the large, electron-dense granules contain neuropeptides. Exocytosis is not limited to synaptic sites. The small vesicles release their contents at morphologically differentiated synapses ('active zones'; Fig. 3.2a), but the larger granules undergo exocytosis either at undifferentiated points of cellular contact ('parasynaptic'; Fig. 3.2b) or at unapposed, undifferentiated membranes ('non-synaptic') (Chase and Tolloczko, 1992).

Electrical coupling of neurons has been reported in the cerebral ganglion (Yoshida and Kobayashi, 1992), the buccal ganglion (Yoshida and Kobayashi, 1991; Altrup and Speckmann, 1994) and the parietal/visceral ganglia (Furukawa and Kobayashi, 1987b). Gap junctions, which are the morphological specializations believed to represent 'electrical synapses', have been found in the procerebrum of *C. aspersus* (Müller) (S. Ratté and R. Chase, unpublished observations), but their occurrence elsewhere in the CNS, as suggested by physiological coupling, requires confirmation.

The 'classical' neurotransmitters in molluscs include acetylcholine, serotonin, several monoamines and several amino acids (S.-Rózsa, 1984). In addition, the application of immunohistochemical methods has indicated the presence of numerous peptides that are candidates for hormones, transmitters or modulators. To date, positive results have been reported for antibodies against at least 16 peptide species. Especially noteworthy is the evidence for APGWamide (Griffond *et al.*, 1992; Li and Chase, 1995), enkephalin (Marchand *et al.*, 1991), FMRFamide (Elekes and Nässel, 1990; Marchand *et al.*, 1991; Li and Chase, 1995), vasoactive intestinal peptide (Kaufmann *et al.*, 1995) and tachykinin (Elekes and Nässel, 1994). Each substance has its own unique distribution among the ganglia, and among the individual cells in each ganglion. It is important to bear in mind, however, that a positive result with immunohistochemisty is not sufficient to establish the presence of a given molecule. More definitive tests require *in situ* hybridization or direct chemical assay, but these methods have rarely been used. There is also a paucity of knowledge concerning the physiological roles of the peptides. FMRFamide, originally described as a molluscan cardioactive peptide, has been investigated the most thoroughly. It produces changes in the membrane potentials of neurons in *Helix*, and its ionotropic receptor in this genus has been cloned (Lingueglia *et al.*, 1995). Also, FMRFamide can modulate the strength of muscle contractions (Lehman and Greenberg, 1987). A role for FMRFamide in the meditation of dart shooting and other aspects of mating behaviour has been proposed (Li and Chase, 1995).

An interesting candidate transmitter molecule is the gas nitric oxide. The enzyme responsible for nitric oxide production is present in specific regions of the PNS and CNS, with particularly high concentrations in the procerebrum (Cooke *et al.*, 1994; Sánchez-Alvarez *et al.*, 1994). A neurotransmitter role for nitric oxide in molluscs is suggested by data from vertebrates and is supported by the finding that electrical oscillations in the procerebrum are dependent on nitric oxide production (Gelperin, 1994; see below).

The Cerebral Ganglion

The paired cerebral ganglia, commonly referred to as 'the brain' (Chase, 2000), are situated just dorsal to the origin of the oesophagus in the buccal mass. These ganglia are united by a commissure arching over the oesphagus and by a subcerebral commissure passing under the oesophagus. The ganglia receive inputs, via peripheral nerves and connective nerves, from the sense organs and the distributed sensory cells of the entire body, but principally from the head region. The afferents from each nerve terminate within a fairly discrete area of the neuropil. The size of the terminal field is roughly proportional to the diameter of the nerve itself (Chase and Tolloczko, 1992, 1993; Zaitseva, 1994). The cerebral ganglia integrate this sensory information and produce motor commands that are often routed directly back into the peripheral nerves. The three bilateral pairs of lip nerves, for example, each contains mixed populations of afferent and efferent fibres (Hernádi *et al.*, 1987). In other cases, cerebral output is relayed to the pleural, pedal and visceral ganglia (Fig. 3.3).

The major regions of the cerebral ganglion are designated the procerebrum, the mesocerebrum and the postcerebrum (Fig. 3.6). Some authors, for example Bullock (1965), recognize the postcerebrum as comprising pleural, pedal and commissural lobes. The term metacerebrum can also be found in the literature in reference to the postcerebral structures. A full discussion of nomenclature from a historical perspective is given in Chase (2000).

Procerebrum

In several respects, the procerebrum is the most unusual region in the entire CNS. Its embryological origins are associated with those of the tentacle ganglion, and it only joins the rest of the cerebrum at a late stage in development (van Mol, 1974). Because it links the peripheral tentacular structures with the CNS, an olfactory function has been assumed. Interestingly, the procerebrum is unique to the pulmonate gastropods, and it is most highly developed in the Stylommatophora (Fig. 3.4; van Mol, 1974). Metabolic and physiological studies support the assumption of olfactory function (Chase, 1985; Gelperin and Tank, 1990). Morphologically, the procerebrum is exceptional because, in the Stylommatophora, it contains a very large number of very small neurons. There are perhaps as many as 50,000 neurons per lobe (see above), but each has a soma diameter of only 6–8 μm. Unlike all other ganglionic regions, in the procerebrum the cells are located adjacent to, rather than exterior to, associated neuropil (Fig. 3.6). Although the entire neuropil is finely textured, one can distinguish a small inner mass and a larger outer (or 'terminal') mass based on histological appearance and immunohistochemistry (Cooke *et al.*, 1994; Sánchez-Alvarez *et al.*, 1994).

The properties of small size and large numbers in the procerebral cell population invite comparison with the Kenyon cells of the insect protocerebrum and the granule cells of the vertebrate olfactory bulb, two structures that are likewise implicated in olfactory function (Chase and Tolloczko, 1993). In all three cases, the morphology of the intrinsic neurons does not include an obvious axon. Silver staining of procerebral cells fails to reveal any processes leaving the region (Zaitseva, 1994). Chase and Tolloczko (1989) described a cluster of about two dozen neurons in the pedal ganglion with cells that send long processes into the procerebrum (Fig. 3.1). Since these fibres were shown to receive synapses, but were seldom presynaptic, they are believed to function as dendritic receptors of procerebral output. A similar pathway has been described in *Limax* in which long dendritic processes from a pair of buccal ganglion cells allow information to flow from the procerebrum to the buccal ganglion (Gelperin and Flores, 1997). Procerebral neurons were injected individually with a visible tracer in *C. aspersus* (Ratté and Chase, 1997). This technique revealed a class of procerebral cells whose processes extended well into the ganglion's core neuropil. Electron microscopic observations of the labelled processes confirmed that output synapses predominate over input synapses in the central neuropil (Ratté and Chase, 2000). Thus, it is now apparent that the procerebrum participates directly in distributing olfactory information within the CNS.

A striking electrophysiological feature of the procerebrum is the presence of oscillating field potentials. These are generated intrinsically, and they occur continuously at a rate of about 0.7 Hz (Gelperin and Tank, 1990). Since they can be recorded with extracellular electrodes, they must arise from the synchronous activity of large numbers of neurons. Gelperin and colleagues have combined field potential and intracellular recording methods to study their origin (Kleinfeld *et al.*, 1994). It appears that two classes of procerebral neurons can be defined by intracellular electrophysiology. One class, which comprises only about 10% of the entire population, exhibits spontaneous bursts of action potentials. The second class does not burst but is strongly and periodically hyperpolarized by the spiking activity of the bursting cells. The field potentials reflect the sum of the currents resulting from membrane events in the two classes of neurons.

The role of the procerebral electrical oscillations in olfactory function, if any, is uncertain. Odours influence the oscillations in a variety of ways, as reported by several investigators, with changes in frequency, amplitude and spatial dynamics being reported (Chase, 2000). Since terrestrial gastropods rely on olfactory learning to regulate their search for food (Croll and Chase, 1980), it is possible that the oscillations are part of a procerebral learning mechanism.

Mesocerebrum

Although the borders of the mesocerebrum can be difficult to define, especially posterior to the commissure, the mesocerebrum possesses two diagnostic morphological features (Fig. 3.6). First, it contains many large neurons, approximately 140 cells with soma diameters larger than 50 μm in *C. aspersus* (Chase, 1986a). Secondly, in mature animals, it is noticeably larger on the right side than on the left side, with the right mesocerebrum containing more and larger neurons than the left mesocerebrum (Chase, 1986a). This asymmetry corresponds to the anatomical localization of the reproductive organs on the right side of the animal. Experimental data support the inference that the functional role of the mesocerebrum is to control reproduction (Chase and Li, 1994; Koene *et al.*, 2000).

The mesocerebral cells predominately innervate the penis and the dart sac in *Cantareus* (Li and Chase, 1995). About 25% of the cells innervate each structure, and some cells innervate both structures. Various peripheral nerves and the pedal ganglion are also targeted. Electrical stimulation of the right mesocerebrum, but not the left mesocerebrum, causes movements of the penis and the dart sac *in vitro*. Even when individual neurons are stimulated, movements can be elicited in one or both structures (Chase, 1986a). A motor function for the mesocerebrum was confirmed recently by electrical stimulation *in vivo* using a fine wire electrode glued to the mesocerebrum (Koene *et al.*, 2000). Use of the same implanted electrode further demonstrated that electrical activity in the mesocerebrum increases significantly during the expression of natural mating behaviour.

The neurons receive afferent inputs from the peripheral nerves, mostly on the right side of the animal regardless of whether a cell lies in the left or right mesocerebrum (Chase, 1986a). Excitation predominates over inhibition, and the strongest inputs enter from the peritentacular nerves and the lip nerves. When tactile stimulation of the skin is applied *in vivo* while recording from the mesocerebrum with an implanted wire electrode, maximal responses are recorded with stimulation of either the ocular tentacles or the area surrounding the genital pore (Koene *et al.*, 2000). The neurons have little spontaneous activity, and they are not easily induced to fire action potentials, suggesting that the excitability of the cells is regulated hormonally and that it varies seasonally or with other factors.

Several peptides have been localized to the mesocerebrum including somatostatin, FMRFamide, APGWamide, insulin, leucokinin I and met-enkephalin. In *C. aspersus*, the innervation of the dart sac is predominately by cells containing FMRFamide, whereas innervation of the penis is predominately by cells containing APGWamide (Li and Chase, 1995). This could indicate a chemical partitioning of function between dart shooting and copulation in this species.

In contrast, *Achatina*, which has a simpler mating behaviour and uses neither a dart nor a spermatophore, has a much less conspicuous mesocerebrum than *Cantareus*. Further comparative studies could take advantage of the highly variable nature of mating and reproduction in stylommatophoran pulmonates.

Postcerebrum

This lobe contains mostly small cells. From backfill studies in Limacidae and Helicidae it is has been determined that the small postcerebral cells send axons into a variety of nerves – namely, the lip nerves (Ierusalimsky *et al.*, 1994), the cutaneous nerves of the pedal ganglion (Ierusalimsky and Zakharov, 1994; Li and Chase, 1995), the cerebrobuccal connectives (Delaney and Gelperin, 1990a) and the penial nerve (Eberhardt and Wabnitz, 1979; Li and Chase, 1995). Some postcerebrum neurons are excited when chemical stimuli are applied to the lips (Hernádi *et al.*, 1987; Delaney and Gelperin, 1990c). The cells that are excited in this manner are not themselves sensory cells, but they receive input from lip sensory neurons either directly or indirectly. Eight of these cells, each sending an axon to the buccal ganglion, have been identified individually in *Limax* (Delaney and Gelperin, 1990a). The neurons evidently are involved in initiating feeding behaviours because their activity can trigger rhythmical firing in the buccal ganglion (Delaney and Gelperin, 1990b). A single neuron of the same type has been described in *A. fulica* (Yoshida and Kobayashi, 1992).

The giant cerebral neuron (GCN)

This is the largest neuron in the cerebral ganglion (soma diameter ~150 μm in *C. aspersus*). It exists as a pair of bilaterally symmetrical cells, situated on the ventral surface of the ganglion (Fig. 3.7). The cell is often referred to as the giant serotonergic neuron, because it contains a high titre of serotonin. The size of the GCN has enabled detailed descriptions of its unique properties which have led, in turn, to the realization that homologous neurons exist in the brains of other pulmonate and opisthobranch gastropods, but not in any prosobranchate gastropod (Sakharov, 1976; Pentreath *et al.*, 1982; Croll, 1987). Thus, the GCN was the first well-documented example of the important idea of cellular homology in molluscs; other examples are described later in this chapter.

The function of the GCN is to facilitate feeding behaviour. Its extensive dendritic arborizations in the cerebral ganglion receive inputs (Figs 3.2a and 3.7) via all the major peripheral nerves, that is the olfactory (tentacle) nerve and the three lip nerves (Kandel and Tauc, 1966; Chase and Tolloczko, 1992). These afferents carry sensory information to signal the presence of food. The efferent projections of the GCN are targeted for the buccal ganglion, where the terminal arborizations are as exuberant as the dendrites (Fig. 3.7). Features of the morphology differ slightly among species, which is a curiosity of unknown significance. Physiological studies (Weiss *et al.*, 1978; Yoshida and Kobayashi, 1991, 1995) have shown that the release of serotonin by the GCN modulates feeding by three separate mechanisms: (i) it depolarizes buccal motoneurons; (ii) it increases the contractile response of the feeding muscles contingent upon excitation via the motoneurons; and (iii) it accelerates the feeding motor programme.

The neuron C3

This is another large, easily identifed, cerebral neuron, best described in *Helix* and *Cantareus*. The soma, about 100 μm in diameter, is situated on the dorsal, anterior surface between the mesocerebrum and the procerebrum. C3 is a motoneuron innervating the tentacle retractor muscle (Cottrell *et al.*, 1982; Zakharov *et al.*, 1982). Mechanical and chemical stimuli, which elicit the tentacle withdrawal reflex when delivered to the tentacle tip, also strongly activate C3 (Chase and Hall, 1996). Quantitative tests of the mediation of the reflex using specific lesions of C3 demonstrate that C3 contributes 85% of the central motor component (Prescott and Chase, 1997). The dominant peptide contained within C3 is FMRFamide, which is present at a concentration of about 0.2 pmol per cell (Cottrell *et al.*, 1992).

The Buccal Ganglion

The small, paired buccal ganglia are sited beneath the oesophagus, on the posterior surface of the buccal mass (pharynx). Anatomically, they supply innervation of the buccal musculature, the salivary glands, oesophagus (including the crop, when present) and stomach (Fig. 3.3). Physiologically, they generate a rhythmical motor output ('feeding motor programme') that drives the various buccal muscles in the coordinated movements necessary for feeding (Gelperin *et al.*, 1978). Although only a few individual neurons are known in any detail, one must consider the buccal ganglion as a miniature control system for executing a limited but sophisticated motor function.

Several of the larger neurons in *Helix* have been characterized morphologically, physiologically and pharmacologically (Altrup and Speckmann, 1994). The most complete cellular description is available for *Achatina* (Yoshida and Kobayashi, 1991, 1992). Five pairs of bilaterally symmetrical motoneurons of the retractor muscles have been identified on the caudal (ventral) surface (Fig. 3.8). One pair of protractor motoneurons has been identified on the rostral surface. As noted above, the GCN can enhance the peripheral actions of both protractor moto-neurons and retractor motoneurons. All of the identified motoneurons fire action potentials in rhythmic bursts, which is an expression of the feeding motor programme and reflects the neurons' cyclical control of the feeding muscles. The patterning of the bursts is determined by a combination of inherent membrane instability in some of the motoneurons and synaptically mediated interactions within the network of motoneurons and associated interneurons (Fig. 3.8). Detailed mechanisms of buccal ganglion pattern generation have been described in basommatophoran freshwater species of *Lymnaea* de Lamarck (Lymnaeidae) (Benjamin and Elliott, 1989), but not in any terrestrial species.

Descriptions of two other functional cell types offer a sampling of the regulatory influences that are likely to be common and diverse. First, a motoneuron of the salivary duct in *Limax* fires continuous bursts of action potentials, and the frequency of the bursts is modulated by stretch of the salivary duct (Beltz and Gelperin, 1980). Thus, activity in the motoneuron is controlled by feedback from the effector organ. The role of this feedback might be to facilitate transport of secreted saliva from the duct to the buccal mass. Secondly, a mechanosensory neuron has been found in *I. fruhstorferi* that is excited by bending of the radula during its retraction phase (Kawahara *et al.*, 1994, 1995). Since its activity exerts significant modulatory effects on the feeding motor programme, its function may be to monitor the hardness of food and to regulate feeding behaviour accordingly.

Visceral and Parietal Ganglia

The nerves from these ganglia innervate the mantle, anus, pneumostome, nephridium, digestive gland, intestine, heart and hermaphroditic duct (Fig. 3.3). The right parietal ganglion is consistently larger than the left parietal ganglion. The size of the left ganglion relative to the unpaired visceral ganglion varies between species. Some homologous neurons, for

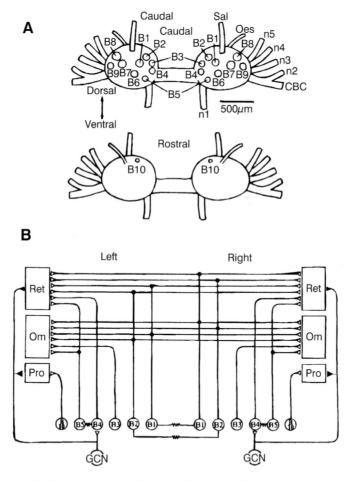

Fig. 3.8. Identified neurons and circuitry underlying feeding behaviour in *Achatina fulica* Bowdich (Achatinidae). (A) Location of identified cells in the buccal ganglion. (B) Schematic circuit, based on electrophysiological data. Open triangles, excitatory synapses; closed triangles, modulatory synapses; zig-zags, electrical coupling. Muscles: Ret, radular retractor; Om, outer muscle; Pro, radular protractor. GCN, giant cerebral neuron. Modified from Yoshida and Kobayashi (1991, 1995).

example the rostral giants described below, seem to shift their positions between the latter two ganglia in different species. In any case, the three ganglia appear to contain overlapping populations of functional neuron types, and the cells of one ganglion commonly send axons out the nerves of another. These arrangements reflect degrees of fusion between the ganglia.

The ganglia contain mostly large, or giant, neurons (Fig. 3.9). In *Helix* and *Cantareus*, each of the parietal ganglia has about ten neurons with soma diameters greater than 100 μm, and the visceral ganglion has about 20 neurons of this size (Kerkut *et al.*, 1975; Ierusalimsky *et al.*, 1994). Two cells, located in the rostral–dorsal region of the right parietal and visceral ganglia of *Achatina*, measure more than 300 μm in soma diameter

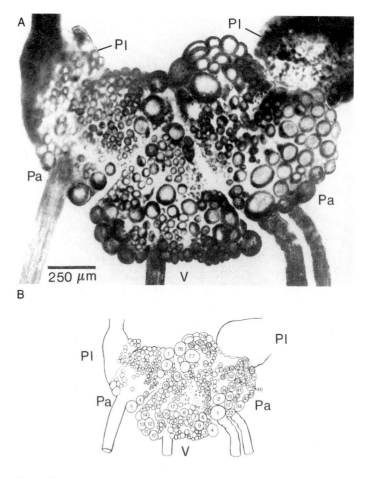

Fig. 3.9. Neurons in the fused complex of suboesophageal ganglia of *Cantareus aspersus* (Müller) (Helicidae). (A) Photograph of a preparation stained with methylene blue, touched up to accentuate the boundaries between cells. (B) Drawing showing a numbering scheme for the identification of individual neurons. Pl, pleural ganglion; Pa, parietal ganglion, V, visceral ganglion. Modified from Kerkut *et al.* (1975).

(Munoz *et al.*, 1983; Pawson and Chase, 1988). Despite these impressive dimensions, the identification of particular cells often is unreliable unless other criteria are employed in addition to size and position of the soma. Adding to the difficulty is the fact that various authors have used different nomenclatures to identify the same cells. Such problems are evident in the numerous cellular 'maps' that have been published for *Helix* and *Cantareus* (e.g. Kerkut *et al.*, 1975; Ierusalimsky *et al.*, 1994).

According to one report (Maksimova and Balaban, 1983, cited in Zakharov, 1994), 90% of the cells respond to tactile stimulation of the skin. In another survey of more than 150 neurons in the left parietal and visceral ganglia of *Helix*, 70% of the cells responded to tactile stimulation of the heart, kidney or pneumostome, or some combination thereof (S.-Rózsa, 1981). The identified neuron V21 has sensory fields in the heart, pericardium, blood vessels, kidney, liver and pneumostome. Other neurons are activated by tactile stimulation of the reproductive organs.

The output functions of the neurons have not been examined so thoroughly. Several of the large neurons are known to have a modulatory influence on cardiac function. In *Helix*, two heart inhibitory motoneurons and four heart excitatory motoneurons have been described, together with three interneurons that also influence heart activity (S.-Rózsa, 1981). In *Achatina*, four excitatory motoneurons and two inhibitory interneurons have been identified (S.-Rózsa, 1979; Furukawa and Kobayashi, 1987a). Many physiological connections among the heart regulatory cells, both within the parietal/visceral ganglia and with identified cells in other ganglia, have been described (S.-Rózsa, 1981; Furukawa and Kobayashi, 1987b). The possibility of homologies of the aforementioned cells in *Helix* and *Achatina* has been addressed but not settled (S.-Rózsa, 1979; Furukawa and Kobayashi, 1987a).

Several giant cells at the rostral end of the dorsal surface have been implicated in withdrawal behaviours. Balaban (1979) identified seven cells in *Helix* that command closure of the pneumostome, retraction of the mantle and contraction of the pedal skin. It is unclear whether they are motoneurons or premotor neurons. Giant cells can be found in an equivalent position in *Achatina*, and they are likely to have the same function (Munoz *et al.*, 1983). Because of fusion, at least one of the giants is found in the visceral ganglion of *Achatina*, whereas its presumed homologue in *Helix* is in the left parietal ganglion.

A convincing case for widespread homology has been obtained for a cell in the right parietal ganglion, which, in isolated CNS preparations, has a distinctive pattern of activity consisting of periodic bursting. Cells with the same electrophysiological signature, and also located in the left caudal part of the ganglion, have been reported in several species, although the nomenclature is idiosyncratic, for example *Helix* (F1, RPa1, Br), *Otala* Schumacher (Helicidae) (Cell 11) and *Achatina* (RPa11, PON). Further, similar cells are known in basommatophoran *Lymnaea* (RPD$_2$) and opisthobranch *Aplysia* (R15). The case for homology has been supported by immunohistochemical evidence showing that all the mentioned cells express a closely related gene (Kerkhoven *et al.*, 1993). There have been numerous proposals for the physiological function of the cell, with the most recent evidence suggesting a role in egg laying (Alevizos *et al.*, 1991).

Pleural Ganglia

These small, symmetrical ganglia have received little attention. The only peripheral nerves to which they give rise in *Helix* are two slender ones that innervate the columellar muscle. By retrograde tracer methodology, it has been found that these nerves are innervated by neurons whose soma lie in either the parietal ganglia or the visceral ganglion, but not in the pleural ganglion (Samygin and Karpenko, 1980). In *Helix*, intracellular stimulation of the largest cell in the ganglion causes withdrawal of the tentacles and the head (Balaban, 1979).

Pedal Ganglia

The pedal ganglion has more peripheral nerves than any other ganglion (Fig. 3.3). In *Helix*, there are three pairs of nerves on the dorsal side of the ganglion that innervate the skin of the lower body (exclusive of the head region) and portions of the male reproductive organs, and ten pairs of nerves that penetrate into the foot to connect with the pedal neural plexus. In addition, a connective nerve links the pedal ganglia to the cerebral ganglia. The majority of the neurons in the pedal ganglia are probably involved, by their connections with the plexus and the skin, in the control of locomotion and posture. However, individual cells having such functional properties have not yet been described because they are small in size. Otherwise, a variety of functions have been attributed to pedal ganglion neurons.

Withdrawal behaviour is controlled or modulated, in part, by cells in the pedal ganglion (Zakharov, 1994). Two large neurons with rostral, lateral placements innervate the columellar muscle, and they contribute directly to withdrawal behaviour (Samygin and Karpenko, 1980). Another identified neuron has properties similar to those of the parietal command neurons, with pronounced motor effects on the pneumostome. A group of cells in the rostral and medial part of the ganglion has received considerable experimental attention (Zakharov *et al.*, 1995). These cells (50 or more on each side) contain serotonin, and they modulate the expression of withdrawal behaviour. They do so, in part, by facilitating the synaptic excitation of parietal command neurons.

Four neurons on the lateral–ventral surface of the right ganglion of *Helix* innervate the penis retractor muscle (Eberhardt and Wabnitz, 1979). Two of these appear to have a purely sensory function, one can produce contractions of the muscle, and one has an unknown function.

Additional large neurons are capable of regulating the heart rate, by either direct or indirect efferent control (S.-Rózsa, 1981; Furukawa and Kobayashi, 1987b). One of these cells, d-LPeLN, is particularly interesting because it is found only on the left side. It is serotonergic, and it might be homologous to certain identified neurons described in other gastropod species, which are also serotonergic and which have a similar, asymmetric location (Croll, 1988).

Acknowledgement

The author thanks the Natural Sciences and Engineering Research Council of Canada for supporting his research.

References

Alevizos, A., Weiss, K.R. and Koester, J. (1991) Synaptic actions of identified peptidergic neuron R15 in *Aplysia*. III. Activation of the large hermaphroditic duct. *Journal of Neuroscience* 11, 1282–1290.

Altrup, U. and Speckmann, E.-J. (1994) Identified neuronal individuals in buccal ganglia of *Helix pomatia*. *Neuroscience and Behavioral Physiology* 24, 23–32.

Amoroso, E.C., Baxter, M.I., Chiquoine, A.D. and Nisbet R.H. (1964) The fine structure of neurons and other elements in the nervous system of the giant African land snail *Archachatina marginata*. *Proceedings of the Royal Society of London, Series B* 160, 167–180.

Bailey, S.E.R. (1981) Circannual and circadian rhythms in the snail *Helix aspersa* Müller and the photoperiodic control of annual activity and reproduction. *Journal of Comparative Physiology* 142, 89–94.

Balaban, P.M. (1979) A system of command neurons in snail's escape behavior. *Acta Neurobiologiae Experimentalis* 39, 97–107.

Bargmann, H.E. (1930) The morphology of the central nervous system in the Gastropoda Pulmonata. *Journal of the Linnaean Society of London, Zoology* 37, 1–59.

Beltz, B. and Gelperin, A. (1980) Mechanosensory input modulates activity of an autoactive, bursting neuron in *Limax maximus*. *Journal of Neurophysiology* 44, 665–674.

Benjamin, P. and Elliott, C.J.H. (1989) Snail feeding oscillator: the central pattern generator and its control by modulatory interneurons. In: Jacklet, J.W. (ed.) *Neuronal and Cellular Oscillators*. Marcel Dekker, New York, pp. 173–214.

Bieler, R. (1990) Haszprunar's 'clado-evolutionary' classification of the gastropoda – a critique. *Malacologia* 31, 371–380.

Brace, R.C. (1977) Anatomical changes in nervous and vascular systems during the transition from prosobranch to opisthobranch organization. *Transactions of the Zoological Society of London* 34, 1–25.

Brandenburger, J.L. (1975) Two new kinds of retinal cells in the eye of a snail, *Helix aspersa*. *Journal of Ultrastructure Research* 50, 216–230.

Bullock, T.H. (1965) Mollusca: Gastropoda. In: Bullock, H. and Horridge, G.A. (eds) *Structure and Function in the Nervous Systems of Invertebrates*, Vol. II. W.H. Freeman, San Francisco, pp. 1283–1386.

Carew, T.J., Castellucci, V.F., Byrne, J.H. and Kandel, E.R. (1979) Quantitative analysis of relative contribution of central and peripheral neurons to gill-withdrawal reflex in *Aplysia californica. Journal of Neurophysiology* 42, 497–509.

Chase, R. (1985) Responses to odors mapped in snail tentacle and brain by ^{14}C-2-deoxyglucose autoradiography. *Journal of Neuroscience* 5, 2930–2939.

Chase, R. (1986a) Brain cells that command sexual behavior in the snail *Helix aspersa. Journal of Neurobiology* 17, 669–679.

Chase, R. (1986b) Lessons from snail tentacles. *Chemical Senses* 11, 411–426.

Chase, R. (2000) Structure and function in the cerebral ganglion. *Microscopy Research and Technique* 49, 511–520.

Chase, R. and Hall, B. (1996) Nociceptive inputs to C3, a motoneuron of the tentacle withdrawal reflex in *Helix aspersa. Journal of Comparative Physiology* 179A, 809–818.

Chase, R. and Kamil, R. (1983) Morphology and odor sensitivity of regenerated snail tentacles. *Journal of Neurobiology* 14, 43–50.

Chase, R. and Li, G. (1994) Mesocerebral neurons and their role in the control of mating behaviour. *Netherlands Journal of Zoology* 44, 212–222.

Chase, R. and Rieling, J. (1986) Autoradiographic evidence for receptor cell renewal in the olfactory epithelium of a snail. *Brain Research* 384, 232–239.

Chase, R. and Tolloczko, B. (1985) Secretory glands of the snail tentacle and their relation to the olfactory organ. *Zoomorphology* 105, 60–67.

Chase, R. and Tolloczko, B. (1986) Synaptic glomeruli in the olfactory system of a snail. *Cell and Tissue Research* 246, 567–573.

Chase, R. and Tolloczko, B. (1987) Evidence for differential DNA endoreplication during the development of a molluscan brain. *Journal of Neurobiology* 18, 395–406.

Chase, R. and Tolloczko, B. (1989) Interganglionic dendrites constitute an output pathway from the procerebrum of the snail. *Journal of Comparative Neurology* 283, 143–152.

Chase, R. and Tolloczko, B. (1992) Synaptic innervation of the giant cerebral neuron in sated and hungry snails. *Journal of Comparative Neurology* 318, 93–102.

Chase, R. and Tolloczko, B. (1993) Tracing neural pathways in snail olfaction: from the tip of the tentacle to the brain and beyond. *Microscopy Research and Technique* 24, 214–230.

Chernorizov, A.M., Shekhter, E.D., Arakelov, G.G. and Zimachev, M.M. (1994) The vision of the snail: the spectral sensitivity of the dark-adapted eye. *Neuroscience and Behavioral Physiology* 24, 59–62.

Chiasson, M.W., Baker, M.W. and Croll, R.P. (1994) Morphological changes and functional recovery following axotomy of a serotonergic neurone in the land snail *Achatina fulica. Journal of Experimental Biology* 192, 147–167.

Cooke, I.R.C., Edwards, S.L. and Anderson, C.R. (1994) The distribution of NADPH diaphorase activity and immunoreactivity to nitric oxide synthase in the nervous system of the pulmonate mollusc *Helix aspersa. Cell and Tissue Research* 277, 565–572.

Cottrell, G.A., Schot, L.P.C. and Dockray, G.J. (1982) Identification and probable role of a single neurone containing the neuropeptide *Helix* FMRFamide. *Nature* 304, 638–640.

Cottrell, G.A., Price, D.A., Doble, K.E., Hettle, S., Sommerville, J. and Macdonald, M. (1992) Identified *Helix* neurons: mutually exclusive expression of the tetrapeptide and heptapeptide members of the FMRFamide family. *Biological Bulletin* 183, 113–122.

Croll, R.P. (1987) Identified neurons and cellular homologies. In: Ali, M.A. (ed.) *Nervous Systems in Invertebrates*. Plenum, New York, pp. 41–59.

Croll, R.P. (1988) Distribution of monoamines within the central nervous system of the juvenile pulmonate snail, *Achatina fulica. Brain Research* 460, 29–49.

Croll, R.P. and Chase, R. (1980) Plasticity of olfactory orientation to foods in the snail *Achatina fulica. Journal of Comparative Physiology* 136A, 267–277.

Delaney, K. and Gelperin, A. (1990a) Cerebral interneurons controlling fictive feeding in *Limax maximus*. I. Anatomy and criteria for re-identification. *Journal of Comparative Physiology* 166A, 297–310.

Delaney, K. and Gelperin, A. (1990b) Cerebral interneurons controlling fictive feeding in *Limax maximus*. II. Initiation and modulation of fictive feeding. *Journal of Comparative Physiology* 166A, 311–326.

Delaney, K. and Gelperin, A. (1990c) Cerebral interneurons controlling fictive feeding in *Limax maximus*. III. Integration of sensory inputs. *Journal of Comparative Physiology* 166A, 327–343.

Dorsett, D.A. (1986) Brains to cells: the neuroanatomy of selected gastropod species. In: Willows, A.O.D. (ed.) *The Mollusca*, Vol. 9, *Neurobiology and Behavior, Part 2*. Academic Press, Orlando, pp. 101–187.

Eberhardt, B. and Wabnitz, R.W. (1979) Morphological identification and functional analysis of central neurons innervating the penis retractor muscle of *Helix pomatia. Comparative Biochemistry and Physiology* 63A, 599–613.

Elekes, K. and Nässel, D.R. (1990) Distribution of FMRFamide-like immunoreactive neurons in the central nervous system of the snail *Helix pomatia. Cell and Tissue Research* 262, 177–190.

Elekes, K. and Nässel, D.R. (1994) Tachykinin-related neuropeptides in the central nervous system of the snail *Helix pomatia*: an immunocytochemical study. *Brain Research* 661, 223–236.

Emberton, K.C. and Tillier, S. (1995) Clarification and evaluation of Tillier's (1989) stylommatophoran monograph. *Malacologia* 36, 203–208.

Emery, D.G. (1992) Fine structure of olfactory epithelia of gastropod molluscs. *Microscopy Research and Technique* 22, 307–324.

Fernandez, J. (1966) Nervous system of the snail *Helix aspersa*. I. Structure and histochemistry of ganglionic sheath and neuroglia. *Journal of Comparative Neurology* 127, 157–182.

Fretter, V. and Graham, A. (1962) *British Prosobranch Molluscs: Their Functional Anatomy and Ecology*. Ray Society, London.

Furukawa, Y. and Kobayashi, M. (1987a) Neural control of heart beat in the African giant snail, *Achatina fulica* Férussac. I. Identification of the heart regulatory neurones. *Journal of Experimental Biology* 129, 279–293.

Furukawa, Y. and Kobayashi, M. (1987b) Neural control of heart beat in the African giant snail, *Achatina fulica* Férussac. II. Interconnections among the heart regulatory neurones. *Journal of Experimental Biology* 129, 295–307.

Gelperin, A. (1994) Nitric oxide mediates network oscillations of olfactory interneurons in a terrestrial mollusc. *Nature* 369, 61–63.

Gelperin, A. and Flores, J. (1997) Vital staining from dye-coated microprobes identifies new olfactory interneurons for optical and electrical recording. *Journal of Neuroscience Methods* 72, 97–108.

Gelperin, A. and Tank, D.W. (1990) Odour-modulated collective network oscillations of olfactory interneurons in a terrestrial mollusc. *Nature* 345, 437–440.

Gelperin, A., Chang, J.J. and Reingold, S.C. (1978) Feeding motor program in *Limax*. I. Neuromuscular correlates and control by chemosensory input. *Journal of Neurobiology* 9, 285–300.

Gillette, R. (1991) On the significance of neuronal giantism in gastropods. *Biological Bulletin* 180, 234–240.

Giusti, F., Manganelli, G. and Schembri, P.J. (1995) *The Non-marine Molluscs of the Maltese Islands*. Monografie XV, Museo Regionale di Scienze Naturali, Turin.

Griffond, B., van Minnen, J. and Colard, C. (1992) Distribution of APGWa-immunoreactive substances in the central nervous system and reproductive apparatus of *Helix aspersa*. *Zoological Science* 9, 533–539.

Haszprunar, G. (1985) The Heterobranchia – a new concept of the phylogeny of the higher Gastropoda. *Zeitschrift für Zoologische Systematik und Evolutionsforschung* 23, 15–37.

Haszprunar, G. (1988) On the origin and evolution of major gastropod groups, with special reference to the Streptoneura. *Journal of Molluscan Studies* 54, 367–441.

Haszprunar, G. and Huber, G. (1990) On the central nervous system of Smeagolidae and Rhodopidae, two families questionably allied with the Gymnomorpha (Gastropoda: Euthyneura). *Journal of Zoology (London)* 220, 185–199.

Hernádi, L. (1992) Relationships between the distribution of serotonergic cell bodies and the running of vascular elements in the central nervous system of the snail, *Helix pomatia*. *Comparative Biochemistry and Physiology* 103A, 85–92.

Hernádi, L., Kemenes, G. and Salánki, J. (1987) Sensory responses and axonal morphology of two different types of cerebral neurones in *Helix pomatia* L. *Comparative Biochemistry and Physiology* 88, 641–646.

Hickmott, P.W. and Carew, T.J. (1991) An autoradiographic analysis of neurogenesis in juvenile *Aplysia californica*. *Journal of Neurobiology* 22, 313–326.

Ierusalimsky, V.N. and Zakharov, I.S. (1994) Mapping of neurons participating in the inneration of the body wall of the snail. *Neuroscience and Behavioral Physiology* 24, 33–39,

Ierusalimsky, V.N., Zakharov, I.S., Palikhova, T.A. and Balaban, P.M. (1994) Nervous system and neural maps in gastropod *Helix lucorum* L. *Neuroscience and Behavioral Physiology* 24, 12–22.

Kandel, E.R. (1979) *Behavioral Biology of Aplysia*. W.H. Freeman, San Francisco.

Kandel, E.R. and Tauc, L. (1966) Input organization of two symmetrical giant cells in the snail brain. *Journal of Physiology* 183, 269–286.

Kaufmann, W., Kerschbaum, H.H., Hauserkronberger, C., Hacker, G.W. and Hermann, A. (1995) Distribution and seasonal variation of vasoactive intestinal (VIP)-like peptides in the nervous system of *Helix pomatia*. *Brain Research* 695, 125–136.

Kawahara, S., Yano, M. and Shimizu, H. (1994) Radular mechanosensory neuron in the buccal ganglia of the terrestrial slug, *Incilaria fruhstorferi*. *Journal of Comparative Physiology* 174, 111–120.

Kawahara, S., Yano, M. and Shimizu, H. (1995) Modulation of the feeding system by a radular mechanosensory neuron in the terrestrial slug, *Incilaria fruhstorferi*. *Journal of Comparative Physiology* 176, 193–203.

Kerkhoven, R.M., Ramkema, M.D., van Minnen, J., Croll, R.P., Pin, T. and Boer, H.H. (1993) Neurons in a variety of molluscs react to antibodies raised against the VD1/RPD2 alpha-neuropeptide of the pond snail *Lymnaea stagnalis. Cell and Tissue Research* 273, 371–379.

Kerkut, G.A., Lambert, J.D.C., Gayton, R.J., Loker, J.E. and Walker, R.J. (1975) Mapping of nerve cells in the suboesophageal ganglia of *Helix aspersa. Comparative Biochemistry and Physiology* 50A, 1–25.

Kleinfeld, D., Delaney, K.R., Fee, M.S., Flores, J.A., Tank, D.W. and Gelperin, A. (1994) Dynamics of propagating waves in the olfactory network of a terrestrial mollusk: an electrical and optical study. *Journal of Neurophysiology* 72, 1402–1419.

Koene, J.M., Jansen, R.F., ter Maat, A. and Chase, R. (2000) A conserved location for the CNS control of mating behaviour in gastropod molluscs: evidence from a terrestrial snail. *Journal of Experimental Biology* 303, 1071–1080.

Koester, J. and Byrne, J.H. (eds) (1980) *Molluscan Nerve Cells: From Biophysics to Behavior.* Cold Spring Harbor Laboratory Press, Cold Spring Harbor, New York.

LaBerge, S. and Chase, R. (1992) The development of mesocerebral neurons in the snail *Helix aspersa maxima. Canadian Journal of Zoology* 70, 2034–2041.

Lehman, H.K. and Greenberg, M.J. (1987) The actions of FMRFamide-like peptides on visceral and somatic muscles of the snail *Helix aspersa. Journal of Experimental Biology* 131, 55–68.

Li, G. and Chase, R. (1995) Correlation of axon projections and peptide immuno-reactivity in mesocerebral neurons of the snail *Helix aspersa. Journal of Comparative Neurology* 353, 9–17.

Lingueglia, E., Champigny, G., Lazdunski, M. and Barbry, P. (1995) Cloning of the amiloride-sensitive FMRFamide peptide-gated sodium channel. *Nature* 378, 730–733.

Marchand C.R., Griffond, B., Mounzih, K and Colard, C. (1991) Distribution of methionine-enkephalin-like and FMRFamide-like immunoreactivities in the central nervous system (including dorsal bodies) of the snail *Helix aspersa* Müller. *Zoological Science* 8, 905–913.

McCarragher, G. and Chase, R. (1985) Quantification of ultrastructural symmetry at molluscan chemical synapses. *Journal of Neurobiology* 16, 69–74.

Mirolli, M. and Talbott, S.R. (1972) The geometrical factors determining the elec-trotonic properties of a molluscan neurone. *Journal of Physiology* 227, 19–34.

Munoz, D.P., Pawson, P.A. and Chase, R. (1983) Symmetrical giant neurons in asymmetrical ganglia: implications for evolution of the nervous system in pulmonate molluscs. *Journal of Experimental Biology* 107, 147–161.

Page, L.R. (1992) New interpretation of a nudibranch central nervous system based on ultrastructural analysis of neurodevelopment in *Melibe leonina.* I. Cerebral and visceral loop ganglia. *Biological Bulletin* 182, 348–365.

Pawson, P. and Chase, R. (1988) The development of transmission at an identified molluscan synapse. I. The emergence of synaptic plasticities. *Journal of Neurophysiology* 60, 2196–2210.

Pentreath, V.W., Berry, M.S. and Osborne, N.N. (1982) Serotonergic cerebral cells in gastropods. In: Osborne, N.N. (ed.) *Biology of Serotonergic Transmission.* John Wiley & Sons, New York, pp. 457–513.

Pin, T. and Gola, M. (1984) Axonal mapping of neurosecretory *Helix* bursting cells. *Comparative Biochemistry and Physiology* 78A, 637–649.

Prescott, S. and Chase, R. (1996) Two types of plasticity in the tentacle withdrawal reflex of *Helix aspersa* are dissociated by tissue location and response measure. *Journal of Comparative Physiology* 179A, 407–414.

Prescott, S. and Chase, R. (1997) The neural circuit mediating tentacle wthdrawal in *Helix aspersa*, with specific reference to the motoneuron C3. *Journal of Neurophysiology* 78, 2951–2965.

Ratté, S. and Chase, R. (1997) Morphology of interneurons in the procerebrum of the snail *Helix aspersa*. *Journal of Comparative Neurology* 384, 359–372.

Ratté, S. and Chase, R. (2000) Synapse distribution of olfactory interneurons in the procerebrum of the snail *Helix aspersa*. *Journal of Comparative Neurology* 417, 366–384.

Régondaud, J. (1964) Origine embryonnaire de la cavité pulmonaire de *Lymnaea stagnalis* L. Considérations particulières sur la morphogenèse de la commissure viscérale. *Bulletin Biologique de la France et de la Belgique* 48, 433–471.

Sakharov, D.A. (1976) Nerve cell homologies in gastropods. In: Salánki, J. (ed.) *Neurobiology of Invertebrates. Gastropoda Brain*. Akadémiai Kiadó, Budapest, pp. 27–40.

Salánki, J. and van Bay, T. (1976) Peripheral and central discrimination of chemoreceptor stimulation in the snail, *Helix pomatia* L. In: Salánki, J. (ed.) *Neurobiology of Invertebrates. Gastropoda Brain*. Akadémiai Kiadó, Budapest, pp. 497–510.

Samygin, F.I. and Karpenko, L.D. (1980) Localization of neurons innervating the columellar muscles of the snail. *Neurophysiology* 12, 424–431.

Sánchez-Alvarez, M., León-Olea, M., Talavera, E., Pellicer, F., Sánchez-Islas, E. and Martinez-Lorenzana, G. (1994) Distribution of NADPH-diaphorase in the perioesophageal ganglia of the snail, *Helix aspersa*. *Neuroscience Letters* 169, 51–55.

S.-Rózsa, K. (1979) Heart regulatory neural network in the central nervous system of *Achatina fulica* (Férussac) (Gastropoda: Pulmonata). *Comparative Biochemistry and Physiology* 63A, 435–445.

S.-Rózsa, K. (1981) Interrelated networks in regulation of various functions in gastropoda. In: Salánki, J. (ed.) *Neurobiology of Invertebrates. Gastropod Brain*. Akadémiai Kiadó, Budapest, pp. 147–169.

S.-Rózsa, K. (1984) The pharmacology of molluscan neurons. *Progress in Neurobiology* 23, 79–150.

Sokolove, P.G. and McCrone, E.J. (1978) Reproductive maturation in the slug, *Limax maximus*, and the effects of artificial photoperiod. *Journal of Comparative Physiology* 125, 317–325.

Suzuki, H., Watanabe, M., Tsukahara, Y. and Tasaki, K. (1979) Duplex system in the simple retina of a gastropod mollusc, *Limax flavus* L. *Journal of Comparative Physiology* 133, 125–130.

Tillier, S. (1989) Comparative morphology, phylogeny and classification of land snails and slugs (Gastropoda: Pulmonata: Stylommatophora). *Malacologia* 30, 1–303.

Van Mol, J.-J. (1974) Evolution phylogenetique du ganglion cerebroide chez les gasteropodes pulmones. *Haliotis* 4, 77–86.

Weiss, K.R., Cohen, J.L. and Kupfermann, I. (1978) Modulatory control of buccal musculature by a serotonergic neuron (metacerebral cell) in *Aplysia*. *Journal of Neurophysiology* 41, 181–203.

Willows, A.O.D. (ed.) (1986) *The Mollusca*, Vol. 9, *Neurobiology and Behavior, Part 2*. Academic Press, Orlando, Florida.

Yoshida, M. and Kobayashi, M. (1991) Neural control of the buccal movement in the African giant snail, *Achatina fulica. Journal of Experimental Biology* 155, 415–433.

Yoshida, M. and Kobayashi, M. (1992) Identified neurones involved in the control of rhythmic buccal motor activity in the snail *Achatina fulica. Journal of Experimental Biology* 164, 117–133.

Yoshida, M. and Kobayashi, M. (1995) Modulation of the buccal muscle contraction by identified serotonergic and peptidergic neurons in the snail *Achatina fulica. Journal of Experimental Biology* 198, 729–738.

Zaitseva, O.V. (1994) Structural organization of the sensory systems of the snail. *Neuroscience and Behavioral Physiology* 24, 47–57.

Zakharov, I.S. (1994) Avoidance behavior of the snail. *Neuroscience and Behavioral Physiology* 24, 63–69.

Zakharov, I.S., Mats, V.N. and Balaban, P.M. (1982) Role of the giant cerebral ganglion neuron in control of defensive behavior of *Helix lucorum. Neurophysiology* 14, 262–266.

Zakharov, I.S., Ierusalimsky, V.N. and Balaban, P.M. (1995) Pedal serotonergic neurons modulate the synaptic input of withdrawal interneurons of *Helix. Invertebrate Neuroscience* 1, 41–52.

Zakharov, I.S., Hayes, N.L., Ierusalimsky, V.N., Nowakowski, R.S. and Balaban, P.M. (1998) Postembryonic neurogenesis in the procerebrum of the terrestrial snail, *Helix lucorum* L. *Journal of Neurobiology* 35, 271–276.

4 Radular Structure and Function

U. Mackenstedt[1] and K. Märkel[2]

[1]Institut für Zoologie, Fachgebiet Parasitologie, Universität Hohenheim, Emil-Wolff-Str. 34, D-70599 Stuttgart, Germany; [2]formerly Lehrstuhl für Spezielle Zoologie, Ruhr-Universität Bochum, Universitätstraße 150, D-44780 Bochum, Germany

Introduction

The radula is a special feature of molluscs. It is a cuticular structure of the ectodermal foregut epithelium and is part of the buccal mass, the mollusc feeding apparatus. The radula itself consists of a flexible membrane provided with numerous teeth arranged in transverse and longitudinal rows (Fig. 4.1A and B). For a long time, the dentition of the radula has been used for the taxonomy of gastropods (e.g. Troschel 1856–1863; Ponder and Lindberg 1996). However, little attention has been paid to the histology and function of the buccal mass, which is essential for the feeding function of the radula, although Troschel (1836) recognized the lick-like movements of the 'Schneckenzunge'. The buccal mass comprises the jaw, the radula, the odontophore and numerous muscles. The muscles protract and retract the odontophore and thus are involved in the movements of the overlying radula (Trappmann, 1916). The neural control of the movements in the buccal mass is provided by the paired buccal ganglia. During feeding, the radula is severely strained. The foremost part is destroyed continually, while at the posterior end new radular membrane and teeth are produced continually. The rates of disintegration and formation of the radula are in balance, and the whole process resembles tooth renewal in sharks (Märkel, 1969).

The radulae differ fundamentally among gastropods. In many prosobranch gastropods, the number of teeth per transverse row remains constant during the life of the animal, and the number of teeth as well as their shape play an important role in classification at high taxonomic rank (cf. Ordinal taxonomic categories Rhipidoglossa and Taenioglossa). The taenioglossate radula, for example, consists of seven teeth per transverse row, that is a middle or rhachis tooth, flanked on both sides by one lateral tooth and two marginal teeth (Fig. 4.2A and C). With growth of the animal, the radula grows in width and length, and the size of the teeth increase but the number of teeth per transverse row remains constant (Mischor and Märkel, 1984). At the anterior of the prosobranch radula, a semicircular cuticular alary process is attached to either side (Lutfy and Demain, 1964) (Figs 4.2A and 4.5). Muscles of the buccal mass are attached to the alary

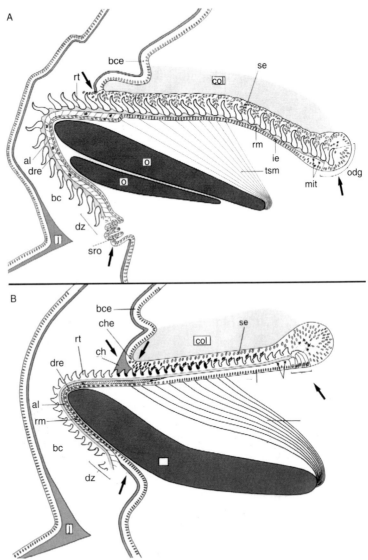

Fig. 4.1. Longitudinal section through the buccal mass. (A) *Pomacea bridgesi* (Reeve) (Ampullariidae), illustrative of the Caenogastropoda. (B) A generalized pulmonate. Note the important differences (arrows). (Reproduced with permission of Springer-Verlag.) al, adhesive layer; bc, buccal cavity; bce, buccal cavity epithelium; che, collostylar hood epithelium; col, collostyle; dre, distal radular epithelium; dz, degeneration zone; ie, inferior epithelium; j, jaw; mit, mitosis; o, odontophore; odg, odontoblast group; rm, radular membrane; rt, radular tooth; se, superior epithelium; sro, subradular organ; tsm, supramedian tensor muscle.

processes and function to protract the radula over the front face of the odontophore during feeding. In pulmonates, the teeth are much smaller and new rows of longitudinal teeth are added on both sides in the course of ontogeny (Fig. 4.4A and B). Alary processes are lacking in pulmonates (Smith, 1987, 1990)

The dentition of the radula to some extent is indicative of the animals' diet, but radulae of the same morphological type may function quite

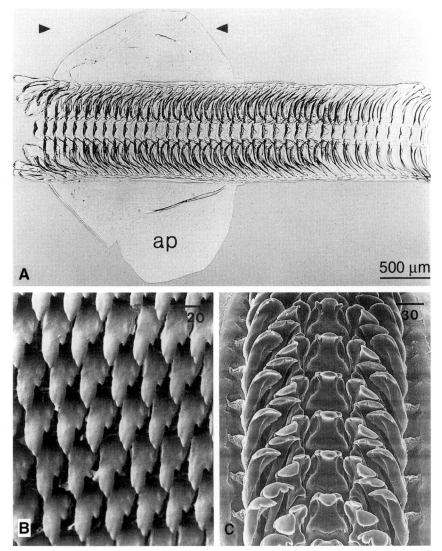

Fig. 4.2. Radulae. (A and C) Taenioglossate radula of the caenogastropod *Pomacea bridgesi* (Reeve) (Ampullariidae). Note the alary processes (ap) which are characteristic for the prosobranchate radula. Arrows indicate the functional part of the radula (A) Micrograph originals by B. Mischor. (B) Dentition from the polyglossate radula of the stylommatophoran *Helix pomatia* Linnaeus (Helicidae). (C) Scanning electron micrograph. ap, alary processes.

differently, and vice versa, due to the movements of the odontophore and the numerous muscles of the buccal mass.

Anatomy of the Buccal Mass

The radula coats the shoehorn-shaped cartilaginous (or cartilaginous-like) odontophore complex located in the centre of the buccal mass (Figs 4.1A and B, 4.3, and 4.4A and B). The odontophoral cartilages in gastropod molluscs vary from a single pair to three or more pairs. In taenioglossate prosobranchs, the odontophore complex is composed of two pairs, with the two dominant pieces overlapping in the centre of the complex to form a trough. The odontophore of prosobranchs consists of strongly vacuolated turgor cells, and the entire odontophore is ensheathed by a layer of collagenous tissue. Muscles are lacking on the odontophore complex of the above-mentioned gastropods, although processes provide anchorage points for the buccal muscles. In pulmonates, the odontophore consists of

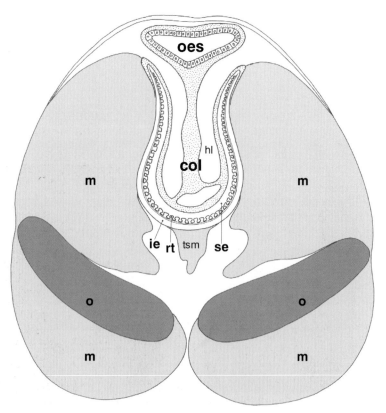

Fig. 4.3. Semi-schematic cross-sectional view through the buccal mass of *Cepaea nemoralis* (Linnaeus) (Helicidae). Note the U-shaped radular sheath and the two pieces of the odontophore forming a trough. col, collostyle; m, muscle; o, odontophore; oes, oesophagus; rt, radular tooth; se, superior epithelium; tsm, supramedian tensor muscle.

two pieces of modified cartilage, which are connected by muscles at the posterior end (Fig. 4.3). The turgor cells are intermingled with muscles, which in Stylommatophora totally replace the turgor cells. Due to its muscular nature, the pulmonate odontophore is able to change shape during feeding.

The functional part of the radula, which is active during feeding, is bent over the anterior face of the odontophore which bulges the radula into the buccal cavity (Figs 4.4A and B and 4.5). The radula is not intimitately connected with the odontophore and can be moved to and fro by the supramedian tensor muscle or the laterally attached protractors. The generative part of the radula lies in the radular sac or radular sheath, which is a pouch of the epithelium of the buccal cavity. The radular sac is U-shaped and the dorsal trough or groove is filled by the collostyle (Fig. 4.3). In *Limax maximus* Linnaeus (Limacidae), the collostyle comprises muscle cells and glycogen-rich fibroplasts (Curtis and Cowden, 1977). Its function is not precisely known, but it may supply nutrients to the epithelia of the radular sac. Additionally, the collostyle may be involved in the disposal of degenerated cells (see below). The radular sac consists of several epithelia that are involved in the formation, tanning and transport of the radula, respectively.

Radula Formation

The matrix of the radula is produced by membranoblasts and odontoblasts which are located in the blind end of the radular sac (Fig. 4.1A and B). The membranoblasts secrete the radular membrane continually,

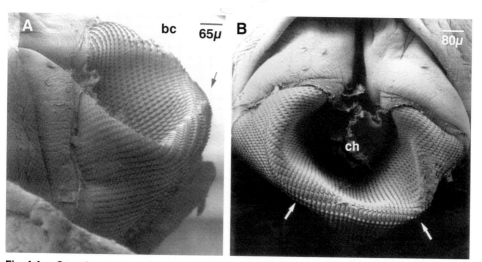

Fig. 4.4. Scanning electron miscroscopical views of the functional part of the radula of *Cepaea nemoralis* (Linnaeus) (Helicidae) spread over the anterior face of the odontophore (arrow). (A) Lateral view; (B) frontal view. Note that the anterior face of the radular sheath is closed by the collostylar hood (ch). bc, buccal cavity.

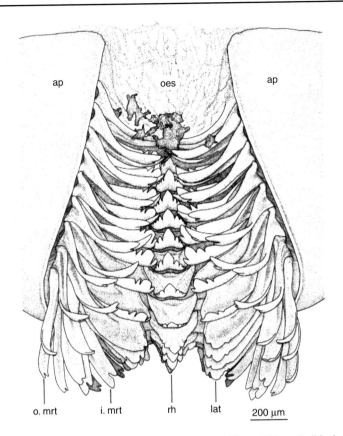

Fig. 4.5.   The taenioglossate radula of *Pomacea bridgesi* (Reeve) (Ampullariidae). Dorsal
view of the frontal face of the odontophore, with alary processes and radular teeth. Note that at
the bent, anterior face of the odontophore, the teeth are elevated into their functional position.
(Original by Dr B. Mischor; reproduced with the permission of Springer-Verlag.) ap, alary
processes; i.mrt, inner marginal tooth; lat, lateral tooth; oes, oesophagus; o.mrt, outer marginal
tooth; rh, rhachis tooth.

whereas the odontoblasts form the matrix of the radular teeth by intermittent secretion. In prosobranchs (as well as in radula-bearing, non-
gastropod molluscs), the odontoblasts are arranged in multicellular
cushions, each of which gives rise to a longitudinal row of teeth (Rössler,
1885; Rottmann, 1901; Gabe and Prenant, 1950a, 1957). The odontoblast
groups comprise several hundred cells, and it is impossible to differentiate between membranoblasts and odontoblasts on ultrastructural
grounds (Fig. 4.6B). In the taenioglossate prosobranch *Pomacea bridgesi*
(Reeve) (Ampullariidae), the matrix of the teeth consists of microvilli,
cytoplasmic protrusions and secretion material from the odontoblasts.
Autoradiographic experiments demonstrated the occurrence of cell
divisions in the multicellular odontoblastic cushions (Mischor and
Märkel, 1984). Although the mitotic activity was very low, the rate of cell
division was shown to be sufficient to replace the spent odontoblasts and
to increase the number of odontoblasts per group. This is significant,

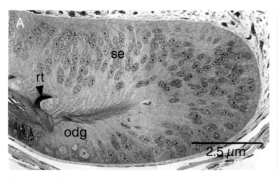
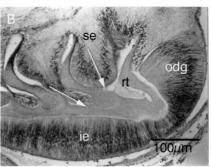

Fig. 4.6. The blind end of the radular sac. (A) Electron micrograph of *Cepaea nemoralis* (Linnaeus) (Helicidae), illustrative of stylommatophoran gastropods. Note the subterminal location of the odontoblast group, whereas the blind terminus is occupied by the proliferation zone of the superior epithelium. (B) Electron micrograph of *Pomacea bridgesi* (Reeve) (Ampullariidae), illustrative of prosobranchate gastropods. The numerous odontoblasts are located at the end of the radular sac. Cells of the superior epithelium immigrate into the space between the radular teeth (arrows). ie, inferior epithelium; odg, odontoblast group; rt, radular tooth; se, superior epithelium.

because the odontoblasts define the size and shape of the radular teeth. In prosobranchs, the size of the teeth correlates with the size of the animal (Mischor and Märkel, 1984).

In pulmonates, the odontoblasts form a subterminal girdle of voluminous cells which are arranged in small groups at the terminus of the radular sac (Fig. 4.6A) (Hoffmann, 1932; Kerth and Krause, 1969; Kerth and Hänsch, 1977; Wiesel and Peters, 1978; Mackenstedt and Märkel, 1987). The matrix of pulmonate radula teeth is composed mainly of microvilli (Mackenstedt and Märkel, 1987). During growth in pulmonates, new groups of odontoblasts are added in the marginal areas of the odontoblast girdle, thus increasing the number of radular teeth per transverse row. The formation of new odontoblasts is a process that is still poorly understood. Once formed, the individual odontoblasts are maintained for the life of the pulmonate animal. So far, neither autoradiographic nor microscopical studies have revealed mitotic activity in the odontoblast groups of pulmonates (Runham, 1963; Mackenstedt and Märkel, 1987). Kerth (1983a,b) discussed the possibility that cells of the adjacent superior epithelium might differentiate into odontoblasts. Mackenstedt and Märkel (1987) described small amoeboid cells in the marginal areas of the odontoblast girdle and believed that these cells might represent pre-odontoblasts which immigrate into the odontoblast girdle, thus adding new odontoblasts on both sides. However, the origin of the so-called pre-odontoblasts remains uncertain. An additional cell type occurs in the odontoblast girdle of pulmonates. This cell type generates the basal plates of the radular teeth (Kerth and Krause, 1969), which play a crucial role in maintaining the spacing of the teeth and supporting them as they engage the food material (Solem, 1973). Basal plates are lacking in prosobranchs (Fig. 4.6A and B). In pulmonates, the number of cells per odontoblast group seems to be species-specific. Kerth and Krause (1969) described 15

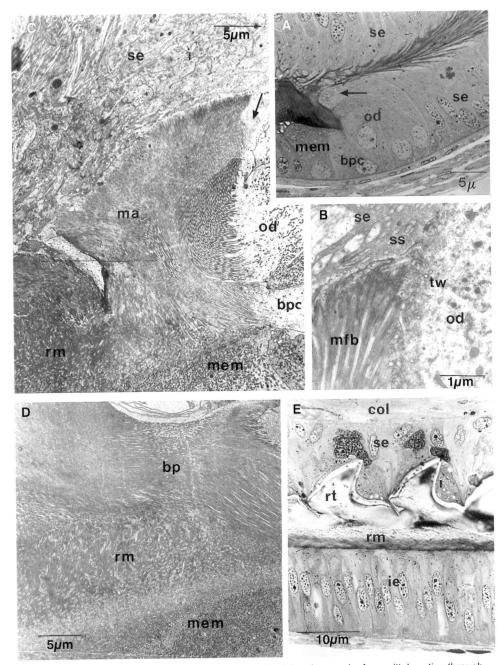

Fig. 4.7. *Cepaea nemoralis* (Linnaeus) (Helicidae). (A) Light micrograph of a sagittal section through the odontoblast group at an advanced stage of radular tooth formation. The apical regions of the real odontoblasts are inclined towards the orifice of the radular sac (arrow). (B) Electron micrograph of the cusp of the newly formed radular tooth. Note the texture of the microfibrils in the outer aspect of the tooth. The tooth is still connected to the surface of the odontoblasts. Substances secreted by the superior epithelium are already incorporated into the tooth matrix. (C) Electron micrograph illustrating the advanced stage of radular tooth formation. The cusp of the near complete tooth is already detached from

odontoblasts in *Limacus flavus* (Linnaeus) (Limacidae), whereas Wiesel and Peters (1978) counted 12 ± 2 odontoblasts in *Biomphalaria glabrata* Say (Planorbidae). D. Coste (1986 and unpublished results) differentiated between membranoblast, basal plate cells and odontoblasts, and described one membranoblast, three basal plate cells and 9–11 real odontoblasts in *Lymnaea stagnalis* (Linnaeus) (Lymnaeidae).

The formation of the tooth and its connection with the radular membrane requires a well-coordinated activity by the different members of the odontoblast group (Fig. 4.7A–D). In pulmonates, the membranoblast is located anteriorly in the odontoblast group and lies adjacent to the inferior epithelium (Fig. 4.7A). The membranoblast secretes the radular membrane. The formation of the membrane is a process that occurs continually and requires a high metabolic activity. The distal surface of the membranoblast is provided with microvilli, which are constantly extended at their bases by newly formed membrane material (Fig. 4.7D). They are orientated in different directions and are finally detached from the surface. The microvilli are embedded in a ground substance, which is produced by the continuous exocytosis of small vesicles at the distal surface of the membranoblast. The microvilli are often detectable by electron microscopy up to the fourth transverse row of teeth. The tooth formation is initiated when the posterior odontoblasts secrete the cuspid head of the tooth (Fig. 4.7B). The odontoblasts located more anteriorly become active only when the cusp of the tooth is almost finished, and they function in adding the matrix of the tooth body. Finally, the odontoblasts responsible for the basal plates generate very long microvilli and connect the tooth matrix with the radular membrane, which is secreted constantly by the membranoblast (Fig. 4.7C and D). In the final phase of tooth formation, the odontoblasts bulge dorsally and their apical region inclines toward the opening of the radular sac (Fig. 4.7A and C). This change in cell shape, together with the tractive power of the constantly secreted radular membrane, pushes the newly formed teeth into the upright and thus final position (Fig. 4.7C). The shape of the odontoblasts at this stage mirrors the shape of the posterior concavity of the tooth. The microvilli are detached progessively from the surface of the odontoblasts and are completely incorporated into the tooth matrix. The tooth thus becomes separated from the odontoblasts, with the microvilli detaching from the surface of the cells, first at the cuspid head region of the tooth and then in the basal plate region.

the surface of the odontoblasts (arrow). (D) Electron micrograph section through the matrix of the radular membrane and the radular tooth basal plate. Note the different orientations of the microvilli. (E) Light micrograph sagittal section through the radular sheath close to the orifice of the radular sheath. Cells of the superior epithelium start to degenerate. Note the number of clumping nuclei. (Reproduced with the permission of Springer-Verlag.) bp, basal plate; bpc, basal plate cell; col, collostyle; ie, inferior epithelium; ma, tooth matrix; mem, membranoblast; mfb, microfibrils; od, real odontoblast; rm, radular membrane; rt, radular tooth; se, superior epithelium; ss, secretion substance; tw, terminal web.

The newly formed parts of the radula are transported continually into the buccal cavity. This migration is accompanied by a hardening of the radular teeth through incorporation of organic compounds and minerals. As a result, the radular teeth are ready for use before they reach the opening of the radular sac. Tanning and mineralization of the radular teeth start immediately after the teeth become separated from the odonto-blasts. The cells of the superior epithelium interdigitate with the radular teeth and their secretions cause the hardening of the teeth (Figs 4.6A and B, and 4.7B and C). Hardening progresses from the surface to the tooth interior, with disappearance of the fibrous texture of the tooth matrix. The newly formed tooth matrix consists of proteins and mucopolysaccharides, mainly chitin, which become cross-linked, possibly by a quinine-tanning process (Runham, 1961). Subsequently, the radular teeth are impregnated by the incorporation of inorganic salts (Gabe and Prenant, 1949, 1950b, 1951a,b, 1952a,b, 1958, 1962; Ducros, 1967; Mann et al., 1986; Kim et al., 1989). Different minerals have been detected in the radulae of molluscs. Iron and silicon (in the form of goethite) are incorporated into the radula of the patellid prosobranchs *Patella* Linnaeus (Runham et al., 1969) and *Lottia* (Gray) Sowerby (Rinkevich, 1993), as well as those in Polyplaco-phora (Lowenstam, 1962; Carefoot, 1965; Towe and Lowenstam, 1967). Calcium is incorporated in the radula of *Helix* Linnaeus (Helicidae) (Sollas, 1907), *Lymnaea* de Lamarck (Lymnaeidae), *Cepaea* Held (Helicidae) (Mackenstedt and Märkel, 1987) and *Pomacea* Perry (Ampullariidae) (Mischor and Märkel, 1984). The intussusception of organic compounds and minerals changes the chemical composition of the teeth, but this is not comparable with a coating because the size and the shape of the radular teeth are not altered. The incorporation of minerals is limited to the teeth, as the radular membrane remains flexible (see below).

Radula Transport

Labelling experiments reveal that the superior epithelium migrates slowly towards the opening of the radular sac, and this migration rate is compara-ble with the migration of the radula itself. The proliferation involves active mitoses in a very restricted zone adjacent to the odontoblast cells (Fig. 4.7A). As these cells migrate towards the opening of the radular sac, they begin to degenerate and sometimes conglomerate into clusters (Fig. 4.7E). In prosobranchs, the degenerated cells are simply extruded into the buccal cavity. In pulmonates, however, the orifice of the radular sac is closed by the collostylar hood, which is generated by a distinct epithelium. Autoradiographic experiments show that the collostylar hood epithelium is, in contrast to the superior epithelium, stationary. In the basommatophoran *Lymnaea*, the collostylar hood is limited to the zone where the radular sac opens to the buccal cavity, and the border between the stationary collostylar hood epithelium and the migrating superior epithelium is very distinct. In contrast, in stylommatophoran *Cepaea* and *Helix*, the collostylar hood extends deeply into the radular sac, and the border between the stationary collostylar hood epithelium and the migrating superior epithelium is lacking. Mackenstedt and Märkel (1987) suggested that the collostylar hood epithelium might be involved in the phagocytosis of the degenerating cells of the superior epithelium. Furthermore, they discussed the possibility that parts of the superior epithelium are ingested by the collostyle.

The radular membrane is closely attached to the inferior epithelium within the radular sac and to the distal radular epithelium, which supports the radula in the buccal cavity. The results of autoradiographic experiments revealed that the inferior epithelium migrates much more slowly than the radula itself, whereas the distal radular epithelium is stationary. The link between the inferior epithelium and the superimposed radular membrane must compensate for strong mechanical forces, because the tensor muscle inserts at the anterior part of the inferior epithelium and pulls the radula into the trough formed by the odontophore. The muscle does not insert directly, but it is connected to an adhesive zone of transverse and longitudinal collagen fibres. At the opening of the radular sac, the inferior epithelium generates the adhesive layer that is limited to the functional part of the radula, that is in the buccal cavity. The adhesive layer compensates for the tensile forces during feeding, so that the connection between the radular membrane and the distal radular epithelium is maintained. The inferior epithelium degenerates at the opening of the radular sac, although a distinct degeneration zone is lacking. The distal radular epithelium carries the radula in the buccal cavity, and for a long time it was believed that this epithelium comprised an extension of the inferior epithelium (Märkel, 1957; Runham, 1963; Kerth, 1976). However, by labelling the cells with [³H]thymidine, there is no doubt that the inferior and the distal radular epithelium represent two different epithelia (Mischor and Märkel, 1984; Mackenstedt and Märkel, 1987).

The transport of the radula has been studied intensively, but the process is still not fully understood. Isarankura and Runham (1968) and Kerth (1973) destroyed the posterior part of the radular sac by means of X-rays, dissection and cold shocks, but found that the radula was still transported into the buccal cavity. Kerth (1973) stated that the epithelia of the radular sac are not involved in the transport of the radula. This is surprising because the distal radular epithelium, as a specialized part of the epithelium of the buccal cavity, does not migrate at all. Therefore the transport of the radula cannot be based on the tractive power of a migrating epithelium. The cells of the distal radular epithelium are linked to the adhesive layer by microvilli-like protrusions, characterized by bundles of microfilaments that traverse the cells and connect with basal hemidesmosomes. The link between the distal radular epithelium and the adhesive layer is not permanent because the adhesive layer and the superimposed radula are transported over the surface of the distal radular epithelium. It is likely that the microvilli-like protrusions go through continuous cycles of engagement and disengagement with the adhesive layer. This process may be described a pseudopodial-like movement, and each cell adds to the tractive power, which transports the radula into the buccal cavity. It is possible that the inferior epithelium is also involved in the transport of the radula.

The radula is constantly destroyed at the anterior end (Fig. 4.8). It is well established that the degradation of the radula is not based on mechanical forces that simply tear the radula apart. Rather, the radula is destroyed in a process which first degrades the adhesive layer (Kerth, 1971; Mischor and Märkel, 1984; Mackenstedt and Märkel, 1987). The anterior part of the distal radular epithelium generates a cuticular layer, which separates the adhesive layer from the surface of the distal radular epithelium. Subsequently, the radular membrane and teeth are shed into the buccal cavity to be swallowed and later passed out with the faeces. In *Pomacea*, a subradular organ is located ventral to the degeneration zone and is apparently responsible for the production of enzymes that digest

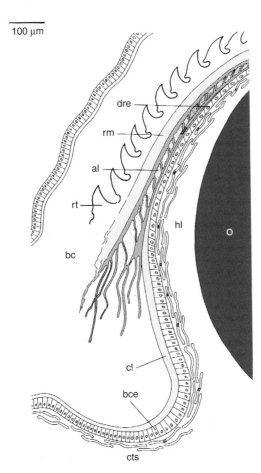

100 μm

Fig. 4.8. Degeneration zone of the radula. Semi-schematic view of a median section through the degeneration zone of a generalized pulmonate gastropod. (Reproduced with the permission of Springer-Verlag.) al, adhesive layer; bc, buccal cavity; bce, buccal cavity epithelium; cl, cuticular layer; cts, connective tissue layer; dre, distal radular epithelium; hl, haemolymph; o, odontophore; rm, radular membrane; rt, radular tooth.

the radular membrane. The subradular organ is absent in pulmonates, but it is likely that special areas of the buccal cavity epithelium secrete enzyme-like substances that similarly destroy the radular membrane.

The rate of complete replacement of the radula has been studied by incorporation of ^{35}S into the radulae of pulmonates and prosobranchs. In the pulmonates, the rate of replacement was 24 and 30–35 days in *L. stagnalis* and *Cepaea nemoralis* (Linnaeus), respectively (Mackenstedt and Märkel, 1987). In *P. bridgesi*, the radula was replaced within 20 days, but Mischor and Märkel (1984) demonstrated that the replacement rate was dependent on the temperature.

Under normal conditions, the processes of radula formation and degeneration are in balance from early in the ontogeny of the animals: Kerth (1971) described the degradation of the radula in embryos of *Viviparus* de Montfort (Viviparidae). Physiological stresses, such as hibernation and starvation, have been shown to reduce the rate of formation of the radula matrix (Isarankura and Runham, 1968; Kerth, 1971; Fujioka, 1985; Smith, 1987; Smith and Russell-Hunter, 1990). During hibernation, for example, the odontoblasts collapse, probably due to water loss, and the formation of the radular matrix ceases. Smith and Russell-Hunter (1990) showed that short exposure to stress resulted in aberrantly shaped teeth in the basommatophoran *Helisoma trivolvis* (Say) (Planorbidae). Longer exposure causes a so-called 'packing' of teeth rows, whereby the distance between the transverse rows was reduced.

Radula Function

In the resting position, the buccal mass lies horizontally within the head and is not in contact with the mouth opening (Fig. 4.1A and B). Feeding is initiated by turning the buccal mass to a nearly vertical position by means of muscles that connect the organ with the body wall. In this manner, the jaw and the foremost part of the radula, which lies upon the U-shaped odontophore, protrude from the mouth. Only the part of the radula that lies outside the radula sac is engaged during feeding. The movement of the radula and the odontophore may be studied in some detail by observing the ventral side of animals placed on a glass plate. The food is gathered into the mouth by movement of the radula over the tip of the odontophore, and by movement of the odontophore itself. Prior to feeding, the radula is in the protracted position and lies loosely upon the odontophore, but, when contact with the substrate is made at the onset of feeding and scraping starts, the radula is pressed against the front face of the odontophore. In the scraping phase of the feeding stroke, the radula without exception traces a posteriad path, whereas the odontophore swivels either forward or backward. At the end of the feeding stroke, the radula is in the retracted position. The swivel action of the odontophore is under neural control from the buccal ganglia. Though the direction of odontophore swivel is in general species-specific, Richter (1962) observed engagement of the radula during both forward and backward strokes or swivel of the odontophore in the prosobranch *Crepidula fornicata* (Linnaeus) (Calyptraeidae).

Details of the feeding process cannot be discerned precisely by direct observation because the components of movement are too rapid. However, much information about the radula's function can be gleaned from analysis of tracks produced by gastropods when feeding on glass slides either thinly coated with a film of flour suspended in agar or coated in grease. Grazing on such artificial pasture results in a pattern of traces produced by the individual radular teeth. Even traces produced by the individual cusps on particular teeth can be discerned. Frequently, these feeding tracks comprise a series of 'bites'. While moving forward, the animal swings its head from side to side and makes repeated bites to produce a zig-zag feeding track (Ankel, 1938; Märkel, 1957, 1964, 1966). The feeding tracks, including the overall shape of the bite, are characteristic at the species level. Animals with the same morphological type of radula will often produce quite different tracks (Fig. 4.11A and B). The salient features of radular function and shape of the tracks may be examined by simple models. The lengthwise cut model (Fig. 4.9A) is based on a radula comprising a single longitudinal row of teeth, the rhachis teeth. The teeth labelled d1–d10 represent those in the functioning part of the radula which come into contact with the pasture during feeding. At the commencement of the bite, the radula is in the protracted position and tooth d1 is the first to come in contact with the pasture. As the radula is retracted over the bent, anterior face of the odontophore, tooth d1, then d2, d3, up to d10 each in turn scrape pieces from the pasture. The traces produced by the single tooth under this model are illustrated in Fig. 4.9B. If the odontophore were not to move at all and feeding was to be effected entirely by movement of the radula, then this model indicates that each tooth would scrape at one location. The result would be the trace labelled a in Fig. 4.9B. The length of that trace is determined by the shape of the leading edge of the odontophore over which the radula moves, the depth to which the teeth cut into the pasture, and the degree of elasticity of the system. In real life, however, each bite results from the combined activities of odontophore and radula. The swivel motion of the odontophore is responsible for the length of the trace, and the extent of the radula retraction at the commencement of the bite determines the number of teeth that contact and scrape the pasture. If the movement of the odontophore is forward (f), the trace d1 lies in the proximal part of the track produced by the bite (i.e. close to the foot). If the odontophore moves backward (b), the trace of that tooth is located in the distal part of the bite. The ratio between the extent of retraction in the radula (vR) and the velocity of the odontophore movement (vO) determines the distances between the traces of the consecutive teeth as well as the length of those traces. In this model, we have held vR constant at 1; whereas vO varies between 3 (very rapid forward movement) to −3 (very rapid backward movement). Increasing vO results in increasing distances between two successive tooth traces, regardless of direction of the odontophore movement. The model shows clearly that the traces are lengthened by the forward movement of the odontophore, depending on the ratio vR : vO. The traces are shortened significantly when the odontophore moves in the opposite direction. At a certain ratio, the retraction of the radula is

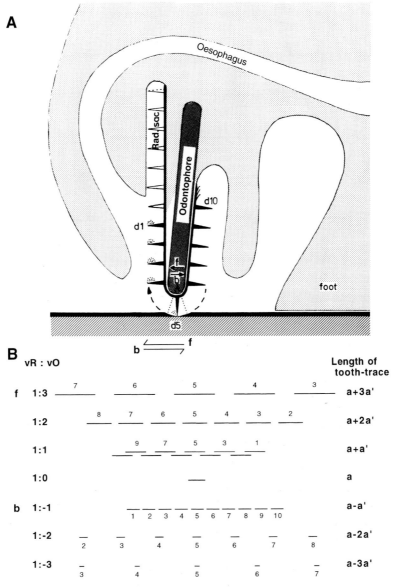

Fig. 4.9. Lengthwise cut model of the buccal mass, showing traces produced by the radular teeth that result from the cooperation of the radula retraction (broken arrow) and the backward (b) or forward (f) movement of the odontophore. a′ is one-third of a (see text for further explanation).

neutralized by the backward movement of the odontophore, but this applies only to the rachis teeth.

The second, broad-front cut, model is based on a more complex radula. Figure 4.10A shows a transverse row of five teeth at the anterior, functional face of the radula model. It comprises a rhachis tooth flanked on either side by one lateral and one marginal tooth. Due to the shape of the odontophore, the lateral and marginal teeth are oriented outwards (position indicated by thin striation) as the radula is pulled over the odontophore in its starting position. They return passively to an inward position (thick striation) when they pass the bent anterior face of the odontophore. Thus the teeth undergo a swinging motion as the transverse row passes over the odontophore face, resulting in non-linear traces of the individual teeth in the feeding track (Fig. 4.10B). Moreover, as the figure shows, the swinging motion results in different parts of the cuspid head of each tooth bearing most of the workload at different phases of the bite. As with the previous model, the movement of the odontophore changes the shape and length of the traces according to the vR : vO ratio. The forward movement of the odontophore results in lengthening and relative diminishing of the curvature of the traces of the marginal teeth (cf. *Littorina littorea* (Linnaeus) (Littorinidae), Fig. 4.11C). The radula functions primarily like a shovel dredge and produces bite-tracks characterized by the well-cleaned longitudinal axis. In contrast, the backward movement of the odontophore results in shortening and an increasingly horizontal arrangement of the marginal traces, that is, collectively, the teeth function to grasp and gather pieces of pasture, rather than to scrap. Bite-tracks of the grasping type have a bipartite appearance with an untouched axis. These traces occur in prosobranchs with rhipidoglossan radulae (Märkel, 1966), but also in some prosobranchs with taenioglossan-type radulae, such as the freshwater *Viviparus contectus* (Millet) (Fig. 4.11D) (Märkel, 1957) and the terrestrial *Pomatias elegans* (Müller) (Pomatiasidae) (K. Märkel, unpublished). In these gastropods, the rhachis teeth hardly touch the pasture, or they do so only at the end of the bite and are involved primarily in the transport of the gathered food.

The two models discussed thus far largely concern gastropods in which the teeth are rather long, and where the odontophore is rigid and thus does not change shape during feeding. The radula and the adjacent alary processes are stretched over the front face of the odontophore. The alary processes are responsible for the extreme lateral tension that spreads the marginal teeth as the radula passes over the tip of the odontophore (Fig. 4.5). Studies in *Littorina* de Férussac indicate that cuticular tendons intensify the spreading of the outermost marginal teeth. The width of the bite-track thus slightly exceeds the width of the radula. Within a single bite-track, differences occur in the length and the inclination of traces produced by individual teeth, and the distances between traces of consecutive rows of teeth. This indicates that the vR : vO ratio is not constant throughout the bite. In *Littorina*, the odontophore movement is forward. The bite-track shows the traces of about 15 transverse rows of teeth. The

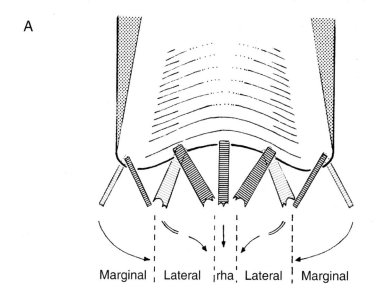

A

Marginal ¦ Lateral ¦rha¦ Lateral ¦ Marginal

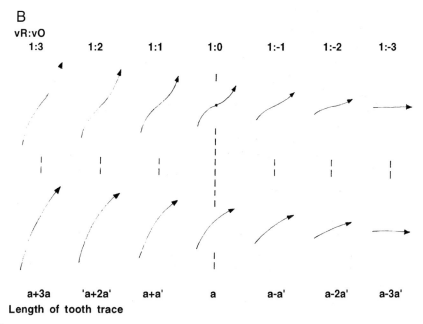

B
vR:vO

| 1:3 | 1:2 | 1:1 | 1:0 | 1:-1 | 1:-2 | 1:-3 |

| a+3a | 'a+2a' | a+a' | a | a-a' | a-2a' | a-3a' |

Length of tooth trace

Fig. 4.10. Broad-front cut model of the buccal mass, showing traces produced by the radular teeth that result from swinging motion of the teeth as the radular moves over the bent, anterior face of the odontophore. (A) As the teeth approach the edge of the odontophore, the teeth are spread (thin striation). As they pass over the bent anterior face, the teeth turn inwards (strong striation) and scrape curved traces (indicated by the arrows). (B) Traces of the left row of lateral (upper row) and marginal (lower row) radular teeth produced by movement of the radular over the odontophore, illustrating modification of the traces by the movement of the odontophore itself. The lateral teeth assume a scrape function at the point when vR : vO = 1 : 0.

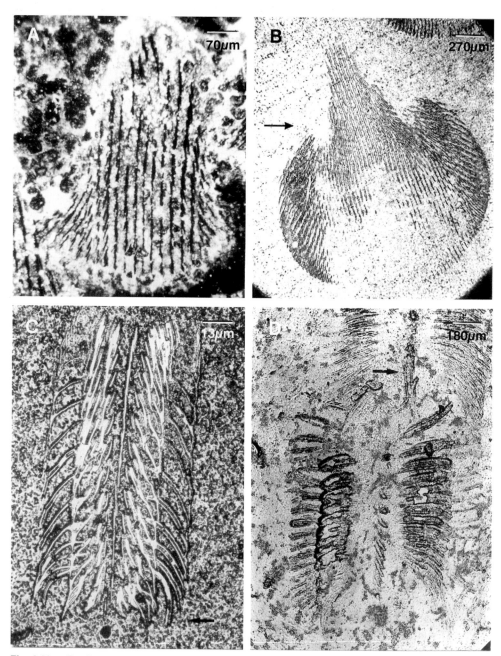

Fig. 4.11. Feeding tracks produced by gastropod molluscs grazing 'pasture' on a glass plate. (A) *Monachoides incarnata* (Müller) (Hygromiidae: Stylommatophora). (B) *Lymnaea stagnalis* (Linnaeus) (Lymnaeidae: Basommatophora). The arrow indicates the modified trace due to changes in the curvature of the anterior face of the odontophore. (C) *Littorina littorea* (Linnaeus) (Littorinidae: Caenogastropoda) (forward movement of the odontophore). (D) *Viviparus contectus* (Millet) (Viviparidae: Caenogastropoda) (backward movement of the odontophore). Arrows indicate the abnormal feeding traces at the beginning of the strokes.

lengths of and the distances between the tooth traces indicate that the stroke starts with a low vO, and the vR : vO ratio is maintained relatively constant throughout the bite (1 : 1.5–1.7). Four traces are visible on each side instead of only three as one would expect from the number of teeth in each transverse row. The lateral teeth are provided with a bipartite cusp, which make a bipartite trace, as the innermost part of the cusp comes into contact with the pasture prior to the outer part of the cusp as the teeth go through their swinging motion (Fig. 4.11C). The interior marginal teeth make flattened sigmoid traces, while the outer marginal teeth produce broadly convex traces. The traces of the interior marginal teeth cross over those of the outer marginal teeth, that is the latter come into contact with the substrate prior to the interior marginal teeth. The transport of the gathered food is by the rhachis teeth. Most tracks reveal that the inclination of the very first traces is not in line with that of the following traces (arrows in Fig. 4.11). This phenomenon is due to the fact that the radula lies loosely upon the odontophore in the initial protractor phase and is pulled tight at the beginning of the scraping sequence.

In _Viviparus_, the odontophore movement is backward. The bite-track shows the traces of about 20 transverse rows of teeth. The shortness of the bite-track, the initially long distances between the traces produced by the rachis teeth and the fact that the traces of the lateral and marginal teeth initially run backward (Fig. 4.11D), indicate that the bite starts with a high vO (e.g. vR : vO 1 : 3). As the bite progresses, the vO continually decreases and the traces of the lateral and marginal teeth turn forward. Thus the bite has a stellar shape. Besides the varying vR : vO ratio, the inclination of the odontophore possibly changes during the bite.

Most pulmonates have numerous, small radular teeth, and the traces made by single teeth are short (Fig. 4.11A and B). In examining the feeding tracks, it is possible to recognize traces produced by the innermost lateral and outermost marginal teeth. However, for much of the track, the distinction between marginals and laterals is not readily apparent as there is a gradual transition in tooth form along each transverse row. As noted earlier, the pulmonate odontophore contains muscle cells and turgor cells, and its shape changes during feeding. In most pulmonates the odontophore moves forward during the bite. In contrast to prosobranchate gastropods, the most marginal teeth do not scrape the pasture, and the width of the radula far exceeds the width of the feeding track. The bite-track is broad at the base and tapers off distally (Fig. 4.11A and B). The innermost lateral teeth produce longitudinal traces, while, more laterally, the traces are inclined due to the swinging motion of the teeth as they pass over the bent, leading edge of the odontophore. The curvature of the odontophore's anterior face increases in the course of the bite. Thus the teeth tend to produce lengthwise traces at the base of the bite-track, and inclined traces more distally. Figure 4.11B shows the well-developed bite-track of _L. stagnalis_, which results from a wide swivel motion of the odontophore. Basically, the track exhibits a delicate circular line of tooth traces, produced by the teeth that scrape the pasture as the radula is pulled tight at the commencement of the bite; the middle area does not exhibit tooth traces because the radula is pulled into the trough of the odontophore. In this example, the marginal teeth are the first to make contact with the pasture. Thereafter, according to the progress of the swivel motion and as the curvature of the odontophore's anterior face increases, the lateral teeth make contact while the marginal teeth lose contact. Not all bite-tracks reveal this full pattern – often the bite tracks show only the marginal teeth traces due to a less extensive swivel motion, which varies from bite to bite. Thus the animal may produce bite-tracks of different shapes.

In the stylommatophoran pulmonates that have been investigated, the middle section of the radular transverse rows are engaged in scraping and food gathering from the very beginning of the bite (Fig. 4.11A). The bite-tracks of these animals often overlap each other – this is especially apparent in larger species; the head sways to and fro, and the axes of succeeding rows of the bite-track alternate from one side to the other. Further, these animals tend to move forward slowly, and the pasture is utilized intensively.

Role of the Jaw

The role of the jaw in feeding has been investigated in pulmonates by Märkel (1957). The jaw is a reinforced part of the foregut cuticle which is located opposite the radula. When pulmonates graze soft substrates like algae, the jaw does not feature predominantly, and it is the radular teeth that scrape, gather and transport the food particles. But pulmonates do not only graze algae but also leaves or tuberous roots. When leaves are fed upon, the leaf is squeezed in between the jaw and the radula, and the radula cuts and extracts pieces. The radula does not comminute the pieces further and the material is transported into the oesophagus. Often the gut contains large pieces of leaves, whose cut surfaces correspond in shape to that of the jaw.

The predaceous pulmonate *Oxychilus glaber* (Rossmässler) (Zonitidae) is not equipped with a well developed jaw but none the less will also feed on leaves. Its head rides the edge of the leaf like a caterpillar and the large marginal teeth graze the leaf from both sides and pull out pieces which are highly damaged (Märkel, 1957). In feeding of tuberous roots, such as carrots, single teeth or groups of teeth are used like a shovel-dredge to remove pieces for swallowing.

Many lineages among Stylommatophora exhibit a trend to reduction in the jaws (often linked to carnivory – see Barker and Efford, 2002), and clearly the radular teeth assume the key role in gathering of food in these animals.

Conclusions

The function of the radula and the uptake of food are components of a very complicated process that involves a well-coordinated cooperation of muscles, the odontophore and the radula. Therefore, it is not possible to interpret the feeding process by looking only at the dentition of the radulae possessed by different animals. Much more information is still necessary if we are to understand fully the function of the organs involved in food gathering. In particular, additional electron microscopical and biochemical studies are needed to provide more information about the function of the collostyle, and the mechanisms of transport and degeneration of the radula.

Acknowledgements

The authors thank D. Coste for allowing quotation of his unpublished results.

References

Ankel, W.E. (1938) Erwerb und Aufnahme der Nahrung bei den Gastropoden. *Verhandlungen der Deutschen Zoologischen Gesellschaft* 40, 223–295.

Barker, G.M. and Efford, M.G. (2002) Predatory gastropods as natural enemies of terrestrial gastropods and other invertebrates. In: Barker, G.M. (ed.) *Natural Enemies of Terrestrial Molluscs*. CAB International, Wallingford, UK.

Carefoot, T.H. (1965) Magnetite in the radula of the *Polyplacophora*. *Proceedings of the Malacological Society of London* 36, 203–212.

Coste, D. (1986) Histologische Untersuchungen an der Radulapapille von *Lymnaea stagnalis* (Gastropoda, Basommatophora). Diplomarbeit, Ruhr-University, Bochum, Germany.

Curtis, S.K. and Cowden, R.R. (1977) Ultrastructure and histochemistry of the supportive structures associated with the radula of the slug, *Limax maximus*. *Journal of Morphology* 151, 187–212.

Ducros, C. (1967) Contribution à l'étude du tannage de la radula chez les gastéropodes. *Annales Histochimie* 12, 243–272.

Fujioka, Y. (1985) Seasonal aberrant radular formation in *Thais bronni* (Dunker) and *T. clavigera* (Küster) (Gastropoda, Muricidae). *Journal of Experimental Marine Biology and Ecology* 90, 43–54.

Gabe, M. and Prenant, M. (1949) Particularités histochimiques de la gaine radulaire chez l'escargot (*Helix aspersa* Müll.) *Comptes Rendus de l'Académie des Sciences Paris* 229, 1269–1270.

Gabe, M. and Prenant, M. (1950a) Recherches sur la gaine radulaire des mollusques. I. La gaine radulaire de *Dentalium entale* Deshayes. *Archives de Zoologie Expérimentale et Générale* 86, 487–498.

Gabe, M. and Prenant, M. (1950b) Recherches sur la gaine des mollusques. II. Données histologiques sur l'appareil radulaire des Hétéropodes. *Bulletin de la Société Zoologique de France* 75, 176–184.

Gabe, M. and Prenant, M. (1951a) Particularités histochimiques de la gaine radulaire chez les Prosobranches Diotocardes. *Bulletin de la Société Zoologique de France* 76, 305.

Gabe, M. and Prenant, M. (1951b) Recherches sur la gaine radulaire des mollusques. III. L'apparail radulaire des Ptéropodes Thécosomes. *Bulletin de la Société Zoologique de France* 76, 315–323.

Gabe, M. and Prenant, M. (1952a) Recherches sur la gaine radulaire des mollusques. IV. L'apparail radulaire d'*Acteon tornatilis* L. *Archives de Zoologie Expérimentale et Générale* 89, 15–25.

Gabe, M. and Prenant, M. (1952b) Recherches sur la gaine radulaire des mollusques. V. L'apparail radulaire de quelques Opistobranches Céphalaspides. *Bulletin du Labaratoire Maritime de Dinard* 37, 13–26.

Gabe, M. and Prenant, M. (1957) Recherches sur la gaine radulaire des mollusques. VI. L'apparail radulaire de quelques Céphalopodes. *Annales de Sciences Naturelles. Zoologie (Series 2)* 19, 587–602.

Gabe, M. and Prenant, M. (1958) Particularités histochimiques de la l'apparail radulaire cheu quelques mollusques. *Annales Histochimie* 3, 95–112.

Gabe, M. and Prenant, M. (1962) Résultats de l'histochimie des polysaccharides: invertébrés. In: Graumann, W. and Neumann, K. (eds) *Handbuch der Histochemie*, Vol. II, *Part I. Polysaccharide*. Fischer-Verlag, Stuttgart.

Hoffmann, H. (1932) Die Radulabildung bei *Lymnaea stagnalis*. *Jenaer Zeitschriften für Naturwissenschaften* 67, 535–550.

Isarankura, K, and Runham, N.W. (1968) Studies on the replacement of the gastropod radula. *Malacologia* 7, 71–91.

Kerth, K. (1971) Radula-Ersatz und Zähnchenmuster der Weinbergschnecke im Winterhalbjahr. *Zoologische Jahrbücher Anatomie* 88, 47–62.

Kerth, K. (1973) Radulaersatz und Zellproliferation in der röntgenbestrahlten Radulascheide der Nacktschnecke *Limax flavus* L. Ergebnisse zur Arbeitsteilung der Scheidengewebe. *Wilhelm Roux Archiv* 172, 317–348.

Kerth, K. (1976) Licht- und elektronenmikroskopische Befunde zum Radulatransport bei der Lungenschnecke *Limax flavus* L. (Gastropoda, Stylommatophora). *Zoomorphology* 83, 271–281.

Kerth, K. (1983a) Radulaapparat und Radulabildung der Mollusken. I. Vergleichende Morphologie und Ultrastruktur. *Zoologische Jahrbücher Anatomie* 110, 205–237.

Kerth, K. (1983b) Radulaapparat und Radulabildung der Mollusken. II. Zahnbildung, Abbau und Radulawachstum. *Zoologische Jahrbücher Anatomie* 110, 239–269.

Kerth, K. and Hänsch, D. (1977) Zellmuster und Wachstum des Odontoblastengürtels der Weinbergschnecke *Helix pomatia*. *Zoologische Jahrbücher Anatomie* 98, 14–28.

Kerth, K. and Krause, G. (1969) Untersuchungen mittels Röntgenbestrahlung über den Radula-Ersatz der Nacktschnecke *Limax flavus* L. *Wilhelm Roux Archiv* 164, 48–82.

Kim, K.S., Macey, D.J., Webb, J. and Mann, S. (1989) Iron mineralization in the radular teeth of the chiton *Acanthopleura hirtosa*. *Proceedings of the Royal Society of London, Series B* 237, 335–346.

Lowenstam, H.A. (1962) Geothite in radular teeth of recent marine gastropods. *Science* 137, 279–280.

Lutfy, R.G. and Demian, E.S. (1964) The histology of the radula and the radular sac of *Marisa cornuarietis* (L.). *Ain Shams of Science Bulletin* 10, 97–118.

Mackenstedt, U. and Märkel, K. (1987) Experimental and comparative morphology of radula renewal in pulmonates (Mollusca, Gastropoda). *Zoomorphology* 107, 209–239.

Mann, S., Perry, C.C., Webb, J., Luke, B. and Williams, R.J.P. (1986) Structure, morphology, composition and organization of biogenic minerals in limpet teeth. *Proceedings of the Royal Society of London, Series B* 227, 179–190.

Märkel, K. (1957) Bau und Funktion der Pulmonaten-Radula. *Zeitschrift für Wissenschaftliche Zoologie* 160, 213–289.

Märkel, K. (1964) Modell-Untersuchungen zur Klärung der Arbeitsweise der Gastropodenradula. *Verhandlungen der Deutschen Zoologischen Gesellschaften in Kiel* 232–243.

Märkel, K. (1966) Über funktionelle Radulatypen bei Gastropoden unter besonderer Berücksichtigung der Rhipidoglossa. *Extrait de 'Vie et Milieu' Seria A, Biologie Qmarine Tome 17*. Fascicule 3-A, 1121–1138.

Märkel, K. (1969) Wie erfolgt der laufende Zahnwechsel bei Schnecken und Knorpelfischen? *Umschau* 69, 477–480.

Mischor, B. and Märkel, K. (1984) Histology and regeneration of the radula of *Pomacea bridgesi* (Gastropoda, Prosobranchia). *Zoomorphology* 104, 42–66.

Ponder, W.F. and Lindberg, D.R. (1996) Gastropod phylogeny–challenges for the 90s. In: Taylor, J. (ed.) *Origin and Evolutionary Radiation of the Mollusca*. Oxford University Press, Oxford, pp. 135–154.

Richter, G. (1962) Die Schnecken 'zunge' als Werkzeug. *Natur und Museum* 92, 391–406.

Rinkevich, B. (1993) Major primary stages in the biomineralization in radular teeth of the limpet *Lottia gigantea*. *Marine Biology* 117, 269–277.

Rössler, R. (1885) Die Bildung der Radula bei den cephalophoren Mollusken. *Zeitschrift für Wissenschaftliche Zoologie* 41, 447–482.

Rottmann, G. (1901) Über die Embryonalentwicklung der Radula bei den Mollusken. *Zeitschrift für Wissenschaftliche Zoologie, Abteilung A* 70, 236–262.

Runham, N.W. (1961) The histochemistry of the radula of *Patella vulgata*. *Quarterly Journal of the Microscopical Society* 102, 371–380.

Runham, N.W. (1963) A study of the replacement mechanism of the pulmonate radula. *Quarterly Journal of the Microscopical Society* 104, 271–277.

Runham, N.W., Thornton, P.R., Shaw, D.A. and Wayte, R.C. (1969) The mineralization and hardness of the radular teeth of the limpet *Patella vulgaris* L. *Zeitschrift für Zellforschung* 99, 608–626.

Smith, D.A. (1987) Functional adaptation and intrinsic biometry in the radula of *Heliosoma trivolvis*. PhD Dissertation, Syracuse University, Syracuse, New York.

Smith, D.A. (1990) Comparative buccal anatomy in *Heliosoma* (Mollusca, Pulmonata, Basommatophora). *Journal of Morphology* 203, 107–116.

Smith, D.A. and Russell-Hunter, W.D. (1990) Correlation of abnormal radular secretion with tissue degrowth during stress periods in *Heliosoma trivolvis* (Pulmonata, Basommatophora). *Biological Bulletin* 178, 25–32.

Sollas, I.B. (1907) The molluscan radula: its chemical composition and some points in its development. *Quarterly Journal of Microscopical Science* 51, 115–136.

Solem, A. (1973) Convergence in pulmonate radula. *Veliger* 15, 165–171.

Towe, K.M. and Lowenstam, H.A. (1967) Ultrastructure and development of iron mineralization in the radular teeth of *Cryptochiton stelleri* (Mollusca). *Journal of Ultrastructure Research* 17, 1–13.

Trappmann, W. (1916) Die Muskulatur von *Helix pomatia*. *Zeitschrift für Wissenschaftliche Zoologie* 115, 489–585.

Troschel, F.H. (1836) Über die Mundteile einheimischer Schnecken. *Archiv für Naturgeschichte* 2, 257–279

Troschel, F.H. (1856–1863) *Das Gebiß der Schnecken zur Begründung einer natürlichen Classification*, Vol. 1. Nicolaische Verlagsbuchhandlung, Berlin.

Wiesel, R. and Peters, W. (1978) Licht- und elektronenmikroskopische Untersuchungen am Radulakomplex und zur Radulabildung von *Biomphalaria glabrata* (Pulmonata). *Zoomorphology* 89, 73–92.

5

Structure and Function of the Digestive System in Stylommatophora

V.K. DIMITRIADIS

School of Biology, Faculty of Sciences, Aristotle University of Thessaloniki, Thessaloniki, Greece 54006

Introduction

A large body of information exists on the structure and function of the digestive system of terrestrial gastropods. The gross morphology of the digestive system is known for a large number of taxa as a result of taxonomic and systematic investigation (e.g. Tillier, 1984, 1989), but studies specifically addressing issues of digestive structure and function have been few in number and highly biased taxonomically. Most of our detailed understanding of the digestive structure and function in Stylommatophora has been derived from studies on a small number of species, primarily in the families Agriolimacidae and Helicidae. It is highly questionable whether *Deroceras reticulatum* (Müller) (Agriolimacidae), the most intensively studied of stylommatophoran species, can be assumed to be indicative of the digestive structure and function of terrestrial gastropods, let alone that of the Stylommatophora: the extent to which *D. reticulatum* is herbivorous on living plant material is not all that common in terrestrial gastropods (possibly not even that common in other Agriolimacidae), yet this species has been studied extensively because of its pest status and availability in modified habitats in many temperate regions. The present chapter relies heavily on recent results, but has been supported by the classical work of Walker (1969, 1970a,b, 1972) on *D. reticulatum*, and the general reviews of Owen (1966), Hyman (1967), Franc (1968) and Runham (1975).

Morphology and Ultrastructure

From the functional perspective, the digestive system of stylommatophorans can be divided into regions with roles in: (i) reception, conduction and storage of food; (ii) digestion and absorption of nutrients; and (iii) formation of faeces. The digestive system comprises the alimentary canal and its associated parts: buccal mass, salivary glands, oesophagus (including the oesophageal crop), gastric pouch or stomach, digestive gland, intestine, rectum and anus (Fig. 5.1).

Buccal mass and salivary glands

The buccal mass, located at the animal anterior, functions primarily in the ingestion of food. The ventrally located mouth, which is surrounded by lips richly supplied with sensory cells (Chase, Chapter 3, this volume), opens to the buccal cavity via a short oral tube. The buccal cavity houses the jaw, and the odontophore with overlying radula, and communicates dorsally with the oesophagus. The jaw protrudes from the anterior–dorsal wall, and is derived from a thickening of the cuticle. The odontophore and overlying radula protrude into the buccal cavity from the posterior.

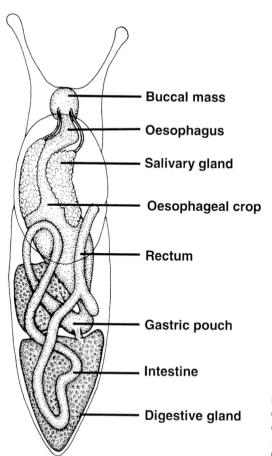

Buccal mass

Oesophagus

Salivary gland

Oesophageal crop

Rectum

Gastric pouch

Intestine

Digestive gland

Fig. 5.1. Semi-schematic drawing of the general morphology of the digestive tract of *Deroceras reticulatum* (Müller) (Agriolimacidae) (original by G.M. Barker).

The structural and functional aspects of the odontophore and radula are described by Mackenstedt and Märkel (Chapter 4, this volume).

The roof of the buccal cavity is open dorsally to the ducts of the paired salivary glands. These glands comprise a sheet of yellow tissues situated on either side of the oesophagus or oesophageal crop (Fig. 5.1).

Using light and electron microscope techniques, Walker (1970b) showed that the salivary glands of *D. reticulatum* are comprised primarily of three types of mucous cell, with smaller numbers of granular and grain cells. In *Achatina fulica* Bowdich (Achatinidae), the salivary glands consist of two types of alveoli, one secreting mucus and the other amylase (Ghose, 1963). Moya *et al.* (1992) found the salivary glands of *Arion ater* (Linnaeus) (Arionidae) to comprise two types of protein cells and four types of mucous cells. Moreno *et al.* (1982), using histochemical and morphological techniques, showed that the salivary glands of *Cantareus aspersus* (Müller) (Helicidae) are composed of two types of mucous cells and one type of protein cell. However, Charrier (1988) reported the presence of two types of mucous cells and four types of serous cells in the salivary glands of the same species.

Using electron microscopical and cytochemical techniques, Dimitriadis and Domouchtsidou (1995) observed two main secretory types in the salivary glands of *Helix lucorum* Linnaeus (Helicidae), mucous cells and granular cells. The morphological differences between various mucous and granular cells were considered to indicate different developmental stages, rather than distinct cell types as supported by previous researchers. In the salivary glands of *H. lucorum*, the mucous granules of the mucous cells display concentric rings in a web-like pattern of periodate-reactive material, which contain sulphated and carboxylated carbohydrates (Dimitriadis and Domouchtsidou, 1995).

The mucous and granular cells of the salivary glands are arranged in clusters or acini around a well-developed network of fine intercellular ducts. Their secretory products are discharged directly into this duct network and are conveyed to the salivary ducts and then to the buccal cavity. In *D. reticulatum*, Walker (1970b) found that the sensory products were conveyed in the intercellular duct network by the ciliary action of the epithelial cells lining the network, while in the main ducts the products were conveyed by peristalsis, as cilia were absent.

Oesophagus and oesophageal crop

The oesophagus is present consistently in Stylommatophora. Its posterior part often is modified into a crop (oesophageal crop; Tillier, 1989) for storage of ingested food. The literature on these structures is confused, however, as many authors have failed to distinguish the primary oesophagus and its modified posterior part. In many Stylommatophora, the oesophageal crop is subdivided into an anterior and posterior portion by a constriction, which is related to organ passage from the head–foot to the visceral cavity. As in *A. fulica*, the posterior part of the oesophageal crop merges with the gastric pouch (stomach), which are not easily distinguished externally (see Tillier, 1989).

The oesophagus is a slender tube. Walker (1969) noted that the oesophageal lumen in *D. reticulatum* is lined with a cuticular epithelium thrown into longitudinal ridges with mucous cells interspersed amongst the ciliated epithelial cells. As the oesophagus broadens into the oesophageal crop, the lumen loses the ridges and cuticle. Roldan and Garcia-Corrales (1988) found that the oesophagus of *Theba pisana* (Müller) (Helicidae) similarly has internal longitudinal ridges. The cell types that constitute the oesophageal epithelium were shown to be a mix of ciliated and unciliated columnar epithelial cells, and mucous cells (Fig. 5.2). In the oesophagus proper, the ciliated cells predominated throughout, but were most abundant on the ridges. In the oesophageal crop, the ciliated cells decreased in abundance and were found only on the ridge crests.

Ultrastructural studies indicate that in *T. pisana* (Roldan and Garcia-Corrales, 1988), *A. ater* (Bowen, 1970; Angulo *et al.*, 1986) and *H. lucorum* (Dimitriadis *et al.* 1992), the ciliated and unciliated cells are similar in possessing small quantities of rough endoplasmic reticulum, a small number of Golgi complexes usually located in the supranuclear cytoplasm, lipid inclusions usually in their middle and basal portion, and many mitochondria mainly located towards the cell apex. A particular characteristic of these columnar cells is the occurrence of many electron-dense granules in their apical cytoplasm. Similar granules have been demonstrated in the oesophageal crop epithelium in marine caenogastropods (Boquist *et al.*, 1971; Lufty and Demian, 1976). In *H. lucorum*, the granules contain periodate-reactive and carboxylated glycoconjugates, as well as acid phosphatases (Dimitriadis and Liosi, 1992), while in *D. reticulatum* they react positively to the lysosomic enzyme non-specific esterase (Walker, 1969). Bowen (1970) found that in *A. ater*, the number of these enzymatic granules, as well as granules containing lipids, increased after a meal, compared with starved animals.

The mucous cells in the oesophageal crop epithelium possess a cytoplasm rich in mucous granules. In *H. lucorum*, the mucous granules are

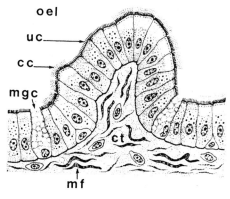

Fig. 5.2. Semi-schematic drawing of a transverse section of the oesophageal wall in *Theba pisana* (Müller) (Helicidae). cc, ciliated cell; ct, connective tissues; oel, oesophageal lumen; mf, muscle fibre; mgc, mucous gland cell; uc, unciliated cell (after Roldan and Garcia Corrales, 1988).

usually spherical in shape, 1–2 μm in diameter and contain a periodate-reactive non-sulphated and non-carboxylated fibrillar matrix, without cores (Dimitriadis *et al.*, 1992). Golgi complexes are numerous in these cells, intimately associated with one or more mucous granules. The rough endoplasmic reticulum of the mucous cells of the oesophageal crop epithelium often have a swollen appearance under electron microscopy and contain a crystalline-like material which reacts negatively to tests for periodate-reactive carbohydrates. The observation of a swollen rough endoplasmic reticulum in starved individuals of *T. pisana* (Roldan Cornejo, 1987) would suggest that the size of the endoplasmic reticulum is unrelated to the feeding cycle.

Underlying the oesophageal crop epithelium, as in the other epithelia of the digestive tract, there are elongated to oval basal cells resting on the basement membrane. These basal cells contain few mitochondria and are poorly supplied with rough endoplasmic reticulum. Dimitriadis *et al.* (1992) regarded these cells as the undifferentiated precursors of the epithelial cell types.

Gastric pouch and digestive gland

The oesophagus passes directly into the gastric pouch or stomach. This region of the digestive tract accepts the ducts from the two lobes of the digestive gland. There have been few studies that have examined the anatomy and ultrastructure of the gastric pouch in Stylommatophora in detail, and in several studies the failure to recognize the distinction between the oesophagus, the oesophageal crop and the gastric pouch makes comparison difficult.

The gastric pouch varies in structure among taxa, comprising what Tillier (1989) calls the gastric crop and gastric pouch. The gastric pouch in *T. pisana* (Roldan and Garcia-Corrales, 1988) is taken as illustrative of the condition in Stylommatophora. Its epithelium comprises ciliated and unciliated columnar cells, and mucous cells. At the intestinal end, the gastric pouch possesses many longitudinal ridges (Fig. 5.3). The ducts opening to the digestive gland (hepatic ducts) are round, and the gastric walls around them show radially arranged ridges. Two unequal intestinal grooves formed by connective tissue and muscle fibres, the major and minor typhlosole, extend from the hepatic ducts into the intestine. The minor typhlosole issues from the duct of the anterior lobe of the digestive gland, and ends a short distance into the pro-intestine. The major typhlosole arises from the duct of the posterior lobe of the digestive gland, and continues along the full length of the pro-intestine.

The digestive gland (also known in the literature as the midgut gland, hepatopancreas and liver) is the largest organ in the stylommatophoran body. It consists of two lobes communicating with the gastric pouch via large ducts, which branch to form smaller ducts, ductules and complex branched blind tubules. The digestive gland is concerned with

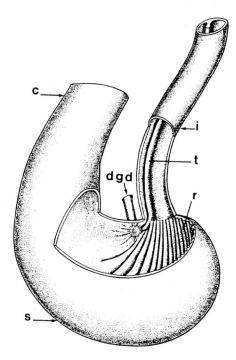

Fig. 5.3. Semi-schematic reconstruction of the gastric pouch and pro-intestine of *Theba pisana* (Müller) (Helicidae). c, oesophageal crop; dgd, duct of the anterior lobe of the digestive gland; i, pro-intestine; s, gastric pouch or stomach; r, ridge; t, major typhlosole (after Roldan and Garcia-Corrales, 1988).

production of digestive enzymes, absorption of nutrients, endocytosis of food substances, food storage, and excretion. In spite of the numerous studies on the morphology and physiology of the digestive gland cells, there is little agreement about the cell types that constitute the digestive gland epithelium of Stylommatophora. At the ultrastructural level, unequivocal interpretation is still difficult because of the cellular changes associated with phases of activity. Most findings are consistent with the hypothesis that the digestive gland epithelium of terrestrial gastropods is composed of three cell types: digestive cells, calcium cells and excretory cells. The presence of a fourth type, the thin cells, has been documented by a number of researchers, but these cells are to be regarded as undifferentiated precursors of the other cell types (see Walker, 1970a). The epithelial cells are bounded by connective tissues and muscle, with interspersed haemocoelic spaces.

   Digestive cells are the most frequent cell type to be found in the digestive gland of Stylommatophora. They are usually columnar in shape, but vary in size. Their apical surface is microvillous. These cells are characterized by numerous cytoplasmic granules and cisternae of varying size, termed 'green' and 'yellow' granules after light microscopic observations of tissues from *C. aspersus* and *D. reticulatum* (Sumner, 1965; Walker, 1970a) or 'apical granules' and 'cisternae with electron-dense cores' after electron microscopical observations of tissues from *H. lucorum* (Dimitriadis and Hondros, 1992; Dimitriadis and Liosi, 1992) (Fig. 5.4). Digestive cells show considerable variation in form related to

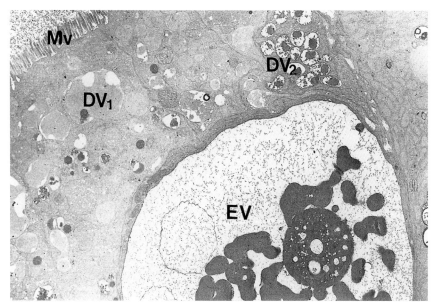

Fig. 5.4. Transmission electron micrograph of the digestive cells in the digestive gland of a hibernating *Helix lucorum* Linnaeus (Helicidae). Note the 'apical granules' (DV1) and 'cisternae with electron-dense cores' (DV2). A 'large cisterna' (EV) occupies an extensive area of cytoplasm in an excretory cell. Mv, microvilli. Magnification ×4000 (after Dimitriadis and Hondros, 1992).

different functional states. There is clear experimental support for their involvement in the digestive process whereby exogenous material is taken up by pinocytotic vesicles. Fusion of pinocytotic vesicles gives rise to so-called 'phagosomes', which have been shown to be degraded subsequently by lysosomal action into 'phagolysosomes', and eventually shed into the hepatic ducts (see 'Digestion Processes', below).

The second cell type in the digestive gland, the calcium cells, are characterized by their pyramidal shape and the presence of calcium granules. The role of calcium in gastropods is multiple, and includes pH homeostasis, reproductive activity such as egg-shell formation, regulation of freeze tolerance, cellular waste metabolism, as a component of mucus of the body wall integument and shell regeneration, and in the formation of the epiphragm during hibernation (for a review, see Dimitriadis and Liosi, 1992). The calcium granules of the calcium cells of *C. aspersus* contain $CaMgP_2O_7$ (Howard *et al.*, 1981) and are regarded as sites of accumulation of a variety of cations that act as sites of detoxification of various dietary metals (Taylor *et al.*, 1988). In *A. ater* and *H. lucorum*, the calcium granules react positively to acid and alkaline phosphatase (Bowen, 1970; Dimitriadis and Liosi, 1992), but it remains uncertain if these phosphatases play an important role in the digestive processes of the digestive gland.

The third cell type in the digestive gland, the excretory cells, are larger in size than the digestive cells and are characterized by one or more vacuoles or 'large cisternae' containing cores or amorphous masses of high density when viewed by electron microscopy (Fig. 5.4). The vacuoles in the excretory cells in *D. reticulatum* react negatively in histochemical tests for neutral polysaccharides but positively for lipofuscin (Walker, 1970a). A positive reaction for lipofuscin is also noted in *A. ater* (Bowen, 1970), while the vacuoles termed 'large cisternae' in the excretory cells of *H. lucorum* (Dimitriadis and Liosi, 1992) exhibit a negative reaction for acid phosphatase and a positive one for sulphated and carboxylated polysaccharides. A negative reaction for acid phosphatase has also been reported for the analogous structures of *Arion hortensis* de Férussac (Bowen and Davies, 1971).

There is a disagreement on the origin and the function of excretory cells in the digestive gland. It has been proposed that these cells are either degenerated calcium cells or differentiated digestive cells (for a review, see Dimitriadis and Liosi, 1992). In addressing prosobranch gastropods, Fretter and Graham (1962) proposed that excretory cells of the digestive gland may be adapted for the uptake of material from the blood, by way of their extensive absorptive basal regions, and for secreting the absorbed material into the lumen of the hepatic ducts. In *D. reticulatum*, Walker (1970a) regards excretory cells as the final phase in the development of the digestive cells, in which remnant food substances are degraded by lysosomal enzymes and subsequently extruded to the hepatic duct lumen.

Intestine and rectum

The intestine is a thin-walled tube of uniform diameter, which in most Stylommatophora makes, firstly, one forwardly directed loop, then one posteriorly directed loop, before again turning forward and running to the anus as the rectum (Fig. 5.1). In carnivorous Stylommatophora, there is a trend towards reduced intestinal length and, as a consequence, the loops are reduced (Tillier, 1989; Barker and Efford, 2002), while in many slug-like taxa the intestine may be lengthened and coiled around the oesophagus and oesophageal crop (Tillier, 1984, 1989).

Three histologically distinct regions have been recognized in the intestine of *D. reticulatum* (Walker, 1969) and *T. pisana* (Roldan and Garcia-Corrales, 1988), named the pro-, mid- and post-intestine, respectively. The major typhlosole extends from the gastric pouch as a prominent fold in the pro-intestinal wall (Fig. 5.5). The intestine epithelium is composed of ciliated and unciliated columnar cells and mucous cells. The epithelium of the lumen in the pro- and mid-intestines is richly ciliated. The mid-intestine is clearly distinguishable due to the presence of folds or ridges in the lumen, the tall stature of the epithelial cells and the presence of numerous granules in the apical cytoplasm of these cells. Towards the posterior end of the mid-intestine, the cilia become restricted to the tips of the folds and this continues along the post-intestine. In addition to the epithelium being comprised of cells of

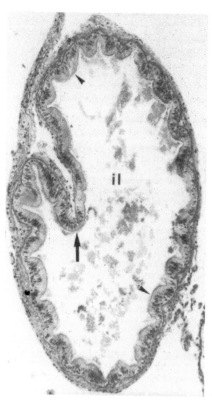

Fig. 5.5. Light micrograph of a transverse section through the pro-intestine of *Theba pisana* (Müller) (Helicidae), demonstrating the major typhlosole (arrow) and the slight internal ridges (arrowheads). il, intestinal lumen. Magnification ×140 (after Roldan and Garcia-Corrales, 1988).

smaller size, the post-intestine is characterized by the presence of a mucous cell type.

Electron microscopical studies have shown that the columnar intestinal cells in *H. lucorum* (Dimitriadis and Domouchtsidou, 1995) possess large numbers of lipid inclusions, usually clustered in the middle and apical region of the cells (Fig. 5.6). Adjacent to or between these lipid inclusions are large numbers of periodate-reactive glycogen granules. Similarly to those in the oesophageal crop, the apical granules of the intestinal columnar cells contain lysosomal enzymes such as esterases and acid phosphatases (Ferreri, 1958; Walker, 1969; Bowen, 1970; Bowen and Davies, 1971). The mucous cells of the intestinal epithelium of *H. lucorum* (Dimitriadis and Domouchtsidou, 1995) produce mucous granules displaying a compact structure when viewed by electron microscopy. The presence of sulphated and carboxylated carbohydrates differentiates these granules from those in the oesophageal crop epithelium of the same species. In *A. ater* (Angulo *et al.*, 1986), sulphated acid and neutral polysaccharides are found in the intestinal mucous cells, as has been demonstrated in other gastropods (e.g. Varute and Patil, 1971) and bivalves (e.g. Zacks, 1955),

The intestine merges imperceptibly with the rectum. The rectum of *D. reticulatum* comprises an epithelium of unciliated and cilated columnar

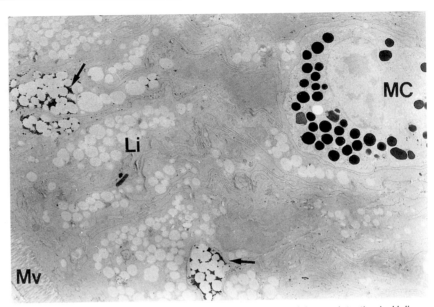

Fig. 5.6.   Transmission electron micrograph of the epithelium of the pro-intestine in *Helix lucorum* Linnaeus (Helicidae). The columnar cells contain large amounts of lipid inclusions (Li), while a mucous cell (MC) exhibits strong reaction for periodate-reactive carbohydrates. A similar reaction is located on glycogen particles closely associated with lipid inclusions (arrow). Magnification ×4000 (after Dimitriadis and Domouchtsidou, 1995).

cells and two types of mucous cells. The epithelium is formed in a number of low folds – very much reduced in height compared with that in the intestine – and a single well-developed lateral fold. While ciliated cells occur widely in the rectal epithelium, they predominate on the lateral fold (Walker, 1969). In *T. pisana*, Roldan and Garcia-Corrales (1988) found the rectum to be morphologically similar to the post-intestine, comprising ciliated and unciliated columnar cells, and mucous cells. The rectum opens to the anus via a sphincter.

Digestion Processes

In the following paragraphs, the participation of the various parts of the digestive system in the digestion processes is discussed.

Salivary glands

Digestion begins with trituration of foods in the buccal cavity and the action of secretions from the salivary glands. The function of the saliva is believed to be lubrication, thus assisting with the removal of food from the radula and its passage into the oesophagus. Chemical analyses of the salivary glands have clearly indicated the presence of amylases, together with trypsin-like proteases (Boers *et al.*, 1967; Walker, 1970b). However, in the salivary glands of *C. aspersus*, histochemical methods show that the amylase is not produced by the cells of the salivary glands (Charrier, 1989). In this species, cellulotic activity has been observed in the connective tissue surrounding the salivary glands and the oesophageal crop, but synthesis of cellulotic enzymes by the salivary gland cells remains to be demonstrated. The secretory activity of the salivary gland cells is associated, with the exception of mucous cells, with the intake of food.

Oesophageal crop and gastric pouch

The oesophagus, and the oesophageal crop when present, functions as an area for food storage, extracellular digestion and absorption. The ingested food is transported rapidly to the oesophagus and oesophageal crop by complex peristaltic rhythms and is there mixed with secretions from the oesophageal epithelial cells (Roach, 1968). Studies with radioisotope-labelled foods have shown that in *D. reticulatum*, ingested food appears in the oesophageal crop almost immediately after being taken into the buccal cavity, and over the following 20–40 min the bulk of the ingested material is transported to the gastric pouch. Usually, it takes several hours for the greater bulk of the meal to pass into the gastric pouch, although very fine material may remain in the oesophageal crop for more than 3 days (Walker, 1972).

In stylommatophorans, the columnar cells of the oesophagus and oesophageal crop epithelium do not show ultrastructural features consistent with extensive secretory function (Roldan Cornejo, 1986, 1987; Dimitriadis *et al.*, 1992). The results are consistent with the view (see, for example, Oxford, 1977) that the digestive activity in the oesophagus/ oesophageal crop of stylommatophorans is mediated by digestive enzymes from the salivary glands and regurgitated from the digestive gland, rather than by enzymes secreted by the columnar cells of the oesophagus.

In *D. reticulatum*, Walker (1969) was able to show that labelled glucose, galactose and glycine are taken up by the plasma membrane of the oesophageal crop epithelium. Bourne *et al.* (1991) demonstrated in the same species that endocytosis of endogenous material labelled with horseradish peroxidase and lanthanum nitrate occurs within 5 min in animals that had been starved for 48 h. In *T. pisana* (Roldan Cornejo, 1986, 1987) and *H. lucorum* (Dimitriadis *et al.*, 1992; Dimitriadis and Domouchtsidou, 1995), the columnar cells of the oesophageal crop are regarded as the sites of nutrient absorption. The cellular features indicative of this function are the very well-developed microvillar border, the numerous mitochondria in the apical cytoplasm and the substantive stores of glycogen and lipids. That the oesophagus and oesophageal crop are important sites for lipid and sugar uptake is supported further by the presence of large amounts of esterases in the apical granules, as well as the presence of cellulases and chitinases (Jeuniaux, 1954; Koopmans, 1967; Flari and Charrier, 1992).

Stylommatophorans are well equipped with a wide array of digestive enzymes, particularly carbohydrases. Of the 30 or more enzymes associated with the digestive tract of Helicidae, more than 20 are carbohydrases: including α-amylase and β-amylase (pH optimum 6.2–6.8 and 4.5, respectively), cellulases and chitinases, as well as a variety of glycosidases (Anker and Vonk 1946; Holden and Tracey, 1950; Myers and Northcote, 1958; Flari and Charrier, 1992; Charrier and Rouland, 1992; Flari and Lazaridou-Dimitriadou, 1996). The oesophageal crop juice and the digestive gland extracts of *A. ater* have also been shown to be capable of digesting a very wide range of carbohydrates, reflecting the fact that most carbohydrases are not substrate specific. A general α-glucosidase could account for the hydrolysis of several substrates containing α-glycosidic bonds, while β-linked polysaccharides and glycosidases may act upon (1,4)β-glycosidase (Evans and Jones, 1962a). The presence of these digestive enzymes is not confined to terrestrial species, however, as a wide range of carbohydrases has also been found in *Tegula funebralis* Adams (Trochidae), a marine species that feeds on a wide diversity of algae (Galli and Giese, 1959).

Due to the contradictory views on the origin of cellulases and chitinases, much research has focused on the presence of these enzymes in the digestive system of terrestrial gastropods. For many years, it was believed that the enzymes are produced by bacteria residing in the gut (Florkin and Lozet, 1949; Jeuniaux, 1954). Other researchers, however, supported an endogenous origin of these enzymes (Holden and Tracey, 1950; Strasdine and Whittaker, 1963) or a mixed endogenous and exogenous origin (Myers and Northcote, 1958; Koopmans, 1967; Flari and Lazaridou-Dimitriadou, 1996). Working with various gastropod species, Jeuniaux (1954, 1963) isolated a large number of bacteria from the oesophageal crop and demonstrated that these bacteria were capable of digesting cellulose and chitin. However, Strasdine and Whittaker (1963) showed that the digestive gland of *Helix pomatia* Linnaeus, which is almost free of bacteria, contains these enzymes. Using scanning electron microscopy, Charrier (1990) showed that *C. aspersus* did not possess an endogenous bacterial flora and that the bacteria were ingested with plant food and more especially with faeces. In *C. aspersus*, as with the isopod *Oniscus asellus* (Linnaeus) (Oniscidae), the bacterial community of the digestive tract was shown to be similar to that found in soil (Watkins and Simkiss, 1990). Parnas (1961) administed various antibiotics to *Levantina hierosolyma* (Boiss.) (Helicidae) and so demonstrated that the cellulase activity persisted only in the digestive gland. He concluded that cellulases were produced within the digestive gland, whereas the cellulotic activity of the oesophageal crop and salivary glands may have resulted either from bacteria or from the passage of enzymes from the digestive gland. From experiments in which animals were reared under aseptic conditions, Jeuniaux (1961, 1963) considered that the bacterial flora made a negligible contribution to the chitinase activity of the digestive system and that the digestive gland was the actual site of chitinase secretion. Moreover, the widespread occurrence of chitinase in the digestive gland of different gastropod clades suggested the biosynthesis of chitinase by the digestive gland to be a general property of the Gastropoda (Jeuniaux, 1963).

The digestive juices of *H. pomatia* exhibit only slight proteolytic activity (Myers and Northcote, 1958). Both in this species (Rosen, 1952), and in *A. fulica* (van Weel, 1950), cathepsin has been detected only intracellularly. In contrast, Evans and Jones (1962b) reported strong activity of a catheptic-type proteinase in the oesophageal crop lumen of *A. ater*, apparently secreted by the digestive gland cells. Lipases have been demonstrated histochemically in the digestive tract epithelium of *H. pomatia* by Ferreri (1958) but, as is the general pattern in gastropods, these enzymes exhibit low activity (see van Weel, 1961).

Digestive gland

Food material remains in the gastric pouch for only a short time. Fine particles and soluble material pass into the digestive gland aided by peristaltic pulses, while coarse material is directed into the intestine by the ciliated folds around the openings of the hepatic duct (Roach, 1968; Walker, 1969). Studies by Walker (1969) indicated that barium sulphate with a particle size of 1–3 μm, and chromopaque of 1 μm, do not enter the digestive gland, while colloidal thorium dioxide of 0.1–0.4 μm particle size readily passes into the ducts. The particles that do enter the hepatic ducts are sufficiently small to pass amongst the cilia, which form an effective filter to the larger particles. Rosen (1952) experimented with many types of particles introduced into the food of *H. pomatia*, and reported that only the very finest (0.1 μm) carbon particles could enter the digestive gland.

There is now ample evidence for the existence of endocytic activity in the digestive gland of Gastropoda (see Owen, 1966; Morton, 1979; Oxford and Fish, 1979). Electron microscopical data (Sumner, 1965; Walker, 1970a; Dimimitriadis and Hondros, 1992) do not indicate the existence of phagocytosis of solid food material by the digestive cells. Fusion of pinocytic vesicles in the digestive cells give rise to so-called 'phagosomes', which subsequently are transformed into 'phagolysosomes' after fusion with small enzyme-rich 'protolysosomes' produced by the Golgi complexes. Digestion of material within the phagolysosomes results in the release of soluble material and the accumulation of indigestible material to form residual bodies. Continuous intracellular digestion results in accumulation of 'phagosomes', which finally are pinched off and shed into the hepatic ducts together with small amounts of apical cytoplasm (Sumner, 1965; Owen, 1970; Walker, 1970a; Bowen and Davies, 1971; Oxford and Fish, 1979; Dimitriadis and Hondros, 1992; Dimitriadis and Liosi, 1992). A portion of the digestive remnants probably leaves the cells basally, as amoebocytes containing them have been detected in the underlying connective tissues (Walker, 1970a).

Chemical analyses indicate that the digestive gland functions as a major store for glycogen (Goddard and Martin, 1966), but this has not been confirmed by histochemical examination (see Runham, 1975). de Jonge-Brink (1973) found, in the freshwater basommatophoran *Lymnaea stagnalis* (Linnaeus) (Lymnaeidae) that the main site of glycogen storage was the connective tissue cells that occur amongst the digestive gland lobules. This seems to be true also for *H. lucorum* (V.K. Dimitriadis, unpublished results).

Intestine and rectum

Radio-opaque markers, introduced with the food, remain in the intestine of *D. reticulatum* for 7–7.5 h (Walker, 1972). In contrast, their passage through the rectum is rapid as faeces are apparently held there only momentarily. The existence of a long intestine in herbivorous animals and the long sojourn of the faeces there could imply that important processes are taking place in this region, in addition to the consolidation of the faecal pellet or string (Carriker, 1946).

As already mentioned above, the features of the columnar epithelium (Roldan Cornejo, 1986, 1987; Dimitriadis *et al.*, 1992; Dimitriadis and Domouchtsidou, 1995) are not suggestive of a role in enzyme secretion, but indicative of an extensive absorptive function of the intestine. Strong intracellular enzymes such as esterases (lipases) and acid phosphates have been detected in the intestine (Ferreri, 1958; Ferreri and Ducato, 1959; Walker, 1969; Bowen, 1970, 1971), apparently localized in the apical granules. From a comparison of fed and starved animals, Walker (1969) concluded that there is an increase in the number of these granules following feeding. Uptake of soluble material from the lumen of the intestine has been clearly demonstrated for minerals (Fretter, 1952, 1953), as well as glycine, galactose, glucose and palmitic acid (Walker, 1972). It seems that low molecular weight materials can enter the cells by diffusion, while large molecules such as palmitic acid are taken up by pinocytosis and then elaborated by a lysosomal mechanism.

The faeces contain two easily recognizable components. The first is the 'liver string', consisting of a membranous sac containing a brown fluid at one end and particulate matter at the other. The liver string is produced from the indigestible or unwanted material discharged by the digestive gland. The second part consists of large particles and represents the material that passes directly through the gastric pouch to the intestine. Pallant (1970) demonstrated that the liver string in *D. reticulatum* is produced periodically, perhaps once a day. It is not known whether this periodic production indicates a discharge of unwanted material only at certain times, or whether the material is accumulated continuously but released periodically.

In *Deroceras panormitanum* (Lessona & Pollonera), the digestive gland cells undergo an endogenous cycle of cytological change over the course of 24 h (Morton, 1979). In the hours of daylight, these cells absorb and intracellularly digest food particles, and at night a process of breakdown releases excretory concretions and fragmentation spherules into the gastric pouch. The release of intracellular enzymes possibly helps the digestion of food in the crop and intestine during the next feeding cycle.

In the rectum, water apparently is absorbed from the faeces. The presence of digestive enzymes in the rectum of *A. fulica* was reported by Ghose (1963), but the rectum is generally believed to have little enzymatic activity and hence little digestive function (Flari and Charrier, 1992).

Nutrition requirements and assimilation efficiency

The nutritional requirements of the stylommatophorans are poorly understood. We do not know their requirements for proteins, fats, carbohydrates, vitamins or minerals. Some species such as *Limacus flavus* (Linnaeus) (Limacidae) and *Arion rufus* (Linnaeus) (Frömming, 1954) are able to live on restricted diets such as wheat (*Triticum aestivum* Linnaeus) (Gramineae) flour. However, experimental studies with basommatophoran gastropods indicate that vitamins are essential for reproductive fitness (Vieira, 1967). Variation in the amounts of lipids, carbohydrates, proteins and fibre in the diet can affect the growth of stylommatophorans (see, for example Jess and Marks, 1989). Growth rate in *A. fulica* was shown by Ireland (1991) to be closely correlated with feeding and excretion rate, which was influenced by dietary calcium concentrations.

The assimilation efficiency of 0.4–0.9 in Stylommatophora is generally high for herbivorous animals (Lamotte and Stern, 1987; Egonmwan, 1991; Bogucki and Helczyk-Kazecka, 1977; Charrier and Daguzan, 1980; Staikou and Lazaridou-Dimitriadou, 1989). Stern (1970) found that when *A. ater* was maintained for its entire life cycle on lettuce (*Lactuca sativa* Linnaeus) (Asteraceae), 70% of the ingested food was utilized. The requirements for respiration amounted to 60% of the total, and growth and reproduction 40%. Similar results have been obtained for other herbivorous stylommatophoran species. In feeding experiments, Stylommatophora generally exhibit higher assimilation efficiencies when provided with foods from the natural environment than when provided with cultivated plants (Mason, 1970; Richardson, 1975; Staikou and Lazaridou-Dimitriadou, 1989). For example, *Eobania vermiculata* (Müller) (Helicidae) fed *Urtica dioica* Linnaeus (Urticaceae) had higher assimilation efficiency than when fed lettuce (Lazaridou-Dimitriadou and Kattulas, 1990). Kornobis and Bogucki (1973) found that in species of *Helix* Linnaeus, foods rich in vegetables were assimilated more efficiently (48.8–88.4%) than artificial diets containing little vegetable matter (39.7%).

Assimilation efficiencies have been found to vary with season. In *H. lucorum*, for example, assimilation was lowest in autumn, just prior to hibernation (Staikou and Lazaridou-Dimitriadou, 1989). Similarly, in *H. pomatia*, assimilation efficiency on a fresh *L. sativa* diet was generally greater in spring than in autumn (Bogucki and Helczyk-Kazecka, 1977). Assimilation rates in *H. lucorum* and *E. vermiculata* were found to be higher in hatchling animals than in adults (Staikou and Lazaridou-Dimitriadou, 1989). A similar result was obtained by Bogucki and Helczyk-Kazecka (1977), working with *H. pomatia*. Richardson (1975) found the assimilation efficiency of *Cepaea nemoralis* (Linnaeus) (Helicidae) not to vary with environmental temperature over the range studied, but this was dependent on the nature of the food.

Conclusion

The incremental gain in our understanding of the structure and function of the digestive tract in stylommatophoran gastropods over the last 20 years has been small compared with the substantive volume of descriptive and experimental data generated in the preceding decades on these or related molluscs. The most obvious gaps in knowledge relate to the unequivocal identification of the cell types that comprise the tissues of the digestive system, the clear understanding of the subcellular changes that these cells undergo during the various phases of both cell development and digestive metabolism, and the clear demonstration of the relative contributions of the various components of the digestive system to food digestion and absorption of nutrients. In addition, more studies are needed that would enable interpretation of the gross morphology and structural data at the cellular level in the light of evolutionary constraints and ecology. Such advances will only be achieved if the new laboratory technologies, such as histochemistry in cryosections and immunocytochemistry, are fully embraced. Studies conducted to date have not sampled terrestrial gastropods adequately, let alone stylommatophorans, across the full spectrum of evolutionary pathways and feeding ecologies. The present data are heavily biased towards large (mainly European) herbivorous taxa, mostly those herbivorous on fresh plant material. There has been little attempt to gather data for species that are primarily detritivorous or those that specialize on feeding on the microbial firms associated with decaying leaves and wood, leaf and tree trunk surfaces – yet it is these species which predominate over herbivores in terrestrial gastropod faunas in most parts of the world.

References

Angulo, E., Moya, J. and de la Vega, I. (1986) Histochemical observation on the intestinal epithelium of the slug *Arion ater*. *Cuadernos de Investigacion Biologica, Bilbao* 9, 59–65.

Anker, L. and Vonk, H.J. (1946) The presence of α- and β-amylase in the saliva of man and the digestive juice of *Helix pomatia*. II. Polarimetric determinations. *Koninklijke Nederlandse Akademie van Wetenshappen, C* 49, 845–851.

Barker, G.M. and Efford, M.G. (2002) Predatory gastropods as natural enemies of terrestrial gastropods and other invertebrates. In: Barker, G.M. (ed.) *Natural Enemies of Terrestrial Molluscs*. CAB International, Wallingford, UK.

Boers, H.N., Bonga, S.E.W. and van Rooyen, N. (1967) Light and electron microscopical investigation on salivary glands of *Lymnaea stagnalis* L. *Zeitschrift für Zellforschung und Mikroskopishe Anatomie* 76, 228–247.

Bogucki, Z. and Helczyk-Kazecka, B. (1977) Efficiency of food assimilation in the Roman snail (*Helix pomatia* L.). *Bulletin de la Societe des Amis des Sciences et des Lettres de Poznan* 17D, 159–167.

Boquist, L., Falkemer, S. and Mehrotra, B.K. (1971) Ultrastructural search for homologues of pancreatic beta-cells in the intestinal mucosa of the mollusc *Buccinum undatum*. *General and Comparative Endocrinology* 17, 236–239.

Bourne, B.B., Jones, G.W. and Bowen, I.D. (1991) Endocytosis in the crop of the slug, *Deroceras reticulatum* (Muller) and the effects of the ingested molluscicides, metaldehyde and methiocarb. *Journal of Molluscan Studies* 57, 71–80.

Bowen, I.D. (1970) The fine structure localization of acid phosphatase in the gut epithelium cells of the slug *Arion ater* (L.). *Protoplasma* 70, 247–270.

Bowen, I.D. (1971) High resolution technique for fine structural localization of acid hydrolase. *Journal of Microscopy* 94, 25–38.

Bowen, I.D. and Davies, P. (1971) The fine structure distribution of acid phosphatase in the digestive gland of *Arion hortensis* (F.). *Protoplasma* 73, 78–81.

Carriker, M.R. (1946) Observations on the functioning of the alimentary system of the snail, *Lymnaea stagnalis appressa* Say. *Biological Bulletin* 91, 88–111.

Charrier, M. (1988) Structure des glandes salivaires d'*Helix aspersa* Muller (Mollusque, Gasteropode, Pulmone). *Haliotis* 18, 171–183.

Charrier, M. (1989) Cycles de secretion et activities enzymatiques dans les cellules des glandes salivaires de l'escargot petit gris *Helix aspersa* Muller (Gasteropode, Pulmonata). *Histoenzymologie, Bulletin de la Societe Zoologique de France* 114, 97–108.

Charrier, M. (1990) Evolution, during digestion, of the bacterial flora in the alimentary system of the *Helix aspersa* (Gastropoda: Pulmonata): a scanning electron microscope study. *Journal of Molluscan Studies* 56, 425–433.

Charrier, M. and Daguzan, J. (1980) Consommation alimentaire: production et bilan energetique chez *Helix aspersa* Muller (Gasteropode pulmone terrestre). *Annales de la Nutrition et de l' Alimentation* 34, 147–166.

Charrier, M. and Rouland, C. (1992) Les osidases digestives de l' escargot *Helix aspersa*: localizations et variations en fonction de l'etat nutrionnel. *Canadian Journal of Zoology* 70, 2234–2241.

de Jonge-Brink M. (1973) Effect of desiccation and starvation upon weight. Histology and ultrastructure of reproductive tract of *Biomphalaria glabrata*, intermediate host of *Schistosoma mansoni*. *Zeitschrift für Zellforschung und Mikroskopische Anatomie* 136, 229–262.

Dimitriadis, V.K. and Domouchtsidou, G.P. (1995) Carbohydrate cytochemistry of the intestine and salivary glands of the snail *Helix lucorum*: effects of starvation and hibernation. *Journal of Molluscan Studies* 61, 215–224.

Dimitriadis, V.K. and Hondros, D. (1992) Effect of starvation and hibernation on the fine structural morphology of digestive gland cells of the snail *Helix lucorum*. *Malacologia* 24, 63–73.

Dimitriadis, V.K. and Liosi, M. (1992) Ultrastructural localization of periodate-reactive complex carbohydrates and acid–alkaline phosphatases in digestive gland cells of fed and hibernated *Helix lucorum* (Mollusca: Helicidae). *Journal of Molluscan Studies* 58, 233–243.

Dimitriadis, V.K., Hondros, D. and Pirpasopoulou, A. (1992) Crop epithelium of normal fed, starved and hibernated snails *Helix lucorum*: a fine structural–cytochemical study. *Malacologia* 34, 343–354.

Egonmwan, R.I. (1991) Food selection in the snail *Limicolaria flammea* Müller (Pulmonata: Achatinidae). *Journal of Molluscan Studies* 58, 49–55.

Evans, W.A.L. and Jones, E.G. (1962a) Carbohydrases in the alimentary tract of the slug, *Arion ater* L. *Comparative Biochemistry and Physiology* 5, 149–166.

Evans, W.A.L. and Jones, E.G. (1962b) A note of proteinase activity in the alimentary tract of the slug *Arion ater* L. *Comparative Biochemistry and Physiology* 5, 223–225.

Ferreri, E. (1958) Ricerche biochimiche ed histochimiche sull' attività lipasica dell' epitelio intestinale di *Helix pomatia*. *Zeitschrift für Vergleichende Physiologie* 41, 373–389.

Ferreri, E. and Ducato, L. (1959) Vergleichende biochemische und histochemische untersuchungen uber die lipolitische tatigkeit des Darmkanalepitheliums von *Planorbis corneus* L. und *Murex trunculus* L. *Zeitschrift für Zellforschung und Mikroskopishe Anatomie* 51, 65–77.

Flari, V. and Charrier, M. (1992) Contribution to the study of carbohydrases in the digestive tract of the edible snail *Helix lucorum* L. (Gastropoda: Pulmonata: Stylommatophora) in relation to its age and its physiological state. *Comparative Biochemistry and Physiology* 102A, 363–372.

Flari, V. and Lazaridou-Dimitriadou, M. (1996) Evolution of digestion of carbohydrates in the separate parts of the digestive tract of the edible snail *Helix lucorum* (Gastropoda; Pulmonata; Stylommatophora) during a complete 24 hour cycle and the first days of starvation. *Journal of Comparative Physiology* 165, 580–591.

Florkin, M. and Lozet, F. (1949) Origine bacterienne de la cellulase du contenu intestinal de l'escargot. *Archives Internationales de Physiology et de Biochimie* 57, 201–201.

Franc, A. (1968) Classe des Gastéropodes (Gastropoda Cuvier 1798). In: Grassé, P.-P. (ed.) *Traité de Zoologie. V. Mollusques*, Fasc. 3. Masson et Cie, Paris, pp 1–893.

Fretter V. (1952) Experiments with ^{32}P and ^{131}I on species of *Arion*, *Helix* and *Agriolimax*. *Quarterly Journal of Microscopical Science* 93, 133–146.

Fretter, V. (1953) Experiments with radioactive strontium (SR-90) on certain molluscs and polychaetes. *Journal of the Marine Biological Association of the United Kingdom* 32, 367–384.

Fretter, V. and Graham, A. (1962) *British Prosobranch Molluscs*. Ray Society, London.

Frömming E. (1954) *Biologie der Mitteleuropäische Landgastropoden*. Duncker and Humblot, Berlin.

Galli, D.R. and Giese, A.C. (1959) Carbohydrate digestion in a herbivorous snail *Tegula funebralis*. *Journal of Experimental Zoology* 140, 415–440.

Ghose, K.C. (1963) The alimentary system of *Achatina fulica*. *Transactions of the American Microscopical Society* 82, 149–167.

Goddard, C.K. and Martin, A.W. (1966) Carbohydrate metabolism. In: Wilbur, M. and Yonge, C.M. (eds) *Physiology of Mollusca*. Academic Press, New York, pp. 275–308.

Holden, M. and Tracey, M.V. (1950) A study of the enzymes that can break down tobacco-leaf components. 2. Digestive juice of *Helix* on defined substrates. *Biochemical Journal* 47, 407–414.

Howard, B., Mitcel, P.C.M., Ritchie, A., Simkiss, K. and Taylor, M.G. (1981) The composition of intracellular granules from the metal accumulating cells of the common garden snail (*Helix aspersa*). *Biochemical Journal* 194, 307–511.

Hyman, L.H. (1967) *The Invertebrates. VI. Mollusca I, 6*. McGraw-Hill, New York.

Ireland, M.P. (1991) The effect of dietary calcium on growth, shell thickness and tissue calcium distribution in the snail *Achatina fulica*. *Comparative Biochemistry and Physiology* 1, 111–116.

Jess, S. and Marks, R.J. (1989) The interaction of diets and substrate on the growth of *Helix aspersa* (Muller) var. *maxima*. In: Henderson, I. (ed.) *Slugs and Snails in World Agriculture*. British Crop Protection Council Monograph No. 41, pp. 311–317.

Jeuniaux, C. (1954) Sur la chitinase et la flore bactérienne intestinale des mollusques gastéropodes. *Bulletin Academie Royale de Belgique, Classe des Sciences* 28, 1–45.

Jeuniaux, C. (1961) Evolution des enzymes chitinolytiques dans le r'egne animal. *Proceedings of the 5th International Congress of Biochemie, Moscow*, Abstract 6.16.1526. Pergamon Press, Oxford.

Jeuniaux, C. (1963) *Chitine et Chitinolyse, un Chapitre de la Biologie Moleculaire*. Mason et Cie, Paris.

Koopmans, J.J.C. (1967) The nature and origin of cellulases in *Helix pomatia*. *Acta Physiologica et Pharmacologica Neerlandica* 14, 349–350.

Kornobis, S. and Bogucki, Z. (1973) Food assimilativeness in some species of the *Helix* L. (Helicidae, Gastropoda) genus. *Bulletin de la Societe des Amis des Sciences et des Lettres de Poznan* 14, 71–75.

Lamotte, M. and Stern, G. (1987) Les bilans energetiques chez les Mollusques Pulmones. *Haliotis* 16, 103–128.

Lazaridou-Dimitriadou, M. and Kattulas, M.E. (1990) Energy flux in a natural population of the land snail *Eobania vermiculata* (Muller) (Gastropoda: Pulmonata: Stylommatophora) in Greece. *Canadian Journal of Zoology* 69, 881–891.

Lufty, R.G. and Demian, E.S. (1976) The histology of the alimentary system of *Marisa cornuarietis* (Mesogastropoda: Ampullariidae). *Malacologia* 5, 375–422.

Mason, C.F. (1970) Food feeding rates and assimilation in woodland snails. *Oecologia* 4, 358–373.

Moreno, F.J., Pineiro, J., Hidalco, J., Navas, P., Aijon, J. and Lopez-Campos, J.L. (1982) Histochemical and ultrastructural studies on the salivary glands of *Helix aspersa* (Mollusca). *Journal of Zoology* 196, 343–354.

Morton, B. (1979) The diurnal rhythm and the cycle of feeding and digestion in the slug *Deroceras carnanae*. *Journal of Zoology* 187, 135–152.

Moya, J., Serrano, M.T. and Angulo, E. (1992) Ultrastructure of the salivary glands of *Arion ater* (Gastropoda, Pulmonata). *Biological Structures and Morphogenesis* 4, 81–87.

Myers, F.L. and Northcote, D.H. (1958) The partial purification and some properties of a cellulase from *Helix pomatia*. *Biochemical Journal* 71, 749–756.

Owen G, (1966) Digestion. In: Wilbur, K.M. and Yonge, C.M. (eds) *Physiology of Mollusca*. Vol. II. Academic Press, New York, pp. 53–96.

Owen, G. (1970) The fine structure of the digestive tubules of the marine bivalve *Cardium edule. Philosophical Transactions of the Royal Society of London, Biological Sciences* 258, 245–260.

Oxford, G.S. (1977) Multiple sources of esterase enzymes in crop juice of *Cepaea* (Mollusca, Helicidae). *Journal of Comparative Physiology* 122, 375–383

Oxford, G.S. and Fish, L.J. (1979) Ultrastructural localization of esterase and acid phosphatase in digestive gland cells of fed and starved *Cepaea nemoralis* (L.). *Protoplasma* 101, 186–196.

Pallant, D. (1970) A note on the faeces of *Agriolimax reticulatus* (Müller). *Journal of Conchology* 27, 111–113.

Parnas, I. (1961) The cellulotic activity in the snail *Levantina hierosolyma* Boiss. *Journal of Cellular Comparative Physiology* 58, 195–201.

Richardson, A.M.M. (1975) Food, feeding rates and assimilation in the snail *Cepaea nemoralis* (L). *Oecologia* 19, 59–70.

Roach, D.K. (1968) Rhythmic muscular activity in alimentary tract of *Arion ater* (L.) (Gastropoda, Pulmonata). *Comparative Biochemistry and Physiology* 24, 865–878.

Roldan Cornejo, C. (1986) Fine structure of the epithelium of the anterior digestive tract in *Theba pisana* (Müller) (Mollusca, Gastropoda, Pulmonata). *Iberus* 6, 269–283.

Roldan Cornejo, C. (1987) Ultrastructural modification of the epithelium in the anterior digestive tract in starved specimens of *Theba pisana* (Mollusca, Gastropada, Pulmonata). *Iberus* 7, 153–164.

Roldan, C. and Garcia-Corrales, B. (1988) Anatomy and histology of the alimentary tract of the snail *Theba pisana* (Gastropoda: Pulmonata). *Malacologia* 28, 119–130.

Rosen, B. (1952) The problem of phagocytosis in *Helix pomatia* L. *Arkiv für Zoologie* 3, 33–50.

Runham, N.W. (1975) Alimentary canal. In: Fretter, V. and Peake, J. (eds) *Pulmonates*, Vol. 1, *Functional Anatomy and Physiology*. Academic Press, London, pp. 53–105.

Staikou, A. and Lazaridou-Dimitriadou, M. (1989) Feeding experiments on and energy flux in a natural population of the edible snail *Helix lucorum* L. (Gastropoda: Pulmonata: Stylommatophora) in Greece. *Malacologia* 31, 217–227.

Stern, G. (1970) Production et bilan énergetique chez la limace rouge. *Terre et la Vie* 117, 403–424.

Strasdine, G.A. and Whittaker, D.R. (1963) On the origin of the cellulase and chitinase of *Helix pomatia. Canadian Journal of Biochemistry and Physiology* 41, 1621–1626.

Sumner, A.T. (1965) The cytology and cytochemistry of the digestive gland cells of *Helix. Quarterly Journal of Microscopical Science* 106, 173–192.

Taylor, G.M., Simkiss, K., Greaves G.N. and Harries, J. (1988) Corrosion of intracellular granules and cell death. *Proceedings of the Royal Society of London* 234, 436–476.

Tillier, S. (1984) Patterns of digestive tract morphology in the limacisation of helicarionid, succineid and athoracophorid snails and slugs (Mollusca: Pulmonata). *Malacologia* 25, 173–192.

Tillier, S. (1989) Comparative morphology and classification of land snails and slugs. *Malacologia* 30, 1–303.

van Weel, P.B. (1950) Contribution to the physiology of the glandula media intestini of the African giant snail, *Achatina fulica* (Fer.) during the first hours of digestion. *Physiol Comparata at Oekologica* 2, 1–19.

van Weel, P.B. (1961) The comparative physiology of digestion in molluscs. *American Zoologist* 1, 245–252.

Varute, A.T. and Patil, V.A. (1971) Histochemical analysis of molluscan stomach and intestinal alkaline phosphatase: a sialoglycoprotein. *Histochemie* 25, 77–90.

Vieira, E.C. (1967) The influence of vitamin E on reproduction of *Biomphalaris glabrata* under axenic conditions. *American Journal of Tropical Medicine and Hygiene* 16, 792–796.

Walker, G. (1969) Studies on digestion of the slug *Agriolimax reticulatus* (Müller) (Mollusca, Pulmonata, Limacidae). PhD Thesis, University of Wales, Cardiff.

Walker, G. (1970a) The cytology, cytochemistry and ultrastructure of the cell types found in the digestive gland of the slug *Agriolimax reticulatus* (Müller). *Protoplasma* 71, 91–109.

Walker, G. (1970b) Light and electron microscopy investigation on the salivary glands of the slug, *Agriolimax reticulatus. Protoplasma* 71, 111–126.

Walker, G. (1972) The digestive system of the slug *Agriolimax reticulatus* (Müller): experiments on phagocytosis and nutrient absorption. *Proceedings of the Malacological Society of London* 40, 33–43.

Watkins, B. and Simkiss, K. (1990) Interactions between soil bacteria and the molluscan alimentary tract. *Journal of Molluscan Studies* 56, 267–274.

Zacks, S.I. (1955) The cytochemistry of the amoebocytes and intestinal epithelium of *Venus mercenaria* (Lamellibranchiata), with remarks on a pigment resembling ceroid. *Quarterly Journal of Microscopical Science* 96, 57–71.

6 Food and Feeding Behaviour

B. SPEISER

Research Institute of Organic Agriculture (FiBL), Ackerstrasse, CH-5070 Frick, Switzerland

Introduction

This chapter is largely dedicated to the relationships between terrestrial gastropods and plants. Ecologists have investigated which plants are eaten by gastropods, why they are eaten, and how gastropods select them as food. Agronomists, on the other hand, have focused on the effects of gastropod feeding on crop plants. In this chapter, I attempt to link the ecologist's viewpoint with that of the agronomist. My fundamental assumption is that the principles of gastropod foraging and feeding are no different, whether wild plants or crops are eaten, and that the differences between wild plants and crops can be accounted for by differences between these two types of food: (i) many crops are highly palatable to gastropods (because they have been bred for low content of secondary compounds), while most wild plants are fairly unpalatable; (ii) wild plants usually occur amidst many other potential gastropod foods, while crops are often cultivated in monocultures, thus, gastropods may be forced to eat crops due to the lack of alternative foods; and (iii) in wild plants, gastropod damage is assessed by the reduction in plant growth or fecundity, while in crops it is expressed as economic loss. Most plants can withstand considerable herbivory before their growth or fecundity is seriously impaired, but very little gastropod feeding may be sufficient to cause economic devaluation of a crop (e.g. 'cosmetic damage' and facilitation of secondary infections).

The chapter concentrates on the best studied gastropod taxa. These are the larger species occurring in temperate habitats of Europe and North America, namely arionids, limacids, agriolimacids and helicids. These are mostly herbivorous or omnivorous. Species from other families may also be herbivorous, but the great majority of terrestrial gastropods are microphagous, feeding on the microorganisms associated with live and decaying vegetation. Others are carnivorous. Reviews including the diets of these taxa can be found in Graham (1955) and Chatfield (1976).

Description of Gastropod Foods

Methods used to determine gastropod food preferences

Direct observation

Direct observations seem to be the most straightforward method to determine gastropod food preferences, but this method has shortcomings. Firstly, it is often difficult to decide whether a gastropod is indeed feeding on a given food item or merely resting on it. Secondly, gastropods feeding in conspicuous places may be recorded disproportionately more often than animals feeding in hidden places such as underneath leaf litter, inside tussocks, etc. Thirdly, observations are most meaningful if they are made during the hours of peak activity (i.e. at night and during rain), but, for reasons of convenience, observations are often made during daytime. Examples of food range determination by observation of feeding gastropods can be found in Richardson (1975), Beyer and Saari (1978), Richter (1979), Chang (1991) and Iglesias and Castillejo (1999).

Time-lapse, infrared video-recordings can monitor feeding activity continuously throughout the night and they can be replayed to determine *a posteriori* what gastropods had been doing before or while they ate a certain food. On the other hand, they are restricted to small surfaces with scarce vegetation. A comprehensive overview of the advantages and disadvantages of video-recording is given by Bailey (1993).

Laboratory experiments with natural foods

Several authors have determined gastropod food choices in the laboratory (e.g. Gain, 1891; Frömming, 1937; Grime *et al.*, 1968; Duval, 1971, 1973; Cates and Orians, 1975; Jennings and Barkham, 1975; van der Laan, 1975; Dirzo, 1980; Whelan, 1982; Wink, 1984; Mølgaard, 1986; Speiser *et al.*, 1992; Baur *et al.*, 1994; Briner and Frank, 1998). Unfortunately, there is no general agreement regarding the methodology to be used in such tests. For example, the animals can be offered a single food, a choice of two foods or many different foods simultaneously. Single presentations show whether a food is palatable, while simultaneous choices indicate food preferences. Single presentations might be representative of agricultural monocultures, but simultaneous choices better simulate the conditions in natural habitats. Entire plants (particularly seedlings), leaves, leaf discs, and plant extracts soaked into filter paper or incorporated in gels, have been used.

Laboratory feeding tests can provide useful data on decision making and regulatory mechanisms in gastropods. For several reasons, however, laboratory tests reveal little about the food range of gastropods in the wild.

Gastropods may show unnatural behaviour under the experimental conditions, and plants may undergo chemical changes very soon after harvesting ('induced defence'). The type of reference food used and the order of presentation may influence the results (Keymer and Ellis, 1978; Richardson and Whittaker, 1982). Most laboratory feeding tests have been made with green leaves, while 'unconventional' food items such as rust-infected, wilted, senescent or decaying plant tissues seldom have been tested. Finally, laboratory tests do not account for properties which affect the likelihood of a food being found, such as phenology, abundance and growth form.

Hard foods such as cereal grains can be fixed to electronic sensors which record individual radular bites ('acoustic pellet' technique; e.g. Wedgewood and Bailey, 1986). Acoustic pellets and video-recordings complement each other nicely (Bailey and Wedgwood, 1991; Bailey, 1993).

Faecal analysis

Faecal analysis is the best method for studying food choices of gastropods under natural conditions. This method has been employed by a number of authors, using very similar techniques (Pallant, 1969, 1972; Grime and Blythe, 1969; Mason, 1970; Wolda *et al.*, 1971; Jennings and Barkham, 1975; Richardson, 1975; van der Laan, 1975; Chatfield, 1976; Williamson and Cameron, 1976; Carter *et al.*, 1979; Speiser and Rowell-Rahier, 1991; Hägele, 1992). Briefly, the method is as follows: gastropods are collected in the field and kept individually until they have defaecated. Then, their faeces are analysed microscopically. Faeces derived from different food types can be distinguished easily because of differences in colour, texture and/or microscopic structure. Using a reference collection, the food plants can be identified by the presence of trichomes or other morphological structures. In some cases, the length of the faecal string of each type of food provides a semi-quantitative measure of the amount of that food eaten.

The strength of faecal analysis is that it provides information on feeding *a posteriori*, without disturbing the animals while they feed. A major drawback of this method is that some plants can only be identified to the genus or family level, while still others may remain undetected, if they lack distinctive anatomic structures. In practice, however, most of the samples can be identified (e.g. Speiser and Rowell-Rahier, 1991). A potential argument against faecal analysis is that foods which have been digested cannot be analysed. It might therefore be argued that faecal analysis is biased towards foods which are ingested, but not digested. Wolda *et al.* (1971) investigated whether such a bias occurs in *Cepaea nemoralis* (Linnaeus) (Helicidae) by comparing the contents of the gut with that of the faeces. Both methods yielded very similar results, thus invalidating the argument of a bias in faecal analysis for this particular case.

Types of food eaten

Almost every kind of organic material has been recorded as a gastropod food: plants of all developmental stages from seedlings to senescing plants, as well as leaf litter, wood and dead animals in different stages of decay with variable shares of microbial and fungal biomass. For simplicity, I have split up this continuous range of food into clear-cut categories below. However, it should be borne in mind that gastropod foods in the wild often may be intermediate between these categories.

Plants

In the diet of the species considered here (see Introduction), plant material usually makes up the largest fraction, followed by fungi, animals and soil. Algae, mosses, lichens and higher plants are sometimes eaten, and roots, shoots, leaves, flowers, anthers, pollen, fruit, seeds and rotting wood have been recorded as food items. Gastropods normally eat only small amounts of grasses (Ingram and Peterson, 1947; Pallant, 1969; Cates and Orians, 1975; van der Laan, 1975; Carter *et al.*, 1979; Barker, 1989; Speiser and Rowell-Rahier, 1991). This may be due to the hard texture of grasses (see below). As an exception, Pallant (1972) found a large proportion of the grass *Holcus lanatus* Linnaeus (Gramineae) in the diet of a grassland population of *Deroceras reticulatum* (Müller) (Agriolimacidae). Among the dicotyledons, there is a continuum from highly palatable to completely unpalatable species or races (see below). In a field study, leaves formed a minor part of the diet of *C. nemoralis*, while stems, fruits, flowers, etc. were eaten frequently (Chang, 1991).

Green plant material often is quantitatively a minor component of the total diet (Grime and Blythe, 1969; Butler, 1976; Bless, 1977; Speiser and Rowell-Rahier, 1991; Hägele, 1992). As an exception, green nettles (*Urtica* Linnaeus: Urticaceae) and mosses often are eaten in large amounts (Wolda *et al.*, 1971; Richardson, 1975). In some gastropod populations and at certain times of the season, green plant material can make up a large proportion of the total food eaten (Hatziioannou *et al.*, 1994; Iglesias and Castillejo, 1999). In many plants, senescent leaves are preferred to both fresh and dead leaves (Richter, 1979). The importance of senescent plant tissues for gastropod nutrition has been described often (Stahl, 1888; Grime and Blythe, 1969; Pomeroy, 1969; Mason, 1970; Wolda *et al.*, 1971; Jennings and Barkham, 1975; Richardson, 1975; Butler, 1976; Chatfield, 1976; Williamson and Cameron, 1976; Bless, 1977; Beyer and Saari, 1978; Carter *et al.*, 1979; Richter, 1979; Wink, 1984; Dussourd and Eisner, 1987; Speiser and Rowell-Rahier, 1991; Hatziioannou *et al.*, 1994). Senescing plant material is probably preferred because of its low toxin content. For example, senescent watercress (*Rorippa nasturtium-aquaticum* (Linnaeus) Hayek) (Brassicaceae) contains only half the amount of nitrogen compared with fresh plant material, but it also contains only 7–10%

of the glucosinolates. Indeed, senescent watercress is favoured over fresh watercress by the physid *Physella* Haldeman sp. (Newman *et al.*, 1992). Air pollution can stimulate senescence in plants (von Sury, 1992); whether this affects gastropod herbivory has not been tested as yet.

Fungi

Many basidiomycetes are eaten avidly by gastropods, including those which are highly toxic to mammals. The utilization of fungi as food varies among gastropod species (Gain, 1891; Frömming, 1937; Kittel, 1956; Pomeroy, 1969; Mason, 1970; Wolda *et al.*, 1971; Jennings and Barkham, 1975; Butler, 1976; Bless, 1977; additional references in South, 1992). Leaves of several plant species infested by rust fungi were preferred to healthy leaves by a number of arionid species (Ramsell and Paul, 1990).

Animals and animal-derived foods

Remains of animals often are detected in gastropod faeces, but in the herbivorous/omnivorous species considered here, they probably constitute only a small fraction of the total food eaten. Earthworm chaetae, mites and remains of insects have been recorded (Pallant, 1969, 1972; Mason, 1970; Wolda *et al.*, 1971; Jennings and Barkham, 1975; Chatfield, 1976; Bless, 1977; Beyer and Saari, 1978; Speiser and Rowell-Rahier, 1991; Hägele, 1992; Hatziioannou *et al.*, 1994; Iglesias and Castillejo, 1999). In most cases, it is not known whether these animals are eaten alive or dead, but Fox and Landis (1973) observed *Deroceras laeve* (Müller) ingest living aphids. Neither do we know whether these animals are devoured accidentally or on purpose, nor how much they contribute to gastropod nutrition. Many gastropods feed on carrion. For example, *Deroceras hilbrandii* Altena was observed to 'steal' animal carcasses which had been trapped on the sticky leaves of the carnivorous plant *Pinguicula vallisneriifolia* Webb (Lentibulariaceae). *D. hilbrandii* itself was not trapped by *P. vallisneriifolia* (Zamora and Gomez, 1996). Living, moribund or dead gastropods may also belong to the diet (e.g. Pomeroy, 1969; Butler, 1976; Chatfield; 1976; Port and Port, 1986). Finally, mammalian faeces are attractive foods for many gastropods, and feeding on faeces can result in the transmission of mesostrongylid nematode parasites from mammals to gastropods.

Soil

Soil, and in particular humic acid, is important for the nutrition of helicid snails in culture (Elmslie, 1998). Soil particles are found regularly in the faeces of gastropods from the wild (Williamson and Cameron, 1976; Godan, 1979, p. 41; Speiser and Rowell-Rahier, 1991; Hägele, 1992), but the role of soil in gastropod nutrition is poorly understood.

Variability in Feeding Behaviour

The food choices of terrestrial gastropods are influenced by the qualitative composition of the food and by its quantitative availability and accessibility, as well as by the nutritional needs of the gastropod. The palatability and availability of almost any given food, as well as the nutritional needs of gastropod species, may be subject to seasonal changes. As a consequence, gastropod diets may vary greatly over the season (Jennings and Barkham, 1975; Chatfield, 1976; Speiser and Rowell-Rahier, 1991; Hatziioannou *et al.*, 1994; Iglesias and Castillejo, 1999).

Many gastropods occur in a range of different habitats, in which completely different food types are available. As a result, the same species may eat a completely different diet at a different place (Pallant, 1969, 1972; Hägele, 1992; Iglesias and Castillejo, 1999). For example, the diet described for *C. nemoralis* at different sites by different authors is very different (Wolda *et al.*, 1971; Richardson, 1975; Williamson and Cameron, 1976; Chang, 1991).

Occasionally, age-specific diet selection has been found in terrestrial gastropods, but it does not seem to be a general phenomenon. A conspicuous example is the age-dependent egg cannibalism observed in many gastropods immediately after hatching (reviewed in Baur, 1992). Newly hatched *Arianta arbustorum* (Linnaeus) (Helicidae) often feed on conspecific eggs. After a few days, this behaviour is lost and the animals thereafter feed mainly on plant material (Baur, 1987). Egg cannibalism speeds up growth and increases fitness (Baur, 1992). Juvenile *Cantareus aspersus* (Müller) (Helicidae) feed more often on fresh plant material than do adults (Iglesias and Castillejo, 1999). Young *C. nemoralis* eat more soil and humus and a larger proportion of grasses than adults, but fewer dicotyledons (Williamson and Cameron, 1976). Individuals of *Achatina achatina* (Linnaeus) (Achatinidae) only begin to eat fruit when they reach the age of 21 months (Hodasi, 1995). On the other hand, different size classes of *D. reticulatum*, *Helminthoglypta arrosa* (Binney) (Helminthoglyptidae), *C. aspersus*, *Bradybaena fruticum* (Müller) (Bradybaenidae), *Helix lucorum* Linnaeus (Helicidae), *Helicella arenosa* (Ziegler) (Hygromiidae), *Monacha cartusiana* (Müller) (Hygromiidae) and *Cepaea vindobonensis* (de Férussac) were found to eat similar diets (Pallant, 1972; van der Laan, 1975; Gallois and Daguzan, 1989; Hatziioannou *et al.*, 1994).

Individual gastropods may exhibit unequal feeding behaviour, if they experienced different feeding histories (Keymer and Ellis, 1978; Wareing 1993; Speiser and Rowell-Rahier 1993). This kind of variability can be reduced by feeding the animals the same diet prior to an experiment (e.g. Burgess and Ennos, 1987; Speiser *et al.*, 1992; Cook *et al.*, 1996). However, Whelan (1982) found that high variation persisted among individuals reared on a uniform diet. In most cases, the reasons for variation between individuals are not known. Genetically determined physiological differences, different learning abilities, different feeding experience, different hunger levels or a different recent feeding history may all cause individual variability in feeding behaviour. Furthermore, apparent variability in feeding behaviour may be caused by heterogeneities in the food which are invisible to the investigator.

Principles of Foraging and Feeding Behaviour

An overview of the behaviour sequencies that control gastropod feeding is given in Fig. 6.1. The diagram incorporates only those regulatory processes known to be of general importance. It includes mechanisms of the pre-ingestive, ingestive and post-ingestive phase. Hunger, which also alters feeding behaviour, is not included in the diagram because its position in the behaviour sequence is poorly understood. In the wild, many gastropod species take several discrete 'meals' each night, which often alternate with locomotor phases (Dobson and Bailey, 1983; Rollo 1988a; Bailey, 1989a,b; Cook *et al.*, 1997; Hommay *et al.*, 1998). A meal is defined as a period of more or less continuous feeding on a single food item. For each meal, the behavioural sequence shown in Fig. 6.1 is passed.

Food finding

Locomotion of terrestrial gastropods is slow, and also costly because a mucus trail is secreted along the entire path covered by the animal (Denny, 1980). This might be the ultimate reason why gastropods have adopted a foraging strategy that minimizes the need for locomotion: generalized feeding. Generalized feeding, in turn, requires sampling behaviour and sophisticated learning mechanisms, which allow the animals to assess a large number of potential food items along their path for palatability (see below).

Most gastropods leave their shelters around sunset, at least partially motivated by the need to find food. Indeed, the likelihood of initiating a meal is highest soon after the beginning of locomotor activity (Bailey and Wedgwood, 1991; Hommay *et al.*, 1998). Some species roost outside of their feeding grounds. As a consequence, they have to move to the food patch before initiating a food search (Bailey, 1989b). For example, 'wildflower strips' are excellent habitats for *Arion lusitanicus* Mabille (Arioniidae) in European agricultural environments, and within 2 m from a wildflower strip, rape (*Brassica napus* Linnaeus) (Brassicaceae) seedlings were much more heavily grazed by *A. lusitanicus* than further inside the field (Frank, 1996).

Smells direct gastropods towards some foods. For example, the stinkhorn *Phallus impudicus* Linnaeus (Basidiomycota) a fungus with a strong odour to the human nose, was detected by *Arion empiricorum* (de Férussac) from a distance of 1.2 m (Kittel, 1956). *Achatina fulica* Bowdich reliably oriented towards the smell of sliced cucumber (*Cucumis sativus* Linnaeus) (Cucurbitaceae) or carrot (*Daucus carota sativus* (Linnaeus) Schuebler &

Martens) (Umbelliferae) (Croll and Chase, 1977, 1980; Chase, 1982). Hexanols are olfactous attractants for *D. reticulatum* (Pickett and Stephenson, 1980), but their role in directing these animals to specific food plants is not clear since hexanols are common to many plants. Gastropods perceive smells with their cephalic tentacles. If signals of different intensity are

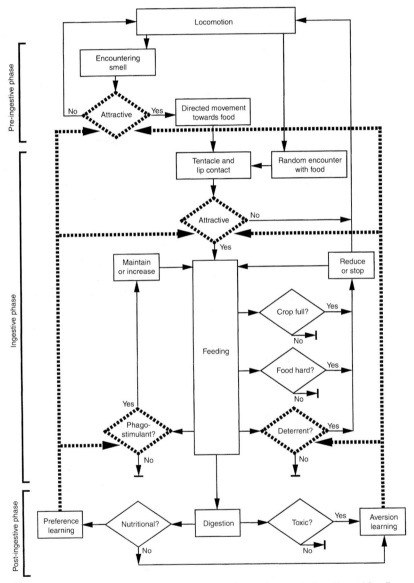

Fig. 6.1. Sequence of events leading from locomotion to the onset of gastropod feeding or its termination. The pre-ingestive phase decides whether or not a food is found. In the ingestive phase, the amount of a particular food item eaten is controlled. The post-ingestive phase influences the feeding behaviour towards a given food on subsequent occasions. Modification of behaviour through learning is shown with dashed arrows and rhombs.

perceived by the two tentacles, the gastropod turns towards the side with the stronger signal (Chase, 1982). However, this orientation takes place only if the difference in signal received by the two tentacles is large, that is, close to a potential food. Comprehensive descriptions of orientation towards odour sources (anemo-, rheo-, klino- and tropotaxis) are given by Croll (1983) and Chase (1986). Orientation is most pronounced towards smells of familiar, favourite foods (i.e. foods which have been found to be palatable in previous meals; see below). 'Negative orientation' away from smells of unattractive food has not been observed (Chase, 1982, 1986). Some foods apparently smell little and are only found if the animal approaches them by chance. Wheat (*Triticum aestivum* Linnaeus) (Gramineae) grains, for example, were detected by agriolimacids from a distance of 2–4 cm only (Bailey and Wedgwood, 1991).

The availability of foods influences the rate with which they are encountered by foraging gastropods. Indeed, several field studies have revealed a broad correlation between food availability and its proportion in the gastropod diet (Speiser and Rowell-Rahier, 1991; Hägele, 1992; Hatziioannou *et al.*, 1994; Iglesias and Castillejo, 1999). In practical terms, availability is difficult to measure. The biomass per unit area or the percentage cover provide crude estimates of availability. However, some of the potential food items present in a habitat may be inaccessible to foraging gastropods. For example, the leaves of several herbs were not grazed apparently because the gastropod species were reluctant to climb up the stems. When the stems were bent down, however, the leaves were heavily grazed (Rathcke, 1985). Some foods are eaten disproportionately more (or less) than predicted from their abundance. These deviations can be explained largely by the presence of nutrients, phagostimulants and deterrents, and by food hardness (see below).

Feeding decisions and the role of hunger

Once a gastropod has found a potential food, it must decide whether or not to initiate feeding. First, the potential food is touched with the inferior tentacles, which possess chemical and mechanoreceptors. Then, the potential food is touched with the lips. At both steps, the food may be rejected. If this is not the case, the gastropod will start to feed (Gelperin, 1975).

Food-deprived ('hungry') individuals exhibit feeding behaviours different from those of well-fed animals. Hungry gastropods reacted more strongly to attractive chemical stimuli than fed ones (Croll and Chase, 1980; Chase, 1982). In a series of experiments with *D. reticulatum*, hungry and fed animals took meals of similar size, but food-deprived animals were more likely to take a second meal than fed animals (Bailey, 1989a). Whether starvation increases selectivity in feeding choices is unclear: hungry individuals of *A. fulica* selected familiar foods more strongly than fed ones (Croll and Chase, 1980). Also, hungry individuals of *D.*

reticulatum were more selective than fed animals with respect to the type of food they accepted (Bailey, 1989a). On the other hand, some food plants were accepted by starved *D. reticulatum*, but rejected by fed individuals (Dirzo and Harper, 1980). Hunger enhanced locomotor activity in *D. reticulatum*, but reduced it in *Tandonia budapestensis* (Hazay) (Airey, 1987). Prolonged periods of food deprivation reduced subsequent food intake, while short periods of starvation had the opposite effect (Rollo, 1988b; Bailey, 1991). Depending on the nutrient which is in short supply, hunger may cause different behaviours. The concept of 'specific hungers' for specific nutrients is discussed below.

Control of intake

Although intermediate behaviours do occur, gastropods often exhibit either of two extreme patterns of feeding behaviour: they may accept a food item and feed until their oesophagus or oesophageal crop is full, or they may reject it (often because of the presence of deterrents). To determine the proportion of all potential foods which are accepted or rejected, I analysed nine laboratory studies (see Table 6.1). In these studies, the feeding tests were performed in dissimilar ways, and different indices were calculated by the respective authors. Therefore, I indicate for each study the threshold value above which I considered a food item as 'readily eaten'. Grime *et al.* (1968) studied the palatability of 52 plant species to the snail *C. nemoralis*. For this analysis, I considered plants as 'readily eaten' if their palatability index was ≥0.5. This was the case for 14 plants. Dirzo (1980) tested the acceptability of 30 plant species to *Deroceras*

Table 6.1. Acceptability of plant species to herbivorous gastropods, estimated from nine laboratory studies

| Gastropod species | Potential food plants | | | Reference |
	No. investigated	No. readily eaten	Percentage readily eaten	
Cepaea nemoralis (Linnaeus) (Helicidae)	52	14	27	Grime *et al.* (1968)
Deroceras caruanae (Pollonera) (Agriolimacidae)	30	6	20	Dirzo (1980)
Arion fasciatus (Nilsson) *Arion subfuscus* (Draparnaud) (Arionidae), *Deroceras reticulatum* (Müller) (Agriolimacidae)	60–61	14–21	27	Rathcke (1985)
Cepaea nemoralis (Linnaeus) (Helicidae)	20	2	10	Chang (1991)
Ariolimax columbianus (Gould), *Arion ater* (Linnaeus) (Arionidae)	100	28–32	30	Cates and Orians (1975)
Helix pomatia Linnaeus (Helicidae), *Arion ater* (Linnaeus) (Ariondidae)	18–28	4–13	34	Mølgaard (1986)
Deroceras reticulatum (Müller) (Agriolimacidae)	26	10	38	Duval (1971)
Arion hortensis de Férussac (Arionidae)	27	4	15	Duval (1973)
Arion lusitanicus Mabille (Arionidae)	78	17	22	Briner and Frank (1998)
20 gastropod species	74–138	0–59	14	Gain (1891)

The number of food items investigated and the number and percentage of readily eaten food items are given. When several gastropod species were studied, the range of foods investigated and readily eaten is given, but the percentages are averaged for all species. The methods for determining which food items are readily eaten are explained in the text. On average, 24% of all food items were readily eaten.

caruanae (Pollonera). Six of these plants had an acceptability index ≥0.5. Rathcke (1985) determined the acceptability of 60–61 plants to *Arion subfuscus* (Draparnaud), *Arion fasciatus* (Nilsson) and *D. reticulatum*. Fourteen, 15 and 21 plants had a relative acceptability ≥0.5. In a study with *C. nemoralis* by Chang (1991), two out of 20 plant species had an acceptability index ≥0.5. Cates and Orians (1975) investigated the palatability of 100 plants from different communities to *Arion ater* (Linnaeus) and *Ariolimax columbianus* (Gould) (Arionidae). Here, I consider all plants with a palatability index ≥1 as readily eaten. This was the case for 28 and 32 plant species, respectively. In his feeding tests, Mølgaard (1986) distinguished the preference groups I, II and III. Here, I consider all plants in preference group I as readily eaten. *Helix pomatia* Linnaeus readily ate four out of 18 plants, and *A. ater* 13 out of 28. Duval (1971, 1973) distinguished four categories of plant acceptability to *D. reticulatum* and *Arion hortensis* de Férussac. For this analysis, I followed his definition of 'readily eaten' plants, and pooled the other three categories of lower acceptability. *D. reticulatum* readily ate ten out of 26 weeds tested, while *A. hortensis* readily ate four out of 27. Brimer and Frank (1998) investigated the palatability of 78 wildflowers to *A. lusitanicus*. Seventeen plant species had a consumption index >5 (criterion used by the authors to distinguish highly palatable species). Gain (1891) investigated the acceptability of 197 food items for 20 gastropod species. Here, only the first 138 food items are considered, with which a gastropod is likely to be confronted in the wild (not all were tested with each gastropod species). I consider all food items with a score of four ('special favourite') or three ('taken freely') as readily eaten. Depending on the gastropod species, 0 out of 131 up to 59 out of 138 food items were readily eaten. Summing up, an average of 24% of all food items were readily eaten by gastropods in the laboratory, while the remainder were eaten reluctantly or not at all.

Deterrents

Deterrency describes the animal's behavioural (pre-ingestive or ingestive) response to food, while toxicity refers to the food's physiological effects after ingestion. Deterrents interact with chemoreceptors, thereby reducing or inhibiting feeding. Foods often are rejected after contact with the inferior tentacles or the lips, but deterrency can also act during feeding. Many deterrents acquire their inhibitory effects through the process of aversion learning (see below), ultimately due to their toxicity. In most cases, deterrency prevents gastropods from ingesting a lethal dose while feeding on plants containing molluscicidal compounds; the potential molluscicidal effects can only be observed if their deterrency is by-passed artificially (e.g. Henderson, 1969).

The best studied deterrents for terrestrial gastropods are cyanogenic glucosides. Gastropods clearly prefer plants which do not contain cyanogenic glucosides over cyanogenic varieties or species (e.g. Jones, 1962; Angseesing, 1974; Keymer and Ellis, 1978; Dirzo and Harper,

1982a,b; Compton *et al.*, 1983; Horrill and Richards, 1986; Burgess and Ennos, 1987; Raffaelli and Mordue, 1990). In the laboratory, animals grew poorly on a pure diet of cyanogenic clover, relative to those maintained on a diet of acyanogenic clover (Dirzo and Harper, 1982a). In a field experiment, cyanogenic white clover (*Trifolium repens* Linnaeus) (Fabaceae) was better protected from gastropods, but not from weevils, sheep or fungal attack, than acyanogenic clover (Dirzo and Harper, 1982b). Monoterpenoids are also well-known deterrents for gastropods. The mint *Clinopodium douglasii* (Bentham) Kuntze (Lamiaceae) contains an array of different monoterpenoids, and chemotypes containing predominantly *p*-menthane monoterpenoids were better protected against grazing by *Ariolimax dolichophallus* Mead than chemotypes containing mainly bicyclic monoterpenoids (Rice *et al.*, 1978). Similarly, individual seedlings of *Thymus vulgaris* Linnaeus (Labiatae) were grazed differently by gastropods, depending on the type of monoterpenoids which they contained (Gouyon *et al.*, 1983; Linhart and Thompson, 1995). A number of monoterpenoids were highly repellent for *D. reticulatum* and therefore show some promise as anti-feedant seed dressings (e.g. Powell and Bowen, 1996). Sesquiterpenes were deterrent for *A. arbustorum, Milax* Gray sp. (Millacidae) and *A. hortensis* (Baig *et al.*, 1989; Speiser *et al.*, 1992; Hägele, 1996; Hägele *et al.*, 1998). Glucosinolates apparently protect crucifers against gastropod grazing (Glen *et al.*, 1990; Giamoustaris and Mythen, 1995; Byrne and Jones, 1996). Many alkaloids were deterrent for gastropods (Wink, 1984), but pyrrolizidine alkaloids were hardly deterrent for *A. arbustorum* (Speiser *et al.*, 1992). Latex was deterrent for *D. reticulatum* (Dussourd and Eisner, 1987). Many crops have been bred for their low content of secondary compounds (e.g. Glen *et al.*, 1990). In such cases, crop plants may be attacked more heavily than their wild relatives.

Phagostimulants

Substances which are nutritionally valuable have been found to stimulate feeding in gastropods. Examples include amino acids, sugars, starch and casein (Senseman, 1977; Dobson and Bailey, 1983). Various alcohols are also phagostimulant (Pickett and Stephenson, 1980). Dobson and Bailey (1983) allowed *D. reticulatum* to feed on agar containing 1% starch until they finished their meal. The authors then presented agar containing 15% starch to these animals, which resumed feeding on the more attractive agar. In *Limax maximus* Linnaeus (Limacidae), phagostimulant concentration has been shown to influence the likelihood of initiating a meal and the duration of the meal, but not the bite frequency (Reingold and Gelperin, 1980).

Food texture

Comparisons of natural foods indicate that soft leaves are eaten more readily than hard leaves (Grime *et al.*, 1968; Dirzo, 1980). Silica crystals have been shown to reduce feeding in *D. reticulatum* (Wadham and

Wynn-Parry, 1981). In contrast, trichomes, which are effective defences against many insect herbivores (reviewed in Levin, 1973), do not seem to prevent gastropod grazing (Grime *et al.*, 1968; Dirzo, 1980; Cook *et al.*, 1996). As an exception, the hairy forms of *Silene dioica* (Linnaeus) Clairville (Caryophyllaceae) were eaten less by gastropods than the glabrous forms (Westerbergh and Nyberg, 1995). Cook *et al.* (1997) observed different feeding patterns for soft and hard foods in *D. reticulatum*: wheat seeds (which are relatively hard) were rasped with the radula, while particles were bitten off from the softer dandelion (*Taraxacum officinale* Weber) (Asteraceae) leaves. In none of these studies was the texture of different foods quantified. The effects of food hardness were demonstrated unambiguously by Reingold and Gelperin (1980), using artificial foods. When *L. maximus* was presented with agar cubes made up of different amounts of agar, it ate more of the softer cubes. These authors also studied isolated lip-brain preparations of *L. maximus*. When the buccal mass was loaded with weight (simulating harder food), biting frequency was reduced.

Oesophageal crop fill

Dobson and Bailey (1983) found that the oesophageal crop of *D. reticulatum* is normally empty at the beginning of a meal and full at the end. This indicates a possible relationship between crop fullness and the termination of meals. Experimental evidence corroborates this finding: when crops in isolated lip-brain preparations of *L. maximus* were inflated, feeding activity declined (Reingold and Gelperin, 1980). Crop fullness is probably the factor which normally terminates meals on palatable foods.

Learning

After digestion, foods may have positive or negative nutritional effects or may be toxic, but this need not be evident to the animal during ingestion. To account for such effects in subsequent meals, gastropods acquire preferences or aversions for particular foods. When gastropods encounter a novel food, they normally eat a small quantity of it. If the food is nutritionally poor or toxic, they reject it in subsequent encounters (e.g. Whelan, 1982; Gouyon *et al.*, 1983; Linhart and Thompson, 1995). However, if the food turns out to be nutritional and non-toxic, gastropods preferentially orientate towards this food and eat larger quantities in subsequent meals (e.g. Delaney and Gelperin, 1986; Desbuquois and Daguzan, 1995; Teyke, 1995). In the field, gastropods experience most of the potential foods early in life, and their feeding behaviour thereafter is guided mainly by learned preferences and aversions. However, preferences and aversions are re-examined constantly and can be modified, if they do not reflect the post-ingestive consequences of a food adequately. This is necessary because many potential food plants undergo changes in their nutritional value or toxicity during the season. If a preferred food deteriorates, this will become evident in the normal feeding process. In contrast, sampling is necessary to find out whether a non-preferred food has become palatable. Sampling might be one reason why a large proportion of non-preferred foods is included in the gastropod diet (Table 6.2). Learning can modify the response towards chemical cues at all steps of the feeding process: food finding (smell), food 'probing' (initial contact with inferior tentacles and lips) and tasting during ingestion (dashed lines in Fig. 6.1). The same training procedure may cause an aversion to the smell of a food, its taste or both (Sahley *et al.*, 1981a).

In the laboratory, aversions for specific foods can be caused by subjecting gastropods to a noxious stimulus while they are feeding on that food, or shortly afterwards. Preferences can be induced if a food or smell is paired with a reward (e.g. Sahley *et al.*, 1990). Sahley *et al.* (1981a, 1990) found that one training run often was sufficient in *L. maximus* to acquire an aversion, while several runs were needed before a preference was established. In *H. pomatia*, a single meal induced a preference for the smell of that food (Teyke, 1995). Most learned aversions lasted for a few days only, but aversions for deficient diets lasted several months (see below). Preferences have been shown to last as long as 120 days (Croll and Chase, 1977). *L. maximus* also exhibits complex forms of learning, which are more adequate for natural situations, as illustrated by a series of experiments by Sahley *et al.* (1981b). In the first experiment, animals were exposed to carrot tissues and quinidine sulphate. This resulted in an aversion to carrots (aversion learning). Thereafter, carrots were paired with potatoes (*Solanum tuberosum* Linnaeus) (Solanaceae) (without quinidine sulphate). As a result, the animals developed aversions to potatoes also, although these had never been paired with the noxious stimulus ('second order conditioning'). In another experiment, *L. maximus* were conditioned in the same way for an aversion to carrots. Thereafter, carrots,

Table 6.2. Major constituents of the diet of gastropods, estimated from six field studies

Gastropod species	No. of plants available as potential food	Major diet constituents		Reference
		No. of plants	Percentage of plants	
Cepaea nemoralis (Linnaeus) (Helicidae)	24	5	21	Richardson (1975)
Arianta arbustorum (Linnaeus) (Helicidae)	14–16	1–3	12	Speiser and Rowell-Rahier (1991)
Arianta arbustorum (Linnaeus) (Helicidae)	14–22	1–4	13	Hägele (1992)
Helix lucorum Linnaeus, *Cepaea vindobonensis* (de Férussac) (Helicidae), *Bradybaena fruticum* (Müller) (Bradybaenidae), *Monacha cartusiana*, *Helicella arenosa* (Ziegler) (Hygromiidae)	23	1–7	12	Hatziioannou *et al.* (1994)
Cantareus aspersus (Müller) (Helicidae)	5–25	1–4	16	Iglesias and Castillejo (1999)
Monacha cantiana (Montagu), *Trichia striolata* (Pfeiffer) (Hygromiidae)	7	2–4	43	Chatfield (1976)

The number of food items available to the animals ('choice') and the number and percentage of food items eaten as major diet constituents are given. When several gastropod species or populations were studied, or when the investigations were repeated on several dates, the range is given for the choice of foods and for the major diet constituents, but the percentages are averaged. The methods for determining which food items are major diet constituents are explained in the text. On average, 19% of all food items are major diet constituents.

potatoes and quinidine sulphate were presented in a triplet. As a result, the aversion to carrots remained, but no aversion to potatoes was formed ('blocking'). In the third experiment, animals were first subjected to quinidine sulphate alone. Then, quinidine sulphate was paired with carrots. In this case, *L. maximus* did not form an aversion to carrots ('unconditioned stimulus pre-exposure effect'). Hopfield and Gelperin (1989) demonstrated that *L. maximus* could even be trained to develop aversions to a combination of two smells, while retaining their initial, positive reaction towards each smell alone.

Delaney and Gelperin (1986) presented *L. maximus* with a diet which contained all essential amino acids except methionine. The first meal taken from this diet was normal, indicating that the animals could not detect the absence of methionine by taste. No symptoms of malnutrition were observed in the animals but, in subsequent feeding trials, the 'amino acid-deficient diet' was completely rejected. However, if methionine was injected into the haemolymph of *L. maximus* shortly after eating the methionine-deficient diet, no aversion was formed. In contrast, a diet deficient in alanine instead of methionine did not lead to the formation of any aversion.

Plants and Gastropod Fitness

What foods do gastropods need?

Feeding serves the purpose of supplying energy and nutrients to an animal. Herbivorous species of gastropods face the same nutritional problems as most other herbivores: on average, their bodies consist of 10% nitrogen, while their food plants only contain 4%. Therefore, it is more difficult for most herbivores to meet the demand for nitrogen than the demand for energy (Crawley, 1983). However, the gastropods considered here are not strict herbivores and they may alleviate the problem of nitrogen uptake by including animal-derived foods in their diet (see above). When given a choice between differently fertilized lettuce (*Lactuca sativa* Linnaeus (Asteraceae)), *D. reticulatum* preferred the leaves with the highest nitrogen contents (Pakarinen *et al.*, 1990).

Other nutrients are also important for gastropods. Delaney and Gelperin (1986) list carbohydrates, cellulose, some amino acids, fatty acids, salts, minerals and vitamins as important components of gastropod diets. According to Howes and Whellock (1937, cited in Carter *et al.*, 1979), the vitamins A, B and D or other sterols were essential for *H. pomatia*. Calcium apparently is also vital for gastropods (e.g. Pomeroy, 1969; Ireland, 1991; Baur, 1992).

Some foods are better suited for gastropod nutrition than others. For example, rock-dwelling gastropods exhibited different growth rates when fed different lichens. When given a choice, each gastropod species selected the species of lichen on which it grew best (Baur *et al.*, 1994).

Optimizing diet composition

The ideal gastropod diet should contain all essential nutrients in pre-determined proportions, and the concentration of secondary compounds should not exceed certain threshold values. It is unlikely that any single food item will meet these requirements. However, a mixed diet of two or more components may come close to the optimal composition. A mixed diet may ensure an intake of nutrients in relatively constant proportions, while the chemical composition of the individual diet constituents varies over the season. Also, in a mixed diet, the secondary compounds of each component will be 'diluted'. One could expect, therefore, that gastropods eat food items which complement the previously eaten foods in such a way that the resulting diet is nutritionally optimal as a whole. There is some evidence that diets composed of several components are superior to diets consisting of one component only: when kept on a mixed diet, *Sarasinula plebeia* (Fischer) (Vaginulidae) grows faster and has a lower mortality than on a pure diet (Herrera, cited in Rueda, 1989). According to Stephenson (1962), mixed rather than pure diets should be used for breeding Agriolimacidae, Milacidae and Arionidae. *A. columbianus* (Gould) apparently grows best on a mixed diet (Richter, 1976, cited in Richter, 1979). In a field study, 89% of all individuals of *A. arbustorum* had eaten more than one type of food (Speiser and Rowell-Rahier, 1993).

Quantitatively, some foods are more important in the diet than others. Information on the quantitative composition of gastropod diets in the wild is summarized in Table 6.2. Richardson (1975) studied the diet of *C. nemoralis* in sand dunes, distinguishing between 24 different food items. If all food items were eaten in equal quantities, each would make up 4.15% of the diet. In the present context, I considered every food item to be a 'major' constituent of the diet if it was eaten at least twice as much as the percentage determined in this way (in this case: ≥8.3%). In Richardson's analysis of oesophageal crop contents and faeces, five out of the 24 food items were major constituents of *Cepaea's* diet. Speiser and Rowell- Rahier (1991) studied the diet of *A. arbustorum* five times over a season. Depending on the sampling date, 14–16 food items were available to the snails. Of these, only 1–3 were major components of *A. arbustorum* diet. Hägele (1992) studied the diet of six populations of *A. arbustorum* at two dates. These animals had a choice of 14–22 food items, 1–4 of which were major diet constituents. Hatziioannou *et al.* (1994) studied the diets of *B. fruticum* (Müller), *H. lucorum*, *H. arenosa*, *M. cartusiana* and *C. vindobonensis* in Greece. From a choice of 23 food items, 1–7 items were major diet constituents, depending on the season. Iglesias and Castillejo (1999) observed three populations of *C. aspersus* during feeding. Four out of 25, three out of 24, and one out of five food items were major components of the diet, respectively. Chatfield (1976) studied the diet of *Monacha cantiana* (Montagu) at various times of the season. In the present context, I consider all foods as major diet constituents which were found in the guts or faeces of half the sampled animals. From the seven food items distinguished, 2–4 were major diet constituents. In summary, 19% of all food items were major diet constituents. In addition to these major diet constituents, a large number of other food items were also eaten, but only in small quantities.

The mechanisms with which gastropods select the individual components of their diet are unknown. The above-mentioned experiments of Delaney and Gelperin (1986) show that *L. maximus* can detect nutrient deficiencies in their diet. This seems a prerequisite for selecting foods which nutritionally complement the previously eaten foods. Selection against recently eaten foods favours the ingestion of sequentially mixed diets. Wareing (1993) kept *D. reticulatum* for several days on a pure diet of either bran, carrot or potato. After this training period, these animals were given a choice between all three food types. In most cases, they ate more of the novel foods than of the food which they had been offered during the training phase. Speiser and Rowell-Rahier (1993) found that *A. arbustorum* ate more of a sequentially mixed diet of two plant species than of a pure diet of one plant species alone, if the two plants belonged to different, chemically distinct families. However, if two plants with a similar secondary chemistry were offered, the mixed diet was eaten no more than the pure diet. On the other hand, *D. reticulatum* has also been observed to eat preferentially foods which it had eaten recently (Cook *et al.*, 1997).

Food limitation in gastropod populations

Herbivores are said to be 'food limited' if their population growth is limited by the amount of available food. Conversely, they are said not to be food limited if population growth is restricted by either abiotic factors or natural enemies. There has been a debate whether herbivores in general are limited by food or by natural enemies. However, it is now clear that both regulatory mechanisms can act to varying degrees on herbivore populations (Crawley, 1983). Thus, instead of asking whether or not gastropods are food limited, it is more appropriate to investigate the conditions under which these animals are food limited, and which foods are limiting. Methodically, however, this is very difficult, and reliable data are scarce.

Some evidence pointing towards food limitation has been obtained by comparing gastropod populations from locations where food is 'abundant' with populations from sites where food is 'scarce' (e.g. Pomeroy, 1969; Butler, 1976; Carter *et al.*, 1979). Such observations are no proof of food limitation or of its absence, since vegetation which provides good food simultaneously may provide shelter and/or a favourable microclimate. The most obvious case of food limitation in terrestrial gastropods was observed in rock-dwelling species on the Baltic island of Öland (Baur and Baur, 1990). *Chondrina clienta* (Westerlund) (Chondrinidae) and *Balea perversa* (Linnaeus) (Clusiliidae) aestivate in fissures. The animals cannot travel far from the fissures and, indeed, the limestone pavements were conspicuously bare of lichens in the vicinity of the fissures. However, as the authors point out, the situation is atypical for terrestrial gastropods first because even low-quality food is extremely scarce in the limestone habitat, and secondly because of the specialized feeding habits in these two species. Other authors found no evidence of food limitation (e.g. Wolda *et al.*, 1971). *B. fruticum, H. lucorum, H. arenosa, M. cartusiana* and *C. vindobonensis* co-occur in many habitats in Greece yet seem to be able to share the available resources. Their favourite food, senescent plant material, could not be demonstrated to be in short supply (Hatziioannou *et al.*, 1994).

Several authors have tackled the problem of food limitation by experimentally manipulating gastropod population density or food availability (Cameron and Carter, 1979; Smallridge and Kirby, 1988; Baur and Baur, 1990). However, the growth depression observed at higher populations does not necessarily indicate food limitation, but may also have been caused by contamination of the food with mucus.

High-quality foods are usually rarer than low-quality foods. Probably, foraging gastropods follow a double strategy. Primarily, they search for high-quality food, but if they fail to find good food, they utilize a poorer food source which is more easy to find. In such a situation of 'relative shortage' of food, the total amount of food is not limiting, but high-quality food is rare and limits population growth. Butler (1976) described this situation for *Cernuella virgata* (da Costa) (Hygromiidae) as follows: 'Each snail has more than it can eat, but the food is not rich. Should it find richer food, (. . .) the snail would grow faster and (. . .) produce more eggs'.

Food supply is very different in natural and agricultural habitats. In agricultural habitats, soil cultivation removes most of the available food, thus starving the gastropod population. When the new crop emerges, it constitutes a superabundant source of potential food with very little diversity in developmental stage and chemical composition. Under these conditions, gastropods would even feed on crops which are not particularly palatable, but the seedlings of many crops are highly palatable to many herbivorous gastropod species. As the crops grow, seedlings often loose their palatability. However, agricultural fields always contain some foods other than the crop plant (crop debris, weeds, dead organic matter, etc.). How much and when such alternative food sources are utilized by gastropod pests is hardly known. It is possible that weeds distract gastropods from feeding on crops. For example, the presence of volunteer plants of barley (*Hordeum vulgare* Linnaeus) (Gramineae) reduced grazing by *D. reticulatum* on rape (Glen *et al.*, 1990). In laboratory experiments, *D. reticulatum* ate less wheat when they were presented with certain palatable weeds as alternative food (Cook *et al.*, 1996). However, if unpalatable weeds were presented to them as an alternative food, the consumption of wheat was unaltered. Indeed, intercropping wheat and dandelion (a palatable alternative food) reduced feeding by *D. reticulatum* on the wheat seeds, but intercropping wheat and chickweed (*Stellaria media* (Linnaeus) Cirillo) (Caryophylloceae) (an unpalatable food) did not (Cook *et al.*, 1997). Over longer time periods, however, weeds might also have an opposite effect. For example, weeds could help gastropods to maintain a nutritionally balanced diet, and non-crop foods might help gastropods to survive until the next crop is sown or planted. Whether alternative foods alleviate or aggravate the feeding pressure on crops depends on the type of the alternative food as well as on the species and the developmental stage of the crop plant. Compared with natural habitats, food availability changes much more quickly and more drastically in cultivated land. When crops are rotated, the pattern of changes on any given field is also very different from year to year. Gastropod populations often are not in synchrony with crop rotation and thus not in equilibrium with the available food resources.

Gastropods and Plant Performance

The performance of wild plants normally is evaluated on the basis of Darwinian fitness. Gastropod grazing can lower plant fitness either through reduced survival or through slower growth. For example, when *D. reticulatum* fed on *Capsella bursa-pastoris* (Linnaeus) Medikus (Brassicaceae), they reduced the size of the rosettes without killing the plants. In the same experiment, the feeding pattern on young *Poa annua* Linnaeus (Gramineae) plants was very different: a few plants were killed while the others were left undamaged (Dirzo and Harper, 1980). Of course, a reduction of plant size may finally also lead to death, especially under conditions of competition or when strong abiotic stress factors prevail (see 'Effects of gastropod feeding on plant communities' below).

The performance of crop plants, on the other hand, usually is evaluated on the basis of economic yield. Crop yield may be reduced by lower survival or slower plant growth in the presence of gastropod feeding. In addition, however, the economic value of the harvested crop may also be lowered by 'cosmetic damage' (e.g. gastropod feeding marks on Brussels sprouts, *Brassica oleracea* Linnaeus), by the facilitation of secondary infections during transport or storage (e.g. mould infections of oranges (*Citrus sinensis* (Linnaeus) Osbeck (Rutaceae), rotting of potatoes) or by the mere presence of gastropods (e.g. *D. reticulatum* in lettuce heads; see Table 6.3).

Direct effects of gastropod feeding on plants

Available information strongly suggests that the bulk of the plant material eaten by herbivorous gastropods is unhealthy, decaying or dead tissue (e.g. rust-infected tissue, senescing leaves, wilted flowers or decaying wood), the removal of which hardly affects plant fitness. Only by eating healthy plant tissue can gastropods potentially exert a direct effect on

Table 6.3.   Mechanisms by which gastropods affect wild plants and crops.

Type of gastropod feeding	Proportion of plant tissue removed	Fitness reduction in wild plants	Economic damage in crops
Seeds or plants killed	substantial	yes	yes
Size reduction, slower growth	substantial	yes	yes
Cosmetic damage	minimal	no	yes
Secondary infections	minimal	no	yes
Presence of gastropods, mucus trails or faeces	none	no	yes

Wild plants are only affected if substantial amounts of biomass are removed. In contrast, crops can also be affected by quantitatively minimal feeding, or by the mere presence of gastropods in harvested produce. Although this last point does not involve feeding, it was included for completeness.

plant fitness. In the following, I discuss the effects of gastropod feeding which palatable plants may experience; less palatable plants suffer less severe consequences.

At any given time in the life cycle of a plant, some organs contribute more to plant fitness than others. The damage done to plants depends not so much on the amount of biomass consumed as on the organs attacked and on the age of the plant (Dirzo and Harper, 1980; Edwards and Gillman, 1987; Hanley *et al.*, 1995a,b). Feeding on seeds is important in many agricultural crops sown into gastropod infested land. In germinating wheat seeds, the embryo is eaten prior to the starch reserves (references in Runham and Hunter, 1970). To what extent seeds of wild plants are attacked by gastropods is unknown.

Consumption of seedlings is probably the most severe way in which gastropods affect plants. For seedlings, gastropod feeding often is lethal. Generally, the younger the seedlings are, the more they are affected by pest feeding (Edwards and Gillman, 1987; Fenner, 1987; Barker, 1989; Hulme, 1994; Hanley *et al.*, 1995b). Several factors may account for this: first, a meal of a given size (i.e. one crop-full) constitutes a larger proportion of the total biomass in a young seedling than in an older seedling. Secondly, it seems that very young seedlings sometimes lack the chemical defences which are present in older plants. For example, at the age of 11 days, seedlings of white clover were palatable irrespective of the plant's cyanogenic potential. At a later age, the seedlings of some cultivars developed cyanogenesis which protected them from gastropod herbivory (Horrill and Richards, 1986). Thirdly, the leaves and meristematic tissue of small seedlings are closer to the ground than those of larger seedlings. Here, they are easily accessible to foraging gastropods. Finally, competition may also be most intense in young seedlings (see below).

Grazing of leaves is common and readily observed, and thus well documented. Slower growth and smaller plant size are typical consequences of leaf grazing (Dirzo and Harper, 1980; Rai and Tripathi, 1985; Cottam, 1986; Kelly, 1989; Rees and Brown, 1992), but only in severe cases does it lead to plant death (Dirzo and Harper, 1982b). In this type of feeding, the reduction in plant fitness sometimes correlates with the amount of biomass eaten by the gastropod. Many plants, however, can compensate for defoliation and some forage plants are even more productive under conditions of periodic defoliation.

Feeding on stems often results in the 'felling' of an entire plant, plant parts or leaves. The parts distal to the site of attack are bound to die, and often the entire plant dies as a consequence of this type of feeding (Dirzo and Harper, 1980; Hanley *et al.*, 1995b). In many cases, the felled plant parts are not eaten. Thus, the consumption of very little biomass can cause severe fitness losses to the plant.

If meristematic tissues are eaten, this may be lethal to the plant or at least cause deformations of its growth form. For example, maize (*Zea mays* Linnaeus) (Gramineae) plants usually die if the meristematic tissues are eaten, while surviving even severe grazing of the leaves (Hommay, 1995). Because of their softness and high nutrient content, growing tips are attractive foods for gastropods. Nevertheless, gastropod feeding on stem tips is not very common because they are often relatively inaccessible (either hidden underground or inside rosettes, or at the top of stems which are difficult to climb and extend into a 'gastropod-unfriendly' microhabitat).

Effects of gastropod feeding on plant communities

At the community level, gastropod feeding has complex effects. On one hand, even the loss of relatively little tissue may put a plant at a competitive disadvantage, thus amplifying the effects of gastropod herbivory. On the other hand, neighbouring plants profit from reduced competition and grow larger. The net result is a shift in community structure towards greater abundance of unpalatable species or varieties (Rai and Tripathi, 1985; Cottam, 1986; Edwards and Gillman, 1987; Oliveira Silva, 1992). Finally, gastropods contribute to the cycling of dead organic material (e.g. Lindquist, 1941; Jennings and Barkham, 1975, 1979; Richter, 1979; Theenhaus and Scheu, 1996).

Monospecific plant populations often experience 'self-thinning', that is some plants die from intraspecific competition in the absence of herbivores. In some cases, the numbers of plants removed from the community by gastropods is similar to or lower than those which are bound to die from self-thinning. In quantitative terms, the effect of gastropod herbivory is therefore negligible (however, herbivory by gastropods may selectively remove certain size classes or unhealthy plants). For example, in the above-mentioned experiment by Dirzo and Harper (1980), feeding by *D. reticulatum* in pure stands of *P. annua* eliminated some of the individual plants. In the ungrazed plots, however, similar numbers of *P. annua* plants died as a result of competition. In pure, ungrazed stands of *C. bursa-pastoris*, competition was intense and led to high plant mortality. In the grazed plots, the size of some plants was reduced by *D. reticulatum*, while the ungrazed plants grew larger. As a result, more *Capsella* plants survived in the grazed plots than in the ungrazed plots. In this last case, the presence of *D. reticulatum* thus improved the survival of the *Capsella* plants! In other cases, gastropods remove more plants from the community than the number which would die from competition. This is especially the case in agricultural crops which are sown at the desired population density. For example, if sugarbeet (*Beta vulgaris* Linnaeus) (Chenopodiaceae) seeds are densely sown and later the crop population reduced by hand-thinning, losses due to pestiferous gastropods can be compensated by thinning less. However, if the seeds are sown at densities equivalent to final plant density, losses due to gastropods cannot be compensated for.

Several descriptive studies have shown that gastropod feeding can affect plant communities. One of the main effects of gastropod feeding on the vegetation is the removal of seedlings (see above). Unfortunately, the disappearance of seedlings often remains unnoticed. Even for seedlings which are recorded to have disappeared between two censuses, it is difficult to establish the cause of their death (i.e. gastropod herbivory versus other causes; Edwards and Gillman, 1987; Kelly, 1989; Rees and Brown, 1992). Edwards and Gillman (1987) estimated that half of the seedlings which disappeared in their study plots were consumed by gastropods. Darwin (1859, cited in Crawley, 1983) reported a death rate of 80% among seedlings and attributed these deaths to 'slugs and insects'. Jennings and Barkham (1975) noted the consumption of various tree seedlings by gastropods and concluded that these animals have an important influence on forest regeneration. From a total of 167 sycamore seedlings (*Acer pseudoplatanus* Linnaeus) (Aceraceae) monitored by Paterson *et al.* (1996), 150 were killed by gastropods within 4 weeks, 17 were eaten by small mammals and none survived. Dirzo and Harper (1982) report that gastropods were the most important consumers of white clover in their study fields. In an analysis of the spatial occurrence of different clover morphs, the less palatable cyanogenic morphs occurred more frequently in areas with high gastropod abundance, and the more palatable acyanogenic morphs were more frequent in areas with low gastropod abundance (Dirzo and Harper, 1982b).

Experimental evidence

The impact of gastropods on plant communities only becomes evident when they are excluded from a habitat. Molluscicides have been used to poison gastropods selectively. Kelly (1989) studied the impact of gastropods on mixed-species seedling populations. When these animals were excluded, the survival of *Linum* Linnaeus (Linaceae) seedlings increased 3-fold, and the number of flowers per plant doubled, because the plants grew larger. By contrast, the molluscicide treatment had no effect on the performance of *Euphrasia* Linnaeus (Scrophulariaceae) seedlings. Barker (1989) reports that molluscicide application to gastropod-infested pastures increased white clover yield by 12–40%. Raffaelli and Mordue

(1990) studied the effects of herbivory by gastropods and insects on cyanogenic and acyanogenic white clover. Gastropods fed more on the acyanogenic varieties, whereas insects caused more damage to cyanogenic plants (at least in spring). In the plots accessible to all herbivores, the net effect of herbivory was greater damage to the acyanogenic plants, indicating that feeding by gastropods was more severe for the plants than that by insects. Rees and Brown (1992) selectively excluded either gastropods or insects from artificial communities of four crucifer species. Excluding gastropods resulted in a 37% increase in plant size, while excluding insects had no significant effect. Hulme (1994) assessed the impacts of herbivory by gastropods, insects and rodents on seedlings of 17 plant species in grassland. Both gastropods and rodents accounted for significant seedling mortality by exploiting about 30% of all seedlings. In contrast, the impact of insects was small. Hanley *et al.* (1995a, 1996) investigated the impact of gastropods on artificial and natural grassland seedling communities. Gastropod grazing altered the composition of the seedling community strikingly: the palatable plants were reduced up to 50-fold, while the unpalatable species dominated in communities developing under the herbivory regime.

In conclusion, gastropods affect the vegetation most strongly by interfering with the growth and survival of seedlings. This is difficult to observe and even more difficult to quantify. Exclusion experiments demonstrate that gastropods can have profound effects on plant communities. These effects can be comparable with, or even larger than, those of insect or rodent herbivores. These findings for wild plants are confirmed for arable crops by farmers' experience that: (i) many crops are most vulnerable at the seedling stage, and (ii) the application of molluscicides may considerably improve crop performance.

Acknowledgements

I warmly thank A. Baur, B. Baur, S.E.R. Bailey, G.M. Barker, L. Tamm and an anonymous reviewer who have improved this chapter with their useful suggestions and critical comments, and the Swiss Federation Office for Education and Science for financial support (grant no. 97.0194, linked with EU FAIR project 97.3355).

References

Airey, W.J. (1987) The influence of food deprivation on the locomotor activity of slugs. *Journal of Molluscan Studies* 53, 37–45.

Angseesing, J.P.A. (1974) Selective eating of the acyanogenic form of *Trifolium repens*. *Heredity* 32, 73–83.

Baig, M.A., Banthorpe, D.V. and Gutowski, J.A. (1989) Accumulation in cultivars of *Santolina chamaecyparissus* of a rare sesquiterpene with gastropod-repellent activity. *Fitoterapia* 60, 373–375.

Bailey, S.E.R. (1989a) Daily cycles of feeding and locomotion in *Helix aspersa*. *Haliotis* 19, 23–31.

Bailey, S.E.R. (1989b) Foraging behaviour of terrestrial gastropods: integrating field and laboratory studies. *Journal of Molluscan Studies* 55, 263–272.

Bailey, S.E.R. (1991) Foraging behaviour of terrestrial gastropods: effects of changing level of food arousal on consumption. *Proceedings of the Tenth International Malacological Congress, Tübingen, 1989.* 421–424.

Bailey, S.E.R. (1993) Terrestrial molluscs. In: Wratten, S.D. (ed.) *Video Techniques in Animal Ecology and Behaviour.* Chapman & Hall, London, pp. 65–88.

Bailey, S.E.R. and Wedgwood, M.A. (1991) Complementary video and acoustic recordings of foraging by two pest species of slugs on non-toxic and molluscicidal baits. *Annals of Applied Biology* 119, 163–176.

Barker, G.M. (1989) Slug problems in New Zealand pastoral agriculture. In: Henderson, I.F. (ed.) *Slugs and Snails in World Agriculture.* British Crop Protection Council Monographs 41, pp. 59–68.

Baur, A., Baur, B. and Fröberg, L. (1994) Herbivory on calcicolous lichens: different food preferences and growth rates in two co-existing land snails. *Oecologia* 98, 313–319.

Baur, B. (1987) Effects of early feeding experience and age on the cannibalistic propensity of the land snail *Arianta arbustorum. Canadian Journal of Zoology* 65, 3068–3070.

Baur, B. (1992) Cannibalism in gastropods. In: Elgar, M.A. and Crespi, B.J. (eds) *Cannibalism. Ecology and Evolution Among Diverse Taxa.* Oxford University Press, Oxford, pp. 102–127.

Baur, B. and Baur, A. (1990) Experimental evidence for intra- and interspecific competition in two species of rock-dwelling land snails. *Journal of Animal Ecology* 59, 301–315.

Beyer, W.N. and Saari, D.M. (1978) Activity and ecological distribution of the slug, *Arion subfuscus* (Draparnaud) (Stylommatophora, Arionidae). *American Midland Naturalist* 100, 359–367.

Bless, R. (1977) Beitrag zur Ernährungsweise ausgewählter Nacktschneckenarten des Naturparks Kottenforst-Ville. *Anzeiger für Schädlingskunde, Pflanzenschutz, Umweltschutz* 50, 73–74.

Briner, T. and Frank, T. (1998) The palatability of 78 wildflower strip plants to the slug *Arion lusitanicus. Annals of Applied Biology* 133, 123–133.

Burgess, R.S.L. and Ennos, R.A. (1987) Selective grazing of acyanogenic white clover: variation in behaviour among populations of the slug *Deroceras reticulatum. Oecologia* 73, 432–435.

Butler, A.J. (1976) A shortage of food for the terrestrial snail *Helicella virgata* in South Australia. *Oecologia* 25, 349–371.

Byrne, J. and Jones, P. (1996) Responses to glucosinolate content in oilseed rape varieties by crop pest (*Deroceras reticulatum*) and non-pest slug species (*Limax pseudoflavus*). *Test of Agrochemicals and Cultivars* 17, 78–79.

Cameron, R.A.D. and Carter, M.A. (1979) Intra- and interspecific effects of population density on growth and activity in some helicid land snails (Gastropoda: Pulmonata). *Journal of Animal Ecology* 48, 237–246.

Carter, M.A., Jeffery, R.C.V. and Williamson, P. (1979) Food overlap in co-existing populations of the land snails *Cepaea nemoralis* (L.) and *Cepaea hortensis* (Müll). *Biological Journal of the Linnaean Society* 11, 169–176.

Cates, R.C. and Orians, G.H. (1975) Successional status and the palatability of plants to generalized herbivores. *Ecology* 56, 410–418.

Chang, H.W. (1991) Food preference of the land snail *Cepaea nemoralis* in a North American population. *Malacological Review* 24, 107–114.

Chase, R. (1982) The olfactory sensitivity of snails, *Achatina fulica. Journal of Comparative Physiology* 148A, 225–235.

Chase, R. (1986) Lessons from snail tentacles. *Chemical Senses* 11, 411–426.

Chatfield, J.E. (1976) Studies on food and feeding in some European land molluscs. *Journal of Conchology* 29, 5–20.

Compton, S.G., Beesley, S.G. and Jones, D.A. (1983) On the polymorphism of cyanogenesis in *Lotus corniculatus* L. IX. Selective herbivory in natural populations at Porthdafarch, Anglesey. *Heredity* 51, 537–547.

Cook, R.T., Bailey, S.E.R. and McCrohan, C.R. (1996) Slug preferences for winter wheat cultivars and common agricultural weeds. *Journal of Applied Ecology* 33, 866–872.

Cook, R.T., Bailey, S.E.R. and McCrohan, C.R. (1997) The potential for common weeds to reduce slug damage to winter wheat: laboratory and field studies. *Journal of Applied Ecology* 34, 79–87.

Cottam, D.A. (1986) The effects of slug-grazing on *Trifolium repens* and *Dactylis glomerata* in monoculture and mixed swards. *Oikos* 47, 275–279.

Crawley, M.J. (1983) *Herbivory. The Dynamics of Animal–Plant Interactions.* Blackwell Scientific Publications, Oxford.

Croll, R.P. (1983) Gastropod chemoreception. *Biological Reviews* 58, 203–219.

Croll, R.P. and Chase, R. (1977) A long-term memory for food odors in the land snail, *Achatina fulica. Behavioural Biology* 19, 261–268.

Croll, R.P. and Chase, R. (1980) Plasticity of olfactory orientation to foods in the snail *Achatina fulica. Journal of Comparative Physiology* 136A, 267–277.

Delaney, K. and Gelperin, A. (1986) Post-ingestive food-aversion learning to amino acid deficient diets by the terrestrial slug *Limax maximus. Journal of Comparative Physiology* 159A, 281–295.

Denny, M. (1980) Locomotion: the cost of gastropod crawling. *Science* 208, 1288–1290.

Desbuquois, C. and Daguzan, J. (1995) The influence of ingestive conditioning on food choices in the land snail *Helix aspersa* Müller (Gastropoda: Pulmonata: Stylommatophora). *Journal of Molluscan Studies* 61, 353–360.

Dirzo, R. (1980) Experimental studies on slug–plant interactions. I The acceptability of thirty plant-species to the slug *Agriolimax caruanae. Journal of Ecology* 68, 981–998.

Dirzo, R. and Harper, J.L. (1980) Experimental studies on slug–plant interactions. II The effect of grazing by slugs on high density monocultures of *Capsella bursa-pastoris* and *Poa annua. Journal of Ecology* 68, 999–1011.

Dirzo, R. and Harper, J.L. (1982a) Experimental studies on slug–plant interactions. III Differences in the acceptability of individual plants of *Trifolium repens* to slugs and snails. *Journal of Ecology* 70, 101–117.

Dirzo, R. and Harper, J.L. (1982b) Experimental studies on slug–plant interactions. IV The performance of cyanogenic and acyanogenic morphs of *Trifolium repens* in the field. *Journal of Ecology* 70, 119–138.

Dobson, D. and Bailey, S.E.R. (1983) Duration of feeding and crop fullness in *Deroceras reticulatum. Journal of Molluscan Studies* 48, 371–372.

Dussourd, D.E. and Eisner, T. (1987) Vein-cutting behavior: insect counterploy to the latex defense of plants. *Science* 237, 898–901.

Duval, D.M. (1971) A note on the acceptability of various weeds as food for *Agriolimax reticulatus* (Müller). *Journal of Conchology* 27, 249–251.

Duval, D.M. (1973) A note on the acceptability of various weeds as food for *Arion hortensis* Férussac. *Journal of Conchology* 28, 37–39.

Edwards, P.J. and Gillman, M.P. (1987) Herbivores and plant succession. In: Gray, A.J., Crawley, M.J. and Edwards, P.J. (eds) *Colonization, Succession and Stability.* Blackwell Scientific Publications, Oxford, pp. 295–314.

Elmslie, L.J. (1998) Humic acid: a growth factor for *Helix aspersa* Müller (Gastropoda: Pulmonata). *Journal of Molluscan Studies* 64, 400–401.

Fenner, M. (1987) Seedlings. *New Phytologist* 106 (Suppl.), 35–47.

Fox, L. and Landis, B.J. (1973) Notes on the predacious habits of the gray field slug, *Deroceras laeve. Environmental Entomology* 2, 306–307.

Frank, T. (1996) Sown wildflower strips in arable land in relation to slug density and slug damage to rape and wheat. In: Henderson, I.F. (ed.) *Slug and Snail Pests in Agriculture.* British Crop Protection Council Monographs No. 66, pp. 289–296.

Frömming, E. (1937) Das Verhalten von *Arianta arbustorum* L. zu den Pflanzen und höheren Pilzen. *Archiv für Molluskenkunde* 69, 161–169.

Gain, W.A. (1891) Notes on the food of some of the British mollusks. *Journal of Conchology* 349–360.

Gallois, L. and Daguzan, J. (1989) Recherches écophysiologiques sur le régime alimentaire de l'escargot petit-gris (*Helix aspersa*, Müller) (mollusque gastéropode pulmoné stylommatophore). *Haliotis* 19, 77–86.

Gelperin, A. (1975) Rapid food-aversion learning by a terrestrial mollusk. *Science* 189, 567–570.

Giamoustaris, A. and Mithen, R. (1995) The effect of modifying the glucosinolate content of leaves of oilseed rape (*Brassica napus* ssp. *oleifera*) on its interaction with specialist and generalist pests. *Annals of Applied Biology* 126, 347–363.

Glen, D.M., Jones, H. and Fieldsend, J.K. (1990) Damage to oilseed rape seedlings by the field slug *Deroceras reticulatum* in relation to glucosinolate concentration. *Annals of Applied Biology* 117, 197–207.

Godan, D. (1979) *Schadschnecken und ihre Bekämpfung.* Ulmer, Stuttgart.

Gouyon, P.H., Fort, P.H. and Caraux, G. (1983) Selection of seedlings of *Thymus vulgaris* by grazing slugs. *Journal of Ecology* 71, 299–306.

Graham, A. (1955) Molluscan diets. *Proceedings of the Malacological Society of London* 31, 144–159.

Grime, J.P. and Blythe, G.M. (1969) An investigation of the relationships between snails and vegetation at the Winnats Pass. *Journal of Ecology* 57, 45–66.

Grime, J.P., MacPherson-Stewart, S.F. and Dearman, R.S. (1968) An investigation of leaf palatability using the snail *Cepaea nemoralis* L. *Journal of Ecology* 56, 405–420.

Hägele, B. (1992) Saisonale Änderungen der Nahrungswahl des generalistischen Herbivoren *Arianta arbustorum* (L.) in verschiedenen, von Pflanzen aus der Tribus Senecioneae dominierten Habitaten. Diploma thesis, University of Tübingen, Germany.

Hägele, B.F., Wildi, E., Harmatha, J., Pavlik, M. and Rowell-Rahier, M. (1998) Long-term effects on food choice of land snail *Arianta arbustorum* mediated by petasin and furanopetasin, two sesquiterpenes from *Petasites hybridus. Journal of Chemical Ecology* 24, 1733–1743.

Hanley, M.E., Fenner, M. and Edwards, P.J. (1995a) An experimental field study of the effects of mollusc grazing on seedling recruitment and survival in grassland. *Journal of Ecology* 83, 621–627.

Hanley, M.E., Fenner, M. and Edwards, P.J. (1995b) The effect of seedling age on the likelihood of herbivory by the slug *Deroceras reticulatum*. *Functional Ecology* 9, 754–759.

Hanley, M.E., Fenner, M. and Edwards, P.J. (1996) The effect of mollusc grazing on seedling recruitment in artificially created grassland gaps. *Oecologia* 106, 240–246.

Hatziioannou, M., Eleutheriadis, N. and Lazaridou-Dimitriadou, M. (1994) Food preferences and dietary overlap by terrestrial snails in Logos area (Edessa, Macedonia, Northern Greece). *Journal of Molluscan Studies* 60, 331–341.

Henderson, I.F. (1969) A laboratory method for assessing the toxicity of stomach poisons to slugs. *Annals of Applied Biology* 63, 167–171.

Hodasi, J.K.M. (1995) Effects of different types of food on the growth of *Achatina achatina*. *Abstracts of the 12th International Malacological Congress, Vigo, 1995.* 488–489.

Hommay, G. (1995) Les limaces nuisibles aux cultures. *Revue Suisse d'Agriculture* 27, 267–286.

Hommay, G., Jacky, F. and Ritz, M.F. (1998) Feeding activity of *Limax valentianus* Férussac: nocturnal rhythm and alimentary competition. *Journal of Molluscan Studies* 64, 137–146.

Hopfield, J.F. and Gelperin, A. (1989) Differential conditioning to a compound stimulus and its components in the terrestrial mollusc, *Limax maximus*. *Behavioural Neuroscience* 103, 329–333.

Horrill, J.C. and Richards, A.J. (1986) Differential grazing by the mollusc *Arion hortensis* Fér. on cyanogenic and acyanogenic seedlings of the white clover, *Trifolium repens* L. *Heredity* 56, 277–281.

Hulme, P.E. (1994) Seedling herbivory in grassland: relative impact of vertebrate and invertebrate herbivores. *Journal of Ecology* 82, 873–880.

Iglesias, J. and Castillejo, J. (1999) Field observations on feeding of the land snail *Helix aspersa* Müller. *Journal of Molluscan Studies* 65, 411–423.

Ingram, W.M. and Peterson, A. (1947) Food of the giant Western slug, *Ariolimax columbianus* (Gould). *Nautilus* 61, 49–51.

Ireland, P. (1991) The effect of dietary calcium on growth, shell thickness and tissue calcium distribution in the snail *Achatina fulica*. *Comparative Biochemistry and Physiology* 98A, 111–116.

Jennings, T.J. and Barkham, J.P. (1975) Food of slugs in mixed deciduous woodland. *Oikos* 26, 211–221.

Jennings, T.J. and Barkham, J.P. (1979) Litter decomposition by slugs in mixed deciduous woodland. *Holarctic Ecology* 2, 21–29.

Jones, D.A. (1962) Selective eating of the acyanogenic form of the plant *Lotus corniculatus* L. by various animals. *Nature* 193, 1109–1110.

Kelly, D. (1989) Demography of short-lived plants in chalk grassland. II Control of mortality and fecundity. *Journal of Ecology* 77, 770–784.

Keymer, R. and Ellis, W.M. (1978) Experimental studies on plants of *Lotus corniculatus* L. from Anglesey polymorphic for cyanogenesis. *Heredity* 40, 189–206.

Kittel, R. (1956) Untersuchungen über den Geruchs- und Geschmackssinn bei den Gattungen *Arion* und *Limax* (Mollusca: Pulmonata). *Zoologischer Anzeiger* 157, 185–195.

Levin, D.A. (1973) The role of trichomes in plant defense. *Quarterly Review of Biology* 48, 3–15.

Lindquist, B. (1941) Experimentelle Untersuchungen über die Bedeutung einiger Landmollusken für die Zersetzung der Waldstreu. *Kungl. Fysiografiska Sällskapets I Lund Förhandlingar* 11, 144–156.

Linhart, Y.B. and Thompson, J.D. (1995) Terpene-based selective herbivory by *Helix aspersa* (Mollusca) on *Thymus vulgaris* (Labiatae). *Oecologia* 102, 126–132.

Mason, C.F. (1970) Food, feeding rates and assimilation in woodland snails. *Oecologia* 4, 358–373.

Mølgaard, P. (1986) Food plant preferences by slugs and snails: a simple method to evaluate the relative palatability of the food plants. *Biochemical Systematics and Ecology* 14, 113–121.

Newman, R.M., Hanscom, Z. and Kerfoot, W.C. (1992) The watercress glucosinolate–myrosinase system: a feeding deterrent to caddiesflies, snails and amphipods. *Oecologia* 92, 1–7.

Oliveira Silva, M.T. (1992) Effects of mollusc grazing on the development of grassland species. *Journal of Vegetation Science* 3, 267–270.

Pakarinen, E., Niemelä, P. and Tuomi, J. (1990) Effect of fertilization, seaweed extracts and leaf-damage on palatability of lettuce to *Deroceras*-slugs. *Acta Oecologia* 11, 113–119.

Pallant, D. (1969) The food of the grey field slug (*Agriolimax reticulatus* (Müller)) in woodland. *Journal of Animal Ecology* 38, 391–398.

Pallant, D. (1972) The food of the grey field slug, *Agriolimax reticulatus* (Müller), on grassland. *Journal of Animal Ecology* 41, 761–769.

Paterson, J.P.H., Binggeli, P. and Rushton, B.S. (1996) Slug- and small mammal-induced seedling mortality in sycamore *Acer pseudoplatanus* L. *Biology and Environment: Proceedings of the Royal Irish Academy* 96B, 49–53.

Pickett, J.A. and Stephenson, J.W. (1980) Plant volatiles and components influencing the behavior of the field slug, *Deroceras reticulatum* (Müll.). *Journal of Chemical Ecology* 6, 435–444.

Pomeroy, D.E. (1969) Some aspects of the ecology of the land snail, *Helicella virgata*, in South Australia. *Australian Journal of Zoology* 17, 495–514.

Port, C.M. and Port, G.R. (1986) The biology and behaviour of slugs in relation to crop damage and control. *Agricultural Zoology Reviews* 1, 255–299.

Powell, A.L. and Bowen, I.D. (1996) The screening of naturally occurring compounds for use as seed treatments for the protection of winter wheat against slug damage. In: Henderson, I.F. (ed.) *Slug and Snail Pests in Agriculture.* British Crop Protection Council Monographs No. 66, pp. 231–236.

Raffaelli, D. and Mordue, A.J. (1990) The relative importance of molluscs and insects as selective grazers of acyanogenic white clover (*Trifolium repens*). *Journal of Molluscan Studies* 56, 37–46.

Rai, J.P.N. and Tripathi, R.S. (1985) Effect of herbivory by the slug, *Mariella dussumieri*, and certain insects on growth and competitive success of two sympatric annual weeds. *Agriculture, Ecosystems and Environment* 13, 125–137.

Ramsell, J. and Paul, N.D. (1990) Preferential grazing by molluscs of plants infected by rust fungi. *Oikos* 58, 145–150.

Rathcke, B. (1985) Slugs as generalist herbivores: tests of three hypotheses on plant choices. *Ecology* 66, 828–836.

Rees, M. and Brown, V.K. (1992) Interactions between invertebrate herbivores and plant competition. *Journal of Ecology* 80, 353–360.

Reingold, S.C. and Gelperin, A. (1980) Feeding motor programme in *Limax*. II. Modulation by sensory inputs in intact animals and isolated central nervous systems. *Journal of Experimental Biology* 85, 1–19.

Rice, R.L., Lincoln, D.E. and Langenheim, J.H. (1978) Palatability of monoterpenoid compositional types of *Satureja douglasii* to a generalist molluscan herbivore, *Ariolimax dolichophallus*. *Biochemical Systematics and Ecology* 6, 45–53.

Richardson, A.M.M. (1975) Food, feeding rates and assimilation in the land snail *Cepaea nemoralis* L. *Oecologia* 19, 59–70.

Richardson, B. and Whittaker, J.B. (1982) The effect of varying the reference material on ranking of acceptability indices of plant species to a polyphagous herbivore, *Agriolimax reticulatus*. *Oikos* 39, 237–240.

Richter, K.O. (1979) Aspects of nutrient cycling by *Ariolimax columbianus* (Mollusca: Arionidae) in Pacific Northwest coniferous forests. *Pedobiologia* 19, 60–74.

Rollo, C.D. (1988a) A quantitative analysis of food consumption for the terrestrial mollusca: allometry, food hydration and temperature. *Malacologia* 28, 41–51.

Rollo, C.D. (1988b) The feeding of terrestrial slugs in relation to food characteristics, starvation, maturation and life history. *Malacologia* 28, 29–39.

Rueda, A. (1989) Artificial diet for laboratory maintenance of the veronicellid slug *Sarasinula plebeja* (Fisher). In: Henderson, I.F. (ed.) *Slugs and Snails in World Agriculture*. British Crop Protection Council Monographs No. 41, pp. 361–366.

Runham, N.W. and Hunter, P.J. (1970) *Terrestrial Slugs*. Hutchinson University Library, London.

Sahley, C., Gelperin, A. and Rudy, J.W. (1981a) One-trial associative learning modifies food odor preferences of a terrestrial mollusc. *Proceedings of the National Academy of Sciences USA* 78, 640–642.

Sahley, C., Rudy, J.W. and Gelperin, A. (1981b) An analysis of associative learning in a terrestrial mollusc. I Higher-order conditioning, blocking and a transient US pre-exposure effect. *Journal of Comparative Physiology* 144A, 1–8.

Sahley, C.L., Martin, K.A. and Gelperin, A. (1990) Analysis of associative learning in the terrestrial mollusc *Limax maximus*. II Appetitive learning. *Journal of Comparative Physiology* 167A, 339–345.

Senseman, D.M. (1977) Starch: a potent feeding stimulant for the terrestrial slug *Ariolimax californicus*. *Journal of Chemical Ecology* 3, 707–715.

Smallridge, M.A. and Kirby, G.C. (1988) Competitive interactions between the land snails *Theba pisana* (Müller) and *Cernuella virgata* (Da Costa) from South Australia. *Journal of Molluscan Studies* 54, 251–258.

South, A. (1992) *Terrestrial Slugs. Biology, Ecology and Control*. Chapman & Hall, London.

Speiser, B. and Rowell-Rahier, M. (1991) Effects of food availability, nutritional value, and alkaloids on food choice in the generalist herbivore *Arianta arbustorum* (Gastropoda: Helicidae). *Oikos* 62, 306–318.

Speiser, B. and Rowell-Rahier, M. (1993) Does the land snail *Arianta arbustorum* prefer sequentially mixed over pure diets? *Functional Ecology* 7, 403–410.

Speiser, B., Harmatha, J. and Rowell-Rahier, M. (1992) Effects of pyrrolizidine alkaloids and sesquiterpenes on snail feeding. *Oecologia* 92, 257–265.

Stahl, E. (1888) Pflanzen und Schnecken. *Jenaer Zeitschrift für Naturwissenschaften* 22, 557–684.

Stephenson, J.W. (1962) A culture method for slugs. *Proceedings of the Malacological Society of London* 35, 43–45.

Teyke, T. (1995) Food-attraction conditioning in the snail, *Helix pomatia. Journal of Comparative Physiology* 177A, 409–414.

Theenhaus, A. and Scheu, S. (1996) The influence of slug (*Arion rufus*) mucus and cast material addition on microbial biomass, respiration, and nutrient cycling in beech leaf litter. *Biology and Fertility of Soils* 23, 80–85.

van der Laan, K.L. (1975) Feeding preferences in a population of the land snail *Helminthoglypta arrosa* (Binney). *Veliger* 17, 354–359.

von Sury, R. (1992) Pflanzen–Pathogen-Beziehungen im städtischen Umfeld: Auswirkungen anthropogener und natürlicher Stressfaktoren auf die Platanen-Anthraknose. PhD thesis, University of Basel, Switzerland.

Wadham, M.D. and Wynn-Parry, D. (1981) The silicon content of *Oryza sativa* L. and its effect on the grazing behaviour of *Agriolimax reticulatus* Müller. *Annals of Botany* 48, 399–402.

Wareing, D.R. (1993) Feeding history – a factor determining food preference in slugs. *Journal of Molluscan Studies* 59, 366–368.

Wedgwood, M.A. and Bailey, S.E.R. (1986) The analysis of single meals in slugs feeding on molluscicidal baits. *Journal of Molluscan Studies* 52, 259–260.

Westerbergh, A. and Nyberg, A. (1995) Selective grazing of hairless *Silene dioica* plants by land gastropods. *Oikos* 73, 289–298.

Whelan, R.J. (1982) Response of slugs to unacceptable food items. *Journal of Applied Ecology* 19, 79–87.

Williamson, P. and Cameron, R.A.D. (1976) Natural diet of the landsnail *Cepaea nemoralis. Oikos* 27, 493–500.

Wink, M. (1984) Chemical defense of lupins. Mollusc-repellant properties of quinolizidine alkaloids. *Zeitschrift für Naturforschung* 39, 553–558.

Wolda, H., Zweep, A. and Schuitema, K.A. (1971) The role of food in the dynamics of populations of the landsnail *Cepaea nemoralis. Oecologia* 7, 361–381.

Zamora, R. and Gomez, J.M. (1996) Carnivorous plant–slug interaction: a trip from herbivory to kleptoparasitism. *Journal of Animal Ecology* 65, 154–160.

7 Haemolymph: Blood Cell Morphology and Function

E. FURUTA AND K. YAMAGUCHI[1]

Department of Histology and [1]Institute for Medical Science, Dokkyo University School of Medicine, Mibu, Tochigi 321–0293, Japan

Introduction

The main body cavity of gastropods is the haemocoel, which forms during embryogenesis within the mesenchyme. In terrestrial gastropods, lining cells are present in the haemocoel, but they are not endothelial in origin or morphologically. The body cavity is filled with haemolymph (blood) returned from outlying venous sinuses. The vascular system is usually open (e.g. Duval and Runham, 1981), with the heart receiving haemolymph from the veins of the pallial region and pumping it to the various organs of the body. From the heart, the haemolymph passes to large arteries which branch to form increasingly smaller arteries. The ultimate ramifications of the arteries, the capillaries, invest the various organs. From the organs, the haemolymph passes to haemocoel and lacunae and thence to large veins, and reaches the pallial cavity to enter the auricle(s).

The haemolymph contains dissolved haemoproteins. The respiratory pigments are haemoglobin in the freshwater planorbids, and haemocyanin in all other gastropods. Haemoglobins are red, while haemocyanins are blue when oxygenized and colourless when deoxygenized. These pigments are synthesized, stored and released into a special type of connective tissue cell, the pore cell or rhogocyte (Sminia, 1972; Sminia and Boer, 1973). The haemolymph also contains several kinds of blood cells, referred to as amoebocytes or haemocytes, but none are morphologically and functionally comparable with vertebrate erythrocytes.

Morphology of Blood Cells

Blood cells have been investigated extensively in aquatic pulmonates (e.g. Kress, 1968; Stang-Voss, 1970; Sminia, 1972; Yoshino, 1976; Dikkeboom *et al.*, 1984). However, morphological and functional information on blood cells of terrestrial gastropods is scarce (Wagge, 1955; Furuta *et al.*, 1986, 1987, 1990).

In gastropods, as in all molluscs, blood cells play a major role in the internal defence system. Although they do not possess immunoglobulins nor a complement system, terrestrial gastropods are able to maintain sterility of the internal body fluids under most circumstances. Gastropods possess at least one type of blood cell that functions as a phagocyte, performing intracellular digestion of non-self materials (Sminia, 1972, 1981; Yoshino, 1976; van der Knaap, 1981; Furuta *et al.*, 1986, 1987, 1990; Yamaguchi *et al.*, 1988).

In the philomycids *Incilaria fruhstorferi* Collinge and *Incilaria bilineata* (Benson), the haemolymph contains three blood cell types, namely type I (macrophage), type II (lymphocyte-like) and type III (fibroblast-like). When a drop of haemolymph is transferred on to a glass slide or a plastic dish, these blood cells appear to be mostly spherical or ovoid during the first few seconds of observation. Thus, it is extremely difficult to distinguish between cell types in fresh preparations. However, if left undisturbed for about 60 min, the three cell types become recognizable. Type I cells are firstly round with short pseudopodia (Fig. 7.1). However, with time, they become more extensive, spreading in all directions over the substrate by extending their pseudopodia (lamellipodia) with supporting ribs (Figs 7.2 and 7.3). These cells measure approximately 20–30 μm in diameter, with the kidney-shaped or lobulated nucleus

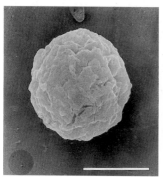 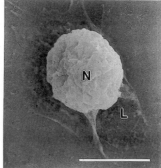 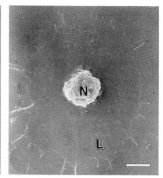

Fig. 7.1. A scanning electron micrograph of a type I cell (macrophage) from *Incilaria fruhstorferi* Collinge (Philomycidae) placed on a plastic dish in SH7 medium (Furuta and Shimozawa, 1983) immediately after harvest from the pallial cavity. Scale = 5 μm.

Fig. 7.2. A scanning electron micrograph of a type I cell (macrophage) from *Incilaria fruhstorferi* Collinge (Philomycidae) placed on a plastic dish in SH7 medium (Furuta and Shimozawa, 1983) 30 min after harvest, showing lamellipodia supported by ribs. L, lamellipodia; N, nucleus. Scale = 5 μm.

Fig. 7.3. A scanning electron micrograph of a type I cell (macrophage) from *Incilaria fruhstorferi* Collinge (Philomycidae) placed on a plastic dish in SH7 medium (Furuta and Shimozawa, 1983) 1 h after harvest, exhibiting flattening and spread, with extending lamellipodia at the sites of contact with the substrate. L, lamellipodia; N, nucleus. Scale = 5 μm.

measuring about 5 μm. The cytoplasm contains numerous mitochondria, rough endoplasmic reticulum, multivesicular bodies, residual bodies, Golgi apparatus and glycogen-like deposits (Figs 7.4 and 7.5).

Type II cells (lymphocyte-like cells) are smaller than type I cells. They are about 5 μm in diameter and the nucleus to cytoplasm ratio is higher. Like type I cells, type II cells also adhere to the substratum, but they remain spherical and, where they are formed, the pseudopodia are weak. Type II cells contain scattered free ribosomes and mitochondria in the cytoplasm surrounding the round nucleus (Fig. 7.6).

Type III cells are fibroblast-like, spindle-shaped cells measuring approximately 75 × 15 μm. They contain microfibrils (12–15 nm in

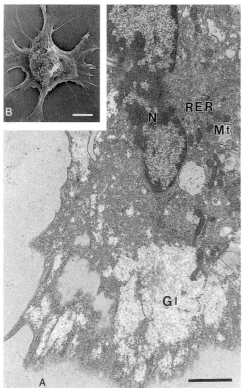

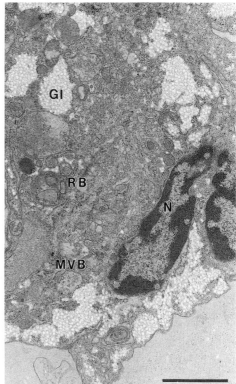

Fig. 7.4. Electron micrographs of a type I cell from *Incilaria fruhstorferi* Collinge (Philomycidae) placed in SH7 medium for 30 min. (A) Transmission electron micrograph illustrating the ultrastructure features of the cell, with the kidney-shaped nucleus (N), small amounts of rough endoplasmic reticulum (RER), sparse mitochondria (Mt) and the abundant glycogen-like deposits (Gl) in the lamellipodia. Scale = 2 μm. (B) Scanning electron micrograph of the cell illustrating the lamellipodia with supporting ribs. Scale = 5 μm.

Fig. 7.5. Transmission electron micrograph of a type I cell from *Incilaria fruhstorferi* Collinge (Philomycidae) placed in SH7 medium for 30 min showing the ultrastructure features of the cell, in particular the presence of multivesicular bodies (MVB), residual bodies (RB) and glycogen-like deposits (Gl). Scale = 2 μm.

diameter) and residual bodies in the cytoplasm, and give rise to collagen-like fibres outside the cytoplasmic membrane (Fig. 7.7).

In addition to blood cells, vast numbers of platelet-like structures are present in haemolymph collected from *I. fruhstorferi* or *I. bilineata*. Most of these platelet-like structures are circular or ovoid in cross-section, with

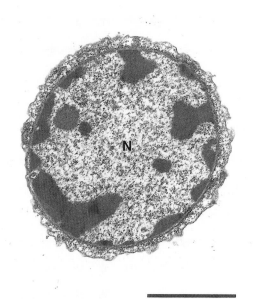

Fig. 7.6. Transmission electron micrograph of a type II cell (lymphocyte-like) from *Incilaria fruhstorferi* Collinge (Philomycidae), incubated for 30 min in SH7 medium, illustrating the spherical nature of the cell, with a relatively large round nucleus (N) and sparse cytoplasm. Scale = 2 μm.

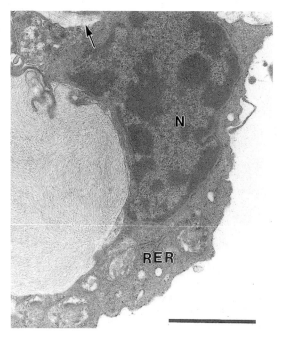

Fig. 7.7. Transmission electron micrograph of a type III cell (fibroblast-like) from *Incilaria fruhstorferi* Collinge (Philo-mycidae), incubated for 30 min in SH7 medium, with small amounts of rough endoplasmic reticulum (RER) in the restricted cytoplasm around the large oval nucleus (N), and the collagen-like fibres (arrow) in the extracellular spaces. Scale = 2 μm.

diameters ranging from 1 to 5 µm. Each is covered by a membrane, but lacks a nucleus (Fig. 7.8).

The number of cell elements in the haemolymph of the helicid *Cantareus aspersus* (Müller) is normally small. However, within about 15 min of wounding, this cell number is increased substantially (Haughton, 1935). *C. aspersus* possesses two cell types in the haemolymph. One is the phagocytic amoebocyte. The other, without phagocytic activity, is a smaller, round cell that makes up about 12–14% of the total cells in the haemolymph (Prowse and Tait, 1969). The amoebocytes possess a large, often kidney- or lobular-shaped nucleus with scattered chromatin blocks. Immediately following haemolymph withdrawal, these cells become adhesive and they rapidly begin to aggregate. After cell aggregation, each amoebocyte gradually begins to spread by extending pseudopodia. In contrast, the non-phagocytic round cells do not form pseudopodia, or they only do so slightly. Compared with spreading amoebocytes, round cells have more extensive, rough endoplasmic reticulum, a greater number of free ribosomes and a cytoplasm that is more opaque in electron microscopy (Sminia, 1981).

Wagge (1955) recognized two different amoebocyte types in *C. aspersus*. Type A cells were found to range in diameter from 10 to 20 µm, depending upon extensions of the cytoplasm. The nucleus was shown to be approximately 10 µm in diameter, and the hyaline cytoplasm often closely contracted around the nucleus. Type B cells contained nuclei of

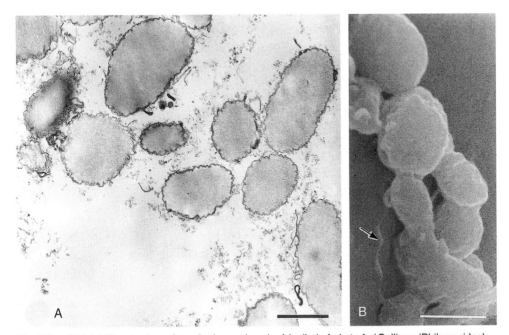

Fig. 7.8. Platelet-like structures from the haemolymph of *Incilaria fruhstorferi* Collinge (Philomycidae), incubated for 30 min in SH7 medium. (A) Transmission electron micrograph of the structures. Scale = 2 µm. (B) Scanning electron micrograph of an aggregate of the structures with fibres (arrow). Scale = 2 µm.

10 μm diameter and the cytoplasm often extends to 50 μm in length. Both cells were shown to originate from the mantle epithelium.

Function of Blood Cells

Invertebrates, like vertebrates, are protected against invading micro-organisms by an internal defence system. Several functions have been attributed to the blood cells in gastropods (Haughton, 1935; Wagge, 1955; Bayne, 1973; Furuta *et al.*, 1987, 1990; Adema *et al.*, 1992), including an important role in defence reactions, such as phagocytosis (Prowse and Tait, 1969; Buchholz *et al.*, 1971; Furuta *et al.*, 1987, 1990), encapsulation and wound healing (Haughton, 1935; Yamaguchi *et al.*, 1989).

Phagocytosis and elimination of foreign materials

Studies involving species of *Incilaria* Benson demonstrate that various types of foreign particles, both biotic (red blood cells and yeast cells) and abiotic (India ink and latex beads of various sizes), are phagocytized by blood cells. The different types of blood cells differentially phagocytize foreign particles, although, on average, smaller particles are phagocytized more readily than larger ones (Figs 7.9 and 7.10). Similar results have been obtained with *Achatina achatina* (Linnaeus), *Achatina fulica* Bowdich

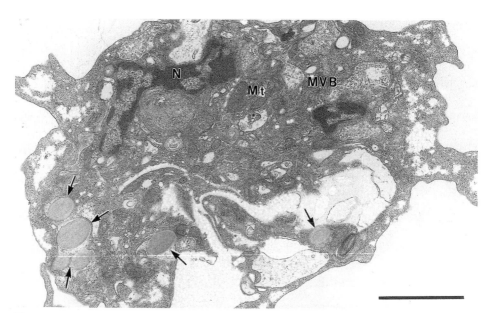

Fig. 7.9. Transmission electron micrograph of a type I cell from *Incilaria fruhstorferi* Collinge (Philomycidae), phagocytizing latex beads *in vivo*. The animal had been injected with latex beads (diameter 0.79 μm) (arrows) 20 h before the collection of the haemolymph. N, nucleus; MVB, multivesicular body; Mt, mitochondria. Scale = 2 μm.

(Achatinidae), *C. aspersus* and *Helix pomatia* Linnaeus (Helicidae) (Haughton, 1935; Wagge, 1955; Brown, 1967; Prowse and Tait, 1969; Buchholz *et al.*, 1971; Bayne, 1973; Renwrantz *et al.*, 1981; Adema *et al.*, 1992), where amoebocytes have been shown to ingest foreign particles (carmine and zymosan) and cells (bacteria, formalized yeast cells, sheep red blood cells and human red blood cells). Morphological and ultra-structural analyses demonstrate that phagocytosis of foreign particles by macrophage-like cells (amoebocytes or type I cells of *Incilaria*) involves the pseudopodia spreading to engulf the particles. With invagination of the cell membrane, the foreign material becomes enclosed in phagosomes. The phagosomes fuse to primary and secondary lysosomes where the enclosed foreign particles are attacked by hydrolytic enzymes. Acid phosphatase and non-specific esterases have been demonstrated in the haemolymph of *I. fruhstorferi* and the freshwater pulmonate *Lymnaea stagnalis* (Linnaeus) (Lymnaeidae) (Sminia, 1972; Furuta *et al.*, 1994). While absent in *Incilaria* (Furuta *et al.*, 1994), peroxidases are present in the lysosomal system of *L. stagnalis* with bactericidal, viricidal and fungicidal function similar to that of mammalian leucocytes (Sminia, 1972).

Phagocytosis is a well-known internal immuno-defence mechanism in all animals. Phagocytosis is considered to be the primary clearance mechanism in gastropods. However, the fate of phagocytized materials may differ. Using X-ray and histological techniques to follow the movement of radio-opaque thorium dioxide, Brown (1967) investigated the elimination of foreign, non-metabolizable particles from the body of

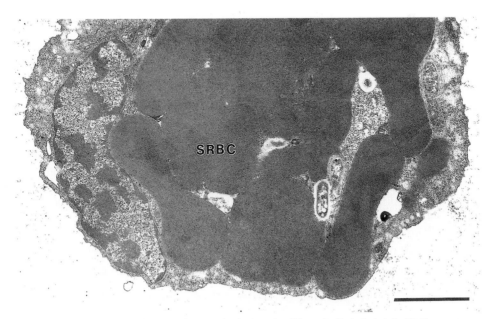

Fig. 7.10. Transmission electron micrograph of a type I cell from *Incilaria fruhstoferi* Collinge (Philomycidae), phagocytizing sheep red blood cells (SRBCs) *in vitro*. The collected haemolymph was incubated for 20 h at 25°C in SH7 medium containing 10% SRBC. Scale = 2 µm.

C. aspersus following injection into the pedal sinus. The radio-opaque thorium dioxide produced intense body shadows under X-ray, with maximum intensity by the fourth or fifth day. This radio-opacity then faded to be barely distinguishable on the tenth day from the shadow cast by the body tissues. Thorium-laden amoebocytes initially accumulated in the distal regions of the reproductive organs, but by day 10 or 11 these cell accumulations had disappeared.

While some investigations support the idea of movement of amoebocytes to sites of foreign particle ingress (e.g. Cheng, 1970; Pauley *et al.*, 1971), others favour a system of fixed phagocytic cells that clear foreign particles from the circulating haemolymph (Reade, 1968; Stuart, 1968; Bayne, 1973; Renwrantz *et al.*, 1981). According to Bayne (1973), bacteria injected into *H. pomatia* are mostly cleared from the haemolymph within 1 h. The bacterial cells do not accumulate in the circulating amoebocytes nor degrade by cell lysis. Rather, the bacteria are eliminated from the circulating haemolymph by phagocytic cells that have a fixed location within the body. Bayne (1973) considered the digestive gland as the primary site where clearance occurs.

Gastropods lack the classical immune recognition molecules of vertebrates, such as immunoglobulins, T-cells, major histocompatibility (MHC) or antigen receptors. However, these animals do possess the capacity to distinguish not only between self and non-self, but also between non-self materials that differ in chemical properties (Bayne, 1990; Cooper *et al.*, 1992).

In invertebrates, no substance comparable with vertebrate immunoglobulin is contained in the haemolymph. However, invertebrates do possess several classes of biologically active humoral factors that functionally mimic antibodies. The molecules that possess agglutinating activity may effect self/non-self recognition in invertebrates (Olafsen, 1986; Couch *et al.*, 1990). Most invertebrate agglutinins are believed to be lectins, some of which possess opsonic activity (Tripp, 1970, 1992a,b; Anderson and Good, 1976; Renwrantz and Mohr, 1978; Sminia *et al.*, 1979; van der Knaap *et al.*, 1983; Furuta *et al.*, 1995). These agglutinating molecules occur, not only within the haemolymph, but also in the mucus secreted by the mucous cells of the body wall (Iguchi *et al.*, 1982, 1985; Miller, 1982; Fountain and Campbell, 1984; Fountain, 1985; Kubota *et al.*, 1985; Furuta *et al.*, 1995), which in part is a haemolymph filtrate (Yamaguchi *et al.*, 2000). Many invertebrates possess lectins or similar factors in their tissues and body surface mucus (Kubota *et al.*, 1985; Kisugi *et al.*, 1992a,b; Yuasa *et al.*, 1998), and such substances surely play an important role in what must be regarded as a primitive defence mechanism.

According to Furuta *et al.* (1995), the body surface mucus from *I. fruhstorferi* contains a water-soluble fraction (WSF) that not only enhances phagocytic activity (Fig. 7.11), but also agglutinates type A and B human red blood cells (HRBCs). The WSF-induced haemagglutination of A and B types of HRBCs is inhibited specifically by *N*-acetyl-galactosamine (GalNAc). Further, WSF causes haemolysis of only B type HRBC 12 h after agglutination (Fig. 7.12). *Incilaria* mucus contains lectin(s) and a haemolysin-like substance, and the *N*-acetyl moiety is essential for binding of the lectin to sugars.

The WSF from *Incilaria* contains three 15 kDa C-type lectins, namely incilarin A, incilarin B and incilarin C, which consist of 150, 149 and 156 amino acids, respectively, and contain signal peptides of 17 amino acids at their N-termini. The three lectins show 44–55% amino acid sequence identity with each other and significant sequence homology with C-type

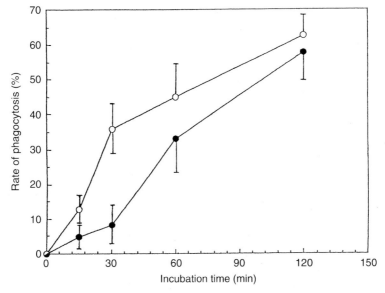

Fig. 7.11. Phagocytosis of sheep red blood cells (SRBCs) by *Incilaria fruhstoferi* Collinge (Philomycidae) macrophages. Filled circles, untreated SRBC; open circles, SRBC treated with the water-soluble fraction (WSF) from the body surface mucus of *Incilaria fruhstoferi*. Untreated and WSF-treated SRBCs were added to a haemolymph cell suspension and incubated at 25°C for 15, 30, 60 and 120 min. After fixation with methanol, the cells were stained with Giemsa. The rate of phagocytosis (%) was calculated from the ratio of SRBC-phagocytized cell number to total cell number. Vertical bars show standard deviation.

Fig. 7.12. Haemolysis of type B human blood cells (HRBCs) (in blood agar) by the water-soluble fraction (WSF) of the body wall mucus of *Incilaria fruhstorferi* Collinge (Philomycidae). Left: the HRBC were agglutinated by WSF, resulting in haemolysis of the erythrocyte and formation of a haemolytic circle (arrow). Right: absence of haemolytic activity around the well due to prior heat treatment (56°C for 30 min) of the WSF.

lectins in other animals. In keeping with the highly conserved C-type lectins in other animal groups, two disulphide bonds are present in incilarin A and incilarin B. However, one of these disulphide bonds is absent in incilarin C (Yuasa *et al.*, 1998). Recently, it has been reported that one of the C-type lectins in echinoderms also lacks one disulphide bond (Hatakeyama *et al.*, 1995); thus two disulphide bonds may not be essential for lectin activity.

Encapsulation

When foreign particles are introduced into animals but are too large to be phagocytized by single cells, they are encapsulated by large numbers of blood cells. Information concerning encapsulation of parasites is available from several studies on freshwater gastropods, but the encapsulation process in terrestrial gastropods is poorly known. According to Sauerländer (1976), in *A. fulica* experimentally challenged with the first stage larvae of the metastrongylid nematode *Angiostrongylus cantonensis* (Chen), large numbers of amoebocytes appear in the haemolymph and the blood spaces of the connective tissue. Within 12 h, the parasites were encapsulated by large numbers of amoebocytes, which assume an elongate shape in the outer regions of the capsule. After 3 days, the nuclei of these amoebocytes had become spindle shaped and the capsule around the parasites had formed concentric layers in its outer region (fibroblastic type of encapsulation). Hori *et al.* (1973) also observed parasites to be surrounded by fibroblast-like cells in *Lehmannia marginata* (Müller) (Limacidae). In the freshwater species *L. stagnalis*, capsules formed around biologically inert implants possess an innermost layer of flattened and closely packed cells. The initially flattened amoebocytes of the middle, and more especially the outer layers become separated and the resulting intercellular spaces become filled with minute fibrils of connective tissue. Clearly, these intercellular substances were synthesized by the amoebocytes that had transformed into fibroblasts (Sminia, 1981).

There are few reports on orthotopic transplantation in terrestrial gastropods. In experiments with the skin of *I. fruhstorferi*, allografts and autografts were initially recognized and attacked by numerous host macrophages. Autografts generally survived in host tissue where a wound-healing response occurred (Yamaguchi *et al.*, 1999). However, allografts were chronically rejected by host macrophages, and destruction of the graft cells was followed by apoptotic cell death (E. Furuta *et al.*, unpublished).

Wound healing

There have been few investigations on the role of blood cells in the process of wound healing in gastropods. Yamaguchi *et al.* (1989) found that when a short incision is made in the skin of *I. fruhstorferi*, the size of the wound opening is reduced immediately by contraction of the body wall muscles. During the first 30 min, the wound region became sealed off by aggregated platelet-like structures (PLSs) sourced from the haemolymph (Fig. 7.13): as PLSs appear to play a role in haemolymph clotting (Furuta *et al.*, 1989), this initial stage of wound repair may prevent further loss of haemolymph. Within 1 day, macrophages were shown to be present at the site of damage, where they assume a flattened shape and spread over the surface to become a provisional wound covering (Fig. 7.14). Within 4 days of the initial tissue damage, undifferentiated epidermal cells had begun to grow over the wound (Yamaguchi *et al.* 1989) (Fig. 7.15).

Shell repair

Amoebocytes are generally thought to play a role in shell repair in terrestrial gastropods by transporting calcium and organic shell matrix substances from the digestive gland to the sites of damage (Abolius-Krogis, 1973, 1976). According to Kapur and Gupta (1970) in the case of *Euplecta indica* (Pfeiffer) (Ariophantidae), within 1 h of the initiation of

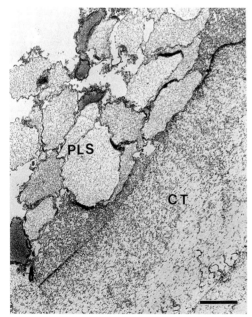

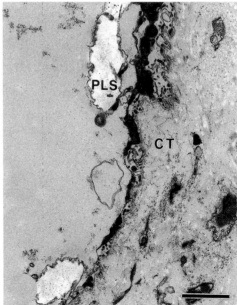

Fig. 7.13. Transmission electron micrograph of the wound-healing process in the foot of *Incilaria fruhstorferi* Collinge (Philomycidae). The wound surface is shown sealed by many aggregated platelet-like structures (PLS) 30 min after inflicting tissue damage. CT, connective tissue.

Fig. 7.14. Transmission electron micrograph of the wound-healing process in the foot of *Incilaria fruhstorferi* Collinge (Philomycidae). Type I (macrophage-like) cells spread over the surface of the wound after 1 day. Note the reduced number of platelet-like structures (PLS).

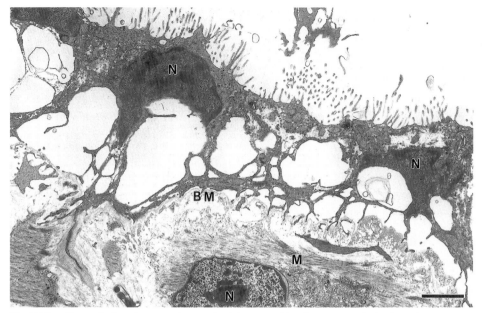

Fig. 7.15. Transmission electron micrograph of the wound-healing process in the foot of *Incilaria fruhstorferi* Collinge (Philomycidae). Undifferentiated epidermal cells begin to cover the wound at day 4. BM, basement membrane; M, smooth muscle cell; N, nucleus. Scale = 2 µm.

shell repair, large numbers of amoebocytes arrive at the site of membrane regeneration. Their cell membrane becomes disorganized and their cytoplasm precipitates, leaving free nuclei. Calcium granules have been detected in the cytoplasm, and these amoebocytes were said to carry histochemically recognizable substances, necessary for calcification, from different organs of the body to the site of regeneration. Wagge (1955) reported that two amoebocyte types in *C. aspersus* were involved in shell repair, one containing calcium granules, another containing brownish-yellow, proteinaceous spherules.

However, calcium is stored in a special type of connective tissue cell, the calcium cell (Richardot and Wautiet, 1972; Sminia, 1981). Sminia (1981) recognized that calcium-rich cells were often mistaken for amoebocytes. It therefore seems unlikely that amoebocytes are involved in shell repair.

Haemopoiesis

Haemopoiesis is the term given to blood cell formation. Several sites of haemopoiesis have been described in gastropods. Wagge (1955) has suggested that blood cells in *C. aspersus* are formed from the epithelium and connective tissue within the mantle. Using a variety of *in vivo* and *in vitro* techniques to investigate responses in *I. fruhstorferi* injected with yeast cells, Furuta *et al.* (1994) could not determine any one organ with a

Surface of Skin

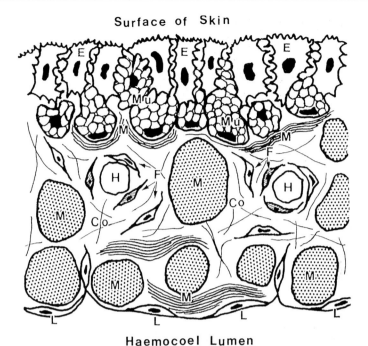

Haemocoel Lumen

Fig. 7.16. A diagram of haemopoiesis in the haemocoel wall of terrestrial gastropods. The connective tissue intermingles with bundles of muscle fibres (M) and collagen fibres (Co), and proliferated fibroblast-like (type III) cells (indicated by diagonal hatching) occur around the haemal spaces (H). E, body surface epithelial cell; L, haemocoel-lining cell (discontinuous); Mu, mucous cell; F, fibroblast-like cell.

special haemopoietic role; blood cell proliferation occurred throughout the body (Figs 7.16 and 7.17). In *in vitro* experiments, Furuta and Shimozawa (1983) found that fibroblast-like amoebocytes originating from the mantle of *I. fruhstorferi* differentiated into macrophages (type I cells). No type II or type III cells were observed after injection of foreign materials. In this species, haemopoiesis was accomplished by two processes: one operating at the foci of challenge by foreign particles such as in microbial infection, another operating throughout the connective tissue and vascular system – the main site is in the cells that line the haemocoel wall, which are derived from fibroblasts.

Summary and Conclusions

In the present chapter, several morphological and functional aspects of blood cells and their formation in terrestrial gastropods have been reviewed and discussed. Terrestrial gastropods lack a specialized haemopoietic organ, and blood cell formation occurs within the connective tissue and vascular system. Their blood cells are usually classified into two types, namely spreading amoebocytes (macrophage-like) and round

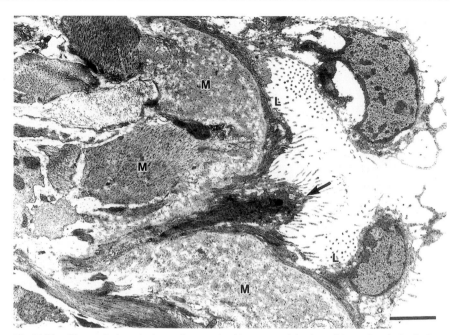

Fig. 7.17. Transmission electron micrograph of haemocoel–body wall boundary in *Incilaria fruhstorferi* Collinge (Philomycidae) 3 h after injection of yeast particles. Fibroblast-like cells (type III) protrude through the connective tissue to line the surface of the haemocoel (arrow). L, lining cell; M, muscle cell. Scale = 2 μm.

cells (lymphocyte-like). The main functions of blood cells are cellular defence (phagocytosis and encapsulation) and tissue repair. At least two types of molecules, foreign body molecules and serum molecules (opsonin and agglutinin), may be present on the plasma membrane of phagocytic cells. These molecules enable blood cells to recognize and ingest foreign particles. Ingested materials are then digested in the lysosomal system.

Acknowledgements

The authors are grateful to Professor E.L. Cooper, Laboratory of Comparative Immunology, Department of Neurobiology, University of California, Los Angeles, California, for review of the manuscript, and to Dr M. Kobayashi, Department of Medical Entomology, The National Institute of Health, Tokyo for his valuable advice.

References

Abolius-Krogis, A. (1973) Fluorescence and histochemical studies of the calcification initiating lipofuscin type pigment granules in shell-repair membrane of the snail, *Helix pomatia* L. *Zeitschrift für Zellforschung Mikroskopische Anatomie* 142, 205–221.

Abolius-Krogis, A. (1976) Ultrastructural study of the shell-repair membrane in the snail, *Helix pomatia* L. *Cell and Tissue Research* 172, 455–476.

Adema, C.E., Harris, R.A. and van Deuteron-Mulder, E.C. (1992) A comparative study of hemocytes from six different snails: morphology and functional aspects. *Journal of Invertebrate Pathology* 59, 24–32.

Anderson, R.S. and Good, R.A. (1976) Opsonic involvement in phagocytosis by mollusk hemocytes. *Journal of Invertebrate Pathology* 27, 57–64.

Bayne, C.J. (1973) Molluscan internal defense mechanism: the fate of C^{14} labelled bacteria in the land snail *Helix pomatia* (L.). *Journal of Comparative Physiology* 86, 17–25.

Bayne, C.J. (1990) Phagocytosis and non-self recognition in invertebrates. *Bioscience* 40, 723–731.

Brown, A.C. (1967) Elimination of foreign particles by the snail, *Helix aspersa*. *Nature* 213, 1154–1155.

Buchholz, K., Kuhlman, D. and Nolte, A. (1971) Aufnahme von Trypanblau und Ferritin in die Blasenzellen des Bindegewebes von *Helix pomatia* und *Cepaea nemoralis* (Stylommatophora, Pulmonata). *Zeitschlift für Zellforschung Mikroskopische Anatomie* 113, 203–215.

Cheng, T.C. (1970) Immunity in mollusca with special reference to reactions to transplants. *Transplantation Proceedings* 2, 226–230.

Cooper, E.L., Rinkevich, B., Uhlenbruch, G. and Valembois, P. (1992) Invertebrate immunity: another viewpoint. *Scandinavian Journal of Immunology* 35, 247–266.

Couch, L., Hertel, L.A. and Loker, E.S. (1990) Humoral response of the snail *Biomphalaria glabrata* to trematode infection: observation on circulating hemagglutinin. *Journal of Experimental Zoology* 255, 340–349.

Dikkeboom, R., van der Knaap, W.P.W., Meulman, E.A. and Sminia, T. (1984) Differences between blood cells of juveniles and adult specimens of the pond snail *Lymnaea stagnalis*. *Cell and Tissue Research* 238, 43–47.

Duval, A. and Runham, N.W. (1981) The arterial system of six species of terrestrial slug. *Journal of Molluscan Studies* 47, 43–52.

Fountain, D.W. (1985) The lectin-like activity of *Helix aspersa* mucus. *Comparative Biochemistry and Physiology* 80B, 795–800.

Fountain, D.W. and Campbell, B.A. (1984) A lectin isolated from mucus of *Helix aspersa*. *Comparative Biochemistry and Physiology* 77B, 419–225.

Furuta, E. and Shimozawa, A. (1983) Primary culture of cells from the foot and mantle of the slug, *Incilaria fruhstorferi* Collinge. *Zoological Magazine* 92, 290–296.

Furuta, E., Yamaguchi, K. and Shimozawa, A. (1986) The ultrastructure of hemolymph cells of the land slug, *Incilaria fruhstorferi* Collinge (Gastropoda: Pulmonata). *Anatomischer Anzeiger* 162, 215–224.

Furuta, E., Yamaguchi, K., Aikawa, M. and Shimozawa, A. (1987) Phagocytosis by hemolymph cells of the land slug, *Incilaria fruhstorferi* Collinge (Gastropoda: Pulmonata). *Anatomischer Anzeiger* 163, 89–99.

Furuta, E., Yamaguchi, K. and Shimozawa, A. (1989) Hemolymph coagulation in the land slug, *Incilaria fruhstorferi* Collinge. *Zoological Science* 6, 1204.

Furuta, E., Yamaguchi, K. and Shimozawa, A. (1990) Hemolymph cells and the platelet-like structures of the land slug, *Incilaria bilineata* (Gastropoda: Pulmonata). *Anatomischer Anzeiger* 170, 99–109.

Furuta, E., Yamaguchi, K. and Shimozawa, A. (1994) Blood cell-producing site in the land slug, *Incilaria fruhstorferi*. *Acta Anatomica Nipponica* 69, 751–764.

Furuta, E., Takagi, T., Yamaguchi, K. and Shimozawa, A. (1995) *Incilaria* mucus agglutinated human erythrocytes. *Journal of Experimental Zoology* 271, 340–347.

Hatakeyama, T., Ohuchi, K., Kuroki, M. and Yamasaki, N. (1995) Amino acid sequence of a C-type lectin CEL-IV from the marine invertebrate *Cucumaria echinata*. *Bioscience, Biotechnology, and Biochemistry* 59, 1314–1317.

Haughton, B.A.I. (1935) Note on the amoeboid elements in the blood of *Helix aspersa*. *Quarterly Journal of Microscopical Science* 77, 157–160.

Hori, E., Kusui, Y., Matui, A. and Hattori, T. (1973) A survey of *Angiostrongylus cantonensis* in the harbor side areas of Tokyo (2) on the intermediate hosts of *Angiostrongylus cantonensis*. *Japanese Journal of Parasitology* 22, 209–217.

Iguchi, S.M.M., Aikawa, T. and Matsumoto, J.J. (1982) Antibacterial activity of snail mucus mucin. *Comparative Biochemistry and Physiology* 72A, 571–574.

Iguchi, S.M.M., Momoi, T.M., Egawa, K. and Matsumoto, J.J. (1985) An *N*-acetylneuraminic acid-specific lectin from the body surface mucus of African giant snail. *Comparative Biochemistry and Physiology* 81B, 897–900.

Kapur, S.P. and Gupta, A.S. (1970) The role of amoebocytes in regeneration of shell in a land pulmonata, *Euplecta indica* (Pfeiffer). *Biological Bulletin* 139, 502–509.

Kisugi, J., Ohye, H., Kamiya, H. and Yamazaki, M. (1992a) Biopolymers from marine invertebrates. XIII Characterization of an antibacterial protein dolabellanin A, from the albumin gland of sea hare, *Dolabella auriculata*. *Chemical and Pharmaceutical Bulletin* 40, 1537–1539.

Kisugi, J., Kamiya, H. and Yamazaki, M. (1992b) Biopolymers from marine invertebrates. XII A novel cytolytic factor from a hermit crab, *Clibanarius longiarsis*. *Chemical and Pharmaceutical Bulletin* 40, 1641–1643.

Kress, A. (1968) Untersuchungen zur Histologie, Autotomie und Regeneration dreir Dotoarten *Doto coronata*, *D. pinnatifida*, *D. fragilis* (Gastropoda, Opisthobranchiata). *Revue de Suisse Zoologie* 75, 235–303.

Kubota, Y., Watanabe, Y., Otsuka, H., Tamiya, T., Tsuchiya, T. and Matsumoto, J.J. (1985) Purification and characterization of an antibacterial factor from snail mucus. *Comparative Biochemistry and Physiology* 82C, 345–348.

Miller, R.L. (1982) A sialic acid-specific lectin from the slug *Limax flavus*. *Journal of Invertebrate Pathology* 39, 210–214.

Olafsen, J.A. (1986) Invertebrate lectins: biochemical heterogeneity as a possible key to their biochemical function. In: Brehelin, M. (ed.) *Immunity in Invertebrates*. Springer-Verlag, Berlin, pp. 94–111.

Pauley, G.B., Krassner, S.M. and Chapman, F.A. (1971) Bacterial clearance in the California sea hare *Aplysia californica*. *Journal of Invertebrate Pathology* 18, 227–239.

Prowse, H.R. and Tait, N.N. (1969) *In vitro* phagocytosis by amoebocytes from the haemolymph of *Helix aspersa* (Müller). 1. Evidence for opsonic factor(s) in serum. *Immunology* 17, 437–443.

Reade, P.C. (1968) Phagocytosis in invertebrates. *Australian Journal of Experimental Biology and Medical Science* 46, 219–229.

Renwrantz, L. and Mohr, W. (1978) Opsonizing effects of serum and albumin gland extracts on the elimination of human erythrocytes from circulation of *Helix pomatia*. *Journal of Invertebrate Pathology* 31, 164–174.

Renwrantz, L., Schäncke, W., Harm, H., Erl, H., Liebsch, H. and Gereken, J. (1981) Discriminative ability and function of the immunobiological recognition system of the snail *Helix pomatia. Journal of Comparative Physiology* 141, 477–488.

Richardot, M. and Wautier, J. (1972) Les cellules à calcium du conjonctif de *Ferrissia wautieri* (Moll. Ancylidae). *Zeitschrift für Zellforschung und Mikroskopische Anatomie* 134, 227–243.

Sauerländer, R. (1976) Histologische Veränderungen bei experimentell mit *Angiostrongylus vasorum* order *Angiostrongylus cantonensis* (Nematoda) infizierten Achatschnecken (*Achatina fulica*). *Zeitschrift für Parasitenkunde* 49, 263–280.

Sminia, T. (1972) Structure and function of blood and connective tissue cells of the fresh water pulmonate *Lymnaea stagnalis* studied by electron-microscopy and enzyme histochemistry. *Zeitschrift für Zellforschung und Mikroskopische Anatomie* 130, 497–526.

Sminia, T. (1981) Gastropods. In: Ratcliffe, N.A. and Rowley, A.F. (eds) *Invertebrate Blood Cells*, Vol. I. Academic Press, London, pp. 190–232.

Sminia, T. and Boer, H.H. (1973) Haemocyanin production in pore cells of the freshwater snail *Lymnaea stagnalis. Zeitschrift für Zellforschung und Mikroskopische Anatomie* 145, 443–445.

Sminia, T., van der Knaap, W.P.W. and Edelenbosch, R. (1979) The role of serum factors in phagocytosis of foreign particles by blood cells of *Lymnaea stagnalis. Developmental and Comparative Immunology* 3, 37–44.

Stang-Voss, C. (1970) Zur Ultrastruktur der Blutzellen wirbelloser Tiere III. Über die Haemocyten der Schnecke *Lymnaea stagnalis* L. (Pulmonata). *Zeitschrift für Zellforschung und Mikroskopische Anatomie* 107, 142–156.

Stuart, A.W. (1968) The reticulo-endothelial apparatus of the lesser octopus, *Eledone cirrosa. Journal of Pathology and Bacteriology* 96, 401–412.

Tripp, M.R. (1970) Defense mechanisms of mollusks. *Journal of the Reticuloendothelial Society* 7, 173–182.

Tripp, M.R. (1992a) Phagocytosis by hemocytes of the hard clam, *Mercenaria mercenaria. Journal of Invertebrate Pathology* 59, 222–227.

Tripp, M.R. (1992b) Agglutinins in the hemolymph of the hard clam, *Mercenaria mercenaria. Journal of Invertebrate Pathology* 59, 228–234.

van der Knaap, W.P.W. (1981) Recognition of foreignness in the internal defence system of the fresh-water gastropod *Lymnaea stagnalis. Developmental and Comparative Immunology* 1, 91–97.

van der Knaap, W.P.W., Sminia, T., Schutte, R. and Boerrigter-Barendsen, L.H. (1983) Cytophilic receptors for foreignness and some factors which influence phagocytosis by invertebrate leukocytes: *in vitro* phagocytosis by amoebocytes of the snail *Lymnaea stagnalis. Immunology* 48, 377–383.

Wagge, L.E. (1955) Amoebocytes. *International Review of Cytology* 4, 31–78.

Yamaguchi, K., Furuta, E. and Shimozawa, A. (1988) Morphological and functional studies on hemolymph cells of land slug, *Incilaria bilineata*, *in vivo* and *in vitro*. In: Kuroda, Y., Kurstak, E. and Maramorosch, K. (eds) *Invertebrate and Fish Tissue Culture*. Japan Scientific Societies Press, Tokyo/Springer-Verlag, Berlin, pp. 247–250.

Yamaguchi, K., Furuta, E. and Shimozawa, A. (1989) Ultrastructural studies on internal reaction of wound-healing in the land slug, *Incilaria fruhstorferi. Zoological Science* 6, 1204.

Yamaguchi, K., Furuta, E. and Nakamura, H. (1999) Chronic skin allografts rejection in terrestrial slugs. *Zoological Science* 16, 485–495.

Yamaguchi, K., Furuta, E. and Seo, N. (2000) Histochemical and ultrastructural analyses of the epithelial cells of the body surface skin from the terrestrial slug, *Incilaria fruhstorferi*. *Zoological Science* 117, 1137–1146.

Yoshino, T.P. (1976) The ultrastructure of circulating hemolymph cells of the marine snail *Cerithidea californica* (Gastropoda: Prosobranchiata). *Journal of Morphology* 150, 485–494.

Yuasa, J.H., Furuta, E., Nakamura, A. and Takagi, T. (1998) Cloning and sequencing of three C-type lectins from body surface mucus of the land slug, *Incilaria fruhstorferi*. *Comparative Biochemistry and Physiology* 119B, 479–484.

8 Structure and Functioning of the Reproductive System

B.J. GÓMEZ

Departamento de Zoología y Dinámica Celular Animal, Facultad de Farmacia, Universidad del País Vasco, Paseo de la Universidad, 7, 01006 Vitoria, Spain

Introduction

Knowledge of reproductive biology is central to management of gastropod populations, whether it be *in situ* conservation of communities or individual threatened species, control of pest species in agriculture, control of helminth parasite transmission to humans and domestic livestock, or captive rearing of edible species. A key component of this knowledge is information on structure and function of the reproductive organs. In addition, morphology of the reproductive organs, and their functional interpretation, has played an increasingly important role in biosystematic studies that seek to provide understanding of the evolutionary history and relationships of terrestrial gastropod lineages. While there is an enormous amount of morphological information on the reproductive organs of terrestrial gastropods, primarily as a result of biosystematic studies over several centuries, detailed structural and functional information is not so plentiful and not representative of the phylogenetic diversity in these animals. Reviews of the morphology and structure of the reproductive system of terrestrial gastropods have been provided by Duncan (1975), Tompa (1984), Nordsieck (1985), Runham (1988) and Barker (Chapter 1, this volume), among others. In this chapter, the focus is on providing a general and brief account of the reproductive system in gastropod groups represented in terrestrial ecosystems, and then a more detailed interpretative summary of the condition in Stylommatophora, the most successful group of terrestrial gastropods.

©CAB *International* 2001. *The Biology of Terrestrial Molluscs*
(ed. G.M. Barker)

General Morphology of the Reproductive System

The reproduction of terrestrial gastropods is characterized by internal fertilization, often elaborate courtship behaviour and direct development by means of cleidoic eggs. Terrestrial prosobranch gastropods are dioecious. The terrestrial pulmonates are simultaneous hermaphrodites, albeit with the gonad predominantly in the male phase initially, and predominantly in the female phase towards the end of the reproductive cycle. Cross-fertilization combined with oviparity is prevalent, but self-fertilization and ovoviviparity are widespread variants that have evolved in several phylogenetic lines independently.

The reproductive system of terrestrial gastropods is complex, with varied function. Some primary functions are: (i) production of sperm and ova; (ii) storage and transport of mature gametes in a suitable medium; (iii) structural and physiological roles in the courtship and copulatory processes; (iv) transference of endogenous sperm (autosperm) to the partner's reproductive ducts; (v) reception of exogenous sperm (allosperm); (vi) supplying a site and proper medium for fertilization of ova; (vii) covering the zygote with nutritive and protective layers; (viii) oviposition; and (ix) resorption of remnant and excess reproductive products.

Terrestrial gastropods exhibit a great diversity with respect to their reproductive systems, reflecting their phylogeny. All of them have only a single gonad located in the visceral mass, usually embedded in the digestive gland. However, in several taxa, notably in the slug forms Athoracophoridae and Rathouisiidae, the gonad is located more anteriorly. The gonad opens to a gonoduct which functions to convey the gametes towards the genital pore and which varies in complexity from one higher taxon to another.

The reproductive system of prosobranch gastropods (Fig. 8.1A and B) consists of tissues differing in embryological origin. The gonad and its associated duct is of mesodermal origin and may be referred to as the coelomic gonoduct. The greater part of the coelomic gonoduct that extends through the visceral cavity to link with the pallial cavity is derived from the kidney and so is termed the renal oviduct or renal gonoduct (Fretter and Graham, 1962). In both males and females, the renal oviduct exhibits little specialization, generally being a narrow, tubular structure. The distal portion of the gonoduct is of ectodermal origin and is termed the pallial gonoduct as it is derived from invagination of the pallial roof epithelium (Fretter and Graham, 1962). It differs considerably between the sexes. In males, it is broad, with thick walls forming a prostatic gland that secretes seminal fluids. In females, the walls of this pallial gonoduct are differentiated into several glandular regions associated with secretion of nutritive and protective coatings of the fertilized zygotes to produce eggs. These female glands are voluminous and include the albumen and the capsule glands, which are placed one next to the other. In the more primitive condition, exemplified by *Pomatias* Studer (Pomatiasidae), the capsule gland opens ventrally to the pallial cavity along most of its length (e.g. Fretter and Graham, 1962; Ibañez and Alonso, 1978). In the more advanced state, achieved independently in a number of prosobranch groups, the female opening of the pallial gonoduct is limited to its distal end and thus is located anterior in the pallial cavity, near the anus (e.g. Tielecke, 1940; Varga, 1984). Two kinds of sperm pouches are present on the female pallial gonoduct, both primarily placed proximally with respect to the albumen and capsule glands. They are the bursa copulatrix, which receives the allospermatozoa and prostatic fluids from mating, and the receptaculum seminis to which these allospermatozoa pass from the bursa copulatrix to be stored and used later to fertilize oocytes. In many prosobranchs, the bursa copulatrix has become secondarily located distally with respect to the albumen and capsule glands. In some cases, the receptaculum seminis is a proper pouch diverticulate on the renal oviduct, as in most Cyclophoridae (Tielecke, 1940), but in others it is simply a slightly enlarged, folded portion of the renal gonoduct (Giusti and Selmi, 1985). In addition to the coelomic and pallial gonoducts, a muscular penis has developed from the anterior body wall, usually on the right side of the head. In some terrestrial prosobranchs, such as the Cyclophoridae, autospermatozoa are transported from the pallial gonoduct to the penis along a ciliated groove that passages the pallial cavity and along the skin

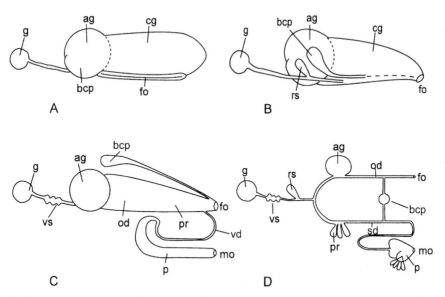

Fig. 8.1. Semi-diagrammatic sketches of the reproductive morphology in selected examples of terrestrial gastropods. (A) *Pomatias* Studer (Pomatiasidae, Prosobranchia) female (after Fretter and Graham, 1962); (B) *Cyclophorus* de Montfort (Cyclophoridae, Prosobranchia) female (after Tielecke, 1940); (C) *Carychium* Müller (Ellobiidae, Pulmonata) (after Martins, 1996); (D) *Vaginula* de Férussac (Vaginulidae, Pulmonata:) (modified from Runham, 1988). ag, albumen gland; bcp, bursa copulatrix; cg, capsule gland; fo, female opening; g, gonad; mo, male opening; od, oviductal gland; p, penis; pr, prostatic gland; rs, receptaculum seminis; sd, spermiduct; vs, vesicula seminalis.

of the dorso-lateral part of the animal's anterior (Tielecke, 1940; Varga, 1984). In others, such as Pomatiasidae, this groove has closed over, giving rise to a vas deferens that opens at or near the tip of penis. The vas deferens together with the penis constitute the cephalopodial portion of the reproductive system in terrestrial prosobranchs.

The distinction between coelomic and pallial gonoducts is also considered to exist in the Pulmonata, although ectodermal elements have also been demonstrated in the gonad (Griffond and Bride, 1987). Further, the components evident in terrestrial prosobranchs are retained in the reproductive system of terrestrial pulmonates, albeit with modification. Often different terminologies are applied to prosobranchs and pulmonates in the literature. The main features of the reproductive system of pulmonates, distinguishing them from terrestrial prosobranchs, are: (i) hermaphroditism, with both male and female gametes within a single gonad; (ii) sinking of the pallial gonoducts into the body cavity; (iii) the separation of the albumen gland from the capsule gland, to open via its own duct to the pallial gonoduct; (iv) the development of a long duct on the bursa copulatrix so that it maintains its distal opening into the distal part of the pallial gonoduct yet its sac retains an association with the pericardium; and (v) the development of an invaginable penis that is retracted into the body cavity when the animal is not sexually active (Duncan, 1975). These represent grades of evolution, rather than a linear phylogenetic sequence, as one or more similar changes have been achieved independently in other gastropod lineages. As applied to the reproductive system of the pulmonates (Nordsieck, 1985), the coelomic gonoduct generally is recognized as comprising the ovotestis and the hermaphodite duct; the latter elaborated into a vesicula seminalis, an enlarged portion where autospermatozoa are stored. The junction of coelomic and pallial gonoducts is termed the carrefour. It generally comprises a sac-like structure that apparently functions as a fertilization chamber, into which open the receptaculum seminis and the duct(s) from the albumen gland. In the majority of stylommatophorans, the receptaculum seminis is a diverticulum, sometimes subdivided into several blind-ended tubules (Lind, 1973; Haase and Baur, 1995). In other Stylommatophora, the receptaculum seminis takes the form of a simple, narrowed hairpin-shaped section in the distal end of the hermaphrodite duct (Fig. 8.2F), without sac-like processes (Runham and Hogg, 1992). This region of the reproductive tract has also been called the fertilization pouch spermatheca complex, or FPSC (Tompa, 1984) but, according to Runham (1988), the absence of any special organ in this region in some pulmonates makes the term carrefour more suitable. Some authors use the term 'talon' for this complex region, others apply the term specifically to the curved or bulbous terminal part of the hermaphrodite duct.

Being hermaphrodite, the pallial gonoducts of terrestrial pulmonates comprise the spermiduct with prostatic gland, and the oviduct with albumen gland and capsule glands. Reflecting the difference in the nature of their secretory produces, the capsule glands in pulmonates generally are referred to as oviductal glands. When spermiduct and oviduct have their lumina joined, the whole organ is called the spermoviduct. The anterior extremity of the female duct, beyond the oviductal glands and leading to the genital orifice, has predominantly muscular walls. Primitively, and in most extant species, the bursa copulatrix connects through a duct termed the pedunculus to this latter section of the female tract. In more advanced forms, the pedunculus can open to the atrium or to the penis.

Distally, the spermiduct continues as a vas deferens to the penis. Frequently, the distal portion of the vas deferens is enlarged and elaborated internally to give rise to a structure known as the epiphallus. If the vas deferens does not insert terminally on the epiphallus, an epiphallial diverticulum, called a flagellum, is formed. According to Nordsieck (1985), the term flagellum should be restricted to this particular appendage. When the vas deferens, or the epiphallus when present, enters the penis subterminally, a penial caecum (also called the penial flagellum) is formed in the proximal portion of the penis.

Auxiliary copulatory organs can be present in the cephalopodial part of the reproductive system and, while taking several forms, primarily comprise one or more glands adjoining and opening through a hard

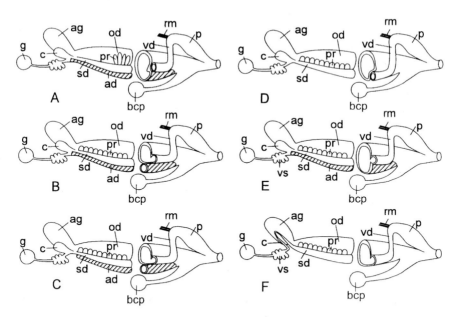

Fig. 8.2. Semi-diagrammatic sketches of the reproductive morphology in selected examples of Stylommatophora. (A) *Succinea* Draparnaud (Succineidae); (B) *Discartemon* Pfeiffer (Streptaxidae); (C) *Gonaxis* Taylor (Streptaxidae); (D) *Achatinella* Swainson (Achatinellidae); (E) *Achatina* de Lamarck (Achatinidae); (F) *Arion* de Férussac (Arionidae). ag, albumen gland; ad, allospermiduct; bcp, bursa copulatrix; c, carrefour; g, gonad; od, oviductal gland; p, penis; pr, prostatic gland; rm, penis retractor muscle; sd, spermiduct; vd, vas deferens; vs, vesicula seminalis (modified from Visser, 1977). Note that Barker (Chapter 1, this volume) treats Succineidae, and the related Athoracophoridae, as sister taxa to the Stylommatophora.

papilla (Nordsieck, 1985). Primitively they are thought to be associated with the male genitalia, but in many extant Stylommatophora reputedly homologous structures are associated with the atrium or with the female genitalia.

The reproductive system of Carychiinae, the most terrestrial of the archeopulmonate Ellobiidae, is monaulic in that one bisexual pallial gonoduct is present. The albumen and oviductal glands are located at the proximal and middle sections of the duct, respectively. The female orifice is directly at the distal extremity of the pallial gonoduct, in the body wall on the right side. The vas deferens runs anteriorly to the penis that in turn opens through the male orifice in the anterior–lateral aspect of the body (Fig. 8.1C). Consequently, the carychine reproductive system possesses two openings, a condition known as diatremy. In this group, the duct of the bursa copulatrix opens to the pallial gonoduct near the female opening (Fig. 8.1C).

The gymnomorph slugs in the families Vaginulidae and Rathouisiidae are diaulic and diatrematic (Fig. 8.1D), with the male and female pallial gonoducts separated from the carrefour up to their respective openings. They are characterized in part by a portion of the vas deferens being embedded in the body wall tissues, and a duct (canalis junctor) communicating between the vas deferens and the bursa copulatrix. The female orifice may be retained mid-laterally (Rathouisiidae) or displaced to the body posterior (Vaginulidae). The auxiliary copulatory organs are well developed and associated with the penial opening.

Most Stylommatophora (Fig. 8.2A–F) are semi-diaulic, in that the section of the pallial gonoduct below the entry of albumen gland duct(s) is incompletely divided into a voluminous oviductal channel and a narrow sperm groove. Some stylommatophorans (e.g. Achatinellidae) have a diaulic reproductive system in that the female and male pallial gonoducts are separated (Fig. 8.2A and D). A third channel, referred to as an allospermiduct (Nordsieck, 1985), is present additionally in the spermoviduct of stylommatophorans such as Achatinidae, Clausiliidae, Helicidae and Subulinidae (Fig. 8.2E). In other groups, this so-called allospermiduct is present as an enclosed duct running separately to the spermoviduct to reach the carrefour, as in the streptaxids *Oophana* Ancey and *Discartemon* Pfeiffer (Fig. 8.2B). A blind-ended appendage on the pallial gonoduct, as in many other Streptaxidae (Fig. 8.2C) and in the Pupilloidea, often is considered part of the allospermiduct system (Steenberg, 1925; Visser, 1973). In other groups (e.g. Arionidae, Agriolimacidae), a channel or duct structurally recognizable as an allospermiduct apparently is absent (Fig. 8.2F), although allospermatozoa still passage the spermoviduct to reach the receptaculum seminis.

In Stylommatophora (Fig. 8.2), the male and female ducts converge distally to a common opening to the body exterior, thus representing a condition known as syntremy. Generally, the male and female genitalia open to a pouch known as the genital atrium from whence they open to the exterior.

Sperm transfer of many stylommatophorans is effected by the use of spermatophores (Fig. 8.3). The spermatophore is formed in the epiphallus, and is transferred via the everted penis to the pedunculus of the bursa copulatrix in the partner. In several families, the pedunculus is equipped with a diverticulum that functions as the site of spermatophore receipt.

Functional Reproductive Anatomy in Stylommatophora

Gonad and renal gonoduct

The lobes of the gonad consist of several to many closely packed acini. Each acinus is a rounded or pear-shaped sac, closed at the basal end but opening to the hermaphrodite duct at the apical end by means of an efferent ductule. The wall of each acinus comprises an inner layer of flattened epithelium, a basement membrane, and an outer connective layer with pigment cells (called Ancel's layer), blood vessels and nerves (Lusis, 1961). Two types of cells become differentiated inside the acini:

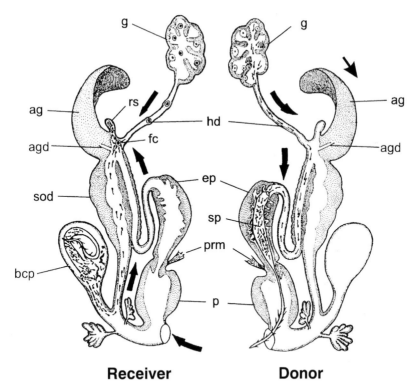

Receiver   **Donor**

Fig. 8.3. Semi-diagrammatic sketches of the reproductive system of a generalized stylommatophoran, illustrating routes of autospermatozoa and allospermatozoa. ag, albumen gland; agd, albumen gland duct; bcp, bursa copulatrix; ep, epiphallus; fc, fertilization chamber; g, gonad; hd, hermaphrodite duct; p, penis; prm, penis retractor muscle; rs, receptaculum seminis; sod, spermoviduct; sp, spermatophore; vs, vesicula seminalis (after Wiktor, 1987).

gametes (spermatozoa and ova) and auxiliary cells (Sertoli and follicular cells, respectively). Spermatozoa and ova develop side by side in the same gonadial acinus. The germinal ring located in the neck of each acinus has been described in *Cantareus aspersus* (Müller) (Helicidae) (Griffond and Bride, 1987) as the only region involved in gamete proliferation.

A cellular gradient occurs in the ovotestis, with oogenesis in the acinar cortex and spermatogenesis in medullar areas. The follicle cells establish a permeable physical barrier between the male and female gametes. Once spermatogenesis has finished, the spermatozoa remain in the acinar lumina before they are transported along the ovotestis efferent ductules and the hermaphrodite duct for storage in the vesicula seminalis. Oocytes undertake their development and maturation entirely within the acini and are transported to the carrefour when fertilization is to take place.

The hermaphrodite duct is lined primarily by a ciliated epithelium, surrounded by connective tissue, circularly arranged smooth muscle and vesicular connective tissue cells (Hodgson, 1996). In its proximal part, formed by the confluence of the ovotestis efferent ductules, the epithelial cells are cuboidal, but they enlarge quickly to become prismatic. However, the mid-section of the hermaphodite duct is differentiated as a vesicula seminalis, and its epithelium consists of two types of cells, ciliated and secretory. The ciliated cells are restricted to a longitudinal band that occupies approximately one-third of the entire circumference. Ciliated and secretory portions can be more or less separated by projecting folds, as in *Achatina fulica* Bowdich (Achatinidae) (Breckenridge and Fallil, 1973). In the Streptaxidae and Enidae, one or several diverticula are present in this organ. When the autospermatozoa are stored in the vesicula seminalis, they are immobile, closely packed and embedded in a seminal fluid apparently secreted by the secretory epithelial cells of this organ. Glycogen secretion from the ciliated cells is used for the maintenance of the autospermatozoa during their storage (Kugler, 1965; Els, 1978; Hodgson and Shachak, 1991; Hodgson, 1996). As the number of spermatozoa in storage increases, the vesicula seminalis becomes more and more distended: the degree of distension in the vesicula seminalis could be used as an indicator of reproductive maturity in semelparous species. Phagocytosis of the autospermatozoa by the secretory cells in the epithelium (Rigby, 1963; Lind, 1973; Hodgson, 1996) indicates that this organ also functions to digest excess autospermatozoa at the end of the reproductive cycle. The structural differences exhibited by spermatozoa in the gonad and hermaphrodite duct suggest that further maturation of autospermatozoa may occur in the vesicula seminalis (Hodgson, 1996).

The carrefour

In Stylommatophora, the structures associated with the carrefour are involved in the storage of exogenous spermatozoa, in the fertilization of oocytes and in the coating of zygotes with the albumen layer. Runham and

Hogg (1992) found that, in *Deroceras reticulatum* (Müller) (Agrio-limacidae), autospermatozoa pass from the hermaphrodite duct into the carrefour within 10 min of the beginning of the stimulatory phase of copu-lation. Within 15 min, the carrefour had become filled with autosperma-tozoa and by 20 min these spermatozoa had all passed to the spermiduct. In *Arion hortensis* de Férussac (Arionidae), these authors found that 90 min after the start of courtship autospermatozoa had entered the carre-four, and by 120 min had passed to the spermoviduct. Lind (1973) showed that, in *Helix pomatia* Linnaeus (Helicidae), autospermatozoa are released from the hermaphrodite duct within 55 s of penial intromission into the partner's vagina, in readiness for incorporation into the spermatophore and transfer to that partner.

In both *D. reticulatum* and *A. hortensis*, allospermatozoa received during mating reach the carrefour within 4 h of copulation (Runham and Hogg, 1992). In *H. pomatia*, the first allospermatozoa reach the receptaculum seminis within 2–3 h of copulation, and the majority reach this site within the first 12 h (Lind, 1973). Lind stated that in *H. pomatia*, only about 0.1% of the allospermatozoa that are transferred at copulation reach the receptaculum seminis.

Usually, allospermatozoa are stored in the receptaculum seminis, with the spermatozoa oriented towards and their apical ends in close con-tact with the epithelial cells (Tompa, 1984). This storage can be extended to more than a year in species with a long life span, as in Achatinidae and Helicidae (Lind, 1973; Bayne, 1973; Baur, 1988; Haase and Baur, 1995). In contrast, no evidence has been obtained for storage of allospermatozoa in the carrefour of *D. reticulatum* and *A. hortensis* (Runham and Hogg, 1992). In these two species, allospermatozoa are instead stored in the proximal portion of the hermaphrodite duct, suggesting that some of them may be transferred to a new partner, together with autogenous spermatozoa, in subsequent matings (Runham and Hogg, 1992). Repeated mating during a single reproductive season is not unusual in terrestrial gastropods (e.g. Haase and Baur, 1995).

The fertilization of oocytes occurs in a specialized region of the carrefour, which is often recognizable externally as a pouch. Secretory cells have been shown to be present in the walls of this fertilization cham-ber (Kugler, 1965; Sirgel, 1973; Gómez, 1991); their secretions are thought to provide a medium for gamete fusion. Up to 50 oocytes were observed in histological sections of the carrefour in *A. hortensis* by Runham and Hogg (1992), and approximately 200 fertilized ova in *Strophocheilus oblongus* (Müller, 1774) (Strophocheilidae) by Tompa (1984).

Albumen gland

The albumen gland in Stylommatophora increases in size with sexual maturation of the animal. The size of this gland, which synthesizes the perivitelline fluid to be added to eggs to provision the developing embryo, seems to be the determinant of the maximum number of eggs that can be produced at any one time (Tompa, 1984). The gland is composed of a large number of branched tubules, lined with voluminous secretory cells whose cytoplasm is filled by large secretion granules. The secretion is released from the secretory cells into the tubules that coalesce to one or two major ducts that open to the carrefour. The tubules and the major ducts are lined with a ciliated epithelium. The main component of the albumen gland secretion has been shown to be galactogen, a highly branched polysaccharide composed predominantly of D-galactose. Duncan (1975) summarized the synthesis process for galactogen. Protein elements, glycosaminoglycans and glycogen have also been showed to be associated intimately with the galactogen (Smith, 1965; Els, 1978; Ramasubramaniam, 1979).

Spermoviduct

The oviductal channel and spermiduct of stylommatophorans are lined with ciliated epithelia, with the cilia being longer and more numerous in the male part. Opening to the lumen among the ciliated cells are several types of secretory cells whose cell bodies are located subepithelially. A thin outer layer of connective tissue, with sparse smooth muscle fibres and interspersed pigment cells, covers the whole pallial gonoduct.

There has been some debate about the exact number of secretory cell types present in the spermiduct of stylommatophorans. Three different types of secretory cells have been demonstrated in the spermiduct of *Arion subfuscus* (Draparnaud) by histochemical and ultrastructural means (Zubiaga *et al.*, 1989, 1990b). The nature of the secretions from spermiduct secretory cells has been investigated in several species (Smith, 1965; Els, 1978; Zubiaga *et al.*, 1990b), showing that they are mostly glycosaminoglycans and neutral carbohydrates. It is considered that these secretions contribute to the seminal fluid that is added to autospermatozoa on their arrival from the hermaphrodite duct (Els, 1978; Zubiaga *et al.*, 1990b), with at least the neutral carbohydrates providing nourishment for the spermatozoa. Secretory cells have not been detected in the spermiduct of *Deroceras caruanae* (Pollonera) or *A. fulica* (Breckenridge and Fallil, 1973; Sirgel, 1973), indicating that their presence may not be a universal feature among stylommatophorans.

Despite some variation in morphology of the prostatic gland, the fine structure of prostatic cells has been shown to be remarkably uniform among investigated stylommatophorans (Zubiaga *et al.*, 1989). The prostatic gland is known to synthesize a mixture of proteins and phospholipids (Smith, 1965; Zubiaga *et al.*, 1990b), with a low concentration of glycosaminoglycans and neutral carbohydrates. The prostatic secretion is the main component of the seminal fluids, providing a fluid medium in which autospermatozoa are embedded, immobile, during their passage through the distal parts of the reproductive tract (Bayne, 1967; Els, 1978).

Histochemically, the prostatic secretions and the spermatophore substances appeared identical in *Arion ater* (Linnaeus) (Runham, 1988). Nevertheless, the prostatic gland continues to be active after mating when male functions have diminished, indicating that this gland may have other functions. It has been shown that prostatic gland secretions can reach the oviduct (Breckenridge and Fallil, 1973; Els, 1978), but it remains uncertain whether prostatic secretions also contribute to egg formation or to activation of received allospermatozoa.

The walls of the oviductal gland are highly folded, as is expected for a duct that distends considerably to allow the passage of the eggs. The wall of the oviductal gland is entirely comprised of large, flask-shaped subepithelial secretory cells that open, individually, into the lumen of the female channel by means of short apical necks passing among the cuboidal, ciliated epithelial cells. Several distinct secretory cell types are present. Els (1978) and Zubiaga *et al.* (1989, 1990b) identified, by histochemical and ultrastructural features, four cell types in *Deroceras laeve* (Müller) and *A. subfuscus*, while Visser (1973) identified five cell types in *Gonaxis gwandaensis* (Preston) (Streptaxidae). The cells of each type are grouped, giving rise to specialized regions along the female pallial gonoduct (Els, 1978; Zubiaga *et al.*, 1990b). Histochemical studies (Smith, 1965; Bayne, 1967; Els, 1974, 1978; Zubiaga *et al.*, 1990b) demonstrate that the oviductal gland, as a whole, secretes mostly glycosaminoglycans, with small amounts of neutral carbohydrate and protein. Its secretions form the egg perivitelline membrane, the jelly or organic matrix of the inner egg cover, and the outer egg cover (Smith, 1965; Bayne, 1968). These layers are deposited sequentially during the descent of zygotes along the oviductal channel. Tompa (1984) proposes the term perivitelline membrane gland for the most proximal glandular region of the oviduct, which synthesizes the perivitelline membrane of eggs.

The calcium content per egg has been found to increase gradually as it passes along the oviduct, showing that the whole organ is involved in calcifying the eggs (Tompa and Wilbur, 1977; Tompa, 1984). The outer egg coat of many stylommatophorans has high concentrations of calcium (often forming a definitive calcareous shell), while the walls of the oviductal gland contain only small quantities of calcium. Rigby (1963), for example, found that a single egg of *Oxychilus cellarius* (Müller) (Zonitidae) contained 76.5–266 µg of calcium, while the spermoviduct contained only 1.3–5.6 µg of this element. Egg calcium is therefore mobilized from elsewhere in the body and transported to the pallial gonoduct via the haemolymph (Tompa and Wilbur, 1977). Ultrastructural studies show that the non-secretory epithelium that lines the oviductal gland has the structural features of transporting epithelia – including basal infolds, highly folded lateral walls, dense microvilli and a well-developed mitochondrial system – suggesting that it is these epithelial cells that transport calcium into the oviductal gland lumen (Tompa, 1984; Zubiaga *et al.*, 1989). In ovoviviparous species (Tompa, 1979), the distal portion of the oviductal gland, together with the adjoining free oviduct, functions as a uterus for brooding of young after their hatching from eggs retained in the female ducts.

The allospermiduct constitutes the route for the allospermatozoa, but there has been much debate about which structures or tissues in the stylommatophoran reproductive system perform this function. The recognition of a functional allospermiduct is readily apparent when a channel or duct provides a connection from the cephalopodial part of the reproductive system to the carrefour (Fig. 8.2A, B and E). The homology of these channels and ducts has been assumed, but ontogenetic data are lacking. In some stylommatophorans, such as *Discartemon* in Streptaxididae (Fig. 8.2C), a function as a route for allospermatozoa is assigned by some authors (e.g. Visser, 1977) to the diverticulum that occurs on the pallial gonoduct, in a position equivalent to the allospermiduct (seminal duct diverticulum *sensu* Visser or cul-de-sac de l'oviducte *sensu* Steenberg). Visser further postulates that this diverticulum is homologous to the diverticulum of the bursa copulatrix pedunculus seen, for example, in *Helix* Linnaeus of Helicidae. That this diverticulum has no apparent connection to the carrefour via a recognizable allospermiduct in the spermoviduct does not negate this hypothesis because many stylommatophorans achieve the function of a route for allospermatozoa without any apparent specialized channel (Fig. 8.2D and F). More problematic for this hypothesis is the occurrence of a diverticulum associated with the female pallial gonoduct in many basal pulmonates and variously retained in Stylommatophora (see Barker, Chapter 1, this volume), without any apparent route function for allospermatozoa.

In the helicid *Theba pisana* (Müller), the allospermiduct (not diverticulum of the bursa copulatrix) has been described as a simple groove, while in streptaxids (Visser, 1973) and in the Pupilloidea *sensu lato* (Steenberg, 1925; Gómez, 1991) it is a duct lined by secretory cells. The secretions elaborated by the allospermiduct have not been investigated extensively, but possible function is as a nutritive medium for activated allospermatozoa as they make their way from the site of deposition in the distal reproductive organs (pedunculus and free oviduct) to the carrefour.

Free oviduct and vagina

Both the free oviduct and vagina have a thick muscular wall, which may be attributed to their role in copulation and oviposition. Ultrastructurally, the epithelium of the free oviduct exhibits the structural features of calcium-transporting tissues. Tompa and Wilbur (1977) demonstrated that in *C. aspersus*, the calcium content of eggs continues to increase as the eggs pass along the free oviduct, but remains constant during passage along the vagina. Subepithelial secretory cells, producing a secretion of probable lubricant function, have been demonstrated in these terminal female ducts in several species (Noyce, 1973; Sirgel, 1973; Els, 1974,

1978), but apparently they are not a universal feature among Stylommato-phora (Smith, 1965).

Vas deferens and epiphallus

The vas deferens is a narrow duct with the luminal epithelium of ciliated cells generally folded extensively. Underlying the luminal folds there is a continuous layer of circularly oriented muscular fibres, and the whole surrounded by connective tissue containing many longitudinal muscular fibres. Peristalsis in the wall, together with ciliary action, contribute to the movement of seminal fluids along the duct, including their expulsion during mating (Runham, 1988).

When present, the epiphallus is usually a highly muscular organ, with the lumen larger and more folded than the vas deferens lumen, often with a finely papillate surface. The structure of the wall is similar to that of the vas deferens, but with the circular and longitudinal muscular layers being more developed. The lining epithelium may be ciliated or non-ciliated, depending on the species, but the presence of long microvilli is a general feature throughout the Stylommatophora. As indicated above, the epiphallus is responsible for the formation of spermatophores.

Studies by Lind (1973) show that, in *H. pomatia*, spermatophore formation is a very rapid process, with complete spermatophores present in the epiphallus within 70 s of the start of copulation. The autosperma-tozoa, embedded in the fluids systhesized by the secretory cells of the spermiduct, together with prostatic secretions, constitute the spermato-phore core. Secretions synthesized by the vas deferens and epiphallus constitute the spermatophore peripheral tunicle and are moulded and hardened within the epiphallus. Only after this peripheral tunicle has been formed is the spermatophore filled with autospermatozoa and seminal fluids from the spermiduct via the vas deferens (Fig. 8.3). Due to the differential structure of the epiphallus and in particular the configuration of the lumen, the hardening process results in formation of spermatophores of species-specific shape and surface characteristics.

Connective interstitial cells, containing large spherules in their cyto-plasm, have been described as occurring in great numbers in the wall of the epiphallus in the Arionidae, Helicidae, Milacidae and Zonitidae (Noyce, 1973; Els, 1974; Runham, 1988). Besides glycosaminoglycans (Lusis, 1961), these cells probably contain calcium salts. The presence of calcium as one of the primary elements of the spherules in the epiphallus interstitial cells has been demonstrated in *Oxychilus atlanticus* (Morelet & Drouet) (Gómez and Rodrigues, 2000). A saccular gland that produces calcium salts is differentiated in the epiphallus of several sytommato-phorans, including Urocyclidae, but its definitive role in spermatophore formation has not been established. Runham (1988) has proposed that calcium is probably associated with the hardening of the spermatophore, but this element apparently is present in only very low concentrations in the tunicle of the *C. aspersus* spermatophore (Koene and Chase, 1998). In other cases, such as in the Milacidae, the spermatophore wall is believed to be of conquioline-like material (Wiktor, 1987). Zubiaga *et al.* (1990b) have found that the vas deferens and epiphallus secrete proteins and neutral polysaccharides in *A. subfuscus*, the two predominant substances of the spermatophore tunicle in this species.

Penis

The penis of the Stylommatophora is a highly muscular organ that is everted and thus extruded during copulation. It functions as a carrier of autospermatozoa, which are usually deposited within the reproductive tract of the partner either in the form of a spermatophore or as a simple seminal mass. In some Stylommatophora, the penis does not penetrate the reproductive tract of the partner and exchange of spermatozoa is external between the intertwined male genitalia; the well-documented copulatory process of *Limax maximus* Linnaeus and *Limax cinereoniger* Wolf (Limacidae) exemplify this type of behaviour (Quick, 1960; Tompa, 1984). In yet further species, the penis is greatly reduced, and in all these cases sperm transmission is achieved by opposition of the genital pores, or by using parts of the terminal female genitalia in a penis-like manner. Probably, as has been observed in Polygyridae by Emberton (1994), the intromission of the penis together with internal deposition of spermatozoa is common in gastropods with rapid gamete exchange, whereas external deposition of sperm masses is typical of species with lengthy courtship and mating. Lind (1973) observed that in *H. pomatia* the transfer of the spermatophore was completed within 5–6 min of intromission into the partner's vagina.

Penial morphology is extremely varied as a result of great evolutionary modifications. The morphology can be observed to be species-specific and thus interpreted as indicating that the penis is the prime species recognition character in mating and particularly in copulation success. The basic structure of the penis of stylommatophorans is a tube consisting of a non-ciliated, non-glandular epithelium, surrounded by a thick muscular wall with inner circular and outer longitudinal layers (e.g. Noyce, 1973). Its proximal lumen often is largely occupied by a terminal part of the vas deferens or epiphallus as a large, conical papilla, or verge. This whole structure may be surrounded by a penial sheath, which mainly comprises muscle fibres embedded within connective tissue. The papilla is completely protruded when the penis is everted during copulation. The contraction of the penis wall effects the hydrostatic pressure necessary to evert the penis, while a retractor muscle that arises from the columellar muscle or diaphragm and inserted into the penis wall effects retraction. Special glandular regions can be present in the penis wall, facilitating such eversion and retraction. The epithelium lining the lumen of the penis often is folded transversely, and several types of raised ridges or folds, known as pilasters, and spines, can function as stimulator or hold-fast surfaces during copulation. Calcium crystals can occur in the extracellular spaces in the penial wall, as demonstrated for *C. aspersus* and *Cepaea nemoralis* (Linnaeus) (Helicidae) by Runham (1988).

In some species, or populations, of stylommatophorans, the male genitalia can be reduced or absent. These animals are referred to as hemiphallic and aphallic, respectively. In some species, euphallic individuals, with the male copulatory apparatus fully developed, can co-exist with hemiphallic and/or aphallic individuals. The female reproductive organs, however, are always fully developed. It is considered that euphallic specimens can reproduce by cross-fertilization, while the hemiphallic and aphallic specimens self-fertilize. Nevertheless, it has also been shown that euphallic individuals may function as males and aphallic individuals as females when copulation occurs between these forms (Pokryszko, 1987). The reduction of the male genitalia is rather common in minute gastropods of the families Pupillidae, Valloniidae and Vertiginidae, leading Boycott (1917) to suggest that self-fertilization is favoured in these animals owing to their limited locomotory abilities and, consequently, their difficulties in finding mates. However, aphally also occurs in gastropod families with larger species, including Chondrinidae, Cochlicopidae, Zonitidae and Agriolimacidae. Furthermore, self-fertilization is common in stylommatophorans across the full range of animal sizes.

Bursa copulatrix

The wall of the bursa copulatrix consists of a columnar, microvillus epithelium, surrounded by a thin layer of connective tissue with embedded muscle fibres and vesicular cells (Sirgel, 1973; Rogers *et al.*, 1980). The epithelial cells produce an apocrine secretion when they are in the active phase (Els, 1974; Stears, 1974; Vorster, 1983). The wall of the pedunculus is often internally folded or ridged, lined by columnar, non-secretory epithelium and surrounded by a thick outer layer of connective tissue with numerous muscle fibres and vesicular cells. The investment in muscle allows for peristaltic movement of the pedunculus wall. Where the diverticulum occurs, its structure is similar to that of the pedunculus, except for a reduced muscle component (Noyce, 1973).

Several studies have shown that, in Stylommatophora, the bursa copulatrix is the site where excess gametes and other reproductive products are digested and resorbed (e.g. Lind, 1973; Gómez *et al.*, 1991). At times other than at copulation, autospermatozoa are expelled periodically from the hermaphrodite duct and pass to the bursa copulatrix where they are digested. Excess secretions from the albumen gland, oviductal glands and seminal channel, as well as the spermatophores and allospermatozoa received during copulation, are also digested in this organ (Lind, 1973; Els, 1978). Initially, there is extracellular digestion by enzymes secreted in apocrine vesicles. Reeder and Rogers (1979) detected hydrolytic enzymes, protease, DNase and RNase in the lumen of the bursa copulatrix. The partially digested materials are taken into the epithelium by endocytosis (Gómez *et al.*, 1991) and digestion products are resorbed intracellularly.

The bursa copulatrix previously has been viewed as the organ where allospermatozoa are stored and activated after mating. However, it has been well established that allospermatozoa and autospermatozoa entering the bursa are immobilized rapidly and permanently (Lind, 1973). Lind (1973) demonstrated that in *H. pomatia* the apical end of the spermatophore reaches the bursa within 3–4 h of copulation, but its tail remains within the pedunculus for 4–5 h. This time is long enough to allow reactivation of the allospermatozoa and their escape via the tail of the spermatophore into the pedunculus and thence along the seminal duct to the receptaculum seminis (Fig. 8.3). During this period of time, the spermatophore tunica has the function of protecting the enclosed allospermatozoa against the digestive fluids of the bursa copulatrix. This method of gamete transfer may allow only the most active allospermatozoa to pass to the receptaculum seminis for use in subsequent fertilization of ova (Haase and Baur, 1995). In *O. atlanticus*, Rodrigues & Gómez (1999) have shown that the spermatophore is released directly into the bursa copulatrix and pedunculus with the penial caecum everting into the bursa copulatrix during copulation. Working with *Arianta arbustorum* (Linnaeus), Haase and Baur (1995) showed that when animals do not want to store spermatozoa from their mate, they are able to direct part or all of the received allospermatozoa to the bursa copulatrix by means of rapid peristaltic activity of the pedunculus. This provides a post-mating strategy for control of genetic diversity among stored spermatozoa, particularly in those species that copulate several times with different partners during the same reproductive cycle.

Different strategies are employed by species that do not produce spermatophores. In *Succinea putris* (Linnaeus) (Succineidae), for example, seminal material is deposited into the vagina of the mate and some allospermatozoa migrate to the receptaculum seminis. The remaining allospermatozoa are captured by the bursa (Rigby, 1965). Tomiyama (1994) suggests that in *A. fulica* the allospermatozoa are introduced slowly into its partner over a period of several hours, thus avoiding capture and digestion by the bursa copulatrix.

Genital atrium

The atrial wall is generally muscular and lined by a columnar or cuboidal epithelium. The atrium often is provisioned with secretory material. This originates in large flask-shaped, subepithelial secretory cells, as in *Sheldonia* Ancey (Urocyclidae) and *Arion* de Férussac, which open directly into the atrial lumen through the epithelium (Smith, 1965; Vorster, 1973; Zubiaga *et al.*, 1990a). The secretion comprises a complex mixture of lipids that are applied to the outside of the eggs during oviposition (Bayne, 1968; Runham *et al.*, 1991), but its function presently is unknown.

Auxiliary copulatory organs

The reproductive system of Stylommatophora is plesiomorphically equipped with an auxiliary copulatory organ that plays an active role in mating. This auxiliary organ is thought to facilitate reciprocal copulation (Nordsieck, 1985) and mutual exchange of male gametes (Tompa, 1984). Reflecting the great diversity in morphology, numerous terminologies have been applied to the auxiliary organ and its components in Stylom-matophora (for a review, see Tompa 1984). In many stylommatophorans, the auxiliary copulatory organ comprises a tubular gland opening through a prominent papilla into the penis (Fig. 8.4A). During copulation, the papilla is protruded from the genital pore and played against the partner's body wall or genitalia, and may even be introduced into the mate's genitalia. This activity can be accompanied by expulsion of secretory material from the auxiliary copulatory organ's gland.

In basal pulmonates (see Barker, Chapter 1, this volume) the papilla is sometimes elaborated into a cartilaginous stylet that, assuming homology,

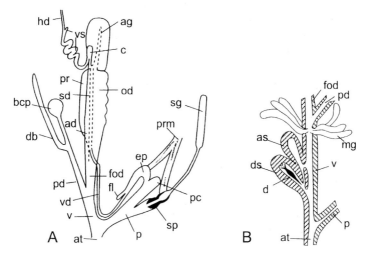

Fig. 8.4. Semi-diagrammatic sketches of the reproductive system of generalized stylom-matophorans. (A) Plesiomorphous condition as exemplified by a generalized orthurethran (after Nordsieck, 1985). (B) The female genitalia of a generalized helicoidean, illustrating the auxillary copulatory organ opening to the vagina (adapted from several sources). ad, allospermiduct; ag, albumen gland; as, accessory-sac of auxiliary copulatory organ; at, atrium; bcp, bursa copulatrix; c, carrefour; d, dart; db, diverticulum of bursa copulatrix; ds, dart-sac of auxillary copulatory organ; ep, epiphallus; fl, epiphallial flagellum; fod, free oviduct; hd, hermaphrodite duct; mg, glands of the auxillary copulatory organ; od, oviductal gland; p, penis; pc, penial caecum; pd, pedunculus of bursa copulatrix; pr, prostatic gland; prm, penis retractor muscle; sd, spermiduct; sg, gland of auxillary copulatory organ; sp, papilla of auxillary copula-tory organ; v, vagina; vd, vas deferens.

suggests that the plesiomorphic condition in Stylommatophora is an auxiliary copulatory organ equipped with a stylet-like structure. Indeed, in many Stylommatophora, the papilla of the auxiliary copulatory organ is equipped with a sharp, calcified or chitinous dart within a so-called dart sac. One or several glands, in a compact glandulous mass or elongate tubules, open to the sac. This configuration is a feature common in helicoidean families, where the organ is placed on the vagina (e.g. Dillaman, 1981; Nordsieck, 1987; Gómez et al., 1996). In addition to the Helicioidea, darts are also to be found in the Ariophantidae, Parmarionidae, Philomycidae, Urocyclidae, Vitrinidae and Zonitidae. Often the courtship in dart-bearing stylommatophorans is characterized by corporal frictions or body wall injuries inflicted by the dart(s) protruded from the genitalia (Tompa, 1984; Gómez, 1991). These mechanical stimulations apparently are combined with pheromone injections (Adamo and Chase, 1990) and serve to synchronize or accelerate the reciprocal copulation in mating pairs. Several authors have pointed out that pre-copulation behaviour contributes to species recognition at mating. Leonard (1991) argues for the idea that pre-copulatory courtship behaviour should enforce reciprocity in exchange of spermatozoa, specifically by preventing individuals from acting only as receivers. Reise (1995) also suggests that this courtship behaviour is important in the physiological recognition of possible mates. In *A. fulica*, which lacks an auxiliary copulatory organ, the duration of this pre-copulation behaviour, which includes injuries by biting, is less than 5 min (Tomiyama, 1994). In contrast, pre-copulatory behaviour in *C. aspersus*, involving dart 'shooting', may last 30 min (Adamo and Chase, 1988).

In Helicidae and Hygromiidae, the dart comprises a bladed, usually hollow spicule of calcium carbonate (e.g. Hunt, 1979). In the case of mating *C. aspersus*, the dart sac is everted rapidly from the vaginal region of the already everted terminal genitalia. The dart is pushed out rapidly so as to pierce the flesh of the partner. The dart carries approximately 2 mg of white mucus secreted by the gland. The dart sac is then withdrawn. The dart is never propelled through the air, but is torn from the dart sac on becoming lodged in the partner's tissues (Chung, 1987; Adamo and Chase, 1988). Adamo and Chase (1996) suggest that such 'dart-shooting' manipulates the endocrine system of the partner. They argue for the introduction of hormonal factors when the dart gains access to the mate's haemocoel, thus influencing mate receptiveness and fertilization success.

A common variation of the auxiliary copulatory organ in Stylommatophora agriolimacids, helicarionids, urocyclids, streptaxids and zonitids among others, is a solid, muscular pad or papilla-like structure located within the penial, atrial or vaginal lumen and with a textured surface (e.g. Sirgel, 1973). The gland of the auxiliary copulatory organ often is reduced in these taxa, comprising a mass of secretory cells opening through tubular connections or opening directly into the lumen of the genitalia. Reise (1995) summarizes the role of this organ in the mating behaviour of agriolimacids. The auxiliary copulatory organ of *Urocyclus*

Gray (Urocyclidae) is elaborated with one or several solid darts (van Mol, 1970), recalling the more plesiomorphic condition described above for helicoids.

The glands of the auxiliary copulatory organ elaborate and store the secretions (Tompa, 1984; Nordsieck, 1987; Panha, 1987; Adamo and Chase, 1990; Prieto *et al.*, 1993; Gómez *et al.*, 1996). The secretion is apocrinous, being composed mainly of glycosaminoglycans and proteins (Bornchen, 1967; Gómez *et al.*, 1996). Both the dart sac and glandular parts of the auxiliary copulatory organ possess thick, mostly muscular walls that facilitate the rapid expulsion of the dart and mucous substances. The dart of helicoidean species is lost during courtship, so that the dart-sac epithelium must renew the dart if this structure is to feature in subsequent courtship. In *C. aspersus*, this dart renewal takes 5–7 days (Dillaman, 1981; Tompa, 1982). The dart is produced by a calcium-secreting epithelium (Runham, 1988) located in the dart-bearing papilla within the dart sac (Tompa, 1982).

Conclusions

There is a wealth of information on the morphology of the reproductive system of terrestrial gastropods in the malacological literature, primarily as a consequence of biosystematic studies. None the less, detailed information on the structure and function of the reproductive system of terrestrial gastropods is limited to rather few species and largely focused on members of Achatinidae, Agriolimacidae, Arionidae, Helicidae, Limacidae and Streptaxidae. On the basis of studies in these families, it is tempting to conclude that those ducts and glands that comprise the essential elements of the reproductive system in stylommatophorans are structurally and functionally well characterized. However, there remains much unstudied structural and functional variation in Stylommatophora. The dearth of information is even more apparent when consideration is given to the archeopulmonate, gymnomorph and prosobranch groups that occupy terrestrial ecosystems.

Significant advances in our understanding of reproduction in terrestrial gastropods have been made in the last decade. Particular contributions have come from ultrastructural analyses of different components of the reproductive system, as well as the identification of the nature of the secretions elaborated. Three-dimensional reconstructions of the reproductive organs, particularly that of the carrefour region, have been obtained from serial histological sections and have shed light on storage of allospermatozoa and fertilization functions. New aspects concerning the copulatory process, transference of spermatozoa, gamete transport and storage in the reproductive system have also been investigated. None the less, further structural and biochemical studies are required to support recent behavioural and ecological studies on the significance of the multiple mating and gamete fitness in the reproductive biology of terrestrial gastropods. It has been shown that the mucus from the gland of the auxiliary copulatory organ in helicids has important physiological and behavioural effects, but the functional significance of 'dart-shooting' is still under debate, and the pheromone injected with the dart remains unidentified. Deposits of calcium have been found in the wall or lumen of the terminal male genitalia in several stylommatophorans, but much more work is required to elucidate the possible function of these calcium salts in the hardening of spermatophores, in the copulatory process or indeed in other aspects of reproduction. More critical studies are needed on the structure of the diversity in auxiliary genitalia present in Stylommatophora, not only to clarify homologies, but also to elucidate functional variation. While detailed field and laboratory studies are now beginning to address the great diversity in reproductive biology, there are still too few examples of the integrated ecological, physiological, genetic, behavioural and structural approaches necessary to achieve a level of understanding that will assist conservation or management of gastropod populations.

References

Adamo, S.A. and Chase, R. (1988) Courtship and copulation in the terrestrial snail *Helix aspersa*. *Canadian Journal of Zoology* 66, 1446–1453.

Adamo, S.A. and Chase, R. (1990) The 'love dart' of the snail *Helix aspersa* injects a pheromone that decreases courtship duration. *Journal of Experimental Zoology* 255, 80–87.

Adamo, S.A. and Chase, R. (1996) Dart shooting in helicid snails: an 'honest' signal or an instrument of manipulation? *Journal of Theoretical Biology* 180, 77–80.

Baur, B. (1988) Repeated mating and female fecundity in the simultaneously hermaphroditic land snail *Arianta arbustorum*. *Invertebrate Reproduction and Development* 14, 197–204.

Bayne, C.J. (1967) Studies on the composition of extracts of the reproductive glands of *Agriolimax reticulatus*, the grey field slug (Pulmonata, Stylommatophora). *Comparative Biochemistry and Physiology* 23, 761–773.

Bayne, C.J. (1968) Histochemical studies on the egg capsules of eight gastropod molluscs. *Proceedings of the Malacological Society of London* 38, 199–212.

Bayne, C.J. (1973) Physiology of the pulmonate reproductive tract: location of spermatozoa in isolated, self-fertilizing succinid snails (with a discussion of pulmonate reproductive tract terminology). *Veliger* 16, 169–175.

Bornchen, M. (1967) Untersuchungen zur Secretion der fingerformigen Drusen von *Helix pomatia*. *Zeitschrift für Zellforschung* 78, 402–426.

Boycott, A. (1917) The genitalia of *Acanthinula aculeata*. *Proceedings of the Malacological Society of London* 12, 221–226.

Breckenridge, W. and Fallil, S. (1973) Histological observations on the reproductive system of *Achatina fulica* (Gastropoda, Pulmonata, Stylommatophora). *Ceylon Journal of Sciences* 10, 85–118.

Chung, D.J.D. (1987) Courtship and dart shooting behaviour of the land snail *Helix aspersa*. *Veliger* 30, 24–39.

Dillaman, R. (1981) Dart formation in *Helix aspersa. Zoomorphology* 97, 247–261.

Duncan, C. (1975) Reproduction. In: Fretter, V. and Peake, J. (eds), *Pulmonates.* Vol. 1, *Functional Anatomy and Physiology.* Academic Press, London, pp. 309–365.

Els, A. (1974) The morphology and histology of the genital system of the pulmonate *Milax gagates* (Draparnaud). *Annale Universiteit van Stellenbosch* 49, 1–39.

Els, W.J. (1978) Histochemical studies on the maturation of the genital system of the slug *Deroceras laeve* (Pulmonata, Limacidae), with special reference to the identification of mucosubstances secreted by the genital tract. *Annale Universiteit van Stellenbosch* 1, 1–116.

Emberton, K.C. (1994) Polygyrid land-snail phylogeny: external sperm exchange, early North American biogography, iterative shell evolution. *Biological Journal of the Linnean Society* 52, 241–271.

Fretter, V. and Graham, A. (1962) *British Prosobranch Molluscs, their Functional Anatomy and Ecology.* Ray Society, London.

Giusti, F. and Selmi, M.G. (1985) The seminal receptacle and sperm storage in *Cochlostoma montanum* (Issel) (Gastropoda: Prosobranchia). *Journal of Morphology* 184, 121–133.

Gómez, B.J. (1991) Morphological and histological study of the genital ducts of *Cryptazeca monodonta* (Pulmonata, Orthurethra), with special emphasis on the auxiliary copulatory organ. *Zoomorphology* 111, 95–102.

Gómez, B.J. and Rodrigues, A. (2000) Calcium phosphate granules in the reproductive system of *Oxychilus atlanticus* (Gastropoda: Pulmonata). *Journal of Molluscan Studies* 66, 197–204.

Gómez, B.J., Angulo, E. and Zubiaga, A. (1991) Ultrastructural analysis of the morphology and function of the spermatheca of the pulmonate slug *Arion subfuscus. Tissue and Cell* 23, 357–365.

Gómez, B.J., Serrano, T. and Angulo, E. (1996) Morphology and fine structure of the glands of the dart-sac complex in Helicoidea (Gastropoda, Stylommatophora). *Invertebrate Reproduction and Development* 29, 47–55.

Griffond, B. and Bride, J. (1987) Germinal and non-germinal lines in the ovotestis of *Helix aspersa*: a survey. *Wilhelm Roux's Archives of Developmental Biology* 196, 113–118.

Haase, M. and Baur, B. (1995) Variation in spermathecal morphology and storage of spermatozoa in the simultaneously hermaphroditic land snail *Arianta arbustorum* (Gastropoda: Pulmonata: Stylommatophora). *Invertebrate Reproduction and Development* 28, 33–41.

Hodgson, A.N. (1996) The structure of the seminal vesicle region of the hermaphrodite duct of some pulmonate snails. *Malacological Review, Supplement* 6, 89–99.

Hodgson, A.N. and Shachak, M. (1991) The spermatogenic cycle and role of the hermaphrodite duct in sperm storage in two desert snails. *Invertebrate Reproduction and Development* 20, 125–136.

Hunt, S. (1979) The structure and composition of the love dart (gypsobellum) in *Helix pomatia. Tissue and Cell* 11, 51–62.

Ibáñez, M. and Alonso, R. (1978) Anatomical observations on *Pomatias sulcatus* (Draparnaud, 1805) (Prosobranchia: Pomatiidae). *Journal of Conchology* 29, 263–266.

Koene, J.M. and Chase, R. (1998) The love dart of *Helix aspersa* Müller is not a gift of calcium. *Journal of Molluscan Studies* 64, 75–80.

Kugler, O.E. (1965) A morphological and histochemical study of the reproductive system of the slug *Philomycus carolinianus* (Bose). *Journal of Morphology* 116, 117–132.

Leonard, J.L. (1991) Sexual conflict and the mating systems of simultaneously hermaphroditic gastropods. *American Malacological Bulletin* 9, 45–58.

Lind, H. (1973) The functional significance of the spermatophore and the fate of spermatozoa in the genital tract of *Helix pomatia* (Gastropoda: Stylommatophora). *Journal of Zoology* 169, 39–64.

Lusis, O. (1961) Postembryonic changes in the reproductive system of the slug *Arion ater rufus* L. *Proceedings of the Malacological Society of London* 137, 433–468.

Martins, A.M.F. (1996) Relationships within the Ellobiidae. In: Taylor, J.D. (ed.) *Origin and Evolutionary Radiation of the Mollusca*. Oxford University Press, Oxford, pp. 285–294.

Nordsieck, H. (1985) The system of the Stylommatophora (Gastropoda), with special regard to the systematic position of the Clausiliidae: I. Importance of the excretory and genital systems. *Archiv für Molluskenkunde* 116, 1–24.

Nordsieck, H. (1987) Revision des System der Helicoidea (Gastropoda: Stylommatophora). *Archiv für Molluskenkunde* 118, 9–50.

Noyce, A.G. (1973) The morphology and histology of the genital system of *Theba pisana* (Müller) (Pulmonata: Helicidae). *Annale Universiteit van Stellenbosch*, 48 (A3), 1–30.

Panha, S. (1987) Histochemical and ultrastructural studies on the amatorial organ of *Hemiplecta distincta* (Pfeiffer) (Pulmonata: Ariophantidae). *Venus* 46, 108–115.

Pokryszko, B.M. (1987) On the aphally in the Vertiginidae (Gastropoda: Pulmonata: Orthurethra). *Journal of Conchology* 32, 365–375.

Prieto, C.E., Puente, A.I., Altonaga, K. and Gómez, B.J. (1993) Genital morphology of *Caracollina lenticula* (Michaud, 1831), with a new proposal of classification of helicodontoid genera (Pulmonata: Hygromioidea). *Malacologia* 35, 63–77.

Quick, H.E. (1960) British slugs (Pulmonata; Testacellidae; Arionidae; Limacidae). *Bulletin of the British Museum (Natural History) Zoology* 6, 103–226.

Ramasubramaniam, K. (1979) A histochemical study of the secretions of reproductive glands of the egg envelopes of *Achatina fulica* (Pulmonata: Stylommatophora). *International Journal of Invertebrate Reproduction* 1, 333–346.

Reeder, R. and Rogers, S. (1979) The histochemistry of the spermatheca in four species of *Sonorella*. *Transactions of the American Microscopical Society* 98, 267–271.

Reise, H. (1995) Mating behaviour of *Deroceras rodnae* Grossu & Lupu, 1965 and *D. praecox* Wiktor, 1966 (Pulmonata: Agriolimacidae). *Journal of Molluscan Studies* 61, 325–330.

Rigby, J.E. (1963) Alimentary and reproductive systems of *Oxychilus cellarius* (Müller) (Stylommatophora). *Proceedings of the Zoological Society of London* 141, 311–360.

Rigby, J.E. (1965) *Succinea putris*: a terrestrial opisthobranch mollusc. *Proceedings of the Zoological Society of London* 144, 445–486.

Rodrigues, A. and Gómez, B.J. (1999) Copulatory process in *Oxychilus* (*Drouetia*) *atlanticus* (Morelet and Drouet, 1857) (Pulmonata: Zonitidae). *Invertebrate Reproduction and Development* 36, 1–3.

Rogers, S., Reeder, R.L. and Shannon, W.A. (1980) Ultrastructural analysis of the morphology and function of the spermatheca of the pulmonate snail *Sonorella santaritana*. *Journal of Morphology* 163, 319–329.

Runham, N.W. (1988) Mollusca. In: Adiyodi, K.G. and Adiyodi, G. (eds) *Reproductive Biology of Invertebrates*, Vol. III, *Accessory Sex Glands*. John Wiley & Sons, Chichester, pp. 113–188.

Runham, N.W. and Hogg, J. (1992) The pulmonate carrefour. In: Gittenberger, E and Goud, J. (eds) *Proceedings of the Ninth International Malacological Congress*. Unitas Malacologica, Leiden, pp. 303–307.

Runham, N.W., East, J. and Hatherly, M.J. (1991) The spongy gland and the function of its lipid secretion. In: Meier-Brook, C. (ed.) *Proceedings of the Tenth International Malacological Congress*. Unitas Malacologica, Tübingen, pp. 115–117.

Sirgel, W.F. (1973) Contributions to the morphology and histology of the genital system of the pulmonate *Agriolimax caruanae* Pollonera. *Annale Universiteit van Stellenbosch* 48 (A2), 1–43.

Smith, B.J. (1965) The secretions of the reproductive tract of the garden slug *Arion ater*. *Annals of the New York Academy of Sciences* 188, 997–1014.

Stears, M. (1974) Contributions to the morphology and histology of the genital system of *Limax valentianus* (Férussac) (Pulmonata: Limacidae). *Annale Universiteit van Stellenbosch* 49 (A3), 1–46.

Steenberg, C. (1925) Études sur l'anatomie et la systematique des maillots. *Videnskabelige Meddelelser Dansk naturhistorisk Forening* 80, 1–211.

Tielecke, H. (1940) Anatomie, Phylogenie und Tiergeographie der Cyclophoriden. *Archiv für Naturgeschichte* 9, 317–371.

Tomiyama, K. (1994) Courtship behaviour of the giant African snail, *Achatina fulica* (Férussac) (Stylommatophora: Achatinidae) in the field. *Journal of Molluscan Studies* 60, 47–54.

Tompa, A.S. (1979) Oviparity, egg retention and ovoviviparity in pulmonates. *Journal of Molluscan Studies* 45, 155–160.

Tompa, A.S. (1982) X-ray radiographic examination of dart formation in *Helix aspersa*. *Netherlands Journal of Zoology* 32, 63–71.

Tompa, A.S. (1984) Land snails (Stylommatophora). In: Tompa, A.S., Verdonk, N.H. and van den Biggelaar, J.A.M. (eds) *The Mollusca*, Vol. 7, *Reproduction*. Academic Press, London, pp. 47–139.

Tompa, A.S. and Wilbur, K.M. (1977) Calcium mobilisation during reproduction in the snail *Helix aspersa*. *Nature* 270, 53–54.

van Mol, J. (1970) Révision des Urocyclidae (Mollusca, Gastropoda, Pulmonata). Anatomie–Systématique–Zoogéographie. *Annales du Musée Royal de l'Afrique Central, Tervuren* 180, 1–234.

Varga, A. (1984) The *Cochlostoma* genus (Gastropoda, Prosobranchiata) in Yugoslavia. I. Anatomical studies. *Miscellanea Zoologica Hungarica* 2, 51–64.

Visser, M. (1973) The ontogeny of the reproductive system of *Gonaxis gwandaensis*, with special reference to the phylogeny of the spermatic conduits of the Pulmonata. *Annale Universiteit van Stellenbosch* 48 (A2), 1–79.

Visser, M. (1977) The morphology and significance of the spermoviduct and prostate in the evolution of the reproductive system of the Pulmonata. *Zoologica Scripta* 6, 43–54.

Vorster, W. (1973) The morphology and histology of the genital system of *Sheldonia cotyledonis. Annale Universiteit van Stellenbosch* 3 (A2), 1–30.

Wiktor, A. (1987) Spermatophores in Milacidae and their significance for classification (Gastropoda, Pulmonata). *Malakologische Abhandlungen Museum für Tierkunde Dresden* 12, 85–100.

Zubiaga, A., Moya, J., Gómez, B.J. and Angulo, E. (1989) Ultrastructural studies on the spermoviduct of *Arion subfuscus* (Draparnaud, 1805) (Gastropoda, Pulmonata). *Biological Structures and Morphogenesis* 2, 124–135.

Zubiaga, A., Gómez, B.J., Mateo, A. and Angulo, E. (1990a) Histochemistry and ultrastructure of the genital atrium of *Arion subfuscus* (Gastropoda: Stylommatophora). *Zeitschrift für Mikroskopische-Anatomische Forschung* 104, 737–750.

Zubiaga, A., Gómez, B.J., Moya, J. and Angulo, E. (1990b) Identification and carbohydrate content of secretory cell types in the spermoviduct of *Arion subfuscus* (Mollusca, Gastropoda) by classical and lectin histochemistry. *Zoologische Jahrbücher für Anatomie* 120, 409–424.

9 Regulation of Growth and Reproduction

A. Gomot de Vaufleury

Laboratoire de Biologie et Ecophysiologie, Faculté des Sciences et des Techniques, Université de Franche-Comté, Place Leclerc, 25030 Besançon Cedex, France

Introduction

The developmental and reproductive biology of terrestrial gastropods is strongly influenced by the environment in which they live. Scientific research over the last few decades has led to markedly increased understanding of the environmental cues, and the endocrine and neuroendocrine response mechanisms, that regulate the growth and reproduction in terrestrial gastropods, albeit in a small number of species. This knowledge has been built up through laboratory experiments involving manipulation of environmental factors and microsurgical manipulation of the animals, in conjunction with cytological and biochemical analyses. The improved understanding of regulation of growth and reproduction has been of particular benefit to heliciculture, where the goal is to optimize the production of edible snail species by maintaining optimum environmental conditions. This knowledge is also pertinent to pest species, where the focus is the development of 'biological' means to reduce populations.

This chapter reviews current knowledge of the environmental cues influencing growth and reproduction, and then reviews knowledge on the endocrine pathways involved.

Effects of Abiotic Environment on Growth and Reproduction

Light, photoperiod and temperature

Growth in many stylommatophorans apparently is not closely linked to photoperiod conditions. Sokolove and McCrone (1978), for example, showed that growth in *Limax maximus* Linnaeus (Limacidae) did not vary with day lengths in the range light:dark (L:D) 8:16 to L:D 16:8. None the

©CAB *International* 2001. *The Biology of Terrestrial Molluscs*
(ed. G.M. Barker)

less, growth in species such as *Cantareus aspersus* (Müller) (Helicidae) is strongly regulated by the prevailing photoperiod (Gomot *et al.*, 1982; Gomot and Deray, 1987). Young of this species, raised under artificially controlled conditions from birth, matured earlier and achieved higher body weights (8 g instead of 2 g) under long-day (L:D 18:6 or L:D 24:0) than under short-day (L:D 8:16 or L:D 12:12) conditions. Laurent *et al.* (1984) showed that progressive lengthening of the photophase at the rate of 30 min week^{-1} promoted more rapid growth than constant L:D 18:6. The inhibitory effect of short-day photoperiods is apparently less pronounced in *C. aspersus* var. *maximus* Taylor than in *C. aspersus aspersus* (Deray and Laurent, 1987), but the level of inhibition has been shown to depend on the prevailing temperature (see below).

In all stylommatophorans that have been investigated to date, reproduction has been shown to be influenced by the prevailing photoperiod. Stephens and Stephens (1966) observed that *C. aspersus* receiving only 9 h of light per day do not lay eggs. Bailey (1981) reported a relationship between photoperiod and annual reproductive activity in *C. aspersus*. In our laboratory, Enée *et al.* (1982) observed that reproductively active *C. aspersus* continued to lay eggs when transferred from the field to controlled long-day environments (L:D 18:6, or permanent lighting at a temperature of 15°C) but, after transfer to short-day environments, reproductive output ceased. The role of photoperiod in *C. aspersus* reproduction was also noted by Le Guhennec and Daguzan (1983). The short-day photoperiod regime (8 h day^{-1}) caused a depression in spermatogenesis and [^{3}H]thymidine incorporation (Gomot and Gomot, 1985). Gomot and Griffond (1987) demonstrated that the rate of spermatogenesis was highest under long-day or continuous light regimes, and that differentiation of the nurse cells and oogenesis were strongly correlated with the rate of development in the male gametes. Bonnefoy-Claudet *et al.* (1983) showed that these benefical effects of long photoperiods in artificial white light were maintained with monochromatic blue, green and yellow light wavelengths (176, 116 and 106% of eggs produced under white light, respectively), although the number of eggs produced was reduced to 60% in green light.

Comparative experiments (Bonnefoy-Claudet and Deray, 1987; Gomot, 1990) have shown that responsiveness to day length varies among closely related species and even populations within species: among helicids, inhibition of reproduction induced by short days occurred in the following order: *Helix pomatia* Linnaeus > *C. aspersus aspersus* > *C. aspersus maximus*. Sokolove and McCrone (1978) found that young *L. maximus* reared under short days (L:D 8:16) grow normally but remained reproductively immature. Subsequent transfer of these animals to a long-day (L:D 16:8) environment induced differentiation of the reproductive organs and sexual activity. In *Deroceras reticulatum* (Müller) (Agriolimacidae), continuous lighting has been shown to stimulate spermatogenesis strongly (Henderson and Pelluet, 1960). In contrast, Hunter and Stone (1986) found that exposure to short days (L:D 8:16) for the entire life span did not prevent or retard the development of functional reproductive organs in *Cepaea nemoralis* (Linnaeus) (Helicidae).

Gametogenesis in Stylommatophora is affected by environmental temperature, as male and female cell lines in the ovotestis of *H. pomatia*, *C. nemoralis* and *Arion ater* (Linnaeus) (Arionidae) exhibit differential sensitivity and responses (Bank, 1931; Bouillon, 1956; Lusis, 1966). Wolda (1967) suggested that the rates of reproduction of certain field populations of *C. nemoralis* were regulated by the temperature conditions prevailing in the habitat.

To analyse the effects of temperature on gametogenesis independently of the other environmental factors, we (P. Gomot *et al.*, 1990) subjected *C. aspersus* to different temperatures (ranging from 5 to 25°C) for a 4-week period commencing at the end of their winter hibernation (6 months at 6°C). Cytological studies and measurement of [³H]thymidine incorporation showed the male cell line to be extremely sensitive to the environmental temperature. At 5°C, DNA synthesis in the ovotestis was negligible, at 10 and 15°C there was modest spermatogenic multiplication but no meiosis, while at 20 and 25°C there was rapid multiplication of spermatogonia and spermatocytes, and differentiation of spermatozoa within 3–4 weeks. Investigations into the effect of various temperatures on oogenesis in the ovotestis of the same animals (Griffond *et al.*, 1992a) indicated that differentiation of young oocytes occurred over the range 5–25°C. However, the vitellogenic oocytes were produced in significant numbers only at 20 and 25°C.

Photoperiod–temperature interference

In laboratory-acclimated animals, the interactive influence of photo-period and temperature has been demonstrated in a number of species. In *C. aspersus maximus*, for example, Gomot (1994) showed that body weights attained at 4 months of age at 15°C under a long-day photoperiod were 2.8 times higher than under short days, whereas body weight at 20°C differed by a factor of 1.6. (Table 9.1 and Fig. 9.1). Four experimental combinations of photoperiod and temperature have been evaluated for effects on the reproductive activity of *C. aspersus aspersus* (Gomot *et al.*, 1989) and *H. pomatia* (Gomot, 1990). The results indicate a compensatory role for long-day photoperiod for the negative effect of low temperature on reproductive output (Table 9.2).

Effects of Abiotic Environment on Hibernation

Stylommatophorans living in temperate regions often hibernate during the winter months. The duration of this hibernation varies depending on the climatic conditions. Bonavita (1972) noted that during the greater part of the natural winter hibernation of *Sphincterochila candidissima*

Table 9.1. Effect of different combinations of photoperiod and temperature on growth of *Cantareus aspersus* (Müller) var. *maximus* Taylor (Helicidae) held in the laboratory from the date of hatching (from Gomot, 1994)

Experimental groups	Mean weight (g) at 4 months of age
Long days (18L:6D)	
15°C	8.9 ± 2.4
20°C	17.1 ± 1.7
23°C	22.2 ± 1.5
Short days (8L:16D)	
15°C	3.1 ± 0.2
20°C	10.7 ± 0.2
23°C	15.8 ± 0.4

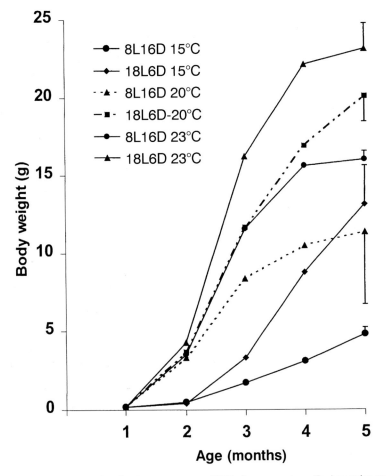

Fig. 9.1. Growth curves for *Cantareus aspersus* (Müller) var. *maximus* Taylor under various combinations of photoperiod and temperature in the laboratory (relative humidity 95%, 60 snails m^{-2}). Error bars = SE of the mean. (From A. Gomot, 1994.)

(Draparnaud) (Sphincterochilidae), cell differentiation/gametogenesis was arrested completely. However, gametogenesis was resumed before hibernation was broken and the snails resumed normal activity. Bonnefoy-Claudet and Deray (1984) demonstrated experimentally that the duration of the hibernation greatly influenced the reproductive performance of *C. aspersus aspersus* on resumption of activity (Table 9.3). Essentially, the longer the hibernation period (up to 18 months evaluated), the sooner the mating behaviour occurred at the break of hibernation and the higher the reproductive output. This suggests the existence of a refractory period.

Deterministic Factors of Endocrine and Neuroendocrine Functions

Knowledge of the endocrinology of pulmonate gastropods has advanced considerably over the last 30 years with the recognition of distinct neurosecretory cell types in the nerve ganglia, based on their stain affinities and ultrastructure, and, in certain cases, with experimental demonstration of the link between the secretory activity of the neurons and

Table 9.2. Effects of different combinations of photoperiod and temperature on reproductive activity in *Helix pomatia* Linnaeus (Helicidae) over an 18-week period in the laboratory (from A. Gomot, 1990)

Experimental groups	Percentage of experimental animals mating	Duration of egg laying (weeks)	Oviposition frequency per animal	Eggs per animal	Mortality (%)
Long days (18 :6D)					
15°C	65.0	17	0.90	45	12.5
20°C	67.5	14	1.30	50	45.0
Short days (8L:16D)					
15°C	25.0	4	0.28	16	22.5
20°C	22.5	8	0.78	35	27.5

These data were obtained for groups of 20 animals each. The data have not been adjusted for the different mortalities among the experimental groups. It is observed that mortality was higher in animals that had a higher rate of reproduction (sexual exhaustion).

Table 9.3. Reproductive activity of *Cantareus aspersus* (Müller) (Helicidae) over a 17-week period in relation to the length of the preceding hibernation period under controlled environmental conditions (from Bonnefoy-Claudet and Deray, 1984)

	Length of hibernation		
	1.5 months	6 months	12 months
Days to first mating	52	17	2
Days to first egg laying	85	52	23
Mating frequency per animal	0.70	0.82	1.76
Oviposition frequency per animal	0.11	0.48	1.40

growth/reproductive physiology. It is now known that neuroendocrine functions of stylommatophorans are regulated by neurosecretory cells in the cerebral ganglia and that endocrine function is regulated by endocrine cells of the dorsal bodies (DB) of the cerebral ganglia connective tissue. An endocrine function has also been detected in the cephalic tentacles and the ovotestis. The progress in elucidation of endocrine and neuro-endocrine physiology in pulmonate gastropods has occurred in parallel with that made in other groups of invertebrates (Durchon and Joly, 1978; Highnam and Hill, 1977). Reviews by Joosse and Geraerts (1983), Joosse (1988) and Geraerts *et al.* (1991) deal especially with basommatophorans in which hormones have been purified and sequenced. In relation to stylommatophorans, information on endocrinology and reproduction was summarized by Griffond and Gomot (1989), and below we focus on recently acquired insights on endocrine and neuroendocrine physiology.

Reproductive strategies in stylommatophorans are diverse, undoubt-edly reflecting species differences in complexity of the reproductive organs and their hormonal controls. Data on the endocrine control of gametogenesis, of reproductive organ function and egg laying are, however, limited to a very small subset of the stylommatophoran species diversity.

The gonad in stylommatophorans is an ovotestis, as male and female gametes occur in the same organ. Ultrastructural study of embryonic *C. aspersus* shows that the ovotestis in these animals has a double origin (Gomot and Griffond, 1993), with two voluminous primordial germinal cells, which later differentiate into male and female gametes, intimately associated with small mesenchymal cells that give rise to the nurse or Sertoli cells and follicle cells, respectively. In post-embryonic life, shortly after the animals hatch from the egg, the gonadal primordium becomes divided on to numerous acini which converge on the proximal part of the hermaphrodite duct. In *C. aspersus*, spermiogenesis and oogenesis are achieved within 3 months if environmental conditions are favourable.

Endocrine control of male gametogenesis

Studies with both *C. aspersus* (Guyard, 1971; Gomot, 1973) and *Arion subfuscus* (Draparnaud) (Wattez, 1980) have shown that when isolated juvenile gonadal tissue is cultured *in vitro* in a hormone-free synthetic medium, the undifferentiated germinal cells evolve only into oocytes, instead of producing both male and female cells. To obtain differentiation of the male line, the cultured gonad must be associated with the cerebral ganglia or the ocular tentacles. These results show that the central nervous system (CNS) and cephalic tentacles can act directly on gameto-genesis by gonadotrophic factors released into the medium.

Several types of *in vivo* experiments have enabled the modes of action of the CNS and the cephalic tentacles to be determined.

The role of the CNS demonstrated by ablation and transplant experiments

In *L. maximus*, implantation of cerebral ganglion tissue from long-day- (L:D 16:8) but not of short-day- (L:D 8:16) treated donors induced mitogenesis of male cells within 1 week (McCrone *et al.*, 1981). The action of the cerebral ganglia was shown to occur via a male gonadotrophic factor (MGF) present in cerebral ganglion homogenates and in the haemolymph of donor animals in the male phase (maintained under long-day conditions) (Fig. 9.2a). The active MGF of protein nature was contained in the 50–100 kDa ultrafiltration fraction (Melrose *et al.*, 1983). The neuroendocrine cells that produce the MGF, located by testing the activity of homogenates prepared from different parts of the ganglia, were shown to be the large neurosecretory cells (area Z in Fig. 9.2a) on the edge of the cerebral commissure (Sokolove *et al.*, 1985).

In *C. aspersus*, the role of the cerebral ganglia (CG) and dorsal bodies (DB, located in the connective sheath) in the control of spermatogonial multiplication (SM) was studied by ablation of the entire CNS, localized extirpations of parts of the cerebral ganglia, and grafting experiments, performed on animals 1–12 months into a hibernation period. SM was evaluated by measurement of the level of [³H]thymidine incorporation into the ovotestis under a temperature favourable for spermatogenesis (25°C). Ablation of the CNS (CG + DB) stimulated SM in animals kept for 1–6 months in hibernation. However, ablation of the CNS from snails kept in hibernation for 12 months had no effect. It seems that during the first 6 months of hibernation, the CNS exerts an inhibiting effect on SM, even at 25°C (Gomot and Gomot 1989a,b). It is therefore apparent that in natural habitat, at the beginning of hibernation, the CNS inhibits the division of male sex cells until spring, when the absence of the inhibitory factor coupled with the external cues favour resumption of gonadal activity. Performing localized extirpations of the CG showed that the factor inhibiting the SM arises from the clusters of neurosecretory cells in the mesocerebrum (Gomot, 1993).

Re-implantation of the whole CNS into the haemocoele of brain-ablated animals failed to correct the effects of CNS extirpation. However, the separate implantation of either DB or CG in to cerebrotomized hosts showed that the DB caused DNA synthesis and SM to return to control cell levels, whereas the CG stimulated these processes further (L. Gomot *et al.*, 1990; P. Gomot *et al.*, 1992). These results indicate the existence of a close functional interrelationship between the CG and the DBs. The action of DB implanted alone probably results from the suppression/absence of the inhibitory nervous control exerted by the neurons of the CG (Vincent *et al.*, 1984b). Thus the regulation of SM in stylommatophorans depends on the interaction between stimulatory and inhibitory factors arising either from the various territories of the CG (in particular from the meso-cerebrum) or from the DB (Fig. 9.3); stimulatory and inhibitory factor release is apparently regulated by the nervous connections that have been

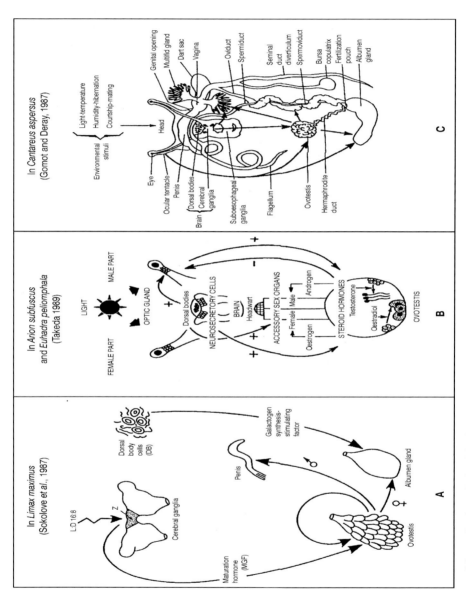

Fig. 9.2. Summary of the endocrine relationships known to be involved in the control of reproduction in terrestrial gastropods:: (A) *Limax maximus* Linnaeus (Limacidae); (B) *Arion subfuscus* (Drapamaund) (Arionidae) and *Euhadra peliomphala* (Pfeiffer) (Bradybaenidae) and (C) *Cantareus aspersus* (Müller) (Helicidae).

demonstrated to occur between these two regions of the CNS (Wijdenes *et al.*, 1983) (Fig. 9.3).

Role of ocular cephalic tentacle ablation on gametogenesis

An endocrine function of the cephalic tentacles has been suggested in stylommatophorans by Pelluet and Lane (1961) and Pelluet (1964), as injections of tentacle extracts stimulated spermatogenesis. Following this, lively debate ensued (Boer and Joosse, 1975), since Gottfried and Dorfman (1970) reported accelerated development of spermatogenesis in immature *Ariolimax californicus* Cooper (Arionidae) deprived of their cephalic tentacles. However, Wattez (1980) presented evidence that the cephalic tentacles of *A. subfuscus* produce a male gamete-promoting substance, while Tadeka (1982a), working with several species, showed induction and stimulation of spermatogenesis by the tentacles.

The mode of action and the origin of the masculinizing effects of the cephalic tentacles are not known accurately. Gottfried and Dorfman

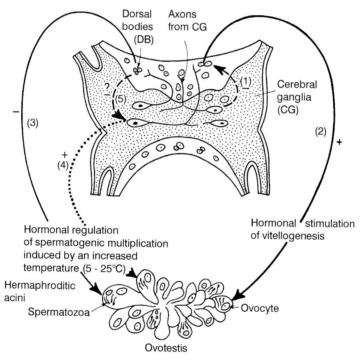

Fig. 9.3.   Roles that are assumed to be played by the various parts of the central nervous system in the regulation of gametogenesis in *Cantareus aspersus* (Müller) (Helicidae). (1) Inhibition of the nervous control of the dorsal bodies (DB) by the cerebral ganglia (CG). (2) Hormonal stimulation of vitellogenesis by the dorsal bodies (DB) no longer inhibited by the cerebral ganglia (CG). (3) Hormonal inhibition of spermatogeneic multiplication (SM) by isolated dorsal bodies (DB). (4) Neuroendocrine stimulation of spermatogenic multiplication (SM) by isolated cerebral ganglia (CG). (5) Inhibition of the stimulating action of the cerebral ganglia (CG) by the dorsal bodies (DB). (From L. Gomot *et al.*, 1990.)

(1970) assume that the cephalic tentacles inhibit steroid biosynthesis by the ovotestis. Such interference between steroids and tentacle factors was also proposed by Bierbauer and Molnar (1972) who observed a stimulating effect of testosterone on spermatogenesis, this stimulation being diminished by about 23% in *H. pomatia* when the cephalic tentacles were removed. According to Takeda (1982a), the active region of the cephalic tentacles, which he called the 'optic gland', is composed of collar cells. Experimentally, a tentacle gonadotrophic factor has been reported to provoke the release of testosterone in the ovotestis of *Euhadra peliomphala* (Pfeiffer) (Bradybaenidae), where it stimulates spermatogenesis (Takayanagi and Takeda, 1985). These authors suggest the occurrence of a feedback control between the ovotestis and the tentacular collar cells (Fig. 9.2B).

From these two lines of experiments, it appears that the CNS (CG + DB) acts on the development of the male cells through the action of both stimulatory and inhibitory factors, whereas the cephalic tentacles more generally have a positive effect on SM. The exact nature of these factors is not known, but it has been suggested that the neurosecretory cells in the CG act via neuropeptides (Marchand and Gomot, 1987). The involvement of steroids in the relationship between the cephalic organs (CNS and cephalic tentacles) and the ovotestis is probable but requires further investigation. Indeed, Le Guellec *et al.* (1987) showed the presence of endogenous steroids in the ovotestis of *C. aspersus* and the ability of the ovotestis to metabolize androstenedione. Moreover, the DB of *C. aspersus* is also known to synthesize steroid hormones (Krusch *et al.*, 1979), and the same tissue in *H. pomatia* is known to contain ecdysteroids (Nolte *et al.*, 1986). Lastly, the ovotestis itself produces a peptide hormone of molecular weight greater than 50 kDa (Sokolove *et al.*, 1987) that accelerates the production of spermatozoa.

Endocrine control of female gametogenesis and ovulation

The *in vitro* explantation of gonadal primordia from *C. aspersus* and *A. subfuscus* shows that oocytes autodifferentiate, while vitellogenesis requires the presence of stimulating factors (Guyard, 1971; Wattez, 1980). Injections of homogenates from the CNS have been shown to increase the number of oocytes (Pelluet and Lane, 1961), thus suggesting the involvement of cerebral factors in oogenesis. This was confirmed from the effects of CNS extracts on *in vitro* culture of ovotestes from *A. ater* (Badino, 1967), *C. aspersus* (Guyard, 1971; Gomot, 1973), *A. subfuscus* (Wattez, 1980) and *E. peliomphala* (Takayanagi and Takeda, 1985). The ablation of the DB in *D. reticulatum* (Wijdenes and Runham, 1976) retards the maturation of the oocytes, whereas *in vitro* culture association of the DB (separated from the CG) with a fragment of the ovotestis favours vitellogenesis in *C. aspersus* (Vincent *et al.*, 1984a). In addition, in *C. aspersus*, an increase in DB size and protein synthesis has been noted in egg-laying animals (Griffond and Vincent, 1985).

The role of the ocular cephalic tentacles in female gametogenesis remains uncertain since the studies of ablation and *in vitro* culture experiments, in the presence of tentacle extracts, have yielded conflicting results. In general, it can be said that tentacle homogenates inhibit oogenesis (Griffond and Gomot, 1989). The endocrine action of the ovotestis in the control of oogenesis was evoked by Bierbauer (1978) as injections of ovotestis homogenates in *H. pomatia* increased the number of oocytes in stages I and II. The involvement of steroids secreted by the ovotestis was proposed by Takeda (1983) as injection of oestradiol stimulated oogenesis in *E. peliomphala*. However, the mode of action of the vertebrate steroids in the control of molluscan oogenesis has not yet been clearly elucidated (Griffond and Gomot, 1989).

Little is known about regulation of ovulation in stylommatophorans, although homogenates of the CNS cause amoeboid movement in oocytes (Saleuddin *et al.*, 1983).

Endocrine control of differentiation of the reproductive tract

Role of the ovotestis

Unlike the situation with basommatophorans, gonadal control of development in the reproductive tract has been clearly demonstrated for Stylommatophora (Joosse, 1988; Griffond and Gomot, 1989). Castration in limacids, agriolimacids and arionids, which is relatively easy to perform since the ovotestis is not embedded extensively in the digestive gland, causes the reproductive tract to remain in an undeveloped state (Abeloos, 1943; Laviolette, 1954a; Runham *et al.*, 1973). Experimental castration in *Limacus flavus* (Linnaeus) (Limacidae) and *E. peliomphala* causes atrophy of the accessory reproductive organs, and the injection of androgens stimulates the development of male tissues (prostatic gland and spermiduct), whereas oestrogens stimulate the development of the albumen gland, oviductal gland and vagina (Takeda, 1985). As observed by Laviolette (1954b) in *A. rufus*, and Hogg and Wijdenes (1979) in *D. reticulatum*, gonad regeneration also occurs after castration in *Lehmannia marginata* (Müller) (Limacidae), *L. flavus*, *E. peliomphala* and *Achatina fulica* Bowdich (Achatinidae) (Takeda and Sugiyama, 1984). Takeda and Sugiyama (1984) found gonadal regeneration to be stimulated by injection of sex steroid hormones, oestradiol and testosterone. Implantation of the hermaphrodite duct from very young into sexually mature *D. reticulatum*, with or without the ovotestis, has demonstrated that the ovotestis produces both male and female tissue-promoting factors (Runham *et al.*, 1973). In *C. aspersus*, castration of juvenile animals results in the cessation of differentiation of the glands of the auxiliary copulatory organ, while transplantation of non-functional fragments of glands from castrated animals into intact, mature animals has been shown to initiate secretory activity of these structures (Gomot and Colard, 1985). The influence of the ovotestis on the reproductive organs (Fig. 9.2c) has also been demonstrated by the ovotestis from mature animals stimulating the development of the spermoviduct and of the albumen gland on trans-plantation into young, immature animals (Berset de Vaufleury *et al.*, 1986). *In vitro* culture experiments with oviduct explants confirm the stimula-tion of polysaccharide synthesis by ovotestis homogenates (Bride and Gomot, 1988). These results suggest that the gonads exert control over the development of the accessory sex organs, possibly mediated by steroid hormones (Fig. 9.2b).

The ovotestis of *L. maximus* is known to release a hormone that causes the development of male accessory reproductive structures under the stimulating effect of the MGF from the CNS (Fig. 9.2a). Development of female reproductive structures has been shown to be stimulated by implantation of small gonadal tissue fragments from sexually mature animals, but it is not known if the hormone (or hormones) stimulates the female organs solely by a direct route, as indicated by *in vitro* studies, or also via the CNS (Sokolove *et al.*, 1987).

Role of the ocular cephalic tentacles

In several species, including *A. fulica*, *Ariolimax columbianus* (Gould) and *Eobania vermiculata* (Müller) (Helicidae) Berry and Chan, 1968; Meenakshi and Scheer, 1969; Nopp, 1971), ablation of the cephalic tenta-cles stimulates the development of the albumen gland. The inhibitory role of the tentacles in the organogenesis of the albumen gland was confirmed in *C. aspersus* in both *in vitro* culture (Gomot and Courtot, 1979) and by *in vivo* injections of tentacle extracts (Bride *et al.*, 1986).

Role of the CNS

In basommatophorans, including *Lymnaea stagnalis* (Linnaeus) (Lym-naeidae), it has been well established that the DB control the maturation of the female part of the reproductive tract (Joosse, 1988). In stylommato-phorans, the role of the DB is more difficult to quantify because the DB cells often are scattered through the cerebral ganglion conjunctive sheath and thus difficult to remove surgically. However, extirpation of the DB considerably slowed the development of the oviductal gland and albumen gland in *D. reticulatum*, and this effect was able to be reversed by trans-plantation of DB from another animal (Wijdenes and Runham, 1976).

The neurohormonal control of galactogen synthesis in the albumen gland of *H. pomatia* was demonstrated by Goudsmit (1975). In this species, CNS homogenate stimulates the synthesis of galactogen in the albumen gland in a dose-dependent manner (Goudsmit and Ram, 1982). A similar effect from a CNS homogenate was obtained in *L. maximus* and the factor responsible for galactogen synthesis (Gal-SF) (Fig. 9.2A) was shown to reside in the DB (Van Minnen and Sokolove, 1984). In *C. aspersus*, the *in vitro* culture method has been employed to show that organogenesis and activity of the albumen gland are stimulated by the

CNS (Gomot and Courtot, 1979) and that it is the DB that stimulate female organ function (Bride and Gomot, 1988) (Fig. 9.2c).

Endocrine control of egg laying

Self-fertilization is a common mode of reproduction in pulmonates. Most widely studied are basommatophorans such as aphallate planorbids, where mating is absent and self-fertilization is the only mode of reproduction (de Larambergue, 1939). Self-fertility has been demonstrated in all basommatophoran species studied (Jarne *et al.*, 1993). Self-fertilization is not uncommon in stylommatophorans, but the great majority of these pulmonates apparently breed primarily by mating and outcrossing (de Larambergue, 1959; Selander *et al.*, 1975; Heller, Chapter 12, this volume). In these animals, mating seems to be an important event in triggering egg laying. *C. aspersus* individuals reared in isolation produce few eggs, and those produced are often abnormal in containing numerous oocytes which do not develop. However, some rare cases of fertile eggs have been observed. *C. aspersus* is thus essentially an obligate outcrossing species. Bride and Gomot (1991) were able to show that mating promotes reproductive tract maturation in previously virgin animals and triggers oviposition within several weeks. The origin of the factors stimulating oviposition in *C. aspersus* presently is not known. Experiments with *D. reticulatum* and *L. flavus* suggest the involvement of the neuroendocrine system, since injection of CNS homogenates stimulates egg laying whereas tentacle homogenates inhibit it (Takeda, 1977, 1989). In addition, Takeda notes the role of steroid hormones, since the injection of oestrogens stimulates egg laying in these species.

Oviposition could also result from the transfer of stimulatory substances, with origins in the reproductive tract, during mating. In *C. aspersus*, injection of prostatic gland homogenates from egg-laying animals stimulated oviposition, whereas injection of bursa copulatrix homogenates caused death of the recipient animals (Lucarz, 1991).

Further experimentation will be necessary to determine the physiological mechanisms that trigger egg laying in stylommatophorans. Immunocytochemical observations using antibodies raised against α-CDCP (caudo-dorsal cell hormone, i.e. egg-laying hormone of *L. stagnalis*) have revealed the presence of immunoreactive neurons in the CNS of *C. aspersus* and *L. maximus* (Griffond *et al.*, 1992b; van Minnen *et al.*, 1992). However, the identification of the egg-laying regulatory system has not been resolved.

Endocrine control of mating

The control of mating behaviour has been investigated extensively for the Helicidae (reviewed by Tompa, 1984; Leonard, 1992). Reproductive development and activity depend on the complementarity of development and function of the various component tissues; removal of any particular tissue, or disruption of its function, affects the whole reproductive system (Bride and Gomot, 1991; Gomot and Griffond, 1993). The secretory products of these tissues most probably are involved (Jeppesen, 1976) and some are thought to have a pheromonal role (Chung, 1986; Adamo and Chase, 1990). In *C. aspersus*, sexual behaviour is controlled by the neurons of the mesocerebrum (Chase, 1986). Mating brings about numerous physiological modifications which can lead, for example, to the activation of the DB (Saleuddin *et al.*, 1991) and an increase of the synthesis of galactogen in the albumen gland (Bride *et al.*, 1991).

A number of stylommatophoran species, representing a number of families, possess specialized cephalic, cutaneous structures that seem to have a function in courtship behaviour (Binder, 1976, 1977; Takeda, 1989). Among those studied to date are the 'frontal organ' in *Gymnarion* Pilsbry (Urocyclidae) (Binder, 1965) (Fig. 9.4), the 'cephalic dimple' in *Achatina* de Lamarck (Chase and Piotte, 1981) and the 'head wart' in *Euhadra* Pilsbry (Bradybaenidae) (Takeda, 1982b). The morphology of these cephalic structures varies among species, but all exhibit the phenomenon of ontogenial development running in parallel to that of the reproductive organs. It is not known presently whether these types of cephalic structures are a common, or indeed a universal, feature in Stylommatophora, as they are, in the majority of species, not readily apparent without histological investigation of the cephalic region of the body.

In *E. peliomphala*, castration led to the atrophy of the head wart, and subsequent injection of an ovotestis homogenate induced the return of the

Fig. 9.4. Photograph of two *Gymnarion* Pilsbry (Urocyclidae) with erection of their 'frontal organ' during courtship. (Courtesy of E. Binder, 1977.)

head wart to its former size (Takeda, 1989). Using *in vitro* tissue culture methodology, the head wart in *Euhadra* has been demonstrated to be the target organ of testosterone (Takeda, 1989).

Endocrine control of growth

The involvement of endocrine factor(s) in growth control in pulmonates was first demonstrated by extirpation and implantation experiments. In the basommatophoran *L. stagnalis*, Geraerts (1976) observed that removal of the neurosecretory light green cells (LGC) from rapidly growing juvenile animals resulted in markedly retarded body growth. Growth was restored by implantation into the haemocoel of cerebral ganglion tissues containing LGC. In the Stylommatophora, removal of the neurosecretory medial cells (MC) from the cerebral ganglia of *D. reticulatum* (Wijdenes and Runham, 1977) and extirpation of the mesocerebrum from young *C. aspersus* (A. Gomot *et al.*, 1992) prevent body growth. Transplants in mesocerebrum-deprived *C. aspersus* show that implantation of CG into the haemocoel partially restores the growth of the animals, indicating that the growth hormone can act by the endocrinal route via the haemolymph (Gomot, 1994). However, chimeric CNS transplants (Gomot and Gomot, 1994) restore growth best, suggesting an essential role in the re-establishment of neuronal connections. While CG grafts from the youngest animals restored growth most efficiently (Gomot and Gomot, 1995), the CG of adults still had a slight growth-stimulating activity. Xenografts of CG between animals of different species within Helicidae were well tolerated, although growth restoration varied depending on the donor species (*C. aspersus maximus* > *H. pomatia* > *C. aspersus aspersus* > *Helix lucorum* Linnaeus). On the other hand, CG grafts from more distantly related stylommatophorans (e.g. *Achatina*) or from basommatophorans (e.g. *Lymnaea*) are not tolerated in helicids; the grafts were destroyed and growth was not re-established (Gomot and Gomot, 1996).

These studies clearly indicate the neurosecretory cells in the cerebral region of the CNS as the source of growth hormone (A. Gomot *et al.*, 1992). The mesocerebrum of *C. aspersus* has a role equivalent to that of the LGC in *L. stagnalis* (Geraerts, 1976), the medial cells in *D. reticulatum* (Wijdenes and Runham, 1977) and the medio-dorsal cells in *Helisoma duryi* Wetherby (Planorbidae) (Saleuddin *et al.*, 1992). Immunocytochemical observations revealed that these groups of cells are indeed homologous and secrete substances related to the insulin of mammals. The chemistry of pulmonate growth hormone is not yet known, although insulin-like molecules are probably involved in the growth of the shell (Saleuddin *et al.*, 1992, 1995), but their mechanism of action remains to be determined.

In summary, we can suggest that the regulation of growth is achieved by the influence of environmental factors on the neurosecretory cells of the CNS, particularly the mesocerebrum (Fig. 9.5).

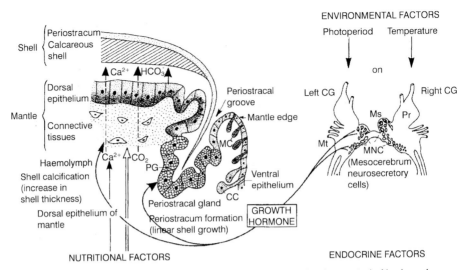

Fig. 9.5. Schematic interpretation of the environmental and endocrine control of body and shell growth in snail forms of Stylommatophora. CC, calcium cells; MC, mucous cells; MNC, mesocerebrum neurosecretory cells; Mt, metacerebrum; Ms, mesocerebrum; PG, periostracal gland; Pr, procerebrum.

Conclusion

The determination of growth and reproduction in terrestrial gastropods is closely linked to the action of environmental factors, integrated by the neuroendocrine system, on bodily functions. Current knowledge was built up through experiments on animals reared in controlled conditions, allowing any of the environmental factors to be varied separately and the consequences on growth and reproduction to be analysed. The data essentially apply to species living in temperate regions, where regular seasonal cycles in day length and temperature occur. Investigations of the same type have not yet been undertaken for species of equatorial and tropical regions, where different climatic regimes, and thus different environmental cues, exist.

Rearing of a number of stylommatophoran species in precisely controlled environments is an established art, and has allowed elucidation of the mechanisms of neuroendocrine control using methods of cell biology, immunocytochemistry and biochemistry, in conjunction with microsurgical techniques. Thus, the roles of the various regions of the CNS in hormone secretion, with effects on growth, gametogenesis, reproductive organ ontogeny and function, and in mating behaviour, have been determined. The endocrine function of the ovotestis in controlling the differentiation of the reproductive organs has been confirmed in several stylommatophoran species – this represents a fundamental physiological difference from freshwater basommatophorans. Although the source locations of the endocrine factors that control the main functions have been established, most of these molecules remain to be isolated and purified. The cells or tissues responsible for these functions often contain substances that are immunologically close to peptides that have been isolated and characterized from vertebrate tissues (e.g. somatostatin, insulin and metenkephalin). However, only two such compounds in stylommatophorans – FMRFamide (Price *et al.*, 1987) and achatin I (Kamatani *et al.*, 1990) – are well known structurally. FMRFamide that is present in the CNS (including the DB) of *C. aspersus* (Marchand *et al.*, 1991) and *L. maximus* (Cooke and Gelperin, 1988) possibly is one of the messengers for regulation of growth and reproduction. The function of certain vertebrate pituitary hormones (follicle-stimulating hormone, thyroid-stimulating hormone and adrenocorticotrophic hormone) in stylommatophoran gametogenesis, suggested by Csaba and Bierbauer (1977) for *H. pomatia*, is yet to be confirmed and is still unexplained.

There is little doubt that progress in our understanding of the determinism of growth and reproduction in terrestrial gastropods will enable, in the future, better management of populations in agricultural crops, parasite transmission and heliciculture. In addition, identifying the factors that regulate reproduction and growth and finding out how to control them has led to *C. aspersus* being permanently available as a laboratory animal, allowing advances in the fields of ecotoxicology (Gomot, 1997; Gomot de Vaufleury and Bispo, 1999; Gomot de Vaufleury, 2000; Gomot de Vaufleury and Kerhoas, 2000; Gomot de Vaufleury and Pihan, 2000).

References

Abeloos, M. (1943) Effets de la castration chez un mollusque *Limax maximus*. *Comptes Rendus de l'Académie des Sciences Paris* 216, 90–92.

Adamo, S. and Chase, R. (1990) The ' love dart ' of the snail, *Helix aspersa*, injects a pheromone that decreases courtship duration. *Journal of Experimental Zoology* 255, 80–87.

Badino, G. (1967) I fattori della gametogenesi di *Arion rufus* studiati con il metodo della cultura *in vitro*. *Archivo Zoologico Italiano* 52, 271–275.

Bailey, S.E.R. (1981) Circannual and circadian rhythms in the snail *Helix aspersa* Müller and the photoperiodic control of annual activity and reproduction. *Journal of Comparative Physiology* 142, 89–94.

Bank, O. (1931) Der einfluss hoher temperatur auf die gonade von *Helix pomatia*. *Biologia Generalis* 7, 429–444.

Berry, A.J. and Chan, L.C. (1968) Reproductive condition and tentacle extirpation in Malayan *Achatina fulica* (Pulmonata). *Australian Journal of Zoology* 16, 849–855.

Berset de Vaufleury, J.P., Bride, J. and Gomot, L. (1986) Mise en évidence du rôle endocrine de la gonade dans le développement de l'appareil génital de l'escargot *Helix aspersa*. *Comptes Rendus de la Société de Biologie* 180, 190–196.

Bierbauer, J. (1978) Effect of hermaphroditic gland homogenate on the regulation of the gametogenesis in *Helix pomatia* (Gastropoda, Pulmonata). *Acta Biologica Academiae Scientiarum Hungaricae* 29, 181–187.

Bierbauer, J. and Molnar, J. (1972) Die experimentelle beeinflussung der regulation der gametogenese bei den lungenschnecken zur zeit des winterschlafes. *Különenyomataz Allattani Kölemények* 59, 36–38.

Binder, E. (1965) Structure de l'organe sexuel frontal des *Gymnarion* des Monts Nimba. *Revue Suisse de Zoologie* 72, 584–593.

Binder, E. (1976) Les *Gymnarion* de l'Afrique de l'ouest, du Sénégal au Togo (Mollusca pulmonata). *Revue Suisse de Zoologie* 83, 705–721.

Binder, E. (1977) La pariade chez le genre *Gymnarion* Pilsbry 1919 (Pulmonata); rôle de l'organe frontal. *Archiv für Molluskenkunde* 107, 249–255.

Boer, H.H. and Joosse, J. (1975) Endocrinology. In: Fretter, V. and Peake, J. (eds) *Pulmonates*, Vol. 1, *Functional Anatomy and Physiology*. Academic Press, London, pp. 245–307.

Bonavita, D. (1972) Eléments pour une monographie écologique et biologique du gastéropode terrestre *Leucochroa candidissima* (Drap.). Thèse doctorat sciences naturelles, Université de Marseille no. A07328, France.

Bonnefoy-Claudet, R. and Deray, A. (1984) Influence de la durée d'hibernation sur l'activité reproductrice de l'escargot *Helix aspersa* Müller. *Comptes Rendus de la Société de Biologie* 178, 442–449.

Bonnefoy-Claudet, R. and Deray, A. (1987) Modalités de reproduction de l'escargot *Helix aspersa maxima* en fonction de la photophase, comparaison avec *Helix aspersa aspersa*. *Haliotis* 16, 69–75.

Bonnefoy-Claudet, R., Deray, A. and Gomot, L. (1983) Action de lumières de longueurs d'onde différentes sur la reproduction de l'escargot *Helix aspersa*. *Comptes Rendus de la Société de Biologie* 177, 504–512.

Bouillon, J. (1956) Influence of temperature on the histological evolution of the ovotestis of *Cepaea nemoralis* L. *Nature* 177, 142–143.

Bride, J. and Gomot, L. (1988) La synthèse polysaccharidique *in vitro* de l'oviducte d'*Helix aspersa*: un bioessai pour éprouver l'activité d'organes endocrines, la gonade et le complexe neuro-endocrine céphalique. *Comptes Rendus de l'Académie des Sciences, Paris* 307, 199–204.

Bride, J. and Gomot, L. (1991) Asynchronisme du développement du tractus génital de l'escargot *Helix aspersa* pendant la croissance et la reproduction. *Reproduction, Nutrition, Développement* 31, 81–96.

Bride, J., Zribi, R. and Gomot, L. (1986) Action de l'extrait de tentacules oculaires d'escargots juvéniles et adultes sur le développement *in vivo* de la glande à albumen d'*Helix aspersa*. *General and Comparative Endocrinology* 61, 383–392.

Bride, J., Gomot, L. and Saleuddin, A.S.M. (1991) Mating and 20-hydroxyecdysone cause increased galactogen synthesis in the albumen gland explants of *Helix aspersa* (Mollusca). *Comparative Biochemistry and Physiology* 98B, 369–373.

Chase, R. (1986) Brain cells that command sexual behavior in the snail *Helix aspersa*. *Journal of Neurobiology* 17, 669–679.

Chase, R. and Piotte, M. (1981) A cephalic dimple in the terrestrial snail, *Achatina achatina*. *Veliger* 23, 241–244.

Chung, D.J.D. (1986) Stimulation of genital eversion in the land snail *Helix aspersa* by extracts of the dart apparatus. *Journal of Experimental Zoology* 238, 129–139.

Cooke, I. and Gelperin, A. (1988) Distribution of FMRF amide-like immunoreactivity in the nervous system of the slug *Limax maximus*. *Cell and Tissue Research* 253, 69–76.

Csaba, G. and Bierbauer, J. (1977) Overlapping effects of different pituitary hormones on the oogenesis and spermatogenesis of Helix pomatia. Acta Biologica Medica Germanica 36, 201–204.

de Larambergue, M. (1939) Etude de l'auto-fécondation chez les gastéropodes pulmonés. Bulletin Biologique de la France et de la Belgique 73, 19–231.

de Larambergue, M. (1959) Sur la reproduction et l'auto-fécondation chez Achatina marginata (gastéropode pulmoné) en élevage. Revue de Pathologie Générale et de Physiologie Clinique 706, 313–316.

Deray, A. and Laurent, J. (1987) Lumière et dynamique de croissance chez l'escargot Helix aspersa maxima (Gros Gris) en conditions hors sol contrôlées. Comptes Rendus de la Société de Biologie 181, 179–186.

Durchon, M. and Joly, P. (1978) L'Endocrinologie des Invertébrés. Collection 'Le biologiste'. Presses Universitaires de France, Paris.

Enée, J., Bonnefoy-Claudet, R. and Gomot, L. (1982) Effets de la photopériode artificielle sur la reproduction de l'escargot Helix aspersa Müller. Comptes Rendus de l'Académie des Sciences, Paris 294, 357–360.

Geraerts, W.P.M. (1976) Control of growth by the neurosecretory hormone of the light green cells in the freshwater snail Lymnaea stagnalis. General and Comparative Endocrinology 29, 61–71.

Geraerts, W.P.M., Smit, A.B., Li, K.W., Vreugdenhil, E. and Van Heerikhuizen, H. (1991) Neuropeptide gene families that control reproductive behaviour and growth in molluscs. In: Osborne, N.N. (ed.) Current Aspects of the Neurosciences, Vol. 3. MacMillan Press, London, pp. 255–305.

Gomot, A. (1990) Photoperiod and temperature interaction in the determination of reproduction of the edible snail, Helix pomatia. Journal of Reproduction and Fertility 90, 581–585.

Gomot, A. (1994) Contribution à l'étude de la croissance d'escargots du genre Helix: influence de facteurs de l'environnement. Nutrition et composition biochimique. Contrôle neuro-endocrine. Thesis no. 398, The University of Franche-Comté, Besançon, France.

Gomot, A. (1997) Dose-dependent effects of cadmium on the growth of snails in toxicity bioassays. Archives of Environmental Toxicology 33, 209–216.

Gomot, A. and Gomot, L. (1994) Cerveaux chimères chez les escargots: croissance et différenciation sexuelle. Comptes Rendus de l'Académie des Sciences, Paris 317, 875–884.

Gomot, A. and Gomot, L. (1995) Brain grafts of cerebral ganglia have effectiveness in growth restoration of damaged Helix aspersa mesocerebrum. Brain Research 682, 127–132.

Gomot, A. and Gomot, L. (1996) Allogeneic and xenogeneic grafts in pulmonate gastropod molluscs: fates of neural transplants. Developmental and Comparative Immunology 20, 193–205.

Gomot, A., Gomot, L., Marchand, C.R., Colard, C. and Bride, J. (1992) Immunocytochemical localization of insulin-related peptide(s) in the central nervous system of the snail Helix aspersa Müller: involvement in growth control. Cellular and Molecular Neurobiology 12, 21–32.

Gomot, L. (1973) Etude du fonctionnement de l'appareil génital de l'escargot Helix aspersa par la méthode des cultures d'organes. Archives d'Anatomie, d'Histologie, d'Embryologie Normales et Expérimentales 56, 131–160.

Gomot, L. and Colard, C. (1985) Etude du contrôle de l'activité sécrétrice des glandes multifides par castration et greffe chez l'escargot Helix aspersa. General and Comparative Endocrinology 58, 159–168.

Gomot, L. and Courtot, A.M. (1979) Etude en culture *in vitro* du contrôle endocrine de la glande à albumen chez l'escargot *Helix aspersa. Malacologia* 18, 361–367.

Gomot, L. and Deray, A. (1987) Les escargots. *La Recherche* 186, 302–311.

Gomot, L. and Griffond, B. (1993) Action of epigenetic factors on the expression of hermaphroditism in the snail *Helix aspersa. Comparative Biochemistry and Physiology* 104A, 195–199.

Gomot, L., Enée, J. and Laurent, J. (1982) Influence de la photopériode sur la croissance pondérale de l'escargot *Helix aspersa* Müller en milieu contrôlé. *Comptes Rendus de l'Académie des Sciences, Paris* 294, 749–752.

Gomot, L., Gomot, P. and Colard, C. (1990) Mise en évidence et analyse de l'activité cérébrale inhibitrice de la multiplication spermatogénétique (acims) induite par la chaleur au cours de l'hibernation d'*Helix aspersa* Müller. In: Loughton, B.G. and Saleuddin, A.S.M. (eds) *Neurobiology and Endocrinology of Selected Invertebrates.* Captus Press, North York, Canada, pp. 55–82.

Gomot, P. (1993) Studies on the control of spermatogenic DNA synthesis by the mesocerebrum in the snail *Helix aspersa. Cellular and Molecular Neurobiology* 13, 517–527.

Gomot, P. and Gomot, L. (1985) Action de la photopériode sur la multiplication spermatogoniale et la reproduction de l'escargot *Helix aspersa. Bulletin de la Société Zoologique de France* 110, 445–459.

Gomot, P. and Gomot, L. (1989a) Inhibition of temperature-induced spermatogenic proliferation by a brain factor in hibernating *Helix aspersa* (Mollusca). *Experientia* 45, 349–351.

Gomot, P. and Gomot, L. (1989b) Etude expérimentale *in vivo* du rôle du cerveau dans l'action de la température sur la spermatogénèse de l'escargot *Helix aspersa* en hibernation. *Comptes Rendus de l'Académie des Sciences, Paris* 308, 135–140.

Gomot, P. and Griffond, B. (1987) Répercussion de la durée d'éclairement journalier sur l'évolution des cellules nourricières et de la lignée mâle dans l'ovotestis d'*Helix aspersa. Reproduction, Nutrition, Développement* 27, 95–108.

Gomot, P., Gomot, L. and Griffond, B. (1989) Evidence for a light compensation of the inhibition of reproduction by low temperatures in the snail *Helix aspersa.* Ovotestis and albumen gland responsiveness to different conditions of photoperiods and temperatures. *Biology of Reproduction* 40, 1237–1245.

Gomot, P., Griffond, B. and Gomot, L. (1990) Action de la température sur la synthèse d'ADN des cellules mâles et la spermatogenèse d'*Helix aspersa* en hibernation. *Journal of Thermal Biology* 15, 267–280.

Gomot, P., Griffond, B., Colard, C. and Gomot, L. (1992) Etude de l'action des ganglions cérébroïdes et des corps dorsaux sur la synthèse d'ADN induite par la chaleur dans l'ovotestis d'*Helix aspersa* en hibernation. *Reproduction, Nutrition, Développement* 32, 55–66.

Gomot de Vaufleury, A. (2000) Standardized growth toxicity testing (Cu, Zn, Pb, and Pentachlorophenol) with *Helix aspersa. Ecotoxicology and Environmental Safety* 46, 41–50.

Gomot de Vaufleury, A. and Bispo, A. (1999) Garden snails as bioindicators for laboratory toxicity assessment of soil contaminated by organic chemicals. In: *SECOTOX 99, Fifth European Conference on Ecotoxicology*

and Environmental Safety. GSF-National Research Center, Neuherberg/
Munich, Germany, 15–17 March, 1999.

Gomot de Vaufleury, A. and Kerhoas, I. (2000) Effects of cadmium on the
reproductive system of the land snail *Helix asperse. Bulletin of Environmental Contamination and Toxicology* 64, 434–442.

Gomot de Vaufleury, A. and Pihan, F. (2000) Growing snails used as sentinels
to evaluate terrestrial environmental contamination by trace elements.
Chemosphere 40, 275–284.

Gottfried, H. and Dorfman, R.J. (1970) Steroids of invertebrates. IV On the
optic tentacle–gonadal axis in the control of the male-phase ovotestis in the
slug (*Ariolimax californicus*). *General and Comparative Endocrinology* 15,
101–119.

Goudsmit, E.M. (1975) Neurosecretory stimulation of galactogen synthesis within
the *Helix pomatia* albumen gland during organ culture. *Journal of Experimental Zoology* 191, 193–198.

Goudsmit, E.M. and Ram, J. (1982) Stimulation of *Helix pomatia* albumen gland
galactogen synthesis by putative neurohormone (galactogenin) and by cyclic
AMP analogues. *Comparative Biochemistry and Physiology* 71B, 417–422.

Griffond, B. and Gomot, L. (1989) Endocrinology of reproduction in stylommatophoran pulmonate gastropods with special reference to *Helix. Comparative
Endocrinology, Life Science Advances* 8, 23–32.

Griffond, B. and Vincent, C. (1985) Etude de l'activité des corps dorsaux
de l'escargot *Helix aspersa* Müller au cours des phases physiologiques de la
vie adulte et sous différentes photopériodes. *Invertebrate Reproduction and
Development* 8, 27–37.

Griffond, B., Gomot, P. and Gomot, L. (1992a) Influence de la température sur le
déroulement de l'ovogenèse chez l'escargot *Helix aspersa. Journal of Thermal
Biology* 17, 185–190.

Griffond, B., Van Minnen, J. and Colard, C. (1992b) Distribution of α CDCP-
immunoreactive neurons in the central nervous system of the snail *Helix
aspersa. Reproduction, Nutrition, Développement* 32, 113–121.

Guyard, A. (1971) Etude de la différenciation de l'ovotestis et des facteurs
contrôlant l'orientation sexuelle des gonocytes de l'escargot *Helix aspersa*
Müller. Thesis University of Besançon, Besançon, France.

Henderson, N.E. and Pelluet, J. (1960) The effect of visible light on the ovotestis
of the slug *Deroceras reticulatum* Müller. *Canadian Journal of Zoology* 38,
173–178.

Highnam, K.C. and Hill, L. (1977) *The Comparative Endocrinology of the
Invertebrates.* Edward Arnold, London.

Hogg, M. and Wijdenes, J. (1979) A study of gonadal organogenesis, and factors
influencing regeneration following surgical castration in *Deroceras
reticulatum* (Pulmonata : Limacidae). *Cell and Tissue Research* 198, 295–307.

Hunter, R.D. and Stone, L.M. (1986) The effect of artificial photoperiod on growth
and reproduction in the land snail *Cepaea nemoralis. International Journal of
Invertebrate Reproduction and Development* 9, 339–344.

Jarne, P., Vianey-Liaud, M. and Delay, B. (1993) Selfing and outcrossing in
hermaphrodite freshwater gastropods (Basommatophora): where, when and
why? *Biological Journal of the Linnean Society* 49, 99–125.

Jeppesen, L.L. (1976) The control of mating behaviour in *Helix pomatia* L.
(Gastropoda: Pulmonata). *Animal Behaviour* 24, 275–290.

Joosse, J. (1988) The hormones of molluscs. In: Laufer, H. and Downer, G.H. (eds) *Endocrinology of Selected Invertebrate Types.* Alan R. Liss, New York, pp. 89–140.

Joosse, J. and Geraerts, W.P.M. (1983) Endocrinology. In: Saleuddin, A.S.M. and Wilbur K.M. (eds) *The Mollusca,* Vol. 4, *Physiology.* Academic Press, New York, pp. 317–406.

Kamatani, Y., Minakata, H., Iwashita, T., Nomoko, K., In, Y., Doi, M. and Ishida, T. (1990) Molecular conformation of achatin-I, an endogenous neuropeptide containing D-amino acid residue. X-ray crystal structure of its neutral form. *Federation of European Biochemical Societies Letters* 276, 95–97.

Krusch, B., Schoenmakers, H.J.N., Voogt, P.A. and Nolte, A. (1979) Steroid synthesizing capacity of the dorsal body of *Helix pomatia* L. (Gastropoda). An *in vitro* study. *Comparative Biochemistry and Physiology* 64B, 101–104.

Laurent, J., Deray, A. and Grimard, A.M. (1984) Influence de la photopériode, du degré d'hétérogénéité de la population sur la dynamique de croissance et la maturation sexuelle de l'escargot *Helix aspersa. Comptes Rendus de la Société de Biologie* 178, 421–441.

Laviolette, P. (1954a) Rôle de la gonade dans le déterminisme humoral de la maturité glandulaire du tractus génital chez les gastéropodes arionidae et limacidae. *Bulletin Biologique de la France et de la Belgique* 88, 310–332.

Laviolette, P (1954b) Etude cytologique et expérimentale de la régénération germinale après castration chez *Arion rufus. Annales des Sciences Naturelles, Zoologie* 11, 16, 427–535.

Le Guellec, D., Thiard, M.C., Rémy-Martin, J.P., Deray, A., Gomot, L. and Adessi, G.L. (1987) *In vitro* metabolism of androstenedione and identification of endogenous steroids in *Helix aspersa. General and Comparative Endocrinology* 66, 425–433.

Le Guhennec, M. and Daguzan, J. (1983) Rôle de la lumière sur la reproduction de l'escargot Petit Gris, *Helix aspersa* Müller. *Comptes Rendus de l'Académie des Sciences, Paris* 297, 141–144.

Leonard, J.L. (1992) The ' love-dart ' in helicid snails: a gift of calcium or a firm commitment? *Journal of Theoretical Biology* 159, 513–521.

Lucarz, A. (1991) Evidence of an egg-laying factor in the prostatic secretions of *Helix aspersa* Müller. *Comparative Biochemistry and Physiology* 100A, 839–843

Lusis, O. (1966) Changes induced in the reproductive system of *Arion ater rufus* by varying environmental conditions. *Proceedings of the Malacological Society of London* 37, 19–26.

Marchand, C.R. and Gomot, L. (1987) Neuroendocrine control of growth and reproduction of the snail *Helix aspersa.* In: Boer, H.H., Geraerts, W.P.M. and Joosse, J. (eds) *Neurobiology Molluscan Models.* North-Holland Publishing Company, Amsterdam, pp. 360–366.

Marchand, C.R., Griffond, B., Mounzih, K. and Colard, C. (1991) Distribution of methionine-enkephalin like and FMRF amide-like immunoreactivities in the central nervous system (including dorsal bodies) of the snail *Helix aspersa* Müller. *Zoological Science* 8, 905–913.

McCrone, E.J., van Minnen, J. and Sokolove, P.G. (1981) Slug reproductive maturation hormone: *in vivo* evidence for long-day stimulation of secretion from brains and cerebral ganglia. *Journal of Comparative Physiology* 143, 311–315.

Meenakshi, V.R. and Scheer, B.T. (1969) Regulation of galactogen synthesis in the slug *Ariolimax columbianus. Comparative Biochemistry and Physiology* 29, 841–845.

Melrose, G.R., O'Neill, M.C. and Sokolove, P. (1983) Male gonadotrophic factor in brain and blood of photoperiodically stimulated slugs. *General and Comparative Endocrinology* 52, 319–328.

Nolte, A., Koolman, J., Dorlöchter, M. and Straub, H. (1986) Ecdysteroids in the dorsal bodies of pulmonates (Gastropoda): synthesis and release of ecdysone. *Comparative Biochemistry and Physiology* 84A, 777–782.

Nopp, H. (1971) Einige wirkungen der amputation der optischen tentakel bei einer landlungen schnecke (*Eobania vermiculata* Müll. Helicidae). *Experientia* 27, 855.

Pelluet, D. (1964) On the hormonal control of cell differentiation in the ovotestis of slugs. *Canadian Journal of Zoology* 42, 195–199.

Pelluet, D. and Lane, N.J. (1961) The relation between neurosecretion and cell differentiation in the ovotestis of slugs (Gastropoda: Pulmonata). *Canadian Journal of Zoology* 39, 789–804.

Price, D.A., Davies, N.W., Doble, K.E. and Greenberg, M.J. (1987) The variety and distribution of the FMRF amide-related peptides in molluscs. *Zoological Science* 4, 395–410.

Runham, N.W., Bailey, T.G. and Laryea, A.A. (1973) Studies of the endocrine control of the reproductive tract of the grey field slug *Agriolimax reticulatus. Malacologia* 14, 135–142.

Saleuddin, A.S.M., Farrel, C.L. and Gomot, L. (1983) Brain extract causes amoeboid movement *in vitro* in oocytes in *Helix aspersa* (Mollusca). *International Journal of Invertebrate Reproduction* 6, 31–34.

Saleuddin, A.S.M., Griffond, B. and Ashton, M.L. (1991) An ultrastructural study of the activation of the endocrine dorsal bodies in the snail *Helix aspersa* by mating. *Canadian Journal of Zoology* 69, 1203–1215.

Saleuddin, A.S.M., Sevala, V.M., Sevala, V.L., Mukai, S.T. and Khan, H.R. (1992) Involvement of mammalian insulin and insulin-like peptides in shell growth and shell regeneration in molluscs. In: Suga, S. and Watabe, N. (eds) *Hard Tissue Mineralization and Demineralization.* Springer-Verlag, Tokyo, pp. 149–169.

Saleuddin, A.S.M., Khan, H.R., Sevala, M.V., Maurella, A. and Kunigelis, S.C. (1995) Hormonal control of biomineralization in selected invertebrates. *Bulletin de l'Institut Océanographique*, Monaco (Spécial Issue 14) 2, 127–140.

Selander, R.K., Kaufman, D.W. and Ralin, R.S. (1975) Self-fertilization in the terrestrial snail *Rumina decollata. Veliger* 16, 265–270.

Sokolove, P.G. and McCrone, E.J. (1978) Reproductive maturation in the slug *Limax maximus* and the effects of artificial photoperiod. *Journal of Comparative Physiology* 125A, 317–325.

Sokolove, P.G., McCrone, E.J., Melrose, G.R., van Minnen, J. and O'Neil, M.C. (1985) Photoperiodic and endocrine control of reproductive maturation in a terrestrial slug. In: Lofts, B. and Holmes, W.N. (eds) *Current Trends in Comparative Endocrinology.* Hong Kong University Press, Hong Kong, pp. 299–302.

Sokolove, P.G., McCrone, E.J. and Albert, G.R. (1987) Role of brain and gonadal hormones in slug reproductive development: a re-examination. In: Boer, H.H.,

Geraerts, W.P.M., and Joosse, J. (eds) *Neurobiology Molluscan Models.* North-Holland Publishing Company, Amsterdam, pp. 350–356.

Stephens, G.J. and Stephens, G.C. (1966) Photoperiodic stimulation of egg laying in the land snail *Helix aspersa. Nature* 212, 1582.

Takayanagi, H. and Takeda, N. (1985) Hormonal control of gametogenesis in the terrestrial snail, *Euhadra peliomphala. Development, Growth and Differentiation* 27, 689–700.

Takeda, N. (1977) Stimulation of egg-laying by nerve extracts in slugs. *Nature* 267, 513–514.

Takeda, N. (1982a) Source of the tentacular hormone in terrestrial pulmonates. *Experientia* 38, 1058–1060.

Takeda, N. (1982b) Notes on the fine structure of the head-wart in some terrestrial snails. *Veliger* 24, 328–330.

Takeda, N. (1983) Endocrine regulation of reproduction in the snail, *Euhadra peliomphala.* In: Lever, J. and Boer, H.H. (eds) *Molluscan Neuro-endocrinology.* North-Holland Publishing Company, Amsterdam, pp. 106–111.

Takeda, N. (1985) Hormonal control of reproduction in some terrestrial pulmonates. In: Lofts, B. and Holmes, W.N. (eds) *Current Trends in Comparative Endocrinology.* Hong Kong University Press, Hong Kong, pp. 303–304.

Takeda, N. (1989) Hormonal control of reproduction in land snails. *Venus* 48, 99–139.

Takeda, N. and Sugiyama, K. (1984) Gonadal regeneration and sex steroid hormones in some terrestrial pulmonates. *Venus* 43, 72–85.

Tompa, A.S. (1984) Land snails (Stylommatophora). In: Tompa, A.S., Verdonk N.H. and van den Biggelaar J.A.M. (eds) *The Mollusca,* Vol. 7, *Reproduction.* Academic Press, New York, pp. 47–140.

Van Minnen, J. and Sokolove, P.G. (1984) Galactogen synthesis stimulating factor in the slug, *Limax maximus*: cellular localization and partial purification. *General and Comparative Endocrinology* 54, 114–122.

Van Minnen, J., Schallig, H.D.F.H. and Ramkema, M.D. (1992) Identification of putative egg-laying hormone containing neuronal systems in gastropod molluscs. *General and Comparative Endocrinology* 86, 96–102.

Vincent, C., Griffond, B., Gomot, L. and Bride, J. (1984a) Etude *in vitro* de l'influence des corps dorsaux sur l'ovogenèse d'*Helix aspersa* Müll. *General and Comparative Endocrinology* 54, 230–235.

Vincent, C., Griffond, B., Wijdenes, J. and Gomot, L. (1984b) Contrôle d'une glande endocrine: les corps dorsaux par le système nerveux central chez *Helix aspersa. Comptes Rendus de l'Académie des Sciences Paris* 299, 421–426.

Wattez, C. (1980) Recherches expérimentales sur le déterminisme de la différenciation sexuelle et du fonctionnement de la gonade chez le pulmoné stylommatophore hermaphrodite *Arion subfuscus* Draparnaud. Thesis University of Lille, Lille France.

Wijdenes, J. and Runham, N.W. (1976) Studies on the function of the dorsal bodies of *Agriolimax reticulatus* (Mollusca, Pulmonata). *General and Comparative Endocrinology* 29, 545–551.

Wijdenes, J. and Runham, N.W. (1977) Studies on the control of growth in *Agriolimax reticulatus* (Mollusca, Pulmonata). *General and Comparative Endocrinology* 31, 154–156.

Wijdenes, J., Vincent, C., Griffond, B. and Gomot, L. (1983) Ultra-structural evidence for the neuro-endocrine innervation of the dorsal bodies and their probable physiological significance in *Helix aspersa.* In: Lever, J. and

Boer, H.H. (eds) *Molluscan Neuroendocrinology*. North-Holland Publishing Company, Amsterdam, pp.147–152.

Wolda, H. (1967) The effect of temperature on reproduction in some morphs of the land snail *Cepaea nemoralis* L. *Evolution* 21, 117–129.

10 Spermatogenesis and Oogenesis

J.M. HEALY

Centre for Marine Studies, The University of Queensland, Brisbane, Queensland 4072, Australia

Introduction

The success of the Gastropoda in colonizing the terrestrial environment owes as much to having appropriate reproductive strategies (internal fertilization, often coupled with hermaphroditism) as it does to the possession of a shell, mucus-coated epidermis and creeping sole. Internal fertilization in gastropods is not, however, the result of selection pressure associated with the transition from water to land – it is already present in marine relatives of all terrestrial groups. Similarly, sperm morphology in each family of terrestrial gastropods has not altered appreciably from that observed in presumed marine antecedents. Even the hermaphroditism of pulmonates is also a feature of virtually all other heterobranchs. Nevertheless, one of the keys to the success of the terrestrial Gastropoda, and in particular the ubiquitous Stylommatophora, remains their ability to reproduce and colonize effectively.

Four superfamilies of prosobranch gastropods occur terrestrially – Neritoidea, Rissooidea, Littorinoidea and Cyclophoroidea – the latter exclusively so. By far the bulk of the land Gastropoda consists of pulmonates, including representatives from the orders Basommatophora (Ellobioidea) and Systellommatophora (Rathouisioidea) and, most importantly, the entire order Stylommatophora, unquestionably one of the most diverse and economically significant groups of living molluscs. This chapter summarizes what is known of the morphology, development and cytochemistry of spermatozoa and oocytes in terrestrial gastropods, with an emphasis on the more intensively studied Pulmonata.

Spermatogenesis

Spermatogenesis includes four developmental stages: (i) derivation of spermatogonia from the germ cells; (ii) mitotic proliferation of spermatogonia; (iii) production of primary and secondary spermatocytes; and (iv) production of spermatids and their eventual transformation into mature spermatozoa (step iv = spermiogenesis). As Roosen-Runge (1977) has emphasized, spermatogenesis is essentially a syncytial process in which generations of cytoplasmically connected cells pass through the same developmental sequence.

In stylommatophorans, spermatogenesis is dependent on several factors other than food and water availability, most notably on the ambient temperature and day length. It has, for instance, been demonstrated that exposure of *Cantareus aspersus* (Müller) (Helicidae) to short photoperiod days not only inhibits spermatogonial proliferation (for up to 3 weeks) but also results in a temporary blockage in the transition of secondary spermatocytes to spermatids (Gomot and Gomot, 1985; Gomot and Griffond, 1987; Griffond and Medina, 1989). Production of primary spermatocytes from spermatogonia and events of spermiogenesis are, however, not affected. Using light and transmission microscopy and ³H-labelled thymidine, Bloch and Hew (1960) determined that during spermatogenesis in *C. aspersus* the lysine-rich nuclear histone present in spermatocytes and early spermatids is gradually replaced by an arginine-rich histone.

Ultrastructural features of spermatogonia and spermatocytes in terrestrial gastropods to date have only been described for a few stylommatophoran pulmonate species (*Euhadra hickonis* Kobelt (Bradybaenidae) by Takaichi and Sawada, 1973; *Arion ater* (Linnaeus) (Arionidae) by Parivar, 1981; *C. aspersus* by Griffond *et al.*, 1991; and *Scutalus tupacii* (d'Orbigny) (Bulimulidae) by Cuezzo, 1995a). Having arisen from the germinal epithelium, spermatogonia become attached via desmosome-like junctions to so-called 'Sertoli cells' to form a cluster. In addition, cytoplasmic bridges develop during spermatogonial proliferative divisions. Spermatogonia are generally pyriform and exhibit an oblong to spherical nucleus (with patchy heterochromatin and 1–3 nucleoli) and, in proportion to the nucleus, a relatively small quantity of cytoplasm containing rounded to oblong mitochondria, many ribosomes and scattered, elongate endoplasmic reticular cisternae. Primary spermatocytes are rounded to pyriform, and are characterized by an increased proportion of cytoplasm relative to the nuclear volume, marked elongation of the mitochondria (cristae now elongate), a more conspicuous Golgi complex (multiple stacks of cisternae, associated with the centrioles) and, within the nucleus, the presence of synaptinemal complexes (during zygotene and pachytene stages) and a single nucleolus. Secondary spermatocytes are reduced in size compared with primary spermatocytes, contain proportionately less cytoplasm and feature clustered stacks of Golgian cisternae. Mitochondrial and endoplasmic reticular features are, however, similar to those observed in primary spermatocytes. Cuezzo (1995a) has found that formation of the centriolar fossa in *S. tupacii* commences in secondary spermatocytes.

Spermiogenesis

The process of spermiogenesis encompasses all those developmental changes, both morphological and cytochemical, which transform spermatids into spermatozoa (Figs 10.1 and 10.2). In terrestrial gastropods, as in other internally fertilizing molluscs, this involves formation of the acrosome, condensation of the nuclear contents and completion of associated nuclear protein transitions, formation of an elongate midpiece (varying degrees of mitochondrial remodelling) and finally the deposition of whatever glycogen reserves may be present around the mitochondria and/or axoneme. Generally, these events overlap, although certain spermiogenic features such as the appearance of the cytoplasmic microtubular sheath occur only late in the process. The vast majority of the literature on spermiogenesis in terrestrial gastropods is devoted to pulmonates, especially stylommatophorans (Beams and Tahmisian, 1954; André, 1956, 1962; Grassé *et al.*, 1955, 1956a,b,c; Rebhun, 1957; Tahmisian, 1964; Rousset-Galangau, 1972; Takahashi *et al.*, 1973; Takaichi and Sawada, 1973; Yasuzumi *et al.*, 1974; Takaichi and Dan, 1977; Takaichi, 1978, 1979; Dan and Takaichi, 1979; Parivar, 1981; Whitehead and Hodgson, 1985; Griffond *et al.*, 1991; Sretarugsa *et al.*, 1991; Cuezzo, 1995a) and, to a lesser degree, systellommatophorans (Lanza and Quattrini, 1965; Quattrini and Lanza, 1964, 1965). Studies treating spermiogenesis in terrestrial prosobranchs include those by Giusti and Mazzini (1973, Truncatellidae), Selmi and Giusti (1980, Cochlostomatidae) and Kohnert and Storch (1984b, Pomatiasidae). The process, as observed in these prosobranchs, is identical to that documented for aquatic, internally fertilizing prosobranchs (for comparisons, see Kohnert and Storch, 1984b; Koike, 1985), and similar to that occurring in heterobranchs, although in the latter group the degree of mitochondrial metamorphosis is considerably more complex than in any prosobranch gastropod.

Pulmonata

In early spermatids of pulmonates, the nucleus is rounded and its contents initially exhibit dense chromatin patches. The nucleus subsequently becomes evenly granular in texture. The cytoplasm contains an extensive endoplasmic reticulum, usually multiple stacks of Golgi cisternae, numerous rounded or elongate mitochondria (exhibiting unmodified cristae and usually one or more intramitochondrial granules) and one or sometimes two centrioles (Figs 10.1A and 10.2A). In some instances, two or more spermatids may develop within a common cytoplasm (Fig. 10.2B). As development proceeds, an electron-dense plaque develops

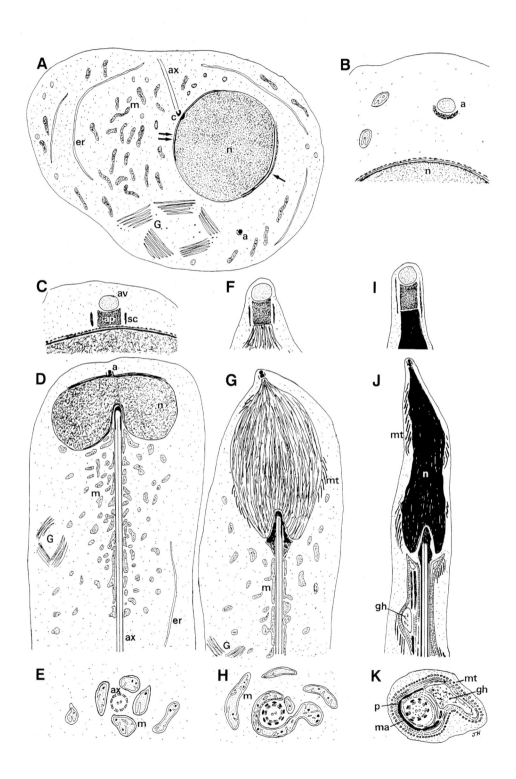

at both the future anterior and posterior poles of the nuclear surface (Fig. 10.1A, B and C). Whereas the anterior plaque is composed of a thin layer of extranuclear material matched by an almost equally thin layer of condensed nuclear material, the posterior plaque consists solely of condensed nuclear material. The Golgi complex secretes a round acrosomal vesicle which migrates to the centre of the anterior nuclear plaque and becomes fixed there (Figs 10.1B and C, and 10.2C and E). During the migration process, the acrosomal vesicle acquires an extravesicular deposit basally and, after attachment to the anterior plaque, also acquires a supporting cylinder (Fig. 10.1B and C). The origins of both of these extravesicular structures appear to lie in cytoplasmic precursors rather than the Golgi complex, but further work is required to confirm this. As these acrosome development events occur, the centriole gives rise to an axoneme with a 9 + 2 microtubular pattern, and becomes lodged in the invaginating base of the nucleus, an area initially defined by the posterior nuclear plaque (Fig. 10.1A). Once the acrosomal and centriolar–axonemal complexes have established their relationship with the anterior and posterior poles of the nucleus, the nuclear contents begin their transformation from fine granules to fibres (Fig. 10.1D). The shape of the spermatid nucleus changes during development, being initially rounded (early spermatid, Fig. 10.1A), then anterio-posteriorly compressed, with spongy fibres (middle stage, Fig. 10.1D), then pyriform, with fibres elongate (advanced stage, Fig. 10.1G) and, finally rod-shaped (usually with helical keels) fusing into lamellae (late stage) (Fig. 10.1J). As transformation of the nuclear contents from granules to fibres takes place, the small cytoplasmic mitochondria cluster along the length of the developing axoneme and begin fusion, thereby heralding the first phase of midpiece formation (Fig. 10.1D and E). During development of the midpiece, the axoneme (now associated with nine coarse fibres) becomes surrounded by

Fig. 10.1. (Oppoiste) Diagram summarizing the basic features of spermiogenesis in a pulmonate gastropod (based on transmission electron microscopy data from several sources including the author's own studies). Only spermatids are depicted (the relationship with supporting or Sertoli cells is not shown). (A) Early spermatid. Note the future polarity of the nucleus marked by the posterior dense plaque (double arrow, site of centriole–axoneme attachment) and the anterior plaque (single arrow, site of acrosome attachment). (B) The acrosomal complex of an early spermatid approaching the anterior plaque of the nucleus. (C) The acrosomal complex and nuclear apex of a middle stage spermatid. (D) Middle stage spermatid. (E) Transverse section of the developing midpiece of a middle stage spermatid. (F) Acrosomal complex and nuclear apex of an advanced stage spermatid. (G) Advanced stage spermatid. Note the presence of perinuclear microtubules. (H) Developing midpiece of an advanced stage spermatid. Note the mitochondrial fusion and periaxonemal wrapping. (I) Acrosomal complex and nuclear apex of a late stage spermatid. (J) Late stage spermatid. Note the presence of perinuclear and perimitochondrial microtubules. (K) Developing midpiece of a late stage spermatid. Note the microtubular sheath, paracrystalline and matrix components of the mitochondrial derivative and formation of a glycogen-filled helix. a, acrosomal complex; ap, acrosomal pedestal; av, acrosomal vesicle; ax, axoneme; cd, centriolar derivative; cf, coarse fibres; er, endoplasmic reticulum; G, Golgi complex; gh, glycogen helix; m, mitochondria; ma, matrix material of the mitochondrial derivative; mt, microtubules; n, nucleus; p, paracrystalline material of the mitochondrial derivative; pns, perinuclear sheath; sc, support cylinder. (A = ×5000; B, C, F, I = ×35,000; D, G, J = ×7500; E, H, K = ×25,000.)

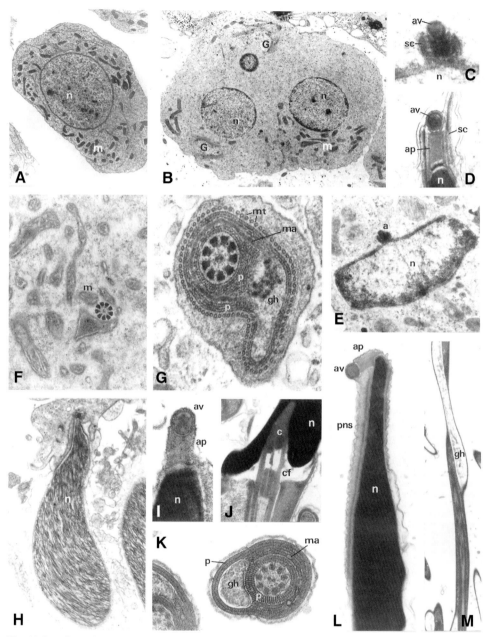

Fig. 10.2. Spermiogenesis and spermatozoa of pulmonates (C–E, Basommatophora; all other figures Stylommatophora). (A) Early spermatid showing the round nucleus and elongate mitochondria of *Cantareus aspersus* (Müller) (Helicidae). (B) Early spermatids within a common cytoplasm in *Pedinogyra hayii* (Griffith & Pidgeon) (Acavidae). Note the large Golgi complexes. (C) Newly attached acrosomal complex on a middle stage spermatid of *Melosidula zonata* (H. & A. Adams) (Ellobiidae). (D) Acrosomal complex of a late spermatid of *M. zonata*. (E) Acrosome and nucleus of a middle stage spermatid of *M. zonata*. (F) Nucleus–midpiece junction of a late spermatid of *P. hayii* showing mitochondrial fusion. (G) Developing mitochondrial derivative and glycogen helix of *Xanthomelon pachystylum* (Pfeiffer) (Camaenidae). Note the associated microtubules. (H) Late stage spermatid

a continuous sheath of mitochondrial material (Figs 10.1F, G and H and 10.2F), only to be followed by a secondary wrapping of mitochondrial material. The mitochondrial material forms one or more helical channels – the glycogen helices (or 'helice secondaire' of André, 1962). As the second wrapping is completed, the mitochondrial sheath begins its transformation into the mature mitochondrial derivative. This involves a fundamental reorganization of the mitochondrial materials into discrete layers of matrix material and highly structured arrays of paracrystalline particles (Figs 10.1J and K, and 10.2G) (André, 1962; Favard and André, 1970). Although glycogen often is not visible in the newly formed glycogen helix, presumably the components for its manufacture are already present in the cytoplasm enclosed by the second mitochondrial wrapping. A helically twisted microtubular sheath is associated with this last phase of midpiece formation (Figs 10.1J and K, and 10.2G). Anderson and Personne (1970a) have demonstrated that glycogen can be synthesized within the glycogen helix of mature pulmonate spermatozoa through the activity of amylophosphorylase present in these cells. Although they conclude that accumulation of glycogen commences only after the completion of spermatozoa maturation, such deposits have been demonstrated in late spermatids and immature spermatozoa of certain basommatophoran pulmonates (Healy, 1983a).

Late in spermiogenesis, both the elongating nucleus and the midpiece become surrounded by a helical sheath of microtubules ('manchette' of various authors) (Figs 10.1G and J, and 10.2G). Almost certainly, these structures play some role in helping to form the helical keel(s) of the nucleus (spiralling fibres now transforming into lamellae – Fig. 10.2H) and the secondary helices within the midpiece sheath. The support cylinder surrounding the acrosomal complex persists until very late in spermiogenesis (Figs 10.1I and 10.2D), but is not observed in mature ovotesticular or hermaphrodite duct spermatozoa (Fig. 10.2I and L). Cytoplasmic debris is removed from the late spermatid through the posterior migration of droplets (containing discarded Golgi and endoplasmic reticular cisternae, membranes, microtubules, mitochondrial materials and vesicular remnants). Morphologically mature spermatozoa, after being released from the supporting or so-called 'Sertoli' cells (during the process of spermiation), move into the lumina of the ovotesticular acini and from there to the hermaphrodite duct (see Hodgson, 1996). In certain

attached to a Sertoli cell in *P. hayii*. (I) Mature acrosomal complex of *Sphaerospira bloomfieldi* (Cox) (Camaenidae). (J) Mature nucleus–midpiece junction of *Sphaerospira yulei* (Forbes) (Camaenidae). (K) Transverse section of a mature midpiece of *S. yulei*. (L) Mature acrosomal complex and nuclear apex of *C. aspersus*. (M) Mature midpiece of *C. aspersus*. a, acrosomal complex; ap, acrosomal pedestal; av, acrosomal vesicle; cd, centriolar derivative; cf, coarse fibres; G, Golgi complex; gh, glycogen helix; m, mitochondria; ma, matrix material of the mitochondrial derivative; mt, microtubules; n, nucleus; p, paracrystalline material of the mitochondrial derivative; pns, perinuclear sheath; sc, support cylinder; SC, Sertoli cell. A = ×1000; B = ×3000; C, D, I = ×30,000; E, F, J = ×15,000; G, K = ×40,000; H = ×7000; L = ×20,000; M = ×4000.)

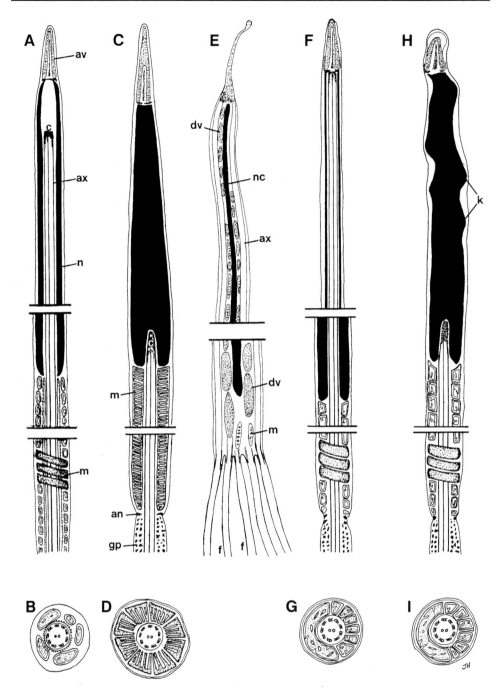

Fig. 10.3. Mature spermatozoa of terrestrial prosobranchs. (A and B) Euspermatozoon of *Waldemaria japonica* (Adams) (Helicinidae, Neritoidea, Neritimorpha; a transverse section of the euspermatozoan midpiece is shown in (B)); based on data of Koike (1985). (C–E) Euspermatozoon (C and D) and multiflagellate paraspermatozoon (E) of a generalized cyclophoroidean (based on data of Selmi and Giusti, 1980, for *Cochlostoma montanum* (Issel) (Cochlostomatidae), and *Liarea ornata* Powell (Liareidae, Healy, unpublished) both of the Cyclophoroidea, Caenogastropoda; a transverse section of

stylommatophorans and basommatophorans, the spermatozoa ultimately are packaged into spermatophores prior to copulation.

Prosobranchia

Ultrastructural features of spermiogenesis (euspermatozoan development) have been investigated in only four terrestrial prosobranchs, all of them caenogastropods (*Pomatias elegans* (Müller), Littorinoidea by Kohnert and Storch, 1984b; *Truncatella subcylindrica* (Linnaeus), Rissooidea by Giusti and Mazzini, 1973; *Cochlostoma montanum* (Issel), by Selmi and Giusti, 1980; and *Liarea ornata* Powell by J.M. Healy, unpublished, (both Cyclophoroidea). These studies clearly show the same sequence of events observed during euspermiogenesis in marine and freshwater caeno-gastropods (for comparisons, see Dohmen, 1983; Maxwell, 1983; Kohnert and Storch, 1984b; Koike, 1985; Healy, 1996). These events include: Golgi production of a conical acrosomal vesicle and subacrosomal deposit; condensation of an initially round spermatid nucleus through granular, fibrillar and lamellar phases into a rod-shaped mature nucleus; transformation of large, round mitochondria of early spermatids into the greatly elongated mitochondria of the definitive midpiece (helical with typical cristae in *Pomatias* Studer and *Truncatella* Risso; straight with parallel cristal plates in *Cochlostoma* Jan); and formation of a glycogen piece posterior to the spermatid annulus. Microtubules surround the late spermatid acrosome, nucleus and midpiece. In *C. montanum*, the microtubules surrounding the late spermatid midpiece are arranged in radiating rows. Presumably the numerous dense intramitochondrial granules observed at this stage give rise to the paracrystalline granules associated with the cristal plates and outer surface of each mitochondrion in the mature euspermatozoon (Selmi and Giusti, 1980) (see Fig. 10.3D). Despite a number of spermiogenic similarities between caenogastropods and heterobranchs – notably the stages of nuclear condensation, the presence of perinuclear and perimitochondrial microtubules in late spermatids and production of a glycogen piece (absent in Stylommato-phora) – the drastic remodelling of mitochondrial material and produc-tion of glycogen helices seen in pulmonates and most other heterobranchs is unique. Similarly, the features of paraspermatozoan development in *C. montanum* (replication of axonemes, generation of proteinaceous dense vesicles, condensation of a greatly reduced-sized nucleus – see Selmi and

euspermatozoan midpiece is shown in D). (F and G) Euspermatozoon of *Pomatias elegans* (Müller) (Pomatiasidae, Littorinoidea, Caenogastropoda; a transverse section of the midpiece is shown in G); based on micrographs of Kohnert and Storch, 1984a, and J.M. Healy, unpublished. (H and I) Euspermatozoon of *Truncatella subcylindrica* (Linnaeus) (Truncatellidae, Rissooidea, Caenogastro-poda; based on data from Giusti and Mazzini, 1973; a transverse section of the euspermatozoan midpiece is shown in I). an, annulus; av, acrosomal vesicle; ax, axoneme; c, centriole; dv, dense vesicle; f, multiple flagella of paraspermatozoon; gp, glycogen piece; k, helical keel of the nucleus; m, mitochondria; n, nucleus; nc, nuclear core. (A, C, F, H = ×15,000; E = ×10,000; B, D, G, I = ×30,000.)

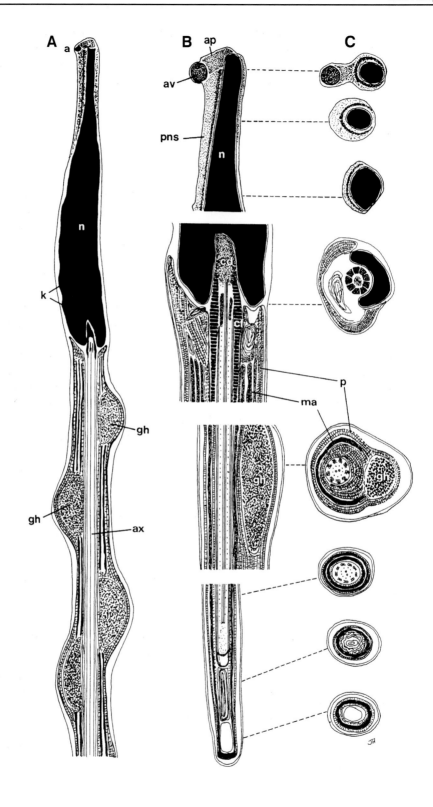

Giusti, 1980), while shared with many other caenogastropods (Healy and Jamieson, 1981; Hodgson, 1997), have no equivalent in any pulmonate in spite of reports of spermiogenic dimorphism in certain limacid and arionid species (Rousset-Galangau, 1972; Parivar, 1981).

Spermatozoa

Pulmonata

Spermatozoa of pulmonates, like those of other heterobranch gastropods, rank among the most complex in the animal kingdom (Thompson, 1973). In certain species of Stylommatophora, the spermatozoa may exceed 1000 µm (1400 µm in *Hedleyella falconeri* (Gray) (Acavidae), 1750 µm in *Pleurodonte acuta* de Lamarck (Camaenidae) – the longest known molluscan spermatozoon) (Thompson, 1973; Maxwell, 1983). Pulmonate spermatozoa show the same key features seen in opisthobranch spermatozoa: a rounded acrosomal vesicle associated with a columnar pedestal, a helical and/or helically keeled nucleus, an axoneme (9 + 2 microtubular pattern) associated with nine periodically striated coarse fibres and an extremely elongate and highly modified midpiece consisting of the axoneme, the mitochondrial derivative featuring matrix and paracrystalline arrays and enclosing one or more glycogen-filled helical compartments (glycogen helices) (Thompson, 1973; Healy, 1983a, 1988a, 1996, and references therein) (Fig. 10.4). Generally, the terminal portion of the sperm 'tail' region of heterobranchs shows a glycogen piece region, preceded by an annulus – as occurs in spermatozoa of caenogastropods (for comparisons and literature, see Healy, 1996). There is, however, a trend toward reduction in the extent of the glycogen piece within the Heterobranchia (Healy and Willan, 1984, 1991; Healy, 1996) and, in stylommatophorans, this portion of the spermatozoon appears to be absent (Takaichi and Sawada, 1973; Dan and Takaichi, 1979; Healy and Jamieson, 1989; Cuezzo, 1995a,b) (see Fig. 10.4). It seems likely that the incorporation of large quantities of glycogen within the mitochondrial derivative of the midpiece may have eliminated the need for a glycogen piece.

Fig. 10.4. (Opposite) Mature spermatozoon of *Cantareus aspersus* (Müller) (Helicidae, Helicoidea, Stylommatophora, Pulmonata). Note the unusual lateral arrangement of the acrosomal complex. (A) Acrosomal complex, nucleus and anterior portion of the midpiece. (B) Longitudinal sections of a spermatozoon showing, in anterior–posterior sequence, details of the acrosomal region, the nucleus–midpiece junction (= 'neck' region), the anterior portion of the midpiece and lastly, the terminal region of the spermatozoon. (C) Corresponding transverse sections of the regions shown in (B). a, acrosomal complex; ap, acrosomal pedestal; av, acrosomal vesicle; ax, axoneme; cd, centriolar derivative; cf, coarse fibres; gh, glycogen helix; k, helical keels of the nucleus; ma, matrix material of the mitochondrial derivative; n, nucleus; p, paracrystalline material of the mitochondrial derivative; pns, perinuclear sheath. (A = ×10,000; B, C = ×25,000.)

The hermaphroditic nature of pulmonates necessitates differentiating between those spermatozoa produced by an individual (autospermatozoa) and those received by that animal during copulation (allospermatozoa or heterospermatozoa). The following summary of the morphology of mature spermatozoa in the Pulmonata is based on studies of autospermatozoa, except where noted.

The acrosomal complex of most investigated terrestrial pulmonates is positioned at the nuclear apex, parallel to the spermatozoan longitudinal axis or slightly tilted in relation to it (Takaichi and Sawada, 1973; Takaichi and Dan, 1977; Shileiko and Danilova, 1979; Atkinson, 1982; Healy, 1983a, 1988a; Giusti et al., 1991; Sretarugsa et al., 1991; Cuezzo, 1995a,b) (Fig. 10.2I). However, in C. aspersus, the acrosomal complex is reflected backwards from the nuclear apex in spermatozoa taken from the hermaphrodite duct and spermatophores (Healy and Jamieson, 1989) (Figs 10.2L and 10.4). Studies of late spermatids and ovotesticular spermatozoa of C. aspersus show an axially aligned acrosomal complex (Griffond et al., 1991; J.M. Healy and B.G.M. Jamieson, unpublished), indicating that the lateral re-positioning of the complex takes place within the hermaphrodite duct. Although the purpose of this phenomenon remains unknown, it does serve to illustrate that important structural changes in pulmonate spermatozoa can occur even after leaving the ovotestis. Whether the acrosomal vesicle of C. aspersus reacts laterally to the egg membrane as it does in marine bivalves of the subclass Anomalodesmata (Hosokawa and Noda, 1994) or undergoes a vertical re-alignment just prior to sperm–egg contact remains to be determined. Unfortunately, no study yet has dealt with the events of the acrosome reaction or sperm–egg interaction in pulmonates, or in fact in any internally fertilizing gastropod.

A number of stylommatophorans exhibit a perinuclear sheath associated with the acrosomal complex and anterior portion of the nucleus (C. aspersus and Helix pomatia Linnaeus (Helicidae) – Healy and Jamieson, 1989; Epiphragmophora tucumanensis (Doering) (Helminthoglyptidae) – Cuezzo, 1995b) (Fig. 10.2I and L), and in Oxyloma elegans (Risso) (Succineidae) it is paracrystalline (Selmi et al., 1989). Although the exact function of this sheath ('paracrosomal body' of Giusti et al., 1991) remains unknown, Selmi et al. (1989) demonstrated that it is lost in the seminal receptacle of O. elegans, and suggested a possible role in spermatozoan capacitation.

Pulmonate spermatozoan nuclei show marked variation in the number and strength of the helical keels between taxa. In some groups such as the Helicidae, the keels may appear almost vestigial. In most cases, however, the keels are dominant sculptural elements of the nucleus (Bayne, 1970; Thompson, 1973; Maxwell, 1975; Atkinson, 1982; Healy, 1983a, and herein; Giusti et al., 1991). Thompson (1973) has suggested that the helical shape of the nucleus and midpiece may provide a means of converting uniplanar flagellar activity into helical forward movement. If this proves to be the case, then variation in keel prominence between pulmonate taxa perhaps may correlate with differing motility demands – possibly to meet differing viscosities of genital duct fluids prior to sperm–egg contact. Results of Bloch and Hew (1960) indicate that a transition in the spermatozoan nucleus of *C. aspersus* from an arginine-rich histone to a protamine occurs within the hermaphrodite duct.

The complexity of the midpiece in stylommatophoran spermatozoa has long attracted research interest, particularly in relation to the sub-structure and biochemistry of the paracrystalline and matrix components of the mitochondrial derivative (Figs 10.2K and M, and 10.4) and the high volume storage of glycogen (André, 1962; Tahmisian, 1964; Anderson, 1970; Anderson and Personne, 1967, 1969a,b, 1970a,b,c, 1976; Anderson *et al.*, 1968; Ritter and André, 1975; Maxwell, 1980; Reger and Fitzgerald, 1982). The paracrystalline material in stylommatophorans (and in other pulmonates and the Opisthobranchia) consists of a three-dimensional, helically coiled, lattice-work composed of proteinaceous, hollow, rodlets or granules with a diameter of approximately 80–90 Å. This material has been confirmed as the site of cytochrome activity (Ritter and André, 1975; Anderson and Personne, 1976). Specific metabolic functions have been ascribed to other components of the midpiece, including: the Krebs tricarboxylic acid cycle in the matrix layers; phosphorylase and dehydrogenase activity in the glycogen helix/helices; and, finally, motility inducement by Mg^{2+}-activated ATPase in the axoneme, coarse fibre–axonemal doublet interface and centriolar region (for further details, see Anderson and Personne, 1976). Aside from variation in the number of 'secondary helices' (helically coiled ridges of the mitochondrial derivative) between taxa, all stylommatophoran spermatozoa exhibit only a single glycogen helix, as do the bulk of the Opisthobranchia and certain allogastropod groups (Healy, 1988a, 1996) (Figs 10.2K and M, and 10.4). This contrasts with the presence of multiple glycogen helices in basommatophoran families such as the Lymnaeidae, Planorbidae and archeopulmonate Ellobiidae (for comparisons and further literature, see Healy, 1983a, 1988a) and in the systellommatophoran family Vaginulidae (Lanza and Quattrini, 1964; Quattrini and Lanza, 1964). Presumably the large quantity of glycogen stored within the midpiece in stylommatophoran spermatozoa plays a significant role in maintaining long periods of viability (Anderson and Personne, 1970c, 1976) – up to 341 days for *Achatina fulica* Bowdich (Achatinidae) and 476 days in *Macrochlamys indica* Godwin-Austen (Ariophantidae), according to Raut and Ghose (1979, 1982); up to 520 days in *Limicolaria martensiana* Smith (Achatinidae) according to Owing (1974). In all investigated species of Stylommatophora, the mitochondrial derivative forms the terminal por-tion of the spermatozoon (see Fig. 10.4). This contrasts with spermatozoa of opisthobranchs and basommatophorans, and possibly some Ellobiidae, which terminate with a glycogen piece preceded by an annulus (Healy, 1996). In the absence of data for many families, it is impossible to state unequivocally whether absence of a glycogen piece is a feature of all stylommatophorans.

In spite of the considerable literature on stylommatophoran sperm-atozoan ultrastructure, the majority of families within this large and eco-logically significant order of gastropods remain unstudied from this perspective. Even species of major economic importance, such as the giant African snail *A. fulica* and related species within the Achatinidae, only recently have been examined using transmission electron micros-copy (Odiete, 1982; Whitehead and Hodgson, 1985; Sretarugsa *et al.*, 1991), as have representatives of certain American families (Bulimulidae, Xanthonychidae – Cuezzo, 1995a,b). The few available comparative studies of stylommatophoran spermatozoa (e.g. Thompson, 1973; Healy and Jamieson, 1989; Giusti *et al.*, 1991) suggest that spermatozoan features may prove taxonomically and phylogenetically informative, as they have in other groups of gastropods (for a recent summary, see Healy, 1996).

Although marked spermatozoan dimorphism (or occasionally polymorphism) occurs widely in internally fertilizing prosobranchs, there is no firm evidence to support the occurrence of this phenomenon in any species of Stylommatophora. Parivar's report (1981) of 'apyrene spermato-zoa' developing in the ovotestis in *A. ater*, together with the normal 'eupyrene spermatozoa', is based on a misunderstanding of the term 'apyrene' which defines a spermatozoon as lacking a nucleus or nuclear material *not* as lacking an acrosome. Micrographs of Parivar not only show that the 'apyrene' spermatids of *A. ater* have a well-developed nucleus, but strongly suggest that these are abnormal, possibly moribund cells. Rousset-Galangau (1972) demonstrated the occurrence of sperma-tids with two axonemes in *Deroceras agrestis* (Linnaeus) (Agrio-limacidae), but concluded that these were possibly the result of meiotic abnormalities. Takaichi (1979) achieved a similar result artificially by exposure of *E. hickonis* to X-rays. This experiment also revealed that X-rays do not interfere with the 9 + 2 axonemal pattern, acrosome devel-opment or the conversion of mitochondrial material into the matrix–paracrystalline derivative, but do lead to significant abnormalities in the final shaping of the midpiece. Occasional biaxonemal spermatids and spermatozoa have been observed by the present author in pulmonate and opisthobranch species (J.M. Healy, unpublished data). Given the scar-city of such cells, it seems likely that these are abnormal products of the gonad. Tompa (1984) has speculated that chemical pollution or deleteri-ous temperature or humidity changes may play some role in generating spermatozoan abnormalities in stylommatophorans. Certainly this aspect of stylommatophoran gametogenesis deserves closer scutiny, as it may prove a useful indicator of environmental degradation.

Prosobranchia

As internally fertilizing prosobranchs frequently show pronounced spermatozoan dimorphism (or polymorphism), it is necessary to dif-ferentiate between the uniaxonemal, fertile type (euspermatozoon) and the infertile, often multiaxonemal type (paraspermatozoon). Of the four superfamilies represented terrestrially, only the Neritoidea and Cyclophoroidea are known to produce dimorphic spermatozoa. At present, it is not known if paraspermatozoa occur in the terrestrial littorinoidean families Pomatiasidae and Chondropomatidae, and, to date, these cells have not been observed in any rissooideans.

Although euspermatozoa of all four superfamilies of terrestrial proso-branchs (Neritoidea, Cyclophoroidea, Littorinoidea and Rissooidea) differ from each other in details, notably in the arrangement of the mitochondria (Fig. 10.3), all possess a conical acrosomal vesicle, a rod-shaped nucleus and an elongate midpiece containing an axoneme surrounded by modified but recognizably cristate mitochondria (see Giusti and Mazzini, 1973; Selmi and Giusti, 1980; Kohnert and Storch, 1984a; Giusti and Selmi, 1985; Koike, 1985; Healy, 1988a, and unpublished). A glycogen piece, composed of the axoneme surrounded by a glycogen sheath, and an annulus (positioned at the junction of the midpiece and glycogen piece) are present in all groups except the Neritoidea (Kohnert and Storch, 1984a; Koike, 1985; Healy, 1988a, 1996) (Fig. 10.3). In the Pomatiasidae (Littorinoidea) and Truncatellidae (Rissooidea), the midpiece is composed of 6–9 mitochondria spiralling around the axoneme (Kohnert and Storch, 1984a; Giusti and Mazzini, 1973; J.W. Healy, unpublished), a pattern repeated in the vast majority of caenogastropod prosobranchs (Healy, 1988a, 1996) (Fig. 10.3F–I). All neritoideans show an elec-tron-dense, spiralling ribbon within the midpiece region, but families differ in the number of mitochondria present (two in Neritidae, 5–6 in Helicinidae; Fig. 10.3A and B) and in the structure of the tail (reflected in Neritidae, straight in Helicinidae) (for comparisons, see Koike, 1985). Cyclophoroidean euspermatozoa resemble those of the freshwater Ampullarioidea and the marine and freshwater Cerithioidea in having the cristae organized as parallel plates (Fig. 10.3C and D). However, cyclophoroideans, alone among the Caenogastropoda, also exhibit intramitochondrial granules linearly arranged along the edges of each angulate mitochondrion and cristal surfaces (Selmi and Giusti, 1980; Healy, 1988a). This represents one of four known incidences of what could be termed 'partial paracrystallization' of the midpiece within the Caenogastropoda, the other three being one member of the marine cypraeoidean family Cypraeidae, and two cases in the estuarine– fresh-water rissooidean family Stenothyridae (Healy, 1983b, 1986a, and unpub-lished). The paracrystalline material of heterobranch spermatozoa differs in organization and substructure from that observed in caenogastropods and probably represents a case of independent acquisition. According to Selmi and Giusti (1980), the substantial deposits of glycogen stored within the euspermatozoa of *C. montanum* 'do not disappear even after long periods after copulation' within the female seminal receptacle. In contrast, euspermatozoan glycogen in *Littorina sitkana* Philippi (Littorinidae) diminishes by half after 3 months storage in the seminal receptacle (J. Buckland-Nicks, personal communication).

Paraspermatozoa of the Neritoidea and Cyclophoroidea differ considerably from each other in morphology. In paraspermatozoa of freshwater and marine neritoideans, no nuclear material is present (the apyrene condition) and glycogen forms the principal deposit associated with the axoneme (Giusti and Selmi, 1982a,b; Selmi and Giusti, 1983; J.M. Healy, unpublished). Tochimoto (1967) reported the occurrence of dimorphic 'typical sperm' (euspermatozoa) in *Waldemaria japonica* (Adams) (Helicinidae), but made no mention of paraspermatozoa being present. Cyclophoroidean paraspermatozoa show closest resemblances to the Ampullarioidea and Cerithioidea: a head region featuring a condensed, rope-like nuclear core and numerous dense vesicles (possibly glycoprotein), glycogen granules and peripheral axonemes, the latter structures emerging posteriorly to form a tail tuft (Selmi and Giusti, 1980) (Fig. 10.3E).

Terrestrial Gastropod Spermatozoa: Some Taxonomic and Phylogenetic Perspectives

The taxonomic position of terrestrial gastropods generally is reflected in ultrastructural features of the developing and mature spermatozoon. Thus, among the terrestrial prosobranchs, the Helicinidae clearly belong within the mostly marine Neritimorpha, whereas the Pomatiasidae and Truncatellidae show spermatozoan features in keeping with their position among the otherwise aquatic Littorinoidea–Rissooidea assemblage (Fig. 10.3A, B and F–I). The parallel cristal plates in the euspermatozoa of Cyclophoroidea (a wholly terrestrial group) are seen elsewhere only in the freshwater Ampullarioidea (Viviparidae only) and the marine or freshwater Cerithioidea, and all three superfamilies share several paraspermatozoan features (Fig. 10.3C–E) (see Healy, 1988a, 1996). A close relationship between these groups seems evident despite Haszprunar's (1988) decision to allocate the Cyclophoroidea + Ampullarioidea (as Architaenioglossa) to the 'Archaeogastropoda' *sensu lato* and the Cerithioidea to the Caenogastropoda (among the Neotaenioglossa).

Among the pulmonates, spermatozoan ultrastructure appears reasonably consistent within families but provides only limited clues regarding relationships between superfamilies and between the many proposed higher groupings. This is partly due to the lack of information for several families or even superfamilies within the Basommatophora *sensu lato* and the Stylommatophora. Sperm ultrastructure is very diverse within the Basommatophora *sensu lato*, suggesting that the order may even be polyphyletic (Healy, 1983a). In this context, it is worth noting that the superfamilies Ellobioidea and Trimusculoidea recently have been removed by Haszprunar and Huber (1990) from the Basommatophora and associated with the Stylommatophora as a superorder Eupulmonata. The Ellobioidea and Stylommatophora possibly may share one spermatozoan apomorphy – the absence of a glycogen piece and annulus (Fig. 10.4) – but this could also prove the end result of convergence because a trend towards marked reduction in size of the glycogen piece is evident in several other groups of heterobranchs including some other pulmonates (Amphiboloidea, Onchidioidea), pyramidelloidean allogastropods and notaspidean and nudibranch opisthobranchs (Healy and Willan, 1984, 1991; Healy, 1983a, 1986b, 1988a,b). Although recent molecular work generally supports the monophyly of the Stylommatophora, it also highlights great uncertainty about subordinal and superfamilial relationships within the group. Tillier *et al.* (1996, p. 268) have even concluded in relation to stylommatophoran classification and evolution that 'confusion at suprafamilial levels is worse than it has ever been since the beginning of the century' and that 'one can doubt the possibility of resolving stylommatophoran phylogeny by morphology alone'. The literature contains many works on stylommatophoran spermatozoa, but the bulk of these papers are non-comparative and concerned primarily with mechanical and chemical aspects rather than taxonomic or phylogenetic implications (see references this chapter). In addition, focus has been largely on the Helicoidea rather than on the vast number of other superfamilies within the Stylommatophora. Hence, at least for the present, it is impossible to use spermatozoan data alone to assess higher level classificatory changes suggested by various authors (e.g. Tillier, 1989; Nordsieck, 1992). Giusti *et al.* (1991) have suggested that the relatively uniform spermatozoan morphology of the Helicoidea may indicate 'excessive fragmentation' of the superfamily in existing classifications which are based purely on gross anatomy.

Oogenesis and Oocytes

In contrast to a substantial spermatozoan and spermatogenic body of literature, very few studies have yet focused on the ultrastructure of oogenesis of terrestrial molluscs. This is most surprising given the ecological and economic importance of groups such as the stylommatophoran pulmonates that contain the bulk of the gastropod agricultural pest species. Aspects of oogenesis have been examined using light and electron microscopy in the systellommatophoran *Vaginulus borellianus* (Colosi) (Vaginulidae) (Quattrini and Lanza, 1964, 1965). The following summary will concern itself only with oogenesis as occurring in the Stylommatophora, drawing principally on the ultrastructural accounts by Hill and Bowen (1976) on the agriolimacid *Deroceras reticulatum* (Müller) and that of Griffond and Bolzoni-Sungur (1986) on the helicid *C. aspersus*.

Oogenesis in stylommatophorans, as in most other heterobranchs, occurs in the same acini of the gonad as does spermatogenesis, and usually simultaneously (Fig. 10.5A). Oogonia are derived from germinal cells lining the acinus closest to the ductule and, during subsequent developmental stages, the oocytes move progressively towards the distal

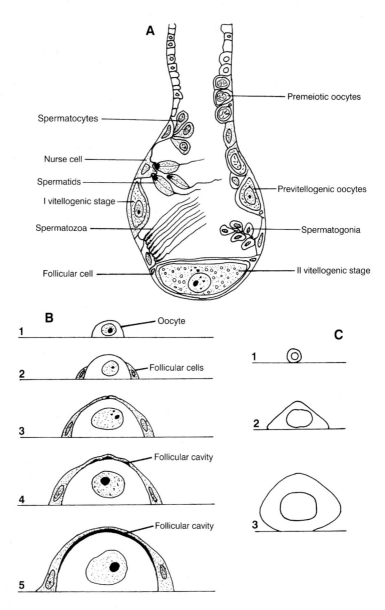

Fig. 10.5. Oogenesis in pulmonates. (A) Diagram showing the distribution of oogenic and spermatogenic activity in a single ovotestis acinus of *Cantareus aspersus* (Müller) (Helicidae, Helicoidea, Stylommatophora, Pulmonata) (redrawn and modified from Griffond and Bolzoni-Sungur, 1986). (B) Diagram showing the relationship of the growing oocyte to the follicular cells occurring within an acinus of the ovotestis of the freshwater planorbid *Biomphalaria glabrata* (Say) (Planorbidae, Basommatophora, Pulmonata). The follicular cells sheath the oocyte, and later a cavity forms between the follicular cells and the oocyte surface (redrawn from Jong-Brink *et al.*, 1976). (C) Diagram indicating the change in volume with growth of oocyte during the vegetative stage in *Deroceras reticulatum* (Müller) (Agriolimacidae, Stylommatophora, Pulmonata) (redrawn from Hill and Bowen, 1976).

region of the acinus (Fig. 10.5A). In *D. reticulatum*, however, no matur-
ational gradient for developing oocytes is evident (Sabelli *et al.*, 1978).
As in other gastropods, oogenesis in stylommatophorans is intimately
associated with a sheath of follicle cells, which on oocyte maturation
release the unfertilized ovum into the acinar lumen (Raven, 1961; Jong-
Brink *et al.*, 1976). Griffond and Bolzoni-Sungur (1986) recognize six
oogenic cell stages in *C. aspersus*: (i) oogonial cells; (ii) young oocytes;
(iii) pre-meiotic oocytes; (iv) pre-vitellogenic oocytes; (v) first stage
vitellogenic oocytes; and (vi) second stage vitellogenic oocytes (for chief
differentiating features, see Table 10.1). The oogonia of *C. aspersus* are
recognizable by their oblong nucleus and particularly well-developed
concentric whorls of rough endoplasmic reticulum (Griffond and Bolzoni-
Sungur, 1986). The three post-oogonial developmental stages described
by Hill and Bowen (1976) in *D. reticulatum* correspond to stages 4–6
of Griffond and Bolzoni-Sungur (1986). Generally, in pre-vitellogenic
oocytes, the plasma membrane shows no surface modifications and
the cytoplasm contains endoplasmic reticular and Golgi cisternae, lipid
vesicles, scattered mitochondria and an abundant supply of free ribosomes. The nucleus is ovoid, decondensed and exhibits a large nucleolus
(or 'eunucleolus' often showing bipartite substructure and peripheral
buds) and frequently a smaller nucleolus ('paranucleolus'). During the
vitellogenic (yolk producing) stages, both the nucleus and cytoplasm
increase dramatically in size, reflecting the activity of both the nucleus
(and nucleoli) and the cytoplasmic organelles (Fig. 10.5B and C). An
additional structure associated with the eunucleolus and paranucleolus,
the so-called 'chromatin nucleolus' has been reported in vitellogenic
oocytes of *Limacus flavus* (Linnaeus) (Limacidae) (for description and
discussion, see Yamamoto, 1977). Yolk vesicles in *C. aspersus* and
many other investigated species of gastropods appear to be derived
largely through the secretory interaction of the Golgi complex and the
endoplasmic reticulum (Jong-Brink *et al.*, 1976; Griffond and Bolzoni-
Sungur, 1986). According to Hill and Bowen (1976), in *D. reticulatum*,
yolk platelets begin to develop in the cytoplasm from either lamellate
bodies of uncertain origin (possibly mitochondrial) or multivesiculate
bodies (of probable Golgian origin) and grow via the incorporation of
probably Golgi-derived cytoplasmic dense vesicles, lipid and glycogen.
The extent to which yolk development may be dependent on extracellular
sources in stylommatophoran oogenesis remains unknown. The internal
structure of yolk vesicles in pulmonates appears to be variable, those of
C. aspersus being uniformly electron dense (Griffond and Bolzoni-
Sungur, 1986), whereas those of *D. reticulatum* contain membranes
(Hill and Bowen, 1976) and those of freshwater planorbids have both
membranes and crystalloid inclusions (Jong-Brink *et al.*, 1976). In late
vitellogenic oocytes (stage 6 of Griffond and Bolzoni-Sungur, 1986; stage
III of Hill and Bowen, 1986), the plasma membrane develops microvilli,
some of which contact the surrounding follicle cells. Throughout
vitellogenesis, a gap between the oocyte and follicle cells becomes

Table 10.1. Summary of the major differentiating features of the six stages of oogenesis in *Cantareus aspersus* (Müller) (Helicidae, Helicoidea, Stylommatophora, Pulmonata) (from Griffond and Bolzoni-Sungur, 1986)

Stage	Shape of the oocyte	Diameter of the oocyte at the end of the stage	Diameter of the nucleus	Diameter of nucleoli	Cytoplasm basophilia	Rough endoplasmic reticulum	Golgi apparatus	Mito-chondria	Annulate lamellae	Cytoplasmic inclusions	Follicular cells
Oogonia	Elongated, variable	10 μm	5 μm	0.8 μm	+	Concentric whorls	+	+	–	–	0
Young oocyte	Round	12 μm	7 μm	0.8 μm	+		+	+	–	–	0 or extensions
Pre-meiotic stage	Oval	15 × 30 μm	18 μm	3 μm	±	Short cisternae	+	+	–	–	(1) Regular intercellular space (2) Enlarged spaces and intermediate junctions
Pre-vitellogenic stage	Oval to elongated	50 × 70 μm	25 μm	8 μm	–	Flattened cisternae	++	++	–	Lipid droplets	Large spaces Intermediate junctions
Vitellogenesis I	Elongated	65 × 160 μm	50 μm	14 μm	–	Clear vesicles	+++	+++	+	Lipid droplets Yolk granules	Large spaces Intermediate junctions
Vitellogenesis II	More rounded	120 × 200 μm	80 μm	20 μm	–	Lamellae, clear vesicles	+++	+++	++	Lipid droplets Yolk granules Calcospherites	Large spaces Intermediate junctions

Greatest height and length indicated for oval-shaped oocytes.

increasingly wider, ultimately resulting in bursting of the follicle and release of the mature oocyte (or ovum) (Fig. 10.5B).

Mature stylommatophoran oocytes range in diameter from 50 to 200 μm (Tompa, 1980, 1984). Those of *C. aspersus* exhibit Ca^{2+}-accumulating vesicles within the ooplasm (Medina *et al.*, 1989). These structures, sometimes known as 'calcium spherules' or 'calcospherites' because of their internal one or two layered spherical mineral deposit, were once believed to be an obvious source of calcium for the embryonic shell (e.g. in *C. aspersus* – for discussion, see Medina *et al.*, 1989; Griffond and Bolzoni-Sungur, 1986). However, calcium for the embryonic shell in stylommatophorans is actually derived largely through re-use of calcium contained in the eggshell (see reviews in Tompa, 1984; Fournié and Chétail, 1984). Recent studies have concluded that the true function of the Ca^{2+}-accumulating vesicles is probably to prevent toxic calcium build-up in the ooplasm by first absorbing and then converting any excess Ca^{2+} (entering from the blood) into osmotically neutral compounds (for further details and discussion, see Medina *et al.*, 1989).

References

Anderson, W.A. (1970) The localization of cytochrome c oxidase during mito-chondrial specialization in spermiogenesis of prosobranch snails. *Journal of Histochemistry and Cytochemistry* 18, 201–210.

Anderson, W.A. and Personne, P. (1967) The fine structure of the neck region of spermatozoa of *Helix aspersa*. *Journal de Microscopie* 6, 1033–1042.

Anderson, W.A. and Personne, P. (1969a) Structure and histochemistry of the basal body derivative, neck and axoneme of spermatozoa of *Helix aspersa*. *Journal de Microscopie* 8, 87–96.

Anderson, W.A. and Personne, P. (1969b) The cytochemical localization of sorbitol dehydrogenase activity in spermatozoa of *Helix aspersa*. *Journal de Microscopie* 8, 97–102.

Anderson, W.A. and Personne, P. (1970a) The localization of glycogen in the spermatozoa of various invertebrate and vertebrate species. *Journal of Cell Biology* 44, 29–51.

Anderson, W.A. and Personne, P. (1970b) Recent cytochemical studies on spermatozoa on some invertebrate and vertebrate species. In: Baccetti, B. (ed.) *Comparative Spermatology*. Academic Press, New York, pp. 431–449.

Anderson, W.A. and Personne, P. (1970c) The cytochemical localization of glycolytic and oxidative enzymes within mitochondria of spermatozoa of some pulmonate gastropods. *Journal of Histochemistry and Cytochemistry* 18, 783–793.

Anderson, W.A. and Personne, P. (1976) The molluscan spermatozoon: dynamic aspects of its structure and function. *American Zoologist* 16, 293–313.

Anderson, W.A., Personne, P. and André, J. (1968) Chemical compartmentalizat-ion in *Helix* spermatozoa. *Journal de Microscopie* 7, 367–390.

André, J. (1956) L'ultrastructure des mitochondries des spermatocytes de l'escargot *Helix pomatia* L. *Compte Rendu Hebdomadaires des Séances de l'Academie des Sciences, Paris, Serie D* 242, 2048–2051.

André, J. (1962) Contribution à la connaissance du chondriome. Étude de ses modifications ultrastructurales pendant la spermiogenése. *Journal of Ultrastructure Research, Supplement* 3, 1–185.

Atkinson, J.W. (1982) An ultrastructural analysis of the mature spermatozoon of *Anguispira alternata* (Say) (Pulmonata, Stylommatophora). *Journal of Morphology* 173, 249–257.

Bayne, C.J. (1970) Organization of the spermatozoon of *Agriolimax reticulatus*, the grey field slug (Pulmonata, Stylommatophora). *Zeitschrift für Zellforschung und Mikroskopische Anatomie* 103, 75–89.

Beams, H.W. and Tahmisian, T.N. (1954) Structure of the mitochondria in the male germ cells of *Helix* as revealed by the electron microscope. *Experimental Cell Research* 6, 87–93.

Bloch, D.P. and Hew, H.Y.C. (1960) Schedule of spermatogenesis in the pulmonate snail *Helix aspersa*, with special reference to histone transition. *Journal of Biophysical and Biochemical Cytology* 7, 515–531.

Cuezzo, M.G. (1995a) Spermatogenesis and sperm structure in the Neotropical pulmonate snail *Scutalus tupacii* (d'Orbigny). *Veliger* 38, 212–222.

Cuezzo, M.G. (1995b) Ultrastructure of the mature spermatozoa of the land snail *Epiphragmophora tucumanensis* (Doering, 1874) (Gastropoda: Helicoidea). *Journal of Molluscan Studies* 61, 1–7.

Dan, J.C. and Takaichi, S. (1979) Spermiogenesis in the pulmonate snail, *Euhadra hickonis*. III. Flagellum formation. *Development, Growth and Differentiation* 21, 71–86.

Dohmen, M.R. (1983) Gametogenesis. In: Verdonk, N.H., van Den Biggelaar, J.A.M. and Tompa, A.S. (eds) *The Mollusca.* Vol. 3, *Development.* Academic Press, New York, pp. 1–47.

Favard, P. and André, J. (1970) The mitochondria of spermatozoa. In: Baccetti, B. (ed.) *Comparative Spermatology.* Academic Press, New York, pp. 415–430.

Fournié, J. and Chétail, M. (1984) Calcium dynamics in land gastropods. *American Zoologist* 24, 857–870.

Giusti, F. and Mazzini, M. (1973) The spermatozoon of *Truncatella* (s. str.) *subcylindrica* (L.) (Gastropoda: Prosobranchia). *Monitore Zoologico Italiano* 7, 181–201.

Giusti, F. and Selmi, M.G. (1982a) The morphological peculiarities of the typical spermatozoa of *Theodoxus fluviatilis* (L.) (Neritoidea) and their implications for motility. *Journal of Ultrastructure Research* 78, 166–177.

Giusti, F. and Selmi, M.G. (1982b) The atypical sperm in the prosobranch molluscs. *Malacologia* 22, 171–181.

Giusti, F. and Selmi, M.G. (1985) The seminal receptacle and sperm storage in *Cochlostoma montanum* (Issel) (Gastropoda: Prosobranchia). *Journal of Morphology* 184, 121–133.

Giusti, F., Manganelli, G. and Selmi, G. (1991) Spermatozoon fine structure in the phylogenetic study of the Helicoidea (Gastropoda, Pulmonata). *Proceedings of the Tenth Malacological Congress, Tübingen, 1989*, 611–616.

Gomot, P. and Gomot, L. (1985) Action de la photopériode sur la multiplication spermatogoniale et la reproduction de l'escargot *Helix aspersa*. *Bulletin de Sociéte Zoologie France* 110, 445–459.

Gomot, P. and Griffond, B. (1987) Répercussion de la durée d'éclairement journalier sur l'evolution des cellules nourricières et de la lignée male dans l'ovotestis d'*Helix aspersa*. *Reproduction, Nutrition, Développement* 27, 95–108.

Grassé, P.P., Carasso, N. and Favard, P. (1955) L'ultrastructure de la spermatide d'escargot *Helix pomatia*: chromosomes, enveloppes nucleaires, centrosome. *Compte Rendu Hebdomadaires des Séances de l'Academie des Sciences, Paris, Serie D* 241, 1430–1432.

Grassé, P.P., Carasso, N. and Favard, P. (1956a) L'ultrastructure des chromosomes et son évolution au cours de la spermiogenèse de l'escargot. I. La spermatide. *Compte Rendu Hebdomadaires des Séances de l'Academie des Sciences, Paris, Serie D* 242, 991–995.

Grassé, P.P., Carasso, N. and Favard, P. (1956b) L'ultrastructure des chromosomes et son evolution au cours de la spermiogenése de l'escargot. II. Le spermatozoïde. *Compte Rendu Hebdomadaires des Séances de l'Academie des Sciences, Paris, Serie D* 242, 1395–1398.

Grassé, P.P., Carasso, N. and Favard, P. (1956c) Les ultrastructures cellulaires au cours de la spermiogenése de l'escargot *Helix pomatia* (L.). Évolution des chromosomes, du chondriome de l'appareil de Golgi etc. *Annales des Sciences Naturelles Zoologie, Biologie Anima II Serie* 18, 339–438.

Griffond, B. and Bolzoni-Sungur, D. (1986) Stages of oogenesis in the snail, *Helix aspersa*: cytological, cytochemical and ultrastructural studies. *Reproduction, Nutrition, Développement* 26, 461–474.

Griffond, B. and Medina, A. (1989) Timing of spermatogenesis and spermiation in snails *Helix aspersa* bred under short photoperiods: a histologic and quantitative autoradiographic study. *Journal of Experimental Zoology* 250, 87–92.

Griffond, B., Dadkhah-Teherani, Z., Medina, A. and Bride, M. (1991) Ultrastructure of *Helix aspersa* spermatogenesis: scanning and transmission electron microscopical contributions. *Journal of Molluscan Studies* 57, 277–287.

Haszprunar, G. (1988) On the origin and evolution of major gastropod groups, with special reference to the Streptoneura. *Journal of Molluscan Studies* 54, 367–441.

Haszprunar, G. and Huber, G. (1990) On the central nervous system of Smeagolidae and Rhodopidae, two families questionably allied with the Gymnomorpha (Gastropoda: Euthyneura). *Journal of Zoology* 220, 185–199.

Healy, J.M. (1983a) An ultrastructural study of basommatophoran spermatozoa. *Zoologica Scripta* 12, 57–66.

Healy, J.M. (1983b) Ultrastructure of euspermiogenesis in the mesogastropod *Stenothyra* sp. (Prosobranchia, Rissoacea, Stenothyridae). *Zoologica Scripta* 12, 203–214.

Healy, J.M. (1986a) An ultrastructural study of euspermatozoa, paraspermatozoa and nurse cells of the cowrie *Cypraea errones* (Gastropoda, Prosobranchia, Cypraeidae). *Journal of Molluscan Studies* 52, 125–137.

Healy, J.M. (1986b) Electron microscopic observations on the spermatozoa of a marine 'pulmonate' slug *Onchidium damelli* (Gastropoda, Onchidiacea). *Journal of Submicroscopic Cytology* 18, 587–594.

Healy, J.M. (1988a) Sperm morphology and its systematic importance in the Gastropoda. In: Ponder, W.F. (ed.) *Prosobranch Phylogeny. Malacological Review Supplement* 4, 252–266.

Healy, J.M. (1988b) The ultrastructure of spermatozoa and spermiogenesis in pyramidellid gastropods, and its systematic importance. *Helgoländer Meeresuntersuchungen* 42, 303–318.

Healy, J.M. (1996) Molluscan sperm ultrastructure: correlation with taxonomic units within the Gastropoda, Cephalopoda and Bivalvia. In: Taylor, J. (ed.)

Origin and Evolutionary Radiation of the Mollusca. Oxford University Press, Oxford, pp. 99–113.

Healy, J.M. and Jamieson, B.G.M. (1981) An ultrastructural examination of developing and mature paraspermatozoa in *Pyrazus ebeninus* (Mollusca, Gastropoda, Potamididae). *Zoomorphology* 98, 101–119.

Healy, J.M. and Jamieson, B.G.M. (1989) An ultrastructural study of spermatozoa of *Helix aspersa* and *Helix pomatia* (Gastropoda, Pulmonata). *Journal of Molluscan Studies* 55, 389–404.

Healy, J.M. and Willan, R.C. (1984) Ultrastructure and phylogenetic significance of notaspidean spermatozoa (Mollusca, Gastropoda, Opisthobranchia). *Zoologica Scripta* 13, 107–120.

Healy, J.M. and Willan, R.C. (1991) Nudibranch spermatozoa: comparative ultrastructure and systematic importance. *Veliger* 34, 134–165.

Hill, R. and Bowen, I. (1976) Studies on the ovotestis of the slug *Agriolimax reticulatus.* I. The oocyte. *Cell and Tissue Research* 173, 465–482.

Hodgson, A.N. (1996) The structure of the seminal vesicle region of the hermaphrodite duct of some pulmonate snails. *Molluscan Reproduction, Malacological Review Supplement* 6, 89–99.

Hodgson, A.N. (1997) Paraspermatogenesis in gastropod molluscs. *Invertebrate Reproduction and Development* 31, 31–38.

Hosokawa, K. and Noda, Y.D. (1994) The acrosome reaction and fertilization in the bivalve, *Laternula limicola*, in reference to sperm penetration from the posterior region of the mid-piece. *Zoological Science* 11, 89–100.

Jong-Brink, de, M., de Wit, A., Kraal, G. and Boer, H.H. (1976) A light and electron microscope study on oogenesis in the freshwater pulmonate snail *Biomphalaria glabrata. Cell and Tissue Research* 171, 195–219.

Kohnert, R. and Storch, V. (1984a) Vergleichend-ultrastrukturelle Untersuchungen zur Morphologie eupyrener Spermien der Monotocardia (Prosobranchia). *Zoologischer Jahrbucher Anatomie* 111, 51–93.

Kohnert, R. and Storch, V. (1984b) Elektronenmikroskopische Untersuchungen zur Spermiogenese der eupyrenen Spermien der Monotocardia (Prosobranchia). *Zoologischer Jahrbücher Anatomie* 112, 1–32.

Koike, K. (1985) Comparative ultrastructural studies on the spermatozoa of the Prosobranchia (Mollusca: Gastropoda). *Science Reports of the Faculty of Education, Gunma University* 34, 33–153.

Lanza, B. and Quattrini, D. (1964) Osservazioni sulla spermiogenesi e sugli spermi di *Vaginulus borellianus* (Colosi) (Gastropoda, Soleolifera). *Bolletino di Zoologia* 31, 1321–1338.

Maxwell, W.L. (1975) Scanning electron microscope studies of pulmonate spermatozoa. *Veliger* 18, 31–33.

Maxwell, W.L. (1980) Distribution of glycogen deposits in two euthyneuran sperm tails. *International Journal of Invertebrate Reproduction* 2, 245–249.

Maxwell,W.L. (1983) Mollusca. In: Adiyodi, K.G and Adiyodi, R.G. (eds) *Reproductive Biology of Invertebrates. II Spermatogenesis and Sperm Function.* John Wiley & Sons, New York, pp. 275–319.

Medina, A., Griffond, B. and Sánchez-Aguayo, I. (1989) Studies on Ca^{2+}-accumulating vesicles in oocytes of the snail *Helix aspersa. Cell and Tissue Research* 257, 597–601.

Nordsieck, H. (1992) Phylogeny and system of the Pulmonata (Gastropoda). *Archiv für Molluskenkunde* 121, 31–52.

Odiete, W.O. (1982) Fine structural studies on the ovotestis of *Archachatina marginata* (Swainson) (Pulmonata, Stylommatophora). *Malacologia* 22, 137–143.

Owing, A.M. (1974) Some aspects of the breeding biology of the equatorial land snail *Limicolaria martensiana* (Achatinidae : Pulmonata). *Journal of Zoology* 172, 191–206.

Parivar, K. (1981) Spermatogenesis and sperm dimorphism in land slug *Arion ater* L. (Pulmonata, Mollusca). *Zeitschrift für Mikroskopische-Anatomie Forschungen* 95, 81–92.

Quattrini, D. and Lanza, B. (1964) Osservazioni sulla ovogenesi e sulla spermatogenesi di *Vaginulus borellianus* (Colosi), Mollusca, Gastropoda, Soleolifera. *Bollettino di Zoologia* 31, 541–553.

Quattrini, D. and Lanza, B. (1965) Ricerche sulla biologia dei Veronicellidae (Gastropoda, Soleolifera). II. Struttura della gonade, ovogenesi e spermatogenesi in *Vaginulus borellianus* (Colosi) e in *Laevicaulis alte* (Ferussac). *Monitore Zoologico Italiano* 73, 1–60.

Raven, C.P. (1961) *Oogenesis*. Pergamon Press, Oxford.

Raut, S.K. and Ghose, K.C. (1979) Viability of sperm in two land snails, *Achatina fulica* Bowdich and *Macrochlamys indica* Godwin-Austen. *Veliger* 21, 486–487.

Raut, S.K. and Ghose, K.C. (1982) Viability of sperms in aestivating *Achatina fulica* Bowdich and *Macrochlamys indica* Godwin-Austen. *Journal of Molluscan Studies* 48, 84–86.

Rebhun, L.I. (1957) Nuclear changes during spermiogenesis in a pulmonate snail. *Journal of Biophysical and Biochemical Cytology* 3, 509–524.

Reger, J.F. and Fitzgerald, M.E.C. (1982) Studies on the fine structure of the mitochondrial derivative in spermatozoa of a gastropod. *Tissue and Cell* 14, 775–783.

Ritter, C. and André, J. (1975) Presence of a complete set of cytochromes despite the absence of cristae in the mitochondrial derivative of snail sperm. *Journal of Experimental Research* 92, 95–101.

Roosen-Runge, E.C. (1977) *The Process of Spermatogenesis in Animals*. Cambridge University Press, Cambridge.

Rousset-Galangau, V. (1972) Présence de deux catégories de spermies chez *Milax gagates* et *Agriolimax agrestis* (Moll. Gast. Pulm. Lim.). Étude comparée des ultrastructures au cours de la spermiogenése. *Annales des Sciences Naturelles, Zoologie, Paris* 38, 319–331.

Sabelli, B., Sabelli Scanabissi, F. and Merloni, M. (1978) Distribution of germ cells in the gonadic acina of *Deroceras reticulatum* (Müller) (Gastropoda Pulmonata Stylommatophora). Monitore.

Selmi, M.G. and Giusti, F. (1980) Structure and function in typical and atypical spermatozoa of Prosobranchia (Mollusca), 1. *Cochlostoma montanum* (Issel) (Mesogastropoda). *Atti della Accademia dei Fisiocritici Siena, IV Congresso Societa Malacologica Italiana, Siena, 1978*, 115–167.

Selmi, M.G. and Giusti, F. (1983) The atypical spermatozoon of *Theodoxus fluviatilis* (L.) (Gastropoda, Prosobranchia). *Journal of Ultrastructure Research* 84, 173–181.

Selmi, M.G., Bigliardi, E. and Giusti, F. (1989) Morphological modifications in stored heterospermatozoa of *Oxyloma elegans* (Pulmonata: Stylommatophora). *Journal of Ultrastructure and Molecular Structure Research* 102, 82–86.

Shileiko, L.V. and Danilova, L.V. (1979) Ultrastructure of spermatozoa in pulmonate molluscs *Trichia hispida* and *Succinea putris*. *Soviet Journal of*

Developmental Biology 10, 392–400. (English translation from *Ontogenez* 10, 437–447.)

Sretarugsa, P., Ngowsiri, U., Kruatrachue, M., Sobhon, P., Chavadej, J. and Upatham, E.S. (1991) Spermiogenesis in *Achatina fulica* as revealed by electron microscopy. *Journal of Medical and Applied Malacology* 3, 7–18.

Tahmisian, T.N. (1964) On orderly domains of particles associated with cytomembranes during spermatogenesis in *Helix aspersa*. *Zeitschrift für Zellforschung und Mikroskopische Anatomie* 64, 25–31.

Takahashi, Y., Nishimura, J. and Yamagishi, N. (1973) Electron microscopic studies on spermiogenesis of *Limax flavus* L. *Journal of the Nara Medical Association* 24, 404–410.

Takaichi, S. (1978) Spermiogenesis in the pulmonate snail *Euhadra hickonis*. II. Structural changes in the nucleus. *Development, Growth and Differentiation* 20, 301–315.

Takaichi, S. (1979) Spermiogenesis in the pulmonate snail *Euhadra hickonis*. IV. Effects of x-rays on the spermiogenesis. *Development, Growth and Differentiation* 21, 87–98.

Takaichi, S. and Dan, J.C. (1977) Spermiogenesis in the pulmonate snail *Euhadra hickonis*. I. Acrosome formation. *Development, Growth and Differentiation* 19, 1–14.

Takaichi, S. and Sawada, N. (1973) An electron microscope study on sperm formation in *Euhadra hickonis*. *Memoirs of the Ehime University, Series B* 7, 17–58.

Thompson, T.E. (1973) Euthyneuran and other molluscan spermatozoa. *Malacologia* 14, 167–206, plus addendum 443–444.

Tillier, S. (1989) Comparative morphology, phylogeny and classification of land snails and slugs (Gastropoda: Pulmonata: Stylommatophora). *Malacologia* 30, 1–303.

Tillier, S., Masselot, M. and Tillier, A. (1996) Phylogenetic relationships of the pulmonate gastropods, from rRNA sequences, and tempo and age of the stylommatophoran radiation. In: Taylor, J. (ed.) *Origin and Evolutionary Radiation of the Mollusca*. Oxford University Press, Oxford, pp. 267–284.

Tochimoto, T. (1967) Comparative histochemical study on the dimorphic spermatozoa of the Prosobranchia with special reference to polysaccharides. *Science Report of the Tokyo Kyoiku Daigaku* 13, 75–109.

Tompa, A.S. (1980) Studies on the reproductive biology of gastropods: part III. Calcium provision and the evolution of terrestrial eggs among gastropods. *Journal of Conchology* 30, 145–154.

Tompa, A.S. (1984) Land snails (Stylommatophora). In: Tompa, A.S., Verdonk, N.H. and van Den Biggelaar, J.A.M. (eds) *The Mollusca*, Vol. 7, *Reproduction*. Academic Press, Orlando, Florida, pp. 47–140.

Whitehead, S.I. and Hodgson, A.N. (1985) A study of the spermatozoa and spermatogenesis of the giant snail *Achatina zebra*. *Proceedings of the Electron Microscopy Society of Southern Africa* 15, 193–194.

Yamamoto, K. (1977) The amphinucleolus of *Limax* oocyte – an electron microscopic study. *Journal of the Nara Medical Association* 28, 341–352.

Yasuzumi, G., Takahashi, Y., Nishimura, Y., Yamagishi, N., Yamamoto, H. and Yamamoto, K. (1974) Spermatogenesis in animals as revealed by electron microscopy. XXVIII. Development of flagella of spermatozoa of *Limax flavus* (L.). *Okajima Folia Anatomica Japonica* 51, 11–28.

11 Population and Conservation Genetics

T. BACKELJAU, A. BAUR[1] AND B. BAUR[1]

Royal Belgian Institute of Natural Sciences, Vautierstraat 29, B-1000
Brussels, Belgium; [1]Department of Integrative Biology, Section of
Conservation Biology (NLU), University of Basel, St Johanns-Vorstadt 10,
CH-4056 Basel, Switzerland

Introduction

Population genetics aims to elucidate the genetic basis of evolutionary
processes. One of its goals is to identify hereditary polymorphisms as
distinct from phenotypic variation due to the environment. Population
genetics also aims to investigate the dynamics of polymorphisms in an
environmental context, that is how phenotype and genotype frequencies
change over time and space, and which forces are responsible for these
changes. This entails the estimation of the threshold between intra- and
interspecific variation. Here we consider these issues, even though less
emphasis is given to aspects of speciation. We also discuss conservation
genetics because this is an issue of increasing importance.

Polymorphisms in Terrestrial Gastropods

Polymorphisms in terrestrial gastropods involve external colour and
pigment patterns, shell morphology (e.g. size, shape and coiling), genital
structures, electrophoretic protein patterns, DNA markers and chromosomes. A number of these polymorphisms represent some of the very
earliest examples of Mendelian traits in animals (Lang, 1904, 1906) and
have provided fundamental insights as illustrated by Cain and Sheppard's
(1950, 1952, 1954) and Cain's (1953) studies on the maintenance of shell
polymorphism in natural populations of *Cepaea nemoralis* (Linnaeus)
(Helicidae) which is one of the classical examples of natural selection.

In this section, we describe and discuss some aspects of polymorphisms in terrestrial gastropod species. More detailed reviews on this
subject can be found in Jones (1973), Murray (1975), Jones *et al.* (1977,

1980), Clarke *et al.* (1978), Cain (1983), Selander and Ochman (1983), Goodhart (1987), Cowie (1992) and Johnson *et al.* (1993).

Shell polymorphisms

Helicidae

The European helicid species *C. nemoralis* is polymorphic for shell colour and the presence, number and appearance of up to five dark bands on the shell. All these characteristics are under genetic control (Murray, 1975). The major colour classes of the shell are, in decreasing order of dominance, brown, pink and yellow. The allele for no bands is dominant to that for the presence of bands. Alleles at other unlinked loci control the numbers of bands and determine the degree of band fusion. The position of bands is indicated conventionally by numbers, with the band closest to the suture being no. 1 and that nearest the umbilicus no. 5 (Fig. 11.1). The absence of a band is indicated by 0. Thus, 12345 represents the full five-banded condition, while 00345 indicates that the two uppermost bands are missing. The inheritance of shell colour and band patterns in *Cepaea hortensis* (Müller) is very similar to that in *C. nemoralis* (Cook and Murray, 1966; Guerrucci-Henrion, 1971, 1973).

The frequency of different morphs of *C. nemoralis* varies greatly from place to place. The complex distribution patterns of morph frequencies might be the result of a variety of factors. Jones *et al.* (1977) listed eight evolutionary factors and stated that several of these affect morph frequency in most *C. nemoralis* populations. Two well-studied forces are visual and climatic selection. Cain and Sheppard (1950) showed that the song thrush *Turdus ericetorum* (Turton) (Muscicapidae) discriminates between morphs and preferentially preys upon animals with conspicuous shells, for example yellow shells in populations living on dark

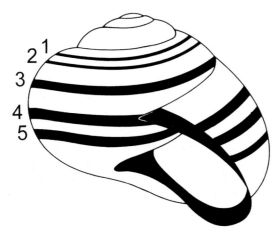

Fig. 11.1.   Numbering of colour band positions in *Cepaea* Held (Helicidae).

backgrounds. Thrushes do not carry snails very far to open the shells on anvils. This allows a comparison of morph frequencies of predated snails with those of individuals that escaped predation, which in turn allows an assessment on frequency-dependent selection. On the other hand, shells of different colours absorb solar radiation differentially, which can result in climatic selection (Heath, 1975). Experiments with *C. nemoralis* showed that when exposed to the sun, brown shells attain a higher temperature than yellow shells (Emberton and Bradbury, 1963; Heath, 1975). Similarly, banded shells attain a higher temperature than unbanded shells.

In *C. nemoralis*, quantitative genetic variation was reported for growth rate, with a heritability of 0.49 (Oosterhoff, 1977), and for shell size, with a heritability of 0.60 (Cook, 1967).

Another helicid morphologically highly variable in shell coloration and banding is *Theba pisana* (Müller). Heller (1981) described *T. pisana* in Israel with a four-banded shell, in which five morphs occur: 1234, 0234, 0034, 0004 and 0000; each band can sometimes be overlaid by up to six dark-brown bandlets. In contrast to Heller (1981), Cowie (1984a) assigned *T. pisana* to the five-banded pulmonate gastropods, as had already been done by some earlier authors (e.g. Taylor, 1912; Zilch, 1960). On the basis of extensive breeding experiments with *T. pisana* from Tenby, South Wales, UK, Cowie (1984a) described four different morphs: (i) unbanded individuals (plain) without any kind of pattern; (ii) unbanded shells with a row of dots at the position of the third band; (iii) five-banded individuals whose bands are broken up longitudinally into five dark-brown lines, which may be broken transversely into dots and dashes; and (iv) five-banded individuals with bands present only as a yellowish-buff background stripe. The bands are broken up transversely. Breeding experiments revealed that unbanded is dominant to five-banded (Cowie, 1984a).

T. pisana is a good example illustrating that the genetic basis of shell colour may be complex and can differ between British and south European populations. For example, the same colour patterns may be produced by different alleles, and the same allele may produce different colour patterns in different populations (Cain, 1984).

Bradybaenidae

Bradybaena similaris (de Férussac) is an East Asian snail which is polymorphic for shell colour (yellow and brown) and the banding pattern (presence/absence of bands) (Komai and Emura, 1955). Shell colour is determined at a single locus, with the allele for brown being dominant to the allele for yellow. The banding pattern of the shell is determined at another locus, with the allele presence of bands dominant to that of no bands (Komai and Emura, 1955). Both loci appear to be closely linked. These findings were confirmed and extended by Asami and Ohba (1982), Shibata *et al.* (1993) and Honda and Asami (1993). Similar colour polymorphisms were reported in *Bradybaena pellucida* Kuroda & Habe (Asami *et al.*, 1993, 1997).

Partulidae

Polymorphism in both shell colour and banding pattern has been found in *Partula taeniata* Mörch which was widely distributed on the island of Moorea, French Polynesia (Johnson *et al.*, 1993). Five different shell colour classes have been defined in this species: white, neutral brown, yellow, pink and orange. All colours can be present at different intensities. In general, alleles for darker colours are dominant to those for lighter colours (Murray and Clarke, 1976a). In some individuals of *P. taeniata*, the spire and lip differ in colour from the rest of the shell. In these specimens, the spire is darker than the rest of the shell and the lip is usually white. Three banding patterns have been recognized (Fig. 11.2): (i) the *frenata* pattern, with two narrow longitudinal lines which are darker than the background; (ii) the *zonata* pattern with a broad central band with lighter areas on both sides; and (iii) the *lyra* pattern with three bands but otherwise as the *frenata* pattern. Two pairs of alleles at two different loci determine the type of banding. One allele produces the *frenata* pattern, the other the *zonata* pattern; when both alleles are expressed together, the *lyra* pattern develops. Banding of the shell is dominant to absence of bands. The genes involved in these patterns show absolute linkage to a supergene (Murray and Clarke, 1976a).

Partula suturalis Pfeiffer co-exists with *P. taeniata* on Moorea. Six different shell morphs have been described by Crampton (1932) and

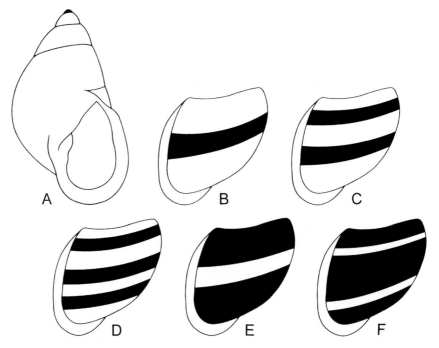

Fig. 11.2.   Some of the shell colour variants in *Partula* de Férussac (Partulidae). (A) General shell morphology; (B) cestata; (C) frenata; (D) lyra; (E) bisecta; (F) zonata.

their genetic basis has been studied by Murray and Clarke (1976b). The different morphs are (Fig. 11.2): (i) *apex* (uniformly whitish or straw-coloured with a darker spire); (ii) *atra* (uniformly dark-brown or black); (iii) *strigata* (unbanded but with cross-striations of brownish-yellow and white); (iv) *bisecta* (dark shell colour with a broad central longitudinal light band); (v) *cestata* (light shell colour with a dark central band); and (vi) *frenata* (light shell colour with two narrow longitudinal dark lines). The morphs described are controlled by at least six alleles at a single locus (Murray and Clarke, 1976b). The alleles for *strigata*, *cestata* and *atra* morphs are given here in increasing order of dominance. Two different alleles can produce the *frenata* pattern, but only one of them is dominant to the *atra* allele, whereas the other allele in combination with the *atra* allele produces the *bisecta* pattern. Finally, the allele for the *apex* morph is most probably dominant to all the other alleles.

Shells of all colour morphs may be either sinistral or dextral. The type of coiling is determined at another locus, with the allele for sinistral coiling being dominant to the one for dextrality. The genotype of the mother determines the direction of coiling among the offspring (maternal effect). Thus, the expression of coiling is delayed by one generation.

There are indications that shell size in *P. taeniata* and *P. suturalis* shows considerable quantitative genetic variation (Murray and Clarke, 1968). In addition, there may be a close relationship between qualitative and quantitative traits. For example, dextral shells on average have a higher spire than sinistral shells (Johnson, 1987).

Achatinidae

Allen (1983, 1985) studied the genetics of colour streak polymorphisms in the shells of three species of giant African snails, namely *Achatina fulica* Bowdich, *Achatina sylvatica* Putzeys and *Limicolaria cailliaudi* (Pfeiffer). In each case, the presence of streaks is controlled by an allele dominant to one for their absence. Similar results were reported for *Limicolaria aurora* (Jay) and *Limicolaria flammulata* (Pfeiffer) (Barker, 1968, 1969). In the latter two species, pink and brown shell colours are dominant to grey.

Colour polymorphisms of the soft body

Although integument colour variation is common and sometimes very conspicuous in stylommatophorans, it is often not simple to disentangle genetic from non-genetic components. Furthermore, individuals may change colour in the course of their lives, due to both genetic and environmental factors (e.g. Gain, 1892; Boettger, 1949). For example, *Arion ater* (Linnaeus) (Arionidae) with a completely white body may become entirely black when kept in captivity (Collinge, 1897; Ant, 1957). In *Arion fasciatus* (Nilsson), the orange–yellow band on the body flanks, generally considered a species-diagnostic character, can be lost irreversibly under laboratory conditions (Simroth, 1885).

Other colour polymorphisms, however, show a simple Mendelian inheritance at single loci, for example the grey–blue body colour dominant to pink, which is dominant to white in *Arion hortensis* de Férussac (Abeloos, 1945). Further examples of Mendelian inheritance are the heavily dark-spotted versus pale mantle in *Trichia striolata* (Pfeiffer) (Hyromiidae) (Cain, 1959), the cream-coloured dorsal integument with dark bands and black mottling (dominant) versus the cream–buff to orange back with indistinct bands (recessive) in *Arion subfuscus* (Draparnaud) (Getz, 1962; Clarke *et al.*, 1978), and the dark (dominant) versus light (recessive) body colour in *Limacus flavus* (Linnaeus) and *Limacus pseudoflavus* (Evans) (Limacidae) (Evans, 1983). The adult mottling and the expression of colour in juveniles in the latter two species is possibly under polygenic control (Evans, 1983).

Deroceras juranum Wüthrich (Agriolimacidae) was described originally as a presumptive endemic of Swiss Jura mountains (Wüthrich, 1993). Yet, breeding experiments (Reise, 1997) showed that *D. juranum* is only a Mendelian phenotype of *Deroceras rodnae* Grossu & Lupu, with the violet colour dominant to cream. The violet morph only occurs at high altitudes in a relatively small area, while the cream-coloured morph has a much wider distribution (e.g. Kerney *et al.*, 1983). In *A. ater*, the occurrence of dark colour morphs at high altitudes is correlated with air temperature and humidity (e.g. Albonico, 1948).

The maintenance of a body colour polymorphism in *Achatina* de Lamarck has been attributed tentatively in part to human selection pressure (Owen and Reid, 1986). In their native range of West Africa, these snails have been shown to possess integument that is either white or black (genetic basis unknown). The white morph is confined to urban areas, while the black morph is found mainly in forests. This difference in distribution might be explained by the fact that in the surveyed regions, only the black morph is regarded as edible (for cultural and medicinal reasons) and thus is collected in preference to the white morphs (Owen and Reid, 1986).

The occurrence of individuals that are completely devoid of body and shell pigmentation is referred to as albinism. Instances of albinism are reported in the malacological literature – for example in *Lehmannia marginata* (Müller) ((Limacidae) Roebuck, 1913). The genetics of albinism has been investigated in *Deroceras agreste* (Linnaeus), *Deroceras reticulatum* (Müller) (e.g. Luther, 1915) and *Philomycus bilineatus* Benson (Philomycidae) (Ikeda, 1937), where a simple Mendelian inheritance was demonstrated, with body pigmentation being dominant to albinism. Williamson (1959) found that body pigmentation in *A. ater* is determined by three loci. One of these carries two alleles for body pigmentation, with black melanin dominant to brown melanin. The other two loci affect the distribution of melanin over the body, that is one locus codes for the presence (dominant) or absence of dark bands in young animals, the other for the spread of melanin over the adult body. Three alleles occur at this latter locus: melanin evenly spread is dominant to melanin confined to the mediodorsal area, which in turn is dominant to white (melanin is only present in the tentacles and the lineolations of the foot fringe).

Genital polymorphisms

Intraspecific polymorphisms in the form of the reproductive tract are common in stylommatophorans, but their mode of inheritance often is unclear.

Currently much work is being focused on phally polymorphisms, that is the co-occurrence in natural populations of hermaphrodite individuals with either normally developed male copulatory organs (euphallics), reduced male parts (hemiphallics) or no male copulatory apparatus at all (aphallics) (Tompa, 1984). All three morphs produce functional spermatozoa, but hemiphallic and aphallic individuals usually cannot transfer their spermatozoa to a partner (Schrag and Read, 1996). Hence they can only act as recipients of spermatozoa. Polymorphism of this nature occurs in species from among several stylommatophoran families (Tompa, 1984; Pokryszko, 1987; Schrag and Read, 1996). Within species, the ratio of hemiphallic and aphallic individuals in natural populations can be highly variable (e.g. Watson, 1934; Pokryszko, 1987; Baur and Chen, 1993; Jordaens *et al.*, 1998), and may (Nicklas and Hoffmann, 1981) or may not (Pokryszko, 1990) show seasonal fluctuations. Breeding experiments with *Vertigo pusilla* Müller (Vertiginidae) showed that both euphallic and aphallic individuals when reared in isolation produce mixed offspring, in which aphallic forms always outnumbered euphallic forms (Pokryszko, 1990). The genetics of phallic polymorphism do not involve a simple single gene system and its expression is, at least partly, mediated by environmental conditions such as temperature and photoperiod regimes (e.g. Nicklas and Hoffmann, 1981; Jarne and Städler, 1995; Doums *et al.*, 1996).

In the helicid *Heterostoma paupercula* (Lowe), there is a functional penis, but polymorphism involves the presence/absence of the epiphallus and flagellum (Lace, 1992). It remains uncertain whether the hemiphallic-like *H. paupercula* has lost male outcrossing ability (Cook and Lace, 1993).

Species of the *Arion hortensis* complex usually have a tripartite free oviduct (Fig. 11.3) consisting of (i) a distal eversible portion containing a 'ligula', which acts as stimulator during copulation; (ii) a firm, musculated portion on to which the retractor muscle inserts; and (iii) a long, narrow proximal portion which merges with the vas deferens (e.g. Davies, 1977). However, in *Arion distinctus* Mabille, the distal eversible portion often is lacking, so that the free oviduct is bipartite. This dimorphism can occur among animals raised from single egg clutches. Pairs of parents with similar oviducts only produce broods of the same parental oviduct type (Davies, 1977). In natural populations, the bipartite morph is more

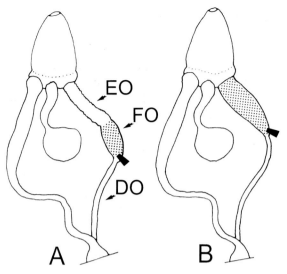

Fig. 11.3. Genital polymorphism in *Arion distinctus* Mabille (Arionidae). (A) Tripartite free oviduct; (B) bipartite oviduct. DO, narrow proximal portion of the oviduct that connects with the vas deferens; EO, eversible portion of the oviduct containing the ligula; FO, firm muscular portion of the oviduct.

common than the tripartite one. Although the dimorphism involves the presence/absence of a major functional part of the copulatory apparatus, it does not seem to affect mating behaviour, as *A. distinctus* mates indiscriminately with respect to free oviduct type (Davies, 1977). Unfortunately, neither the genetic basis nor the biological meaning of this oviduct dimorphism is known.

In *Bradybaena fruticum* (Müller), there is polymorphism in the numbers and configuration of the mucous gland lobes in the auxiliary copulatory organ (Falniowski *et al.*, 1993).

Protein polymorphisms

Protein polymorphisms in terrestrial gastropods are usually investigated by means of electrophoretic techniques (reviewed by Selander and Ochman, 1983; Brown and Richardson, 1988; Jarne, 1995). Most often, protein polymorphisms involve metabolic enzymes, although there are also reports on polymorphisms of haemocyanin in *Cepaea* Held (Manwell and Baker, 1968), oviductal gland proteins in *Partula* de Férussac (Schwabl and Murray, 1970) or albumen gland proteins in *Arion* de Férussac (Backeljau *et al.*, 1988, 1997b; Backeljau, 1989).

Several studies have shown that most electrophoretic 'alleles' are coded by single loci and inherited in a simple co-dominant Mendelian way (e.g. Brussard and McCracken, 1974; Johnson, 1979; McCracken and Brussard, 1980a,b; Crook, 1982; Johnson *et al.*, 1984; Hillis, 1989). Nevertheless, at least in *Cepaea*, non-genetic electrophoretic variation induced by food or environmental conditions was found in esterases (Oxford, 1973, 1975, 1978) and other enzymes (Gill, 1978a,b). Moreover, the degree of protein variation may depend on the techniques used (hidden variation; e.g. Backeljau, 1989). The interpretation of electro-phoretic data may be complicated further by the possible occurrence of null alleles (i.e. alleles that give no detectable enzyme activity), as reported by Burnet (1972) for *A. ater*. Hence, although electrophoretic analyses may provide a wealth of population genetic data, the results must be interpreted with caution and will by definition underestimate the actual amount of genetic variation.

Few population genetic analyses are available on terrestrial proso-branchs. Gofas (1994) showed a strong (micro)geographical genetic differ-entiation between populations of *Cochlostoma* Jan (Diplommatinidae) from adjacent valleys or mountains in north-western Spain. This implies ample opportunities for speciation in these gonochoristic diplomma-tinids. Yet, morphological differentiation in *Cochlostoma* is very limited. This contrasts strongly with patterns observed within several groups of morphologically highly distinctive stylommatophoran taxa, which are genetically very similar or even indistinguishable from one another, for example *Cerion* Röding (Cerionidae) (Gould *et al.*, 1975; Woodruff and Gould, 1980), *Samoana* Pilsbry (Partulidae) (Johnson *et al.*, 1986a), *Partula* (Johnson *et al.*, 1977, 1986b) and *Liguus* de Montfort (Bulimulidae) (Hillis *et al.*, 1991). Hence, several studies on protein poly-morphisms show that geographical or ecological patterns in morpho-logical characters and protein polymorphisms are not concordant. This has, for example, been demonstrated further in *Cantareus aspersus* (Müller) (Helicidae) by Guiller and Madec (1993), *Chondrina clienta* (Westerlund) (Chondinidae) by Baur (1988) and Baur and Klemm (1989) and *C. nemoralis* by Wilson (1996). Moreover, the degree of electropho-retic differentiation between populations is not a good index of taxonomic status (e.g. Johnson *et al.*, 1984).

Polymorphisms at the DNA level

Stylommatophoran DNA has been investigated by restriction fragment length (RFLP) analyses, random amplified polymorphic DNA (RAPD) fingerprinting and nucleotide sequencing.

The maternal, non-recombining inheritance of mitochondrial DNA (mtDNA), as well as its high intraspecific variability and evolutionary pace, make this molecule particularly useful for inferring geographic population structures and colonization events (e.g. Avise *et al.*, 1987). For example, RFLP analyses of mtDNA supported the isozyme data of Johnson *et al.* (1984), which suggested an English origin for introduced *C. nemoralis* from Lexington, Virginia (Stine, 1989). Similarly, mtDNA RFLP data from Moorean *Partula* largely confirmed the geographic patterns revealed by isozyme analyses, except for the Mount Ahutau populations of *P. taeniata*, which, according to isozyme data, are strongly aberrant, but not so according to mtDNA data (Murray *et al.*, 1991). Interestingly, Murray *et al.* (1991) found 32 different haplotypes using only four restriction enzymes. Twelve out of 31 populations surveyed were polymorphic within species. Despite this high degree of intraspecific polymorphism, different species can share the same haplotypes. For example, the haplotype A is shared by *P. taeniata*, *P. suturalis*, *Partula tohiveana* Crampton and *Partula mooreana* Hartman on Moorea, and by *Partula nodosa* Pfeiffer and *Partula hyalina* Broderip on Tahiti.

Another surprising finding is the length heteroplasmy of mtDNA in *C. nemoralis* (i.e. the condition in which two or more mtDNA genotypes co-exist within an individual) (Stine, 1989; Terrett *et al.*, 1994). If mtDNA length polymorphism is common among stylommatophorans, and if it has the same high turnover rate as in pectinid bivalves (e.g. Gjetvaj *et al.*, 1992), then the use of mtDNA RFLP data in population genetics may be limited (Terrett *et al.*, 1994). Yet, heteroplasmy was reported neither in *Partula* (Murray *et al.*, 1991) nor in *Albinaria* Vest (Clausiliidae) (Douris *et al.*, 1995; Hatzoglou *et al.*, 1995).

The *C. nemoralis* mtDNA haplotypes are estimated to have diverged as long ago as 20 million years (Thomaz *et al.*, 1996). This may indicate that the population structure of stylommatophorans favours the persistence of (co-existing) ancient haplotypes. Together with the occurrence of mtDNA gene rearrangements and the occurrence of unusually high amino acid substitution rates, it also suggests that stylommatophoran mtDNAs may evolve with an accelerated pace (Yamazaki *et al.*, 1997). The usefulness of mtDNA sequences for phylogenetic inference in the Gastropoda may therefore be limited (Yamazaki *et al.*, 1997). Moreover, with lengths from 14,100 to ±14,500 bp and with their particular tRNA secondary structures, stylommatophoran mtDNA genomes are among the smallest and most aberrant within the Metazoa (Terrett *et al.*, 1996; Yamazaki *et al.*, 1997). The small size is due to the compact gene organization, the short gene sequences and the absence of long non-coding regions (Lecanidou *et al.*, 1994; Terrett *et al.*, 1994; Hatzoglou *et al.*, 1995). This raises the question as to where transcription and replication are initiated in these compact genomes, as well as which part of the mtDNA genome is responsible for the heteroplasmy indicated in *Cepaea*.

Geneticists have also started to apply Mendelian nuclear DNA markers in stylommatophoran population studies. For example, RAPDs revealed population differentiation in *Trochoidea geyeri* (Soós) (Hygromiidae) (Pfenninger *et al.*, 1995, 1996), *Arion* species (Noble and Jones, 1996), *Cochlicopa lubricella* (Porro) (Cochlicopidae) (Armbruster, 1997) and *Balea perversa* (Linnaeus) (Clausiliidae) (Wirth *et al.*, 1997).

Chromosomal polymorphisms

General accounts of gastropod chromosomes can be found in Patterson (1969, 1973) and Patterson and Burch (1978). Stylommatophoran haploid chromosome numbers range from five in *Catinella* Pease (Succineidae) to 44 in *Aneitea* Gray (Athoracophoridae). The numbers usually are conservative within families, genera and species, so that intraspecific polymorphism for chromosome numbers is rare. Polyploidy, for example, is unknown in stylommatophorans.

The most thorough report on intraspecific variation in chromosome numbers in Stylommatophora is that of Evans (1960), who observed 1–6 supernumerary chromosomes in peripheral populations of *Helix pomatia* Linnaeus (Helicidae). These extra chromosomes are smaller than normal ones, but behave regularly during meiosis. Evans (1960) speculated that they arose from karyological changes after a period of genotype instability caused by inbreeding in these outcrossing snails.

Because of their small size and large number, stylommatophoran chromosomes are difficult to observe. This makes it highly probable that many inversion polymorphisms remain undetected. Nevertheless, Husted and Burch (1953) reported evidence of a paracentric inversion in *Allogona profunda* (Say) (Polygyridae), and Arrenbrecht *et al.* (1970) observed a pericentric inversion in a *C. hortensis* population in which both the normal and aberrant karyotypes co-existed. Clarke *et al.* (1978) wondered whether 'supergenes' for colour and banding pattern polymorphisms might be associated with chromosomal inversions, while Cain (1983) stressed the need for data on chiasma frequencies in order to study the release of variation in relation to the location of supergenes on large chromosomes (e.g. Cook, 1967). Bantock (1972), Price (1974a,b, 1975a,b,c) and Price and Bantock (1975) found that chiasma frequencies on the large chromosomes of *C. hortensis* and *C. nemoralis* varied between populations and species. In *C. hortensis*, this variation was associated with altitude, whereas in *C. nemoralis* it was negatively correlated with population density. The latter results were interpreted as a compensation for inbreeding at low densities.

Polymorphisms and breeding systems

Breeding experiments using colour markers of the integument have shown that a number of stylommatophorans regularly reproduce by non-apomictic selfing (e.g. Luther, 1915; Ikeda, 1937; Reise, 1997), even if some of the offspring are produced by outcrossing, as in *A. ater* (Williamson, 1959). Using protein and RAPD data, Reise *et al.* (1997) showed that in *D. rodnae* selfing can occur in the presence of mutual exchange of spermatozoa. In addition, RAPD data of parent–offspring comparisons indicated that in a natural population of *B. perversa*, the outcrossing rate was 10–16% (Wirth *et al.*, 1997).

Electrophoretic enzyme surveys of natural populations have provided evidence for the occurrence of (facultative) selfing in many other stylommatophorans (e.g. Selander and Kaufman, 1973; Johnson *et al.*, 1986a; Selander and Hudson, 1976; McCracken and Selander, 1980; Foltz *et al.*, 1982a, 1984; Boato, 1988; Brown and Richardson, 1988; Baur and Klemm, 1989; Hillis, 1989; Backeljau and De Bruyn, 1991; Hillis *et al.*, 1991;

Backeljau *et al.*, 1992; Cook and Lace, 1993; Armbruster and Schlegel, 1994; Backeljau *et al.*, 1997a). The high frequency of selfing is deduced mainly from the absence of variation (in terms of numbers of alleles and observed heterozygosities) in these populations. Consequently, most local populations of facultatively selfing species consist of a single mono-morphic multilocus genotype or strain, although the species as a whole may consist of a number of different strains (Selander and Ochman, 1983). Cases in point are *Rumina decollata* (Linnaeus) (Subulinidae) with more than 30 strains (Selander and Kaufman, 1973; Selander and Hudson, 1976), *C. clienta* with two strains (Baur and Klemm, 1989), *Arion intermedius* Normand with at least four strains (McCracken and Selander, 1980; Backeljau and De Bruyn, 1991; Backeljau *et al.*, 1992), species of the *A. fasciatus* complex with more than 20 strains (Jordaens *et al.*, 1996) and *Zonitoides nitidus* (Müller) (Zonitidae) with at least nine strains (Jordaens *et al.*, 1998). The number of strains is expected to increase further if more individuals from different areas are surveyed and if the number of poly-morphic markers scored is extended.

Selander and Ochman (1983) used data on *R. decollata* to construct a model describing interactions between selfing strains. This model states that slight differences in fecundity between strains may lead to the even-tual elimination of the less fecund ones, so that most selfing populations consist of a single strain. Recently, Backeljau *et al.* (1997a) observed that at least in *Arion silvaticus* Lohmander and *Arion fasciatus* (Nilsson), a considerable proportion of populations (64 and 33%, respectively) con-sist of more than one strain.

Hitherto, selfing terrestrial gastropods have been assumed to be autogamous (*sensu* Mogie, 1986). This was suggested for *Deroceras laeve* (Müller) (McCracken and Selander, 1980; Foltz *et al.*, 1982b). Nicklas and Hoffmann (1981), however, reported the occurrence of apomictic parthenogenesis in this species based on breeding experiments showing that heterozygous individuals, reared in isolation, only produced hetero-zygous offspring identical to the parent. Moreover, Nicklas and Hoffmann (1981) were unable to find spermatozoa in the reproductive tract of their specimens. Yet later, Hoffmann (1983) suggested that the data for *D. laeve* are also compatible with automixis. Unfortunately, the issue has not been dealt with since these initial studies, although the claim for apomixis, if correct, calls for a re-examination of the breeding system of many other selfing stylommatophorans (Runham, 1993).

It is usually believed that inbreeding, and thus selfing, has a long-term disadvantage because of the release of recessive lethal genes and the loss of genetic variation. Yet, Jarne (1995) found that the relative loss of alleles caused by bottlenecks is smaller in selfers than in outcrossers, so that genetic variation in selfing populations may be less eroded by bottlenecks than in outcrossers. Nevertheless, it appears difficult to predict the relative amount of genetic variation that can be maintained by selfing and outcrossing metapopulations, without precise knowledge of migration rates and the effective size of the subpopulations. On the other hand, high genetic variability (in terms of heterozygosity) often is assumed to be associated with higher fitness and developmental stability (Allendorf, 1983). Yet, Booth *et al.* (1990) could not detect any significant association between heterozygosity and phenotypic traits in *Cerion bendalli* (Pilsbry & Vanatta), and Elliott and Pierce (1992) found no relationship between growth-rate and heterozygosity in *Otala lactea* (Müller) (Helicidae).

Spatial Patterns of Genetic Variation

Factors causing spatial patterns in genetic variation

The genetic structure of natural populations is thought to be influenced by the combined effects of random genetic drift, restricted gene flow and differential selection pressures. For fitness-related characters, natural selection is an important factor influencing variation among and within groups (Endler, 1986). Selection can stabilize characters such that the genetic variation within populations is reduced, for example in *C. nemoralis* (see above). Disruptive selection may lead to genetic divergence and therefore may eventually increase genetic variation among populations. Not in all cases is genetic differentiation correlated with population subdivision. Frequency-dependent selection can maintain genetic variation within populations in situations where minority genotypes have an adaptive advantage, for example if they can escape predators (Clarke, 1962a,b; Allen, 1976). Frequency-dependent selection for direction of shell coiling in *P. suturalis* resulted in spatial patterns, which were maintained by behavioural isolation (Johnson, 1982).

Large and temporally constant heterogeneity in selection pressures can lead to genetic differences over very short distances even in the face of some gene flow. Several studies provided evidence that the spatial patterns of colour morph frequencies in *C. nemoralis* are a result of natural selection, and often change over distances of less than 100 m (Cain and Sheppard, 1952; Cain and Currey, 1963; Wolda, 1969; Jones *et al.*, 1977; Clarke *et al.*, 1978). If selection pressures change continuously with distance, then clinal variation may evolve, for example altitudinal variation in shell size as observed in *Arianta arbustorum* (Linnaeus) (Helicidae) (Burla and Stahel, 1983; Baur, 1984). However, besides selection, genetic diversity within and among populations in some situations can be due to the balance between genetic drift (which causes the local population to lose genetic diversity, but leads to an increase in among-population differentiation) and gene flow (which brings new genetic diversity into the local population and reduces genetic differentiation among populations).

Because isozymes usually are presumed to be selectively neutral, they may be poor predictors of genetic variation in characters of more direct relevance to fitness, in particular quantitative characters (e.g. life history traits). Nevertheless, in terrestrial gastropods, there is some evidence suggesting that electrophoretic alleles of certain metabolic enzymes may also have a direct adaptive significance in relation to changing environmental conditions (e.g. heavy metal pollution) (for reviews, see Posthuma and Van Straalen, 1993; Riddoch, 1993).

Effective population size, neighbourhood size and gene flow

In the island and stepping-stone models, the genetically relevant size of a gene can be measured by the so-called effective population size (N_e). This is the size an ideal population would have if its genetic composition were influenced by random processes in the same way as the real population of size N (Wright, 1931). Nunney and Elam (1994) reviewed genetic and ecological methods that have been used to estimate N_e. Genetic estimates of N_e rely on frequency data from enzyme electrophoresis or molecular techniques. A general problem of genetic methods is the need to eliminate confounding effects of migration and population subdivision, and the possibility that selection is acting at the marker loci or on loci linked to them. Ecological methods for estimations of N_e depend on theories that link demographic models and behavioural factors to changes in N_e (Nunney, 1993). N_e is almost always lower than the population number of adult individuals.

In the continuum model, the rate of dispersal is small compared with the area occupied by the entire population. Thus, isolation by distance will prevent panmixis, and the effective population size will again be less than the actual number. Wright (1943, 1946) introduced the idea of a neighbourhood and defined it as an area from which the parents of central individuals may be treated as if drawn at random. Individuals are considered to be distributed at a uniform density either along a linear range, for example a roadside verge or a river bank, or throughout an area.

Using the data of Lamotte (1951), Wright (1969) estimated a neighbourhood size of 2800 for *C. nemoralis* and a corresponding neighbourhood area consisting of a circle with 40 m diameter. A neighbourhood size of 5600 was estimated for a two-dimensional population of *A. arbustorum* in central Sweden and a size of 388 individuals for a linear population in the eastern Swiss Alps (Baur, 1993). The neighbourhood area of the two-dimensional population corresponds to a circle with a diameter of 32 m (Baur, 1993). Pfenninger *et al.* (1996) estimated a neighbourhood size of 70–208 for *T. geyeri* in an area of 13–21 m^2.

The effective population size (N_e) may depart from the neighbourhood number (N) under certain conditions (Murray, 1964; Greenwood, 1974; Cowie, 1984b). These include asymmetry in investment in female versus male reproductive function, non-random fertilization of eggs, non-random dispersal, changing population size and non-random variation in lifetime reproductive success. Estimates of effective population size that take into account some of these factors are available for *C. nemoralis* ($N_e = 95–6000$; Greenwood, 1974, 1976), *T. pisana*

(Müller) (N_e = 115–4130; Cowie, 1984b), *C. aspersus* (N_e = 15–215; Crook, 1980) and *Albinaria corrugata* (Bruguière) (N_e = 22–36; Schilthuizen and Lombaerts, 1994).

In *P. taeniata*, both direct and indirect methods have been used to estimate gene flow (Murray and Clarke, 1984). Gene flow, as assessed by recording movements of marked individuals (direct method), agrees well with estimates of gene flow obtained by patterns of isozyme variation (indirect method) in this species. Yet, Schilthuizen and Lombaerts (1994) found in *A. corrugata*, which is more patchily distributed, that direct (marked individuals) and indirect (F_{st} values) estimates of gene flow were of the same order of magnitude only at the lowest spatial levels (<10 m). At higher spatial levels, there were substantial discrepancies between the two methods.

Unsuitable areas (lakes, rivers, desert or mountain ridges) can interrupt gene flow. Artificial habitats such as buildings, roads, railways and agricultural areas can also act as partial or complete barriers to gene flow for many species of terrestrial gastropods (Selander and Kaufman, 1975; Baur and Baur, 1993). Populations separated by paved roads with high traffic density might be isolated from each other (Baur and Baur, 1990). Furthermore, gene flow can be limited to particular types of habitats. Stream embankments serve as gene flow paths for *A. arbustorum* on mountain slopes in the Swiss Alps (Arter, 1990). Adults preferentially move along streams, resulting in a mean displacement of 8 m per year (Baur, 1986). The spatial structure of allozyme variation could be best explained by a functional isolation-by-distance model (Arter, 1990), assuming gene flow along the drainage system. The genetic structure of apparent continuous populations of *T. geyeri* can also be described by the isolation-by-distance model (Pfenninger *et al.*, 1996).

Temporal Variation of Genetic Diversity

Studies of genetic change over time in natural populations can give many insights into the processes of evolution. Studies on changes in morph frequencies in *C. nemoralis* and *C. hortensis* range over periods from a few to 60 years (see review by Cameron, 1992). Most of the studies found little temporal variation in morph frequencies (Wolda, 1969; Williamson *et al.*, 1977; Cain and Cook, 1989; Cain *et al.*, 1990; Arthur *et al.*, 1993). A gradual change in morph frequencies resulting from gene flow was observed in one study (Goodhart, 1973), while in two studies systematic changes in morph frequencies were observed in several populations (Clarke and Murray, 1962; Murray and Clarke, 1978; Wall *et al.*, 1980). Cameron (1992) examined changes in morph frequencies in 71 *C. hortensis* populations over 25 years, corresponding to approximately ten generations. Some populations were stable in shell colour and banding patterns. However, populations in valley bottoms showed a consistent and significant decline in the frequency of yellow shells. Cameron (1992) interpreted this change in morph frequency as a possible case of climatic selection.

Selander and Ochman (1983) described a temporal change in the genetic structure of natural populations of the facultatively selfing species *R. decollata* in southern France. Between 1973 and 1979, one genetic strain (light) gradually replaced another one (dark), probably due to slight differences in fecundity between the strains.

Terrestrial gastropods provide an important resource in long-term studies on genetic variation (Cain, 1983). Using subfossil samples, Currey and Cain (1968) and Cain (1971) showed a major change in shell morph frequencies in *C. nemoralis* over the last 6000 years in southern England. The change in morph frequency was associated with known changes in climate, from unbanded to five-banded shells in valleys and lowlands, and, to a minor extent, to banded shells on uplands. During the same period, populations of *C. hortensis* exhibited changes in the proportion of unbanded and banded morphs, but the species also expanded its range (Cain, 1971). Besides frequency-dependent selection, other factors may contribute to the long-term maintenance of polymorphism in a population, for example overdominance of the heterozygous condition as shown in *B. similaris* (Komai and Emura, 1955).

Interspecific Hybridization

Closely related species sometimes hybridize in natural sympatric conditions. Hybrid zones sometimes reveal unexpected electrophoretic variants which are referred to as 'hybrizymes' (Woodruff, 1989), that is alleles which are outside hybrid zones at frequencies of a few per cent or less (Barton and Hewitt, 1985). Hybrizymes hitherto were detected in *Cerion* (Woodruff, 1989), *Albinaria* (Kemperman and Degenaars, 1992; Schilthuizen and Gittenberger, 1994; Schilthuizen and Lombaerts, 1995) and *Mandarina* Pilsbry (Camaenidae) (Chiba, 1993). Hybrid zones may thus be a source of evolutionary novelties and reticulate evolution (Chiba, 1993). For example, Chiba (1997) found novel shell colour polymorphisms in a hybrid zone of *Mandarina*. On the other hand, it has been suggested that mutation rates are increased in hybrids, resulting in hybrid disadvantage since mutations usually are deleterious (Schilthuizen and Gittenberger, 1994). Yet, currently, there is no evidence supporting this suggestion. Hence, the evolutionary significance of hybrizymes is still unclear.

Conservation Genetics

Habitat loss and fragmentation, environmental stresses (air pollution, soil contamination, increasing temperatures, etc.) and environmental degradation, through introduction of alien predators or competitors, are generally considered as the major threats to indigenous communities, including those of terrestrial gastropods (e.g. Hadfield and Mountain, 1980; Hadfield, 1986; Coppois and Wells, 1987; Baur and Baur, 1993). Within the realm of conservation biology, conservation genetics plays an important role as they focus on inbreeding in small populations, the assessment of spatial patterns of genetic variation, as well as on problems of gene flow, hybridization, outbreeding depression and systematics (e.g. Allendorf, 1983; Simberloff, 1988; Thornhill, 1993; Avise and Hamrick, 1996).

In small populations (<100 individuals), allele frequencies may undergo large and unpredictable fluctuations due to random genetic drift. Large changes of allele frequencies may also result from founder effects and bottlenecks (Johnson, 1988). The loss of genetic variation, that is decreasing heterozygosity levels and the loss of rare alleles, may significantly limit the adaptability of populations in changing environments in future generations. The loss of genetic variation may have stronger effects in outcrossing species than in predominantly selfing species (Jarne, 1995). These aspects should be considered in issues of terrestrial gastropod conservation, but hitherto have received limited attention (Triggs and Sherley, 1993; Daniell, 1994).

Conserving and managing biodiversity also require the identification of species and populations with independent evolutionary histories. However, these concepts are controversial (e.g. Endler, 1989; O'Brien and Mayr, 1991; Rojas, 1992), and intraspecific taxonomy is often confused. Therefore, it is increasingly accepted that genetically divergent populations should be recognized as appropriate units for conservation, regardless of their taxonomic status (Moritz, 1994). This has led to the introduction of the concept of an evolutionarily significant unit (ESU), a set of populations with a distinct, long-term evolutionary history, and the more restrictive concept of a management unit (MU), which involves sets of populations which show significant allelic differences. These concepts aim to reach the main objective of conservation biology, that is the preservation of unique ecological adaptations and the maintenance of evolutionary potential (e.g. Moritz, 1994; Vogler and DeSalle, 1994). Population genetics provide a major tool to identify ESUs and MUs. Existing data sets can be used to give an intuitive and simple illustration of the principle.

Arion owenii Davies is a rare, endemic member of the *A. hortensis* complex in the British Isles, where it is distributed disjunctly over two regions, that is northern Ireland and north-west Scotland on the one hand, and southern England on the other. Electrophoretic analyses of albumen gland proteins revealed a striking differentiation between specimens from the two regions (Backeljau *et al.*, 1988). Irrespective of the taxonomic implications of this result, it suggests that two ESUs or MUs may be involved. Similarly, Baur and Klemm (1989) found that *C. clienta* populations from Öland (Sweden) and Austria differ consistently at four out of 17 enzyme loci. Considering the species' disjunct distribution, these data suggest that *C. clienta* may also consist of at least two ESUs or MUs. The geographically very restricted endemic Australian camaenids (Woodruff and Solem, 1990) and the endangered endemic *Tropidophora* Troschel (Pomatiasidae) radiation in Madagascar (Emberton, 1995) may be more complex examples of the ESU concept.

Finally, population genetic analyses will also play an important role in captive breeding and reintroduction programmes of threatened species. Currently such programmes have been developed for *Partula* (e.g. Murray, 1993; Tonge and Bloxam, 1991; Pearce-Kelly *et al.*, 1995). One of the important issues when reinforcing endangered populations by introducing specimens from other populations is to avoid possible outbreeding depression. This is the disruption of local, highly co-adapted genomes by introgression of foreign, but conspecific genes. This can decrease individual fitness in the recipient population despite genetic variability and population size being increased. Population genetic surveys may provide the data needed to minimize such risks.

Epilogue

This short overview provides evidence that population genetics might be important for various aspects of applied malacology (e.g. captive breeding programmes, spread of pest species, etc.), as well as for the general understanding of the evolution and adaptative radiation of gastropods. Therefore, we believe that population genetic work on terrestrial gastropods should be intensified, particularly since current molecular techniques allow some exciting new possibilities for long-term genetic studies. Indeed, polymerase chain reaction (PCR)-based DNA methods can theoretically can be applied to almost any kind of biological sample (Arnheim *et al.*, 1990). Thus it may become feasible for genetic surveys to include old museum specimens (in alcohol, dried, embedded in paraffin, etc.) that often were collected hundreds of years ago. This allows the direct estimation of genetic change over historical times and thus the assessment of effects of industrialization, pollution, deforestation, colonization, etc., while also providing for genetic data of contemporary rare and endangered species represented by specimens preserved in collections. The importance of these kind of data must be obvious in a time of accelerating environmental change.

Acknowledgements

We are indebted to S. Chiba, H. De Wolf, P. Jarne, K. Jordaens, J.J. Murray and H. Reise for their comments on the manuscript. We thank H. Van Paesschen for preparing the illustrations. Financial support for our own research was received from the Flemish Science Foundation and the Swiss National Science Foundation.

Note added in proof

By the time this chapter is published there will be many more noteworthy additions to the literature in this fast-evolving field. As this chapter was submitted in its final form in March 1998, it has not been possible to include relevant literature from later dates or to adapt the text accordingly. Nevertheless it is expected that the general scope and conclusions presented will remain current for some years.

References

Abeloos, M. (1945) Sur la génétique du gastéropode *Arion hortensis* de Fér. *Comptes Rendus des Séances de la Société de Biologie et ses Filiales* 139, 619–620.

Albonico, R. (1948) Die Farbenvarietäten der grossen Wegschnecke, *Arion empiricorum* Fér., und deren Abhängigkeit von den Umweltbedingungen. *Revue Suisse de Zoologie* 55, 347–425.

Allen, J.A. (1976) Further evidence for apostatic selection by wild passerine birds – 9:1 experiments. *Heredity* 36, 173–180.

Allen, J.A. (1983) The inheritance of a shell colour polymorphism in *Achatina fulica* Bowdich from East Africa. *Journal of Conchology* 31, 185–189.

Allen, J.A. (1985) The genetics of streaked and unstreaked morphs of the snails *Achatina sylvatica* Putzeys and *Limicolaria cailliaudi* (Pfeiffer). *Heredity* 54, 103–105.

Allendorf, F.W. (1983) Isolation, gene flow, and genetic differentiation among populations. In: Schonewald-Cox, C.M., Chambers, S.M., MacBryde, B. and Thomas, L. (eds) *Genetics and Conservation*. Benjamin/Cummings, Menlo Park, California, pp. 51–65.

Ant, H. (1957) Westfälische Nacktschnecken. *Natur und Heimat* 17, 1–20.

Armbruster, G. (1997) Evaluations of RAPD markers and allozyme patterns: evidence for morphological convergence in the morphotype of *Cochlicopa lubricella* (Gastropoda: Pulmonata: Cochlicopidae). *Journal of Molluscan Studies* 63, 379–388.

Armbruster, G. and Schlegel, M. (1994) The land-snail species of *Cochlicopa* (Gastropoda: Pulmonata: Cochlicopidae): presentation of taxon-specific allozyme patterns, and evidence for a high level of self-fertilization. *Journal of Zoological Systematics and Evolutionary Research* 32, 282–296.

Arnheim, N., White, T. and Rainey, W.E. (1990) Application of PCR: organismal and population biology. *BioScience* 40, 174–182

Arrenbrecht, S., Hauschteck-Jungen, E. and Jungen, H. (1970) Normaler und aberranter Karyotyp von *Cepaea hortensis* (Müller) (Gastropoda). *Experientia* 15, 1279–1281.

Arter, H.E. (1990) Spatial relationship and gene flow paths between populations of the alpine snail *Arianta arbustorum* (Pulmonata: Helicidae). *Evolution* 44, 966–980.

Arthur, W., Phillips, D. and Mitchell, P. (1993) Long-term stability of morph frequency and species distribution in a sand-dune colony of *Cepaea*. *Proceedings of the Royal Society of London, Series B* 251, 159–163.

Asami, T. and Ohba, S. (1982) Shell polymorphism in the land snail *Bradybaena similaris* in the Kanto District. *Scientific Reports of the Takao Museum of Natural History* 11, 13–28.

Asami, T., Fukuda, H. and Tomiyama, K. (1993) The inheritance of shell banding in the land snail *Bradybaena pellucida*. *Venus* 52, 155–159.

Asami, T., Ohbayashi, K. and Seki, K. (1997) The inheritance of shell color in the land snail *Bradybaena pellucida*. *Venus* 56, 35–39.

Avise, J.C. and Hamrick, J.L. (eds) (1996) *Conservation Genetics – Case Histories from Nature*. Chapman & Hall, New York.

Avise, J.C., Arnold, J., Ball, R.M., Bermingham, E., Lamb, T., Neigel, J.E., Reeb, C.A. and Saunders, N.C. (1987) Intraspecific phylogeography: the mitochondrial DNA bridge between population genetics and systematics. *Annual Review of Ecology and Systematics* 18, 489–522.

Backeljau, T. (1989) Electrophoresis of albumen gland proteins as a tool to elucidate taxonomic problems in the genus *Arion* (Gastropoda, Pulmonata). *Journal of Medical and Applied Malacology* 1, 29–41.

Backeljau, T. and De Bruyn, L. (1991) Preliminary report on the genetic variability of *Arion intermedius* in Europe (Pulmonata). *Journal of Medical and Applied Malacology* 3, 19–29.

Backeljau, T., Davies, S.M. and De Bruyn, L. (1988) An albumen gland protein polymorphism in the terrestrial slug *Arion owenii*. *Biochemical Systematics and Ecology* 16, 425–429.

Backeljau, T., Brito, C.P., Tristao da Cunha, R.M., Frias Martins, A.M. and De Bruyn, L. (1992) Colour polymorphism and genetic strains in *Arion intermedius* from Flores, Azores (Mollusca: Pulmonata). *Biological Journal of the Linnean Society* 46, 131–143.

Backeljau, T., De Bruyn, L., De Wolf, H., Jordaens, K., Van Dongen, S. and Winnepenninckx, B. (1997a) Allozyme diversity in slugs of the *Carinarion* complex (Mollusca, Pulmonata). *Heredity* 78, 445–451.

Backeljau, T., Jordaens, K., De Wolf, H., Rodriguez, T. and Winnepenninckx, B. (1997b) Albino-like *Carinarion* identified by protein electrophoresis (Pulmonata: Arionidae). *Journal of Molluscan Studies* 63, 559–563.

Bantock, C.R. (1972) Localization of chiasmata in *Cepaea nemoralis* L. *Heredity* 29, 213–221.

Barker, J.F. (1968) Polymorphism in West African snails. *Heredity* 23, 81–98.

Barker, J.F. (1969) Polymorphism in a West African snail. *American Naturalist* 103, 259–266.

Barton, N.H. and Hewitt, G.M. (1985) Analysis of hybrid zones. *Annual Review of Ecology and Systematics* 16, 113–148.

Baur, A. and Baur, B. (1990) Are roads barriers to dispersal in the land snail *Arianta arbustorum*? *Canadian Journal of Zoology* 68, 613–617.

Baur, A. and Baur, B. (1993) Daily movement patterns and dispersal in the land snail *Arianta arbustorum*. *Malacologia* 35, 89–98.

Baur, B. (1984) Shell size and growth rate differences for alpine populations of *Arianta arbustorum* (L.) (Pulmonata: Helicidae). *Revue Suisse de Zoologie* 91, 37–46.

Baur, B. (1986) Patterns of dispersion, density and dispersal in alpine populations of the land snail *Arianta arbustorum* (L.) (Helicidae). *Holarctic Ecology* 9, 117–125.

Baur, B. (1988) Microgeographical variation in shell size of the land snail *Chondrina clienta*. *Biological Journal of the Linnean Society* 35, 247–259.

Baur, B. (1993) Population structure, density, dispersal and neighbourhood size in *Arianta arbustorum* (Linnaeus, 1758) (Pulmonata: Helicidae). *Annalen des Naturhistorischen Museums Wien* 94/95B, 307–321.

Baur, B. and Baur, A. (1993) Climatic warming due to thermal radiation from an urban area as possible cause for the local extinction of a land snail. *Journal of Applied Ecology* 30, 333–340.

Baur, B. and Chen, X. (1993) Genital dimorphism in the land snail *Chondrina avenacea*: frequency of aphally in natural populations and morph-specific allocation to reproductive organs. *Veliger* 36, 252–258.

Baur, B. and Klemm, M. (1989) Absence of isozyme variation in geographically isolated populations of the land snail *Chondrina clienta*. *Heredity* 63, 239–244.

Boato, A. (1988) Microevolution in *Solatopupa* landsnails (Pulmonata Chondrinidae): genetic diversity and founder effects. *Biological Journal of the Linnean Society* 34, 327–348.

Boettger, C.R. (1949) Zur Kenntnis der grossen Wegschnecken (*Arion* s.str.) Deutschlands. *Archiv für Molluskenkunde* 78, 169–186.

Booth, C.L., Woodruff, D.S. and Gould, S.J. (1990) Lack of significant associations between allozyme heterozygosity and phenotypic traits in the land snail *Cerion*. *Evolution* 44, 210–213.

Brown, K.M. and Richardson, T.D. (1988) Genetic polymorphism in gastropods: a comparison of methods and habitat scales. *American Malacological Bulletin* 6, 9–17.

Brussard, P.F. and McCracken, G.F. (1974) Allozymic variation in a North American colony of *Cepaea nemoralis*. *Heredity* 33, 98–101.

Burla, H. and Stahel, W. (1983) Altitudinal variation in *Arianta arbustorum* (Mollusca, Pulmonata) in the Swiss Alps. *Genetica* 62, 95–108.

Burnet, B. (1972) Enzyme protein polymorphism in the slug *Arion ater*. *Genetical Research, Cambridge* 20, 161–173.

Cain, A.J. (1953) Visual selection by tone of *Cepaea nemoralis*. *Journal of Conchology* 23, 333–336.

Cain, A.J. (1959) Inheritance of mantle colour in *Hygromia striolata* (C. Pfeiffer). *Journal of Conchology* 24, 352–353.

Cain, A.J. (1971) Colour and banding morphs in subfossil samples of the snail *Cepaea*. In: Creed, R. (ed.) *Ecological Genetics and Evolution*. Blackwell, Oxford, pp. 65–92.

Cain, A.J. (1983) Ecology and ecogenetics of terrestrial molluscan populations. In: Russell-Hunter, W.D. (ed.) *The Mollusca*, Vol. 6, *Ecology*. Academic Press, London, pp. 597–647.

Cain, A.J. (1984) Genetics of some morphs in the land snail *Theba pisana*. *Malacologia* 25, 381–411.

Cain, A.J. and Cook, L.M. (1989) Persistence and extinction in some *Cepaea* populations. *Biological Journal of the Linnean Society* 38, 183–190.

Cain, A.J. and Currey, J.D. (1963) Area effects in *Cepaea*. *Philosophical Transactions of the Royal Society of London, Biological Sciences* 246, 1–81.

Cain, A.J. and Sheppard, P.M. (1950) Selection in the polymorphic land snail *Cepaea nemoralis*. *Heredity* 4, 275–294.

Cain, A.J. and Sheppard, P.M. (1952) The effects of natural selection on body colour in the land snail *Cepaea nemoralis*. *Heredity* 6, 217–223.

Cain, A.J. and Sheppard, P.M. (1954) Natural selection in *Cepaea*. *Genetics* 39, 89–116.

Cain, A.J., Cook, L.M. and Currey, J.D. (1990) Population size and morph frequency in a long-term study of *Cepaea nemoralis*. *Proceedings of the Royal Society of London, Series B* 240, 231–250.

Cameron, R.A.D. (1992) Change and stability in *Cepaea* populations over 25 years: a case of climatic selection. *Proceedings of the Royal Society of London, Series B* 248, 181–187.

Chiba, S. (1993) Modern and historical evidence for natural hybridization between sympatric species in *Mandarina* (Pulmonata: Camaenidae). *Evolution* 47, 1539–1556.

Chiba, S. (1997) Novel colour polymorphisms in a hybrid zone of *Mandarina* (Gastropoda: Pulmonata). *Biological Journal of the Linnean Society* 61, 369–384.

Clarke, B. (1962a) Balanced polymorphism and the diversity of sympatric species. *Systematic Association Publication* 4, 47–70.

Clarke, B. (1962b) Natural selection in mixed populations of two polymorphic snails. *Heredity* 17, 319–345.

Clarke, B. and Murray, J. (1962) Changes of gene frequency in *Cepaea nemoralis* (L.). *Heredity* 17, 445–465.

Clarke, B., Arthur, W., Horsley, D.T. and Parkin, D.T. (1978) Genetic variation and natural selection in pulmonate molluscs. In: Fretter, V. and Peake, J. (eds) *Pulmonates*. Vol. 2A, *Systematics, Evolution and Ecology*. Academic Press, New York, pp. 219–270.

Collinge, W.E. (1897) On some European slugs of the genus *Arion*. *Proceedings of the Zoological Society of London* 3, 439–450.

Cook, L.M. (1967) The genetics of *Cepaea nemoralis*. *Heredity* 22, 397–410.

Cook, L.M. and Lace, L.A. (1993) Sex and genetic variation in a helicid snail. *Heredity* 70, 376–384.

Cook, L.M. and Murray, J. (1966) New information on the inheritance of polymorphic characters in *Cepaea hortensis*. *Journal of Heredity* 57, 245–247.

Coppois, G. and Wells, S. (1987) Threatened Galapagos snails. *Oryx* 21, 236–241.

Cowie, R.H. (1984a) Ecogenetics of *Theba pisana* (Pulmonata: Helicidae) at the northern edge of its range. *Malacologia* 25, 361–380.

Cowie, R.H. (1984b) Density, dispersal and neighbourhood size in the land snail *Theba pisana*. *Heredity* 52, 391–401.

Cowie, R.H. (1992) Evolution and extinction of Partulidae, endemic Pacific island land snails. *Philosophical Transactions of the Royal Society of London, Biological Sciences* 335, 167–191.

Crampton, H.E. (1932) Studies on the variation, distribution and evolution of the genus *Partula*. The species inhabiting Moorea. *Carnegie Institute of Washington Publication* 410, 1–335.

Crook, S.J. (1980) Studies on the ecological genetics of *Helix aspersa* (Müller). PhD thesis, Dundee University, UK.

Crook, S.J. (1982) Genetic studies of allozyme variation in leucine aminopeptidase in the land snail *Helix aspersa* (Müller). *Journal of Molluscan Studies* 48, 362–365.

Currey, J.D. and Cain, A.J. (1968) Studies on *Cepaea* IV. Climate and selection of banding morphs in *Cepaea* from the climatic optimum to the present day. *Philosophical Transactions of the Royal Society of London, Biological Sciences* 253, 483–498.

Daniell, A. (1994) Genetics and terrestrial mollusc conservation. *Memoirs of the Queensland Museum* 36, 47–53.

Davies, S.M. (1977) The *Arion hortensis* complex, with notes on *A. intermedius* Normand (Pulmonata: Arionidae). *Journal of Conchology* 29, 173–187.

Doums, C., Bremond, P., Delay, B. and Jarne, P. (1996) The genetical and environmental determination of phally polymorphism in the freshwater snail *Bulinus truncatus. Genetics* 142, 217–225.

Douris, V., Rodakis, G.C., Giokas, S., Mylonas, M. and Lecanidou, R. (1995) Mitochondrial DNA and morphological differentiation of *Albinaria* populations (Gastropoda: Clausiliidae). *Journal of Molluscan Studies* 61, 65–78.

Elliott, A.C. and Pierce, B.A. (1992) Size, growth rate, and multiple-locus heterozygosity in the land snail (*Otala lactea*). *Journal of Heredity* 83, 270–274.

Emberton, K.C. (1995) Cryptic, genetically extremely divergent, polytypic, convergent, and polymorphic taxa in Madagascan *Tropidophora* (Gastropoda: Pomatiasidae). *Biological Journal of the Linnean Society* 55, 183–208.

Emberton, L.B.B. and Bradbury, S. (1963) Transmission of light through shells of *Cepaea nemoralis* (L.). *Proceedings of the Malacological Society of London* 35, 211–219.

Endler, J.A. (1986) *Natural Selection in the Wild.* Princeton University Press, New Jersey.

Endler, J.A. (1989) Conceptual and other problems in speciation. In: Otte, D. and Endler, J.A. (eds) *Speciation and its Consequences.* Sinauer Associates, Sunderland, Massachusetts, pp. 625–661.

Evans, H.J. (1960) Supernumerary chromosomes in wild populations of the snail *Helix pomatia* L. *Heredity* 15, 129–138.

Evans, N.J. (1983) Notes on self-fertilization and variation in body colour in *Limax flavus* L. and *L. pseudoflavus* Evans. *Irish Naturalists' Journal* 21, 37–40.

Falniowski, A., Kozik, A., Szarowska, M., Rapala-Kozik, M. and Turyna, I. (1993) Morphological and allozymic polymorphism and differences among local populations in *Bradybaena fruticum* (O.F. Müller, 1777) (Gastropoda: Stylommatophora: Helicoidea). *Malacologia* 35, 371–388.

Foltz, D.W., Ochman, H., Jones, J.S., Evangelisti, S.M. and Selander, R.K. (1982a) Genetic population structure and breeding systems in arionid slugs (Mollusca: Pulmonata). *Biological Journal of the Linnean Society* 17, 225–241.

Foltz, D.W., Schaitkin, B.M. and Selander, R.K. (1982b) Gametic disequilibrium in the self-fertilizing slug *Deroceras laeve. Evolution* 36, 80–85.

Foltz, D.W., Ochman, H. and Selander, R.K. (1984) Genetic diversity and breeding systems in terrestrial slugs of the families Limacidae and Arionidae. *Malacologia* 25, 593–605.

Gain, W.A. (1892) Some remarks on the colour changes in *Arion intermedius* Normand. *Conchologist* 2, 55–56.

Getz, L.L. (1962) Color forms of *Arion subfuscus* in New Hampshire. *Nautilus* 76, 70–71.

Gill, P.D. (1978a) Non-genetic variation in isoenzymes of lactate dehydrogenase of *Cepaea nemoralis. Comparative Biochemistry and Physiology* 59B, 271–276.

Gill, P.D. (1978b) Non-genetic variation in isoenzymes of acid phosphatase, alkaline phosphatase and α-glycerophosphate dehydrogenase of *Cepaea nemoralis. Comparative Biochemistry and Physiology* 60B, 365–368.

Gjetvaj, B., Cook, D.I. and Zouros, E. (1992) Repeated sequences and large-scale size variation of mitochondrial DNA: a common feature among scallops (Bivalvia: Pectinidae). *Molecular Biology and Evolution* 9, 106–124.

Gofas, S. (1994) Les *Cochlostoma* (Gastropoda, Prosobranchia) des Pyrénées et monts Cantabriques: systématique et évolution. PhD thesis, Muséum National d'Histoire Naturelle, Paris, France.

Goodhart, C.B. (1973) A 16-year survey of *Cepaea* on the Hundred Foot Bank. *Malacologia* 14, 327–331.

Goodhart, C.B. (1987) Why are some snails visibly polymorphic, and others not? *Biological Journal of the Linnean Society* 31, 35–58.

Gould, S.J., Woodruff, D.S. and Martin, J.P. (1975) Genetics and morphometrics of *Cerion* at Pongo Carpet: a new systematic approach to this enigmatic land snail. *Systematic Zoology* 23, 518–535.

Greenwood, J.J.D. (1974) Effective population numbers in the snail *Cepaea nemoralis. Evolution* 28, 513–526.

Greenwood, J.J.D. (1976) Effective population number in *Cepaea*: a modification. *Evolution* 30, 186.

Guerrucci-Henrion, M.-A. (1971) Etude de la transmission de quelques caractères de la pigmentation chez *Cepaea hortensis. Archives de Zoologie Expérimental et Général* 112, 211–219.

Guerrucci-Henrion, M.-A. (1973) Etude de la transmission du caractère péristome coloré chez *Cepaea hortensis. Archives de Zoologie Expérimental et Général* 114, 313–316.

Guiller, A. and Madec, L. (1993) A contribution to the study of morphological and biochemical differentiation in French and Iberian populations of *Cepaea nemoralis. Biochemical Systematics and Ecology* 21, 323–339.

Hadfield, M.G. (1986) Extinction in Hawaiian achatinelline snails. *Malacologia* 27, 67–81.

Hadfield, M.G. and Mountain, B.S. (1980) A field study of a vanishing species, *Achatinella mustelina* (Gastropoda, Pulmonata), in the Waianae Mountains of Oahu. *Pacific Science* 34, 345–358.

Hatzoglou, E., Rodakis, G.C. and Lecanidou, R. (1995) Complete sequence and genome organization of the mitochondrial genome of the land snail *Albinaria coerulea. Genetics* 140, 1353–1366.

Heath, D.J. (1975) Colour, sunlight and internal temperatures in the land-snail *Cepaea nemoralis* (L.). *Oecologia* 19, 29–38.

Heller, J. (1981) Visual versus climatic selection in the land snail *Theba pisana* in Israel. *Journal of Zoology* 194, 85–101.

Hillis, D.M. (1989) Genetic consequences of partial self-fertilization on populations of *Liguus fasciatus* (Mollusca: Pulmonata: Bulimulidae). *American Malacological Bulletin* 7, 7–12.

Hillis, D.M., Dixon, M.T. and Jones, A.L. (1991) Minimal genetic variation in a morphologically diverse species (Florida tree snail, *Liguus fasciatus*). *Journal of Heredity* 82, 282–286.

Hoffmann, R.J. (1983) The mating system of the terrestrial slug *Deroceras laeve. Evolution* 37, 423–425.

Honda, C. and Asami, T. (1993) Temporal and spatial variations of genetic structure of populations in the land snail *Bradybaena similaris. Journal of the Tachikawa College, Tokyo* 26, 103–112.

Husted, L. and Burch, P.R. (1953) The chromosomes of the polygyrid snail *Allogona profunda. Virginia Journal of Science* 4, 62–64.

Ikeda, K. (1937) Cytogenetic studies on the self-fertilization of *Philomycus bilineatus* Benson. (Studies of hermaphroditism in Pulmonata II.) *Journal of Science of the Hirosima University, Series B, Division* 1, 5, 67–123.

Jarne, P. (1995) Mating system, bottlenecks and genetic polymorphism in hermaphroditic animals. *Genetical Research, Cambridge*, 65, 193–207.

Jarne, P. and Städler, T. (1995) Population genetic structure and mating system evolution in freshwater pulmonates. *Experientia* 51, 482–497.

Johnson, M.S. (1979) Inheritance and geographic variation of allozymes in *Cepaea nemoralis*. *Heredity* 43, 137–141.

Johnson, M.S. (1982) Polymorphism for direction of coil in *Partula suturalis*: behavioural isolation and positive frequency dependent selection. *Heredity* 49, 145–151.

Johnson, M.S. (1987) Adaptation and rules of form: chirality and shape in *Partula suturalis*. *Evolution* 41, 672–675.

Johnson, M.S. (1988) Founder effects and geographic variation in the land snail *Theba pisana*. *Heredity* 61, 133–142.

Johnson, M.S., Clarke, B. and Murray, J. (1977) Genetic variation and reproductive isolation in *Partula*. *Evolution* 31, 116–126.

Johnson, M.S., Stine, O.C. and Murray, J. (1984) Reproductive compatibility despite large-scale genetic divergence in *Cepaea nemoralis*. *Heredity* 53, 655–665.

Johnson, M.S., Murray, J. and Clarke, B. (1986a) High genetic similarities and low heterozygosities in land snails of the genus *Samoana* from the Society islands. *Malacologia* 27, 97–106.

Johnson, M.S., Murray, J. and Clarke, B. (1986b) Allozymic similarities among species of *Partula* on Moorea. *Heredity* 56, 319–327.

Johnson, M.S., Murray, J. and Clarke, B. (1993) The ecological genetics and adaptive radiation of *Partula* on Moorea. In: Futuyma, D. and Antonovics, J. (eds) *Oxford Surveys in Evolutionary Biology*, Vol. 9. Oxford University Press, Oxford, pp. 167–238.

Jones, J.S. (1973) Ecological genetics and natural selection in molluscs. *Science* 182, 546–552.

Jones, J.S., Leith, B.H. and Rawlings, P. (1977) Polymorphism in *Cepaea*: a problem with too many solutions? *Annual Review of Ecology and Systematics* 8, 109–143.

Jones, J.S., Selander, R.K. and Schnell, G.D. (1980) Patterns of morphological and molecular polymorphism in the land snail *Cepaea nemoralis*. *Biological Journal of the Linnean Society* 14, 359–387.

Jordaens, K., De Wolf, H., Verhagen, R. and Backeljau, T. (1996) Possible outcrossing in natural *Carinarion* populations (Mollusca, Pulmonata). In: Henderson, I.F. (ed.) *Slug and Snail Pests in Agriculture*. British Crop Protection Council Symposium Proceedings No. 66, 13–20.

Jordaens, K., Backeljau, T., De Wolf, H., Ondina, P., Reise, H. and Verhagen, R. (1998) Allozyme homozygosity and phally polymorphism in the land snail *Zonitoides nitidus* (Gastropoda, Pulmonata). *Journal of Zoology (London)* 276, 95–104.

Kemperman, T.C.M. and Degenaars, G.H. (1992) Allozyme frequencies in *Albinaria* (Gastropoda Pulmonata: Clausiliidae) from the Ionian islands of Kephallinia and Ithaka. *Malacologia* 34, 33–61.

Kerney, M.P., Cameron, R.A.D. and Jungbluth, J.H. (1983) *Die Landschnecken Nord- und Mitteleuropas*. Verlag Paul Parey, Hamburg.

Komai, T. and Emura, S. (1955) A study of population genetics of the polymorphic land snail *Bradybaena similaris*. *Evolution* 9, 400–418.

Lace, L.A. (1992) Variation in the genitalia of the land snail *Heterostoma paupercula* (Lowe, 1831) (Helicidae) in Madeira. *Biological Journal of the Linnean Society* 46, 115–129.

Lamotte, M. (1951) Recherches sur la structure génétique des populations naturelles de *Cepaea nemoralis* (L.). *Bulletin Biologique de la France et de la Belgique, Supplément*, 35, 1–239.

Lang, A. (1904) Ueber Vorversuche zu Untersuchungen über die Varietäten-bildung von *H. hortensis* und *H. nemoralis*. *Denkschriften der Medizinisch-Naturwissenschaftlichen Gesellschaft zu Jena* 11, 439–506.

Lang, A. (1906) Ueber die Mendelschen Gesetze, Art- und Varietätbildung, Schnecken und Variation, insbesondere bei unseren Hain- und Gartenschnecken. *Verhandlungen der Schweizerischen Naturforschenden Gesellschaft Luzern* 88, 209–254.

Lecanidou, R., Douris, V. and Rodakis, G.C. (1994) Novel features of metazoan mtDNA revealed from sequence analysis of three mitochondrial DNA segments of the land snail *Albinaria turrita* (Gastropoda: Clausiliidae). *Journal of Molecular Evolution* 38, 369–382.

Luther, A. (1915) Zuchtversuche an Ackerschnecken (*Agriolimax reticulatus* Müll. und *Agr. agrestis* L.). *Acta Societatis Pro Fauna et Flora Fennica* 40, 1–42.

Manwell, C. and Baker, C.M.A. (1968) Genetic variation of isocitrate, malate and 6-phosphogluconate dehydrogenases in snails of the genus *Cepaea* – introgressive hybridization, polymorphism and pollution? *Comparative Biochemistry and Physiology* 26, 195–209.

McCracken, G.F. and Brussard, P.F. (1980a) The population biology of the white-lipped land snail *Triodopsis albolabris*: Genetic variability. *Evolution* 34, 92–104.

McCracken, G.F. and Brussard, P.F. (1980b) Self-fertilization in the white-lipped land snail *Triodopsis albolabris*. *Biological Journal of the Linnean Society* 14, 429–434.

McCracken, G.F. and Selander, R.K. (1980) Self-fertilization and monogenic strains in natural populations of terrestrial slugs. *Proceedings of the National Academy of Sciences USA* 77, 684–688.

Mogie, M. (1986) Automixis: its distribution and status. *Biological Journal of the Linnean Society* 28, 321–329.

Moritz, C. (1994) Applications of mitochondrial DNA analysis in conservation: a critical review. *Molecular Ecology* 3, 401–411.

Murray, E. (1993) The sinister snail. *Endeavour* 17, 78–83.

Murray, J. (1964) Multiple mating and the effective population size in *Cepaea nemoralis*. *Evolution* 18, 284–291.

Murray, J. (1975) The genetics of the Mollusca. In: King, R.C. (ed.) *Handbook of Genetics*, Vol. 3, *Invertebrates of Genetic Interest*. Plenum Press, New York, pp. 3–31.

Murray, J. and Clarke, B. (1968) Inheritance of shell size in *Partula*. *Heredity* 23, 189–198.

Murray, J. and Clarke, B. (1976a) Supergenes in polymorphic land snails. I. *Partula taeniata*. *Heredity* 37, 253–269.

Murray, J. and Clarke, B. (1976b) Supergenes in polymorphic land snails. II. *Partula suturalis*. *Heredity* 37, 271–282.

Murray, J. and Clarke, B. (1978) Changes of gene-frequency in *Cepaea nemoralis* over 50 years. *Malacologia* 17, 317–330.

Murray, J. and Clarke, B. (1984) Movement and gene flow in *Partula taeniata*. *Malacologia* 25, 343–348.

Murray, J., Stine, O.C. and Johnson, M.S. (1991) The evolution of mitochondrial DNA in *Partula*. *Heredity* 66, 93–104.

Nicklas, N.L. and Hoffmann, R.J. (1981) Apomictic parthenogenesis in a hermaphroditic terrestrial slug, *Deroceras laeve* (Müller). *Biological Bulletin* 160, 123–135.

Noble, L.R. and Jones, C.S. (1996) A molecular and ecological investigation of the large arionid slugs of North-West Europe: the potential for new pest species. In: Symondson, W.O.C. and Liddell, J.E. (eds) *The Ecology of Agricultural Pests: Biochemical Approaches.* Chapman & Hall, London.

Nunney, L. (1993) The influence of mating system and overlapping generations on effective population size. *Evolution* 47, 1329–1341.

Nunney, L. and Elam, D.R. (1994) Estimating the effective population size of conserved populations. *Conservation Biology* 8, 175–184.

O'Brien, S.J. and Mayr, E. (1991) Bureaucratic mischief: recognizing endangered species and subspecies. *Science* 251, 1187–1188.

Oosterhoff, L.M. (1977) Variation in growth rate as an ecological factor in the landsnail *Cepaea nemoralis* (L.). *Netherlands Journal of Zoology* 27, 1–132.

Owen, D.F. and Reid, J.C. (1986) The white snails of Africa: the significance of man in the maintenance of a striking polymorphism. *Oikos* 46, 267–269.

Oxford, G.S. (1973) The genetics of *Cepaea* esterases. I. *Cepaea nemoralis*. *Heredity* 30, 127–139.

Oxford, G.S. (1975) Food induced esterase phenocopies in the snail *Cepaea nemoralis*. *Heredity* 35, 361–370.

Oxford, G.S. (1978) The nature and distribution of food-induced esterases in helicid snails. *Malacologia* 17, 331–339.

Patterson, C.M. (1969) Chromosomes of molluscs. *Proceedings of the Symposium on Mollusca held at Cochin*, Part II, 635–686.

Patterson, C.M. (1973) Cytogenetics of gastropod mollusks. *Malacological Review* 6, 141–150.

Patterson, C.M. and Burch, J.B. (1978) Chromosomes of pulmonate molluscs. In: Fretter, V. and Peake, J. (eds) *Pulmonates*, Vol. 2A, *Systematics, Evolution and Ecology.* Academic Press, New York, pp. 171–217.

Pearce-Kelly, P., Mace, G.M. and Clarke, D. (1995) The release of captive bred snails (*Partula taeniata*) into a semi-natural environment. *Biodiversity and Conservation* 4, 645–663.

Pfenninger, M., Frye, M., Bahl, A. and Streit, B. (1995) Discrimination of three conchologically similar Helicellinae (*Helicella*, Gastropoda) species using RAPD-fingerprinting. *Molecular Ecology* 4, 521–522.

Pfenninger, M., Bahl, A. and Streit, B. (1996) Isolation by distance in a population of a small land snail *Trochoidea geyeri*: evidence from direct and indirect methods. *Proceedings of the Royal Society of London, Series B* 263, 1211–1217.

Pokryszko, B.M. (1987) On the aphally in the Vertiginidae (Gastropoda: Pulmonata: Orthurethra). *Journal of Conchology* 32, 365–375.

Pokryszko, B.M. (1990) Life history and population dynamics of *Vertigo pusilla* O.F. Müller, 1774 (Gastropoda: Pulmonata: Vertiginidae), with some notes on shell and genital variability. *Annales Zoologici* 43, 407–432.

Posthuma, L. and Van Straalen, N.M. (1993) Heavy-metal adaptation in terrestrial invertebrates: a review of occurrence, genetics, physiology and ecological consequences. *Comparative Biochemistry and Physiology* 106C, 11–38.

Price, D.J. (1974a) Variation in chiasma frequency in *Cepaea nemoralis*. *Heredity* 32, 211–217.

Price, D.J. (1974b) Differential staining of meiotic chromosomes in *Cepaea nemoralis* (L.). *Caryologia* 27, 211–216.

Price, D.J. (1975a) Chiasma frequency variation with altitude in *Cepaea hortensis* (Müll.). *Heredity* 35, 221–229.

Price, D.J. (1975b) Chiasma frequency variation in *Cepaea hortensis* (Müll.) and a comparison with *C. nemoralis* (L.). *Genetica* 45, 497–508.

Price, D.J. (1975c) Position and frequency distribution of chiasmata in *Cepaea nemoralis* (L.). *Caryologia* 28, 261–268.

Price, D.J. and Bantock, C.R. (1975) Marginal populations of *Cepaea nemoralis* (L.) on the Brendon Hills, England. II. Variation in chiasma frequency. *Evolution* 29, 278–286.

Reise, H. (1997) *Deroceras juranum* – a Mendelian colour morph of *D. rodnae* (Gastropoda: Agriolimacidae). *Journal of Zoology* 271, 103–115.

Reise, H., Backeljau, T. and Lieckfeldt, E. (1996) Verhaltensbeobachtungen und molekulargenetische Methoden zur Untersuchung von Taxonomie und Reproduktionsbiologie terrestrischer Nacktschnecken. *Abhandlungen und Berichte des Naturkundemuseums Görlitz* 69, 191–198.

Riddoch, B.J. (1993) The adaptive significance of electrophoretic mobility in phosphoglucose isomerase (PGI). *Biological Journal of the Linnean Society* 50, 1–17.

Roebuck, W.D. (1913) Perfect albinism in *Limax arborum*. *Journal of Conchology* 14, 92.

Rojas, M. (1992) The species problem and conservation: what are we protecting. *Conservation Biology* 6, 170–178.

Runham, N.W. (1993) Mollusca. In: Adiyodi, K.G. and Adiyodi, R.G. (eds) *Reproductive Biology of Invertebrates*, Vol. VI, *Part A. Asexual Propagation and Reproductive Strategies*. John Wiley & Sons, Chichester, pp. 311–383.

Schilthuizen, M. and Gittenberger, E. (1994) Parallel evolution of an sAat-'hybrizyme' in hybrid zones in *Albinaria hippolyti* (Boettger). *Heredity* 73, 244–248.

Schilthuizen, M. and Lombaerts, M. (1994) Population structure and levels of gene flow in the Mediterranean land snail *Albinaria corrugata* (Pulmonata: Clausiliidae). *Evolution* 48, 577–586.

Schilthuizen, M. and Lombaerts, M. (1995) Life on the edge: a hybrid zone in *Albinaria hippolyti* (Gastropoda: Clausiliidae) from Crete. *Biological Journal of the Linnean Society* 54, 111–138.

Schrag, S.J. and Read, A.F. (1996) Loss of male outcrossing ability in simultaneous hermaphrodites: phylogenetic analyses of pulmonate snails. *Journal of Zoology* 238, 287–299.

Schwabl, G. and Murray, J. (1970) Electrophoresis of proteins in natural populations of *Partula* (Gastropoda). *Evolution* 24, 424–430.

Selander, R.K. and Hudson, R.O. (1976) Animal population structure under close inbreeding: the land snail *Rumina* in southern France. *American Naturalist* 110, 695–718.

Selander, R.K. and Kaufman, D.W. (1973) Self-fertilization and genetic population structure in a colonizing land snail. *Proceedings of the National Academy of Sciences USA* 70, 1186–1190.

Selander, R.K. and Kaufman, D.W. (1975) Genetic structure of populations of the brown snail (*Helix aspersa*). I. Microgeographic variation. *Evolution* 29, 385–401.

Selander, R.K. and Ochman, H. (1983) The genetic structure of populations as illustrated by molluscs. In: Rattazzi, M.C., Scandalios, J.C. and Whitt, G.S. (eds) *Isozymes: Current Topics in Biological and Medical Research*, Vol. 10, *Genetics and Evolution*. Alan R. Liss, New York, pp. 93–123.

Shibata, S., Asami, T. and Shimamura, R. (1993) Is the shell polymorphism controlled by a supergene in the land snail *Bradybaena similaris*? *Journal of the Tachikawa College, Tokyo* 26, 95–102.

Simberloff, D. (1988) The contribution of population and community biology to conservation science. *Annual Review of Ecology and Systematics* 19, 473–511.

Simroth, H. (1885) Versuch einer Naturgeschichte der deutschen Nacktschnecken und ihrer europäischen Verwandten. *Zeitschrift für Wissenschaftliche Zoologie* 42, 203–366.

Stine, O.C. (1989) *Cepaea nemoralis* from Lexington, Virginia: the isolation and characterization of their mitochondrial DNA, the implications for their origin and climatic selection. *Malacologia* 30, 305–315.

Taylor, J.W. (1912) *Monograph of the Land and Freshwater Mollusca of the British Isles*, Vol. 3, Part 19. Taylor Brothers, Leeds, pp. 369–416.

Terrett, J., Miles, S. and Thomas, R.H. (1994) The mitochondrial genome of *Cepaea nemoralis* (Gastropoda: Stylommatophora): gene order, base composition, and heteroplasmy. *Nautilus, Supplement* 2, 79–84.

Terrett, J., Miles, S. and Thomas, R.H. (1996) Complete DNA sequence of the mitochondrial genome of *Cepaea nemoralis* (Gastropoda: Pulmonata). *Journal of Molecular Evolution* 42, 160–168.

Thomaz, D., Guiller, A. and Clarke, B. (1996) Extreme divergence of mitochondrial DNA within species of pulmonate land snails. *Proceedings of the Royal Society of London, Series B* 263, 363–368.

Thornhill, N.W. (ed.) (1993) *The Natural History of Inbreeding and Outbreeding: Theoretical and Empirical Perspectives*. University of Chicago Press, Chicago.

Tompa, A.S. (1984) Land snails (Stylommatophora). In: Tompa, A.S., Verdonk, N.H. and Van Den Biggelaar, J.A.M. (eds) *The Mollusca*, Vol. 7, *Reproduction*. Academic Press, Orlando, Florida, pp. 47–140.

Tonge, S. and Bloxam, Q. (1991) A review of the captive-breeding programme for Polynesian tree snails. *International Zoo Yearbook* 30, 51–59.

Triggs, S.J. and Sherley, G.H. (1993) Allozyme genetic diversity in *Placostylus* land snails and implications for conservation. *New Zealand Journal of Zoology* 20, 19–33.

Vogler, A.P. and DeSalle, R. (1994) Diagnosing units of conservation management. *Conservation Biology* 8, 354–363.

Wall, S., Carter, M.A. and Clarke, B. (1980) Temporal changes of gene frequencies in *Cepaea hortensis*. *Biological Journal of the Linnaean Society* 14, 303–317.

Watson, H. (1934) Genital dimorphism in *Zonitoides*. *Journal of Conchology* 20, 33–42.

Williamson, M. (1959) Studies on the colour and genetics of the black slug. *Proceedings of the Royal Physical Society of Edinburgh* 27, 87–93.

Williamson, P., Cameron, R.A.D. and Carter, M.A. (1977) Population dynamics of the landsnail *Cepaea nemoralis* (L.): a six-year study. *Journal of Animal Ecology* 46, 181–194.

Wilson, I.F. (1996) Application of ecological genetics techniques to test for selection by habitat on allozymes in *Cepaea nemoralis* (L.). *Heredity* 77, 324–335.

Wirth, T., Baur, A. and Baur, B. (1997) Mating system and genetic variability in the simultaneously hermaphroditic terrestrial gastropod *Balea perversa* on the Baltic island of Öland, Sweden. *Hereditas* 126, 199–209.

Wolda, H. (1969) Fine distribution of morph frequencies in the snail, *Cepaea nemoralis* near Groningen. *Journal of Animal Ecology* 38, 305–327.

Woodruff, D.S. (1989) Genetic anomalies associated with *Cerion* hybrid zones: the origin and maintenance of new electrophoretic variants called hybrizymes. *Biological Journal of the Linnaean Society* 36, 281–294.

Woodruff, D.S. and Gould, S.J. (1980) Geographic differentiation and speciation in *Cerion* – a preliminary discussion of patterns and processes. *Biological Journal of the Linnean Society* 14, 389–416.

Woodruff, D.S. and Solem, A. (1990) Allozyme variation in the Australian camaenid land snail *Cristilabrum primum*: a prolegomenon for a molecular phylogeny of an extraordinary radiation in an isolated habitat. *Veliger* 33, 129–139.

Wright, S. (1931) Evolution in Mendelian populations. *Genetics* 16, 97–159.

Wright, S. (1943) Isolation by distance. *Genetics* 28, 114–138.

Wright, S. (1946) Isolation by distance under diverse systems of mating. *Genetics* 31, 39–59.

Wright, S. (1969) *The Theory of Gene Frequencies*, Vol. 2. University of Chicago Press, Chicago.

Wüthrich, M. (1993) *Deroceras (Plathystimulus) juranum* n. sp., eine endemische Nacktschnecke aus dem Schweizer Jura. *Archiv für Molluskenkunde* 122, 123–131.

Yamazaki, N., Ueshima, R., Terrett, J.A., Yokobori, S.-i., Kaifu, M., Segawa, R., Kobayashi, T., Numachi, K.-i., Ueda, T., Nishikawa, K., Watanabe, K. and Thomas, R.H. (1997) Evolution of pulmonate gastropod mitochondrial genomes: comparisons of gene organizations of *Euhadra*, *Cepaea* and *Albinaria* and implications of unusual tRNA secondary structures. *Genetics* 145, 749–758.

Zilch, A. (1960) *Gastropoda Euthyneura*. Handbuch der Paläozoologie Band 6. Gebrüder Borntraeger, Berlin-Nikolassee.

12 Life History Strategies

J. HELLER

Department of Evolution, Systematics and Ecology, The Hebrew University, Jerusalem 91904, Israel

Introduction

Terrestrial gastropods constitute a highly diverse group of animals. Early estimates (Solem, 1978), of about 24,000 species (of which about 20,500 are pulmonate stylommatophorans and about 3650 belong to other groups, mainly the caenogastropods and Neritopsina) are almost certainly an underestimate, because recent studies have revealed a previously unknown high diversity in tropical forests. De Winter and Gittenberger (1998), for example, have recently found 97 species within a single square kilometre of rainforest in Cameroon. Terrestrial gastropods also occupy a wide range of terrestrial habitats, from soils and subterranean caverns, litter and low vegetation, to rock surfaces, epiphytes and tree canopies. Variation in life history traits of terrestrial gastropods is enormous, and some aspects of this diversity of life cycles have been discussed in previous reviews. Peake (1978) concentrated on the distribution and ecology of stylommatophorans and included aspects of self-fertilization, variation in life cycles and egg/clutch size. Cain's (1983) review focused on the ecology and ecogenetics of terrestrial gastropods and referred to variation in life history and population dynamics. Calow (1983) presented an overview of life cycle patterns in molluscs, with emphasis on marine and freshwater (rather than land-) gastropods. Tompa (1984) reviewed various aspects of reproduction in Stylommatophora including courtship and mating, egg laying, and developmental strategies (oviparity versus ovoviviparity). South (1992) has presented a review on the biology of European slug forms of terrestrial Gastropoda, with reference to life cycles. Baur (1994a) reviewed parental care in terrestrial gastropods, including such aspects as egg laying, egg provisioning, egg cannibalism and egg retention.

This chapter provides a synthesis of the accumulated knowledge concerning terrestrial gastropod life history, mainly in reference to such aspects as outcrossing versus selfing, random versus non-random mating, egg size versus egg numbers, egg survival, oviparity versus ovoviviparity, semelparity versus iteroparity and short versus long life spans.

Outcrossing Versus Selfing

Terrestrial prosobranchiates are gonochoristic and reproduce by outcrossing (Creek, 1951, 1953; Kasinathan, 1975; Fretter and Graham, 1994), whereas terrestrial pulmonates are simultaneous hermaphrodites and most breed by cross-fertilization. In most terrestrial pulmonates, courtship and copulation are reciprocal, with both animals acting as males and females at the same time (Tompa, 1984). However, in some species, the two animals might have slightly different courtship roles: one 'male-behaving' (active) and the other 'female-behaving' (the passive role) (Tompa, 1984; see also Lipton and Murray, 1979). Yet other stylommatophorans have developed the ability to self-fertilize. Heller (1993) lists 18 genera (of 12 families) as capable of self-fertilization. The ability to self-fertilize is of considerable value as a last resort when, due to population decline, chances for cross-fertilization decrease: the animal is then able to contribute to perpetuation of the species, by selfing. Selfing can also increase the colonizing ability of the species by assuring reproduction at low population density, reduce the cost of male allocation and keep co-adapted genes together. The disadvantages may include an immediate fitness reduction caused by inbreeding depression, and low genetic diversity resulting from reduced recombination (Chen, 1993, 1994, and references therein). For simultaneous hermaphrodites, the option to reproduce through self-fertilization or cross-fertilization is, therefore, one of the most important life history traits.

In general, self-fertilization is concluded to occur either from observation of reproduction in animals isolated from birth, or by use of allozyme electrophoresis, on the assumption that genetic diversity reflects variation in the mode of reproduction. The frequency of selfing varies greatly among species. In some species, it is absent or very rare, in others it occurs occasionally, whereas in some species self-fertilization occurs regularly.

One case of self-fertilization involves *Rumina decollata* (Linnaeus) (Subulinidae). In its native Mediterranean range, it forms a complex of monogamic strains maintained, at least in part, by a breeding system of facultative self-fertilization. One single strain colonized North America and now occupies much of the southern USA (Selander and Kaufman, 1973). Another case of self-fertilization, in the genus *Arion* de Férussac (Arionidae) in the British Isles, was studied by Foltz *et al.* (1982). They found that members of this genus fall into three categories on the basis of their breeding systems. *Arion lusitanicus* Mabille, *Arion hortensis* de Férussac, *Arion distinctus* Mabille and *Arion owenii* Davies reproduce predominantly, if not exclusively, by outcrossing; *Arion circumscriptus* Johnston, *Arion silvaticus* Lohmander and *Arion intermedius* Normand reproduce by frequent self-fertilization; whereas *Arion ater* (Linnaeus) and *Arion subfuscus* (Draparnaud) include both outcrossing and self-fertilizing forms, that locally hybridize. Foltz *et al.* (1984) found that selfing is more widespread in the Arionidae than in the Limacidae and Agriolimacidae. They also found that seven out of eight self-fertilizing species of the British fauna have colonized eastern North America. This suggests that 'selfers' may be better colonizers than 'outcrossers'. Foltz *et al.* (1984) found no relationship between the breeding system and extent of ecological amplitude: outcrossing species (such as *Deroceras reticulatum* Müller) (Agriolimacidae) did not occupy a greater diversity of habitats than self-fertilizing species, such as *Arion fasciatus* (Nilsson). Chen, Jiaxiang and Jicheng (in South, 1992) found *Deroceras agreste* (Linnaeus) to be two to four times more fecund, and the growth rate of progeny higher (and the life span longer) for animals breeding by self-fertilization rather than by outcrossing. Also, *Deroceras laeve* (Müller) can reproduce in the absence of a mate; Hoffmann's (1983) suggestion that this may probably be by parthenogenesis rather than by self-fertilization was criticized by Foltz *et al.* (1984). *Deroceras rodnae* Grossu & Lupu is capable of uniparental reproduction by self-fertilization rather than partheno-genesis, as animals which mutually exchange spermatozoa may produce outcrossed and/or selfed eggs (Reise and Backeljau, 1995). In Pacific island *Partula* de Férussac (Partulidae), at least five species are capable of self-fertilization, but this is a rare event. In breeding experiments with *Partula taeniata* Mörch, Murray and Clarke (1966, 1976) found that self-ferilization is responsible for about 20% of the progeny produced in the early part of reproductive life, but its contribution declined to undetect-able levels later, and averaged about 2% throughout an individual's life. Species of Partulidae, and individuals within species, may differ in their capacity to self (Murray and Clarke, 1966; Johnson *et al.*, 1977, 1986; Kobayashi and Hadfield, 1996). In the polygyrid *Triodopsis albolabris* (Say), populations generally breed by outcrossing. However, if the proba-bility of finding a mate is low (assessed by experimental isolation for several months), then lone individuals may self-fertilize, though the reproductive output of isolated virgin animals is much lower than that of paired individuals who have mated (McCracken and Brussard, 1980). Helicids were long believed to be self-incompatible, but a recent study has revealed uniparental reproduction in *Arianta arbustorum* (Linnaeus) under laboratory conditions (Chen, 1993, 1994). Over a period of 3 years after maturation, 39% of an experimental group of unmated animals produced fertile eggs, mostly in the second and third years. The number of hatch-lings produced was 1–2% of that from mated animals, although survival was similar for progeny from mated and unmated animals. These results indicate that *A. arbustorum* can self-fertilize, but with a great fitness reduction (Chen, 1993, 1994). In general, among terrestrial gastropods, selfing is less common than in freshwater gastropods and bivalves (see

Heller, 1993; Jarne *et al.*, 1993). That many succineids are capable of selfing (Patterson, 1970) is a noteworthy point, because members of this family frequently live in permanently wet places (such as marshes and the margins of lakes and rivers) and many species are virtually amphibious (Kerney and Cameron, 1979); their habitat is close to that of freshwater gastropods.

Mating

Mating may be indiscriminate (random), or may involve mate choice, in which case individuals are more likely to mate with certain members of their species than with others. In theory, random mating should occur where there is little variance in mate quality or when search costs for mates are high (Parker, 1983). In terrestrial gastropods, there is considerable variance in mate quality, partly reflected in the positive relationship between fecundity and animal size, found within some species (e.g. *A. arbustorum*, Baur and Raboud, 1988). Within populations, the spatial distribution of individuals tends to be aggregated. Thus, on average, search costs for mates may be low when local abundance is high, but may vary considerably between populations with different abundances.

Many terrestrial gastropod species apparently mate at random. Mating in *Cepaea nemoralis* (Linnaeus) (Helicidae) is random with respect to size, colour and banding patterns of shells of conspecifics (Wolda, 1963). In *Helix pomatia* Linnaeus (Helicidae), in one natural population, some animals showed a slight (but non-significant) tendency towards size-assortive mating (Baur, 1992) while, in another, mating between resident animals and a newly introduced animal was random (Woyciechowski and Lomnicki, in Tompa, 1984). In *A. arbustorum*, both in a natural population and in laboratory mate-choice tests, Baur (1992) found that pairs formed randomly with respect to size; in pairs with large size differences, courtship was neither hindered nor prolonged, and when a large animal was placed close to two courting, smaller conspecifics, it did not displace one of them. Baur (1992) proposed that, because of the time-constrained activity and the high costs of locomotion, the best strategy for terrestrial gastropods is to mate with any conspecific adult encountered, so as to minimize the risk of complete mating failure and/or to avoid desiccation during mate search.

Other stylommatophoran species mate non-randomly with respect to age and size. *Achatina fulica* Bowdich (Achatinidae) has indeterminate growth and is protandrous, with smaller, still-growing 'young adults' producing only spermatozoa, and larger, fully-grown 'old adults' producing both spermatozoa and ova (Tomiyama, 1993). A field study in the Bonin Islands revealed that most of the animals that began courtship were rejected by their opponents (Tomiyama, 1994), thereby suggesting that animals may be choosing their mates (Tomiyama, 1996). This study also revealed age-dependent mate choice, in that young adults preferred old adults as mating partners (Tomiyama, 1996). Some mating between young adults did occur, however, indicating that *A. fulica* can store allosperma-tozoa (see Raut and Chose, 1982) for subsequent fertilization of ova.

In several families (e.g. Enidae, Clausiliidae and Helicidae), growth is determinate. Thus the helicid *Theba pisana* (Müller) is usually consid-ered to be mature or sexually active only when fully grown and with a thickened lip at the shell aperture. However, on some (very infrequent) occasions, one of the mating partners may lack a shell lip and be of a very small size and early age (Cowie, 1980); such individuals might be considered as corresponding to the 'young adult' stage of *A. fulica.*

Partula suturalis Pfeiffer is somewhat unusual among gastropod snails in being polymorphic for direction of shell coiling. Polymorphic populations of *P. suturalis* occur in narrow clines between areas of monomorphic dextral or sinistral populations (Clarke and Murray, 1969). In no-choice experiments, pairs with opposite coil mated only 20% as frequently as did pairs with the same coil configuration, and produced fewer young (Johnson, 1982). The rarity of copulation in mixed pairs was due not to infrequent courtship, but to incompatible locations of genital apertures (Lipton and Murray, 1979).

Some species of terrestrial gastropods mate repeatedly in the course of a reproductive season or reproductive life span, even though the supply of spermatozoa stored from a previous mating may not have been depleted (Tompa, 1984). In *C. nemoralis*, up to 20 matings have been observed in pairs kept in the laboratory (Wolda and Kreulen, 1973). Multiple mating has also been observed in *Caracolus caracollus* (Linnaeus), a camaeid of rainforests in Puerto Rico (Heatwole and Heatwole, 1978). In *Cantareus aspersus* (Müller) (Helicidae) reared in the laboratory, mating occurs two or three times per season (Madec and Daguzan, 1993) and in *H. pomatia* five to six times per year (Chen and Baur, 1993). Multiple mating with different partners results in multiple parentage, as shown in *C. nemoralis* (Murray, 1964) and *A. arbustorum* (Baur, 1994c), and may provide a reservoir of genetic diversity in situations of population depletions (Murray, 1964). Additionally, repeated mating may increase fecundity. In *A. arbustorum*, animals allowed to mate repeatedly within a season tended to lay more eggs than animals allowed to mate only once; however, the eggs laid by the repeatedly mating animals had a lower hatching suc-cess (Chen and Baur, 1993). In the wild, *A. arbustorum* mates repeatedly throughout the season even though each individual receives sufficient spermatozoa in one mating to fertilize all the ova produced. Baur and Baur (1992) found that the first clutch laid after a second or repeated copulation contained more eggs than in the first clutch of the season, suggesting that repeated copulation in the course of a reproductive season may lead to an increased number of eggs produced.

Mate choice is only one of the many factors that may determine the distribution of mating among individuals in a breeding population. Another important variant is the propensity to mate. In *C. aspersus*, high mate propensity is associated with large shell size and high fecun-dity (Fearnley, 1996). In mate-choice experiments between resident and non-resident individuals, Fearnley (1996) showed that animals from different sites varied in their readiness to mate – animals of one population started mating much earlier than those from other populations. We do not know how generally applicable these findings are, but variation in mating propensity has also been found in several other helicid species, such as *C. nemoralis, H. pomatia* and *A. arbustorum* (Fearnley, 1996).

Terrestrial gastropods have such poor dispersal abilities that it is reasonable to assume that at the level of a local population, mating is predominantly by inbreeding. This strategy may provide for maintenance of adaptations to local conditions and may hold co-adapted gene complexes together (Partridge, 1983).

Oviposition and Egg Survival

Most terrestrial gastropod species are oviparous. After ovulation, the ova are fertilized and each zygote is embedded in a coating of nutritive albumen, and receives a protective coating, before being laid. More ova are usually shed than are laid in eggs. Thus in *Vaginula* de Férussac (Vaginulidae), approximately 400 oocytes are shed from the ovotestis but only 25–225 are actually utilized in deposited eggs. The excess ova, some of which may be abnormal, are resorbed (Runham and Hunter, 1970). Egg resorption, as a means of retaining nutrients, may well be a common strategy in terrestrial gastropods.

Sites of egg laying vary considerably. Many zonitids drop their eggs singly amongst moss or herbage (Stanlen, 1917). *Zonitoides arboreus* (Say) lays single eggs scattered in forest litter, and Baur (1994a) speculated that in this facultative predatory species, this egg-laying strategy might perhaps have evolved to minimize cannibalism among offspring. For oviposition, some gastropod species utilize natural holes or crevices in the soil or under stones and pieces of wood (e.g. South, 1992). Stanlen (1917) noted that *Testacella haliotidea* Draparnaud (Testacellidae), a subterranean predator, deposits its eggs in subterranean galleries, sometimes a metre or more beneath the surface. In others, the animals first enlarge or excavate the holes in the soil, and a few species actively make a nest (Runham and Hunter 1970; Tompa, 1984; South, 1992). For example, by movements of the foot, *C. nemoralis* hollow out a more or less globular cavity in the soil into which, after some hours of rest, eggs are deposited, over a period lasting up to 3 days (Wolda, 1963). After the last egg of the clutch has been produced, the animal covers the 'nest' with soil, by movements of its foot. Similarly, *H. pomatia* digs a hole of about 6 cm, inserts its body deep into the hole for oviposition and, on completion, the hole is covered with soil and abandoned (Pollard, 1975). *Sphincterochila zonata* (Bourguignat) (Sphincterochilidae) digs a 3–4 cm deep hole and lays its eggs in a mucus sac that hangs in the cavity without touching its walls (Shachak and Phillips, 1971; Yom Tov, 1971). Yom Tov (1971) suggested that this saccule assists embryo survival by reducing egg desiccation and infection by soil fungi. *Archachatina marginata* (Swainson) (Achatinidae), which may reach a shell length of 130–160 mm, may dig a 10–15 cm deep cavity for oviposition in the course of a single night (Plummer, 1975). In some species, each animal digs several holes, but abandons most before one hole eventually is considered suitable for oviposition (Wolda, 1963, 1965; Pollard, 1975). For example, only 10% of the holes excavated by *T. pisana* may be used for oviposition (Heller, 1982). The fastidious nest hole requirements can lead to migratory activity. *H. pomatia*, for example, may move up to 15 m in search of soft soil that is suitable for excavation and egg laying (Pollard, 1975). In the closely related species, *Helix texta* Mousson, however, the animals move only a few metres at most (Heller and Ittiel, 1990).

On the other hand, some gastropods are tree dwellers and consequently lay their eggs in various arboreal sites. For example, the neotropical *Pseudachatina downesi* (Sowerby) (Achatinidae) and *Polydontes acutangula* (Burrow) (Camaenidae) deposit their eggs in the axils of branches of the trees on which they dwell, at microsites which tend to collect water in the base (Heatwole and Heatwole, 1978; Tompa, 1984). *Placostylus miltochelus* (Reeve) (Bulimulidae) places eggs into detritus that collects in the leaf axils of certain plants, while *Cryptaegis pilsbryi* Clapp (Camaenidae) deposits its eggs on the surface of leaves (Tompa, 1984; Baur, 1994a). A more specialized oviposition behaviour is exhibited by *Helicostyla* de Férussaac (Camaenidae). The Philippines species *Helicostyla mindroensis* Broderip deposits its eggs upon a leaf in parallel rows, each standing perpendicularly on end, attached at the base by a glutinous substance (Stanlen, 1917). *Helicostyla pithogaster* (de Férussac), also of the Philippines, deposits eggs in leaves rolled into cornucopia, stuck together and lined with mucus; one to three leaves are used for each nest. Nest construction and egg laying occur simultaneously as the snail moves backwards towards the leaf tips, depositing eggs in the cavity formed by the pulling and cementing together of the leaf margins. The first and last groups of eggs laid are small, misshapen and non-viable, and dry to form plugs to protect the viable eggs from desiccation (Auffenberg and Auffenberg, 1984). Some arboreal species, such as *Liguus fasciatus* (Müller) (Bulimulida), have been found to come down from the tree to deposit their eggs in the soil or litter (Tuskes, 1981).

Some endodontoids, especially in the Pacific genus *Libera* Garrett (Endodontidae), have the peculiar habit of ovipositing into the wide umbilicus of their own shell, which is constricted to form a pouch: the eggs are retained in the umbilicus by a temporary plate which covers over the umbilical opening. In *Libera fratercula* (Pease), during narrowing of the umbilical brood chamber, the animal gradually vacates the upper whorls of the shell, filling in behind itself with deposits of calcium. After the young hatch within the umbilical chamber, they chew their way into the sides of the parent's shell and eventually emerge through a hole which they create at the apex of the parental shell (Stanlen, 1917; Solem, 1976). Beyond this case (and also an observation on *Vallonia pulchella* (Müller)

(Valloniidae) where adults moved with their head and radula over the surface of an egg, perhaps to remove fungi), parental care of eggs or of young hatchlings is not known to occur among terrestrial gastropods (Baur, 1994a).

Some terrestrial gastropods deposit calcium carbonate crystals in the outer layers of the egg. Such calcium carbonate may consist of discrete crystals dispersed in the jelly matrix (e.g. *H. pomatia*) or occur in such quantities as to form a hard, brittle shell of fused crystals of $CaCO_3$, much like an avian egg (e.g. *C. nemoralis*). Shelled eggs are known to occur in at least 36 stylommatophoran families (Tompa, 1976). As slug forms have much less calcium available in their body than snail forms, it might perhaps be expected that slug eggs would be not calcified. This is only partly so. Most Arionidae do indeed form non-calcified or partly calcified eggs, but those of *A. ater* are heavily calcified. Similarly, Limacidae lay non-calcified eggs, but, among the closely related Agriolimacidae, eggs of *D. agreste* and *D. reticulatum* are speckled with calcium carbonate crystals. At least some members of Milacidae, Philomycidae and Parmacellidae form eggs that are partly calcified, and all Testacellidae deposit heavily calcified eggs (Tompa, 1976). In the Succineidae, and in freshwater basommatophoran pulmonates, calcified eggs are absent (Tompa, 1976). Beyond the pulmonates, calcified deposits are found in the eggs of the prosobranchiate families Helicinidae and Cyclophoridae (Tompa, 1976). In the pomatiasid *Pomatias elegans* (Müller), each egg is laid in a large spherical capsule which the foot covers with a coating of soil, before depositing it beneath the soil surface. The capsule wall is permeable to water and salts, which pass in from the surrounding soil (Creek, 1951).

A calcified egg shell functions to support the large volume of egg albumen. Accordingly, the largest eggs are all heavily calcified, while the non-calcified and partly calcified shells are usually between 1 and 8 mm (Tompa, 1976). The shell also functions to supply the developing embryo with enough calcium to form the embryonic shell, by the time of hatching. This requires the parent animal to access and mobilize calcium for deposition in the eggs. *Anguispira alternata* (Say) (Endodontidae) mobilizes 10–25 mm of calcium, from its own shell, for one egg clutch in less than a day, and *D. reticulatum* loses 65% of the total calcium in its body tissues to produce a clutch of 100 eggs (Baur, 1994a). Heavily calcified eggs occur in putative primitive families such as Partulidae, Endodontidae and Zonitidae, and this led Tompa (1976) to view the calcified egg as a primitive stylommatophoran trait, associated with terrestrial adaptation.

One important environmental criterion for successful egg development is a permanently moist site. Drought is the most common mortality factor for eggs (Tompa, 1984), and oviposition strategy principally revolves around minimizing moisture stress in the developing embryo. Thus, when given a choice of substrates, *D. reticulatum* deposits the maximum number of eggs in soil of about 75% water saturation; no eggs are laid when the soil is at 10% saturation (Runham and Hunter, 1970). Likewise, in helicids such as *H. pomatia* and *C. aspersus*, the soil must be thoroughly moistened for oviposition to occur (Pollard, 1975; Potts, 1975).

The dependence of egg laying on damp conditions is not a simple one and may be modified by interactions with other environmental factors. In the desert-dwelling *Trochoidea simulata* (Ehrenberg) (Hygromiidae), for example, rigid reproductive seasonality (Ward and Slotow, 1992) meant that the animals were incapable of increasing the number of eggs to capitalize on unusual but nevertheless favourable moisture conditions. In *C. nemoralis*, optimal conditions for egg production seem to be short periods of alternating rainy and dry weather with relatively high temperatures (Wolda and Kreulen, 1973). Accordingly, average oviposition frequencies at 12°C are less than half of those at 20°C, and experimental lengthening of the period without rainfall reduces the frequency of oviposition (Wolda, 1965). The influence of temperature on egg laying frequency varies further among animals in populations polymorphic for shell coloration and banding. In animals with yellow-banded shells, oviposition frequency is about twice that of animals with pink-banded shells (Wolda, 1967).

Laboratory observations suggest that desiccation may be a major cause of egg mortality in *H. pomatia* (Pollard, 1975). In field studies in northern Greece, using estimates of differences in abundances of juvenile animals and eggs, 38% mortality occurred in the egg stage of *Helix lucorum* Linnaeus, 35% in *Bradybaena fruticum* (Müller) (Bradybaenidae), and 25% in *Monacha cartusiana* (Müller) (Hygromiidae), presumably because of desiccation (Staikou *et al.*, 1988, 1990; Staikou and Lazaridou-Dimitriadou, 1990). Similarly, in the Negev Desert, 100% of *T. simulata* eggs on a northern slope hatched, but only 46% of those on a drier southern slope (Yom Tov, 1971). Further implication of desiccation as a factor in the survival of eggs come from studies of the desert-dwelling sphincterochilid *S. zonata*. Shachak and Phillips (1971) excavated egg holes periodically to observe the conditions of the eggs inside. They found egg survival to decline markedly with time after oviposition, such that by 3 months, only 12% of the clutches contained viable eggs. In an observation plot in which 518 eggs were laid, only 0.6% developed to yield hatchling animals that emerged at the soil surface (Shachak *et al.*, 1975).

In a comparison of eggs from five terrestrial gastropod species, Bayne (1968) found that differences in desiccation rate were due to differences in size of the clutch – a larger clutch having the advantage of a slower drying rate per unit weight. In *D. reticulatum*, eggs laid at 25% soil water saturation failed to hatch, but embryos could survive weight losses of 60–80% without hatching being delayed (Bayne, 1968).

It may perhaps be expected that terrestrial gastropod eggs are heavily predated. However, data concerning egg predation in the real world are very meagre. Staphilinid and phorid larvae occasionally have been recorded feeding on eggs of *Deroceras* Rafinesque Schmaltz (reviewed in South, 1992). In western Europe, the carnivorous *Oxychilus cellarius* (Müller) (Zonitidae) is found repeatedly in nests of *C. nemoralis*, where it feeds on the eggs (Wolda, 1963). However, in other studies the losses to egg predation have been exceedingly low (e.g. in *S. zonata*; Shachak and Phillips, 1971). Similarly, while fungal attack on eggs is rather common under laboratory culture conditions, there is little evidence that this type of disease occurs extensively in nature. South (1992) reported that, on average, only 8–9% of *D. reticulatum* eggs and 6–7% of *A. intermedius* eggs taken over 2 years from permanent pasture in the UK contained fungal mycelium or spores. South concluded that the fungus was most probably saprophagous rather than a pathogen, developing only in eggs where the embryo had already died. Shachak and Phillips (1971) found that *S. zonata* eggs supported no fungal growth while the embryos were alive but, when dried clutches were incubated at room temperature, they were quickly covered by hyphae.

Egg Cannibalism

Egg cannibalism has been documented in various stylommatophoran families. Our most detailed knowledge concerns *A. arbustorum* (Baur and Baur, 1986; Baur, 1990a,b, 1992; Bulman, 1995). In this species, egg laying occurs from late spring to autumn, when small holes of about 20 mm depth are dug into the soil or under logs, grass tufts and stones (Baur and Baur, 1986). Those young earliest to hatch first eat their own egg shells and may then devour the still unhatched eggs within the same clutch (Baur, 1987c). Cannibalistic feeding begins only 12 h or more after hatching, perhaps because beforehand the hatchlings are satiated from eating their own egg shell (Baur, 1993). The cannibals consume approximately one egg within 4 days (Baur, 1988a), and in the field up to three hatchlings have been observed feeding on a single egg (Baur, 1993). Clutch size and egg abundance do not affect the degree of cannibalism (Baur and Baur, 1986). Hatchlings occasionally cannibalize eggs from neighbouring clutches (Baur, 1987c), but gravid animals do not react to the presence of conspecific eggs when starting to oviposit and therefore do not select sites which reduce the likelihood of between-clutch cannibalism (Baur, 1988a,b, 1994a). Also, it might perhaps be expected that a hatchling may be 'pre-conditioned' for the taste of its own clutch, via the eating of its own egg shell. However, in choice experiments, hatchlings do not discriminate between sib eggs and eggs from neighbouring clutches (Baur, 1987a). They also do not discriminate between fertilized and unfertilized eggs, or between eggs with well-developed and less advanced embryos (Baur, 1993). Hatchlings preferentially eat large eggs, but detailed choice experiments reveal that this size preference may be due to a higher encounter probability and/or stronger attraction by chemical cues of larger eggs, rather than prey choice based on size (Baur, 1993). Cannibalism seems to be species-specific, since laboratory studies show that cannibalistic hatchlings of *A. arbustorum* do not attack eggs of other species (Baur and Baur, 1986; Baur, 1988a).

The extent of within-clutch egg cannibalism by hatchlings depends primarily on the hatching asynchrony of the eggs. In *A. arbustorum*, when hatching asynchrony was extended experimentally by manipulation of temperature, the extent of egg cannibalism increased (Baur and Baur, 1986).

Hatchlings from different parents and, above all, from different populations, vary enormously in their propensity for cannibalism (Baur, 1987a). In populations of *A. arbustorum* in Switzerland, the percentage of cannibalistic attacks ranged from 50% in a subalpine forest to 88% in a lowland forest (Baur and Baur, 1986; Baur, 1994b). Hatchlings from populations with a high frequency of cannibals also began to eat eggs at an earlier age. This propensity for egg cannibalism was not correlated with egg size, and Baur (1994b) suggested that this variation may be related to the fact that lowland animals produce more but (in relation to their size) smaller eggs that contain less energy and nutrients than mountain animals. Consequently, lowland hatchlings emerge less nourished than do mountain ones. Further, intrinsic hatching success in the lowland was lower. This combination of undernourished hatchlings and of availability of unhatched, heavily yolked eggs would favour a higher tendency for sib cannibalism in the lowland population.

Egg cannibalism in *A. arbustorum* occurs exclusively during the hatchling stage, not in juvenile or adult stages of life. In choice tests, hatchlings with no prior feeding experience chose eggs exclusively, 4-day-old animals ate eggs and lettuce (*Lactuca sativa* Linnaeus) (Asteraceae) in equal proportions, 16-day-old animals preferred vegetable food, while individuals older than 4 weeks fed exclusively on lettuce (Baur, 1987c, 1992). Egg cannibalism in *H. pomatia*, *C. nemoralis* and *D. laeve* also occurs exclusively during the hatchling stage (Baur, 1992; Shen, 1995). This age specificity of oophagy could perhaps have evolved because hatchlings of some species may need some time to build up certain enzymes (such as cellulase) that a herbivore requires for digesting vegetable food – and it is during this period that they depend upon oophagy (Baur, 1987c).

Egg cannibalism may be considered as a reproductive strategy of the parent, whereby some descendants are nourished from parental sources during the first days of their lives. To the cannibalistic individual, this strategy is obviously advantageous. Eggs are rich in energy and nutrients that include proteins, mucopolysaccharides and calcium; the latter is necessary for shell growth (Tompa, 1984). In theory, cannibalism should be favoured at the point in the life cycle where the benefit is at its maximum, namely when the nutritional gains are high and when age-specific mortality is high (Eikwort, 1973). Age-specific mortality in *A. arbustorum* is indeed highest in the egg and hatchling stage (Baur, 1990b), as in various other terrestrial gastropod species (Wolda, 1963; Cain, 1983). The nutritional and energetic benefits of a cannibalistic diet during the hatchling stage become manifest in an accelerated juvenile growth rate, in a tendency to complete shell growth more rapidly and in higher survival than non-cannibalistic species. Thus, newly hatched *A. arbustorum* doubled their weight within 6 days when fed exclusively on conspecific eggs, whereas siblings given vegetarian food increased their weight by only 18% within this period (Baur, 1987a,c). Similarly, the net weight of *H. pomatia* hatchlings fed upon one single conspecific egg increased within 6 days by 87%, whereas those fed on lettuce needed 20 days to achieve an equal increase in weight (Baur, 1990a).

If a diet rich in protein and calcium confers advantage in early life, why do terrestrial gastropods not produce fewer, but larger eggs? In theory, if parents are unable to increase investment in young through increasing egg size, an alternative strategy is to increase clutch size and allow some siblings to consume others (Alexander, 1974). In *A. arbustorum*, clutch size varies considerably. Further, not all eggs within a batch hatch, since some eggs are not fertilized and in others the embryos die during development. Thus, under natural conditions, eggs that fail to hatch quickly serve as food for earlier hatchlings.

Restriction of the cannibalistic propensity to the hatchling stage and differences between clutches suggest that egg cannibalism is a genetically determined trait (Baur, 1987a,c, 1994a).

Egg Size

The size of the egg determines the size of the hatchling and hence subsequent growth and survivorship properties. In a wide variety of animal and plant groups, minor differences in egg size may have proportional effects on fitness. Hatchlings from large eggs generally are more resistant to starvation than hatchlings from small eggs, and embryo size usually is positively related to the survival, growth and eventual breeding success of offspring. However, large egg size can have disadvantages, including lengthy incubation times and higher rates of mortality before hatching (Clutton-Brock, 1991).

Egg size varies enormously among species. In both snail (Fig. 12.1) and slug forms (Fig. 12.2), a significant correlation exists between animal size and egg size. The smallest eggs are of about 0.4 mm (Table 12.1). The largest eggs (51 × 35 mm) belong to *Megalobulimus popelairianus* (Nyst) (Acavidae), a species with a shell height of 150–230 mm that occurs in Colombia and Ecuador (Table 12.1).

Egg size also varies within species. In *Megalobulimus oblongus* (Nyst), for example, egg size varies considerably, and the larger the egg laid, the larger the hatchling (Tompa, 1984). Hatchlings from small eggs (1 g) can survive starvation for only about 18 days, whereas hatchlings that are larger because they hatch from larger eggs (2.5 g) can survive as long as 80 days on their more extensive stored food (Tompa, 1984). In *A. arbustorum* of the Swiss Alps, egg size decreases with increasing altitude, from 3.3 mm at 600 m to 2.4 mm at 2400 m (Baur, 1984). An increase in egg size was found generally to be accompanied by an increase in the nutrient content (A. Baur, 1994), and hatchlings emerging from larger eggs have a longer developmental period than those from smaller ones (Baur, 1994b), with hatchling size positively correlated with egg size (Baur, 1994b). In *A. marginata*, where growth is indeterminate, there is a continuous increase in the size of the eggs produced. Young adults of 9 months of age (shell height 90 mm) produce eggs of 14 mm, whereas adults of 3 years (160 mm) produce eggs of 26 mm (Plummer, 1975). Likewise, in *Limax maximus* Linnaeus (Limacidae), larger eggs are produced as the reproductive season progresses, mean egg size in late summer being 27%

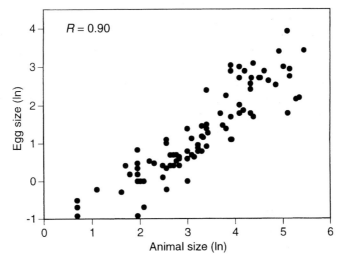

Fig. 12.1.   Egg size versus animal size (maximum shell dimension) in snail forms of terrestrial gastropods.

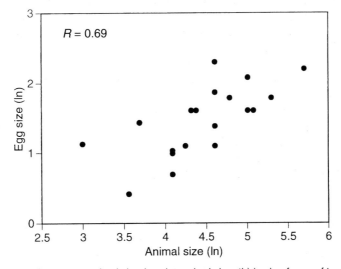

Fig. 12.2.   Egg size versus animal size (maximum body length) in slug forms of terrestrial gastropods.

Table 12.1.   Egg size and animal size of terrestrial gastropod species (maximum shell dimension for snail forms, maximum body length in slug forms)

Family Species	Animal size (mm)	Egg size (mm)	Reference
Cyclophoridae			
Cochlostoma septemspirale (Razoumowsky)	8	1.0	Baur (1989)
Cyclophorus jerdoni (Benson)	31	3.6	Kasinathan (1975)
Microlaux scabra (Theobald)	15	2.0	Kasinathan (1975)
Opisthostoma retrovertens Tomlin	2	0.43	Berry (1964)
Pterocyclus bilabiatus (Sowerby)	20	2.2	Kasinathan (1975)
Theobaldius ravidus (Benson)	27	2.2	Kasinathan (1975)
Theobaldius shiplayi (Pfeiffer)	7	2.3	Kasinathan (1975)
Pomatiasidae			
Pomatias elegans (Müller)	16	2	Creek (1951)
Ellobiidae			
Carychium tridentatum (Risso)	2	0.4 × 0.3	Baur (1989)
Cochlicopidae			
Cochlicopa lubrica (Müller)	7	1.2 × 1.1	Baur (1989)
Vertiginidae			
Vertigo pusilla Müller	2	0.6 × 0.3	Pokryszko (1990)
Valloniidae			
Vallonia costata (Müller)	3	0.8 × 0.5	Frömming (1954)
Vallonia pulchella (Müller)	2	0.5	Baur (1989)
Enidae			
Zebrina detrita (Müller)	25	2.4	Frömming (1954)
Clausiliidae			
Cochlodina laminata (Montagu)	17	1.5	Bulman (1995)
Cristataria genezarethana (Tristram)	20	1.0	Heller and Dolev (1994)
Macrogastra lineolata (Held)	13	2 × 1	Frömming (1954)
Partulidae			
Partula canalis Mousson	30	4.5 × 3.5	Stanlen (1917)
Bulimulidae			
Eucalodium decollatum (Nyst)	75	11.2 × 7.2	Stanlen (1917)
Liguus fasciatus (Müller)	60	7.5 × 4.0	Tuskes (1981)
Orthalicus undata Bruguière	65	6.5 × 5	Stanlen (1917)
Placostylus bollonsi Suter	100	18 × 13	Stanlen (1917)
Placostylus hongi (Lesson)	75	6 × 5	Stanlen (1917)
Plecochilus distortus Bruguière	60	6 × 5.5	Stanlen (1917)
Ferussaciidae			
Cecilioides acicula (Müller)	5	0.75	Frömming (1954)
Cecilioides genezarethensis Forcart	7	0.4	Heller et al. (1991)
Subulinidae			
Euonyma platyacme Melvill & Ponsonby	50	3	Stanlen (1917)
Rumina decollata (Linnaeus)	30	2.5	Stanlen (1917)
Subulina octona Bruguière	13	2.75	Stanlen (1917)
Achatinidae			
Achatina achatina (Linnaeus)	194	8.7 × 7	Bequaert (1950)
Achatina bicarinata Bruguière	150	20 × 14	Stanlen (1917)
Achatina fulica Bowdich	194	8.7 × 7	Bequaert (1950)
Achatina panthera de Férussac	164	6 × 5.5	Stanlen (1917)
Achatina reticulata Pfeiffer	208	9 × 7.5	Bequaert (1950)
Achatina ventricosa (Swanson)	128	17 × 12.5	Bequaert (1950)
Archachatina marginata (Swanson)	170	20.1 × 15.7	Plummer (1975)
Cochlitoma varicosa Pfeiffer	90	15 × 10	Stanlen (1917)

Family Species	Animal size (mm)	Egg size (mm)	Reference
Limicolaria martensiana (Smith)	51	3	Owen (1965)
Perideris auripigmentum (Reeve)	80	5.5 × 4.5	Stanlen (1917)
Acavidae			
Acavus haemastoma (Linnaeus)	50	18 × 15	Stanlen (1917)
Acavus phoenix Pfeiffer	60	20 × 15	Stanlen (1917)
Acavus waltoni Reeve	50	21 × 14	Stanlen (1917)
Caryodes dufresni Leach	30	11 × 8	Stanlen (1917)
Megalobulimus bronni Pfeiffer	93	15 × 11	Stanlen (1917)
Megalobulimus capillaceus Pfeiffer	77	15 × 12	Stanlen (1917)
Megalobulimus oblongus Müller	137	30 × 20	Stanlen (1917)
Megalobulimus popelairianus Nyst	230	51 × 28	Stanlen (1917)
Megalobulimus rosaceus King	67	18 × 11	Stanlen (1917)
Megalobulimus terrestris (Spix)	110	14 × 9	Stanlen (1917)
Stylodon studeriana (de Férussac)	60	15 × 12	Stanlen (1917)
Rhytididae			
Paryphanta busbyi (Gray)	75	13 × 11	Stanlen (1917)
Powelliphanta superba (Powell)	80	22 × 16	Stanlen (1917)
Schizoglossa novoseelandica (Pfeiffer)	20	4 × 3	Stanlen (1917)
Testacellidae			
Testacella haliotidea (Draparnaud)	120	6 × 4	Frömming (1954)
Punctidae			
Punctum pygmaeum (Draparnaud)	2	0.5 × 0.4	Baur (1987b)
Endodontidae			
Anguispira alternata (Say)	13	3	Elwell and Ulmer (1971)
Discus rotundatus (Müller)	7	1	Taylor (1894–1921), Baur (1989)
Succineidae			
Oxyloma elegans (Risso)	12	1.1	Frömming (1954)
Succinea putris (Linnaeus)	17	1.9	Frömming (1954)
Succinella oblonga Draparnaud	8	0.5	Frömming (1954)
Vitrinidae			
Vitrina pellucida (Müller)	6	1.2 × 1.1	Frömming (1954)
Vitrinobrachium breve (de Férussac)	5.5	1.5 × 1	Frömming (1954)
Zonitidae			
Aegopinella nitidula (de Lamarck)	10	1.6 × 1.4	Mordan (1978)
Aegopis verticillus (de Férussac)	30	4	Frömming (1954)
Oxychilus alliarius (Miller)	7	1.4 × 0.7	Frömming (1954)
Oxychilus cellarius (Müller)	14	1.5	Frömming (1954)
Oxychilus draparnaudi (Beck)	16	1.7	Frömming (1954)
Oxychilus villae (Strobel)	13	1.4	Frömming (1954)
Milacidae			
Milax gagates (Draparnaud)	60	2 × 1.5	Taylor (1894–1921)
Tandonia sowerbyi (de Férussac)	75	5	Taylor (1894–1921)
Tandonia rustica (Millet)	100	6.5 × 5	Frömming (1954)
Limacidae			
Limacus flavus (Linnaeus)	100	10 × 6	Frömming (1954)
Bielzia coerulans (Bielz)	160	5 × 4	Frömming (1954)
Limax cineroniger Wolf	300	9 × 5.5	Frömming (1954)
Limax maximus Linnaeus	200	6 × 4.5	Frömming (1954)
Lehmannia marginata (Müller)	80	5 × 4	Frömming (1954)
Malacoilmax tenellus (Müller)	40	4.2 × 2.9	Frömming (1954)

Table 12.1. *Continued.*

Family Species	Animal size (mm)	Egg size (mm)	Reference
Agriolimacidae			
Deroceras agreste (Müller)	60	2.0 × 1.5	Carrick (1938)
Deroceras caruane (Polonera)	35	1.5	South (1992)
Deroceras laeve (Müller)	60	2.8	Frömming (1954)
Deroceras reticulatum (Müller)	60	2.7	Frömming (1954)
Arionidae			
Arion ater (Linnaeus)	150	5 × 4	South (1992)
Arion flagellus Collinge	100	3	Davies (1987)
Arion intermedius Normand	20	3.1 × 2.5	Frömming (1954)
Arion lusitanicus Mabille	100	4	Davies (1987)
Arion rufus (Linnaeus)	150	8 × 6.5	Frömming (1954)
Arion subfuscus (Draparnaud)	70	3 × 2.5	Frömming (1954)
Sphincterochilidae			
Sphincterochila zonata (Bourguignat)	27	4.3 × 3.8	Yom Tov (1971)
Bradybaeindae			
Bradybaena fruticum (Müller)	23	1.9	Staikou *et al.* (1990)
Hygromiidae			
Ashfordia granulata (Alder)	9	1	Taylor (1894–1921)
Cernuella virgata (da Costa)	25	1.5	Taylor (1894–1921)
Chilostoma cingulatum (Studer)	27	3.3	Frömming (1954)
Helicella itala (Linnaeus)	15	1.5	Frömming (1954)
Monacha cantiana (Montagu)	20	1.8	Chatfield (1968)
Monacha cartusiana (Müller)	17	1.8	Chatfield (1968)
Monachoides incarnatus (Müller)	14	2	Frömming (1954)
Pseudotrichia rubiginosa (Rossmaessler)	7	1.6	Frömming (1954)
Perforatella bidentata (Gmelin)	9	1.7 × 2	Frömming (1954)
Trichia hispida (Linnaeus)	12	1.5	Frömming (1954)
Trichia sericea (Draparnaud)	7.5	1	Frömming (1954)
Trochoidea simulata (Ehrenberg)	25	2.6 × 2.1	Yom Tov (1971)
Xerolenta obvia (Menke)	20	1.5	Frömming (1954)
Zenobiella subrufescens (Miller)	10	1.5 × 1	Taylor (1894–1921)
Helicidae			
Arianta arbustorum (Linnaeus)	28	3.2	Baur (1988a)
Cantareus aspersus (Müller)	45	4	Bonnet *et al.* (1990)
Cepaea hortensis (Müller)	22	2	Frömming (1954)
Cepaea nemoralis (Linnaeus)	22	3.1 × 2.6	Frömming (1954)
Eobania vermiculata (Müller)	30	4.1 × 3	Frömming (1954)
Helix lucorum Linnaeus	42	4.4	Staikou *et al.* (1988)
Helix pomatia Linnaeus	40	8.6 × 7.2	Frömming (1954)
Helix texta Mousson	50	5.5	Heller and Ittiel (1990)
Theba pisana (Müller)	25	2.2	Heller (1982)
Camaenidae			
Polydontes acutangula (Burrow)	45	9.6 × 7.6	Heatwole and Heatwole (1978)

References concern egg size; nomenclature is updated, from many sources. Of the data in Stanlen (1917), only species for which I found shell measurements are entered.

smaller than in early autumn. As the animals age, they produce smaller batches of larger eggs, and at the end of their breeding life they produce some very large eggs and some large masses of unpackaged albumen (Rollo, 1983). Further, *L. maximus* was found to produce larger eggs in the presence of *Ariolimax columbianus* (Gould) (Arionidae) than in the presence of the more competitive *A. ater* or in the presence of conspecifics (Rollo, 1983). This suggests that egg size is also influenced by intra- and interspecific competition.

It frequently is assumed that within a species, egg size is a reliable indicator of the amount and quality of resources invested in each off-spring (i.e. larger eggs are supposed to contain more nutritive material), but this assumption may not necessarily be true. In *A. arbustorum*, for example, clutches of larger eggs were found to contain more nutrients than clutches of smaller eggs but, at the within-clutch level, this relationship between egg size and nutrient content could be detected in only 60% of the clutches. Egg size varies more than the nutrient content, perhaps because of imprecise control of egg size during vitellogenesis. Extremely small and large eggs of *A. arbustorum* were found to have significantly lower hatching success than eggs of intermediate size from the same clutch when kept under 'optimal' temperature and humidity conditions in the laboratory (A. Baur, 1994).

There may be considerable differences in egg size, and thus reproductive strategies, among species occupying similar terrestrial habitats.

Egg Numbers

Because the resources available to parents for gamete production are finite, parents may produce either few large eggs, or many small ones. In various animal groups, there is a negative correlation between egg size and number of eggs produced.

Terrestrial gastropods of minute size are constrained to low fecundity – they produce only a few eggs at any one time, and deposit their eggs singly. Consequently, they tend to produce only few eggs throughout their life (e.g. six in *Punctum pygmaeum* (Draparnaud) (Punctidae) during an average life span of 170 days; Baur, 1989). Larger animals produce many more eggs: *D. reticulatum* may produce 500 eggs (Carrick, 1938), *A. arbustorum* 800 eggs (Baur and Raboud, 1988) and *Vaginulus borelliana* (Colosi) (Vaginulidae) over 1300 eggs (Lanza and Quattrini, in Runham and Hunter, 1970).

Within species, there is often considerable variation in number and size of egg clutches, regulated by the size and age of the parent animal, but also strongly modified by environmental factors such as intraspecific and interspecific competition, and seasonality in climate. In *C. nemoralis* there is a positive correlation between parental size (as indicated by shell size) and clutch size: clutches increase by about three eggs for every 1 mm increase in parental size (Wolda, 1967; Wolda and Kreulen, 1973). In this species, young adults tend to lay fewer but larger clutches than older animals of the same size and from the same population (Wolda and Kreulen, 1973; Carter and Ashdown, 1984). Further, the larger animals also tend to lay eggs more frequently (Wolda and Kreulen, 1973), at an earlier age and the eggs produced have a higher hatching success than those from small animals (Oosterhoof, 1977). In some *C. nemoralis* populations, larger animals produce more eggs than smaller individuals of similar age simply because they produce more clutches, whereas in other populations differences in fecundity among animals relate to clutch size (Carter and Ashdown, 1984). In addition, there are among-population differences in egg production. Carter and Ashdown (1984) demonstrated, for example, that animals of the same size drawn from different populations can lay different numbers of differently sized clutches when placed under the same conditions, and that animals of the same size range drawn from one population in different years can lay different numbers of eggs. The allometric scaling of fecundity with parental size is a common feature in terrestrial gastropods (e.g. Cowie, 1984; Baur and Raboud, 1988), but it is by no means universal. In *Cochlicella acuta* (Müller) (Hygromiidae), for example, there is no evidence to suggest that large animals produce more eggs than smaller animals (Baker and Hawke, 1991).

High population density of conspecifics may have an adverse effect on oviposition frequency in *C. nemoralis* (Wolda and Kreulen, 1973). This would result in reduced fecundity per individual, because oviposition frequency is strongly correlated with mean clutch size in natural populations of this species (Wolda and Kreulen, 1973). The effects of crowding apparently do not depend on food abundance, on lack of calcium or on competition for oviposition sites (Carter and Ashdown, 1984). Similar effects have been observed in *T. simulata* (Yom Tov, 1972).

Rollo (1983) demonstrated competitive interactions among sympatric slugs in British Columbia. Under conditions of scarce shelter sites, *L. maximus* in particular exhibited aggressive territorial behaviour against conspecifics and other species. In a field cage experiment, *L. maximus* drastically reduced egg production of the non-aggressive species *A. ater* and *A. columbianus*, and *A. ater* strongly reduced egg production by *A. columbianus*. Rollo concluded that the overall reproductive life strategy of *L. maximus* is strongly linked to density-dependent regulation of behaviour.

Baur and Raboud (1988) found that the number of eggs laid by *A. arbustorum* decreased with increasing altitude in the Swiss Alps, both in absolute terms (life time fecundity of 830 eggs in the valley as compared with 40 eggs at the highest elevation of 2600 m) and for a given length of the breeding season. In terms of energy investment in eggs, an animal from the valley population was found to invest about five times its body dry weight in egg production, in contrast to animals from the summit population where investment was approximately 50% of body weight. Animals translocated from high to low altitudes laid considerably more eggs than those retained at their original site, suggesting that environmental correlates of altitude, such as temperature, explained most of the variation in egg number over the elevational gradient (Baur and Raboud, 1988).

Egg Retention: Ovoviviparity and Viviparity

The vast majority of terrestrial gastropods are oviparous: they lay eggs that contain only a single-celled zygote (or the earliest cleavage stages, of 2–4 cells). As a modification of simple oviparity, some species can retain their eggs inside the reproductive tract for various parts of the embryonic period so that the eggs, when eventually laid, contain embryos advanced in development. In extreme cases of egg retention (termed ovoviviparity), the eggs are retained within the reproductive tract for the entire embryonic period, and the young hatch from the egg inside the parent, followed by birth; the young of these ovoviviparous species are not provisioned from sources other than the egg. In some specialized cases of ovoviviparity (termed viviparity), nutrients are passed from the parent to the developing embryos, which are retained in the reproductive tract until extrusion as free young. Some species may be flexible in their developmental strategy, in that they may extrude eggs, embryos or young as a function of environmental conditions (Tompa, 1979a,b, 1984; Baur, 1994a). As indicated above, most terrestrial pulmonate species are oviparous. However, in 30 of the approximately 80 of terrestrial pulmonate families there is at least one species with either egg retention or viviparity; the total number of species involved is rather small (Tompa, 1979a; Baur, 1994a; Heller *et al.*, 1997).

Egg retention frequently occurs when the environmental conditions are unfavourable for eggs and juveniles. The achatinid *Limicolaria martensiana* Smith of Uganda, for example, potentially can breed throughout the year but retains its eggs during dry periods, when the adults aestivate. At the beginning of the rainy season, eggs immediately are deposited, ensuring them the best prospects of survival (Owen, 1965). The clausiliid *Lacinaria biplicata* (Draparnaud) is also usually oviparous, but under favourable environmental conditions may lay eggs with well-developed embryos (Fechter and Falkner, 1990, in Baur, 1994a). Another partly ovoviviparous species is the pupillid *Pupilla muscorum* (Linnaeus), which can hibernate while gravid, and deposit eggs with partly developed embryos early in the reproductive season, providing the emerging young with a prolonged period of growth (Tompa, 1984). The pupillid *Lauria cylindracea* (da Costa), however, is exclusively ovoviviparous. Adults contain four embryos during the breeding season, but the subsequent number of juveniles in the population is never as high as could be expected from the number of embryos (Heller *et al.*, 1997). The possibility of mutual growth inhibition between embryos, leading to some reaching birth and others perishing, cannot be excluded (Heller *et al.*, 1997).

Opinions differ as to the benefits of egg retention (including ovoviviparity). Peake (1978) noted that among the tornatellinine Achatinellidae of the Pacific Islands, oviparous genera tend to have narrow distributions (in that they were found only on single islands), whereas ovoviviparous genera are very widespread. He suggested that ovoviviparity may contribute to the reduction of the level of mortality and thereby enhance the probability of successful colonization of a new habitat, particularly in species with very low reproductive rates. Tompa (1984) suggested that ovoviviparity may be advantageous among terrestrial gastropods, either in those areas of the world where the initiation of rainy seasons is unpredictable – selective pressure favouring those offspring whose parents do not release offspring immediately – or where releasing of live young at the beginning of the growing season would provide a competitive advantage over young of oviparous animals. Baur (1994a) suggested that the main benefit of egg retention and ovoviviparity might be a minimization of potential egg mortality caused by drought and predators. He noted that ovoviviparity is common in species living in habitats with extreme environmental conditions, such as exposed rock walls or stone walls which provide no suitable oviposition sites, and in species living in tropical regions. Heller *et al.* (1997) suggested that ovoviviparity may be advantageous in minute orthurethran snails that can neither lay many, big eggs, nor dig deep egg cavities. In an animal constrained to (minute size and hence to) low fecundity, one reproductive strategy for increasing fitness would be to retain its few developing eggs inside its body and produce hatchlings. Active hatchlings are advantageous, compared with eggs, in that mobility enables some capacity to escape from unfavourable microclimatic conditions. The phylogenetic tendency to shift from oviparity via facultative ovoviviparity to obligate ovoviviparity seems stronger among certain groups of the Orthurethra than among other stylommatophoran suborders. The reasons for this have not been elucidated but clearly relate to post-embryonic life. The orthurethran egg and early embryo perhaps may be less well adapted than those of other suborders to withstand lengthy periods of immersion in water, which may occur in regions subject to high rainfall.

Viviparity is so extremely rare among pulmonates that Tompa (1979a) considered it non-existent. However, some pieces of evidence suggest that it may occur, in a very few cases. In the vaginulid *Pseudoveronicella zootoca* Hoffmann of Africa, the developing young may reach 38% the length of the parent before birth, and this large size suggested to Solem (1972) that some form of gradual embryonic nutrition may exist. In the achatinellinid *Tekoulina pricei* Solem of the Cook Islands, each embryo is enclosed in a sac from which two stalks are linked to the inner, glandular side of the oviduct in the parent animal (Solem, 1972). These anatomical data indicated to Solem (1972) that in *T. pricei* there is a gradual transfer of nutritive material from the parent to the embryo. Direct evidence of viviparity concerns the acavid *Stylodon studeriana* (de Fèrussac) of the Seychelles, where there is a long-term calcium transfer from parental tissues to developing embryos, through a placenta-like podocyst (Tompa, 1979b, 1984).

Semelparity Versus Iteroparity

The two basic modes of reproductive strategy in terrestrial gastropods comprise semelparity, in which animals reproduce during one season only, after which they die; and iteroparity, in which animals reproduce during one season and then live on to reproduce again. While the theoretical advantage of iteroparity is that if the progeny in one generation are lost due to an environmental catastrophy the parents can create a new generation (Stearns, 1976; Stearns and Crandal, 1981), that of semelparity is not so clear (Calow, 1978; Stearns, 1992). Many models have been suggested to describe the habitats in which semelparity is advantageous, as opposed to those in which iteroparity is (Gadgil and Bossert, 1970; Charnov and Schaffer, 1973; Schaffer, 1974, 1979; Stearns, 1976).

According to these models, the most important factor is adult survival relative to juvenile survival (Partridge and Harvey, 1988). Thus, at sites where survival of adults is unpredictable (a habitat that is unstable and changing), selection will favour a short life, maximum reproductive effort and semelparity. In habitats where high survival of the adult stage is expected (a habitat that is stable and competitive), selection will favour a long life, late maturity, low to medium reproductive effort and iteroparity (Gadgil and Bossert, 1970; Charnov and Schaffer, 1973; Schaffer, 1974, 1979; Stearns, 1976; Reznik *et al.*, 1990).

It might perhaps be expected that semelparity in terrestrial gastropods would be closely associated with an annual life span. In nature, however, the close dependence of terrestrial gastropod life history on climate, with aestivation when it is too hot or dry and hibernation when it is too cold, may cause a semelparous cycle to be uncoupled from the annual cycle and extend over more than 1 year. Further, this uncoupling may vary on a geographic basis. Another factor that may lengthen the life cycle is intraspecific variation in growth rate within a single cohort, perhaps in response to density-dependent factors. This uncoupling of semelparity from the annual cycle is exemplified by *T. pisana*, a sand dune- and coastal scrubland species of the Mediterranean region; in dry weather, it climbs up plant stems to aestivate, sometimes in large aggregates. Where the summer is very hot and dry but the (brief) spring is moist and warm (as in Israel), most individuals of *T. pisana* are annual. Juveniles that emerge in winter grow to subadult size in spring, aestivate in summer, mature in autumn and die after breeding in early winter (Heller, 1982). However, about 5–10% of the population within each cohort may grow so slowly that they do not attain adult size in the following autumn. These individuals live on and enter a second summer of aestivation (Arad and Avivi, 1993). In Wales, where the temperatures are lower, growth of *T. pisana* is slowed and the cycle is more consistently biennial, with most of the animals attaining maturity after 2 years (Cowie, 1984). In southern Australia, where *T. pisana* is an introduced pest, both annual and biennial life cycles are present (Baker and Vogelzang, 1988). This variability in the life cycle of *T. pisana* is not genetically controlled, since animals from many populations are annual in the laboratory (Cowie, 1984). Other semelparous species, in which similar variations in life cycles have been found, include *Cernuella virgata* (da Costa) (Hygromiidae) (Lazaridou-Dimitriadou, 1981; Baker, 1988), *C. acuta* (Baker, 1991), *Xeropicta vestalis* (Pfeiffer) (Hygromiidae) (Heller and Volokita, 1981), *M. cartusiana* (Staikou and Lazaridou-Dimitriadou, 1990), *Helicella itala* (Linnaeus) (Hygromiidae) (Lazaridou-Dimitriadou and Sgardelis, 1995), *Vitrina pellucida* (Müller) (Vitrinidae) and *Semilimax kotulai* (Westerlund) (Vitrinidae) (Uminski, 1975), as well as various species of the genera *Arion*, *Deroceras*, *Milax* Gray (Milacidae) and *Tandonia* Lessona & Pollonera (Milacidae) (South, 1989, 1992).

Semelparous species are largely excluded from deserts. Even *X. vestalis*, with its highly successful physiological adaptations to severely desiccating conditions, has not been able to penetrate desert environments (Heller, 1988; Arad *et al.*, 1992). Deserts are unpredictable environments, susceptible to year-to-year climatic fluctuations to such an extent that one single rainless year would wipe out all the desert populations of a semelparous/annual species, no matter how efficient their water-preserving mechanisms are. Beyond the lack of semelparous species in deserts, there is no clear biogeographic pattern in semelparity among terrestrial gastropods. Many environments are home to both semelparous and iteroparous species, side by side. For example, in Mediterranean climatic regions of Israel, both the semelparous *X. vestalis* and *Monacha haifaensis* (Pallary) and the iteroparous *Helix engaddensis* Bourguignat, *Levantina hierosolyma* (Mousson) (Helicidae) and *Buliminus labrosus* (Olivier) (Enidae) may coexist within the same 0.01 ha plot (Heller, unpublished observations).

One example of an iteroparous species is *Cristataria genezarethana* (Tristram), a clausiliid of Mediterranean habitats. Heller and Dolev (1994) found that it spends 95–98% of its life time within crevices, emerging to the surface of the rock during brief periods of activity to feed and mate. No more than 15–20% of the population is active simultaneously, and an average individual is active only 6–12 days per year (during rainy days or immediately after). The population studied by Heller and Dolev (1994) consisted of two well-defined groups, adults and young; a third group, of juveniles appeared during winter, but died soon afterwards. During the period of study, there was no shift from the young to the adult size group, which suggests a broadly stable population, with virtually no recruitment and final growth to adulthood inhibited by the presence of adults. Growth of the animals is very slow, with maturity reached in about 11 years and individuals living for at least 16 years. Mortality in the adult stage was estimated at only about 5% in each year. The sheltering crevices, possible lack of predators, and lack of competing gastropod species may be factors enabling this long-lived species to reach very high population densities, in which low mating frequencies and growth inhibition may be important population regulatory factors (Heller and Dolev, 1994).

Another iteroparous species of Mediterranean habitat is the helicid *H. texta*. Buried in the soil during late winter, spring and summer, it emerges to the surface during autumn to feed, mate and lay eggs. A population studied by Heller and Ittiel (1990) comprised predominantly, as in *C. genezarethana*, two well-defined groups, adults and young. In the first year of study, there was no recruitment from the young to the adult size group. In the second year, however, immediately after an event of massive predation by wild boar (*Sus scrofa* Linnaeus) (within a few days 50% of the adults were eaten), the young animals grew very rapidly in size (Heller and Ittiel, 1990). This suggests that the extent of recruitment of new adults into the population may be partly controlled by the presence of adults, probably through a growth-inhibiting factor in the mucus. After predation, this inhibiting factor dissipated, enabling subsequent rapid growth of the young. This provides a good illustration of the advantage provided by the iteroparous life strategy in the face of disturbance events or unpredictability.

Other iteroparous species that have been studied in some depth include *C. nemoralis* in England (where a natural population was studied for 23 years; Cain *et al.*, 1990; see also Cowie, 1984) and in The Netherlands (Wolda, 1963); *H. lucorum*, *B. fruticum* (Müller) and *Helicella pappi* (Schütt) in Greece (Staikou *et al.*, 1988, 1990; Lazaridou-Dimitriadou, 1995; Lazaridou-Dimitriadou and Sgardelis, 1995), *Brephulopsis bidens* (Krynicki) (Enidae) in Crimea (Livshits, 1985), *Vestia elata* (Rossmaessler) (Clausillidae) in the Karpathians (Piechocki, 1982), *T. simulata*, *S. zonata* and *Sphincterochila fimbriata* (Bourguignat) in Israel (Yom Tov, 1971, 1972; Shachak *et al.*, 1975; Ben-Yehuda, 1995) and *Rhagada convicta* (Cox) (Camaenidae) in Western Australia (Johnson and Black, 1991).

Collectively, we have good life strategy data concerning species in about 35 genera, of which 15 genera are semelparous and 20 iteroparous. This sample amounts to about 2% of the estimated 1700 genera of terrestrial gastropods (data from Vaught, 1989). It is noteworthy that many of the agricultural pest species, for which information is available for most, are semelparous. This suggests that semelparous species are well adapted to human modified environments.

A classification of life history strategies closely similar to that of semelparity versus iteropariy is that of the *r* and *K* selection (MacArthur and Wilson, 1967). According to this paradigm, *r*-selected species have high reproductive effort, with the combination of early maturity, large litters, small young and a short life, while *K*-selected species show the opposite traits (Stearns, 1976). These two strategies, like semelparity and iteropariy, are at opposing ends of a continuum. Consequently, the terms *r* and *K* are comparative (Stearns, 1976). It has been suggested that *r* selection may occur in fluctuating environments where the ability to capitalize on transient albeit favourable conditions is advantageous. On the other hand, *K* selection should operate in stable environments where population size is near the habitat carrying capacity, and production of a few but fit offspring is more advantageous (MacArthur and Wilson, 1967).

Attempts broadly to apply the *r*–*K* theory to terrestrial gastropods do not yield firm conclusions (Peake, 1978). On a more narrow taxonomic scale, Cowie (1984) compared life history tactics of two helicid species that occur in Britain. He found *C. nemoralis*, a more *K*-selected species, to occupy rather varied habitats (woodland rides, hedges, grassland and dunes), whereas *T. pisana*, a more *r*-selected species, occupies more ephemeral and more patchy habitats (early successional habitats, particularly the more open part of dunelands and waste ground close to the sea). Their different strategies thus agree with theoretical predictions. In another study, Yom Tov (1983) compared life history tactics of two very broadly sympatric desert gastropods. He found that *S. zonata* is closer to the K 'syndrome' end of a theoretical *r*–*K* continuum, as it lays small clutches with large eggs, while *T. simulata* is closer to the *r* 'syndrome' end point in that it is short-lived, lays large clutches of small eggs and suffers from high juvenile mortality. This coexistence of two (similar sized) species with quite different life strategies in an inhospitable habitat conflicts with predictions from MacArthur and Wilson's (1967) theory. To conclude, classification of terrestrial gastropods into *r* versus *K* strategists has not, as yet, yielded broad generalizations.

A further way of classifying life strategies in terrestrial gastropods is to examine their longevity.

Longevity

A survey of the literature (Heller, 1990) has revealed longevity records of 75 species belonging to 57 genera and 30 families of terrestrial gastropods. With an overall estimate of 1700 genera, this is 4%. In analysing these data, all species were assigned to two categories: short-lived species (those living up to 2 years, or those living longer but reproducing during only one season); and long-lived species (those species that live for more than 2 years and breed over at least two seasons). This approach disentangled the fact that in some semelparous species the life span is variable, being annual in one habitat but stretching over several years in a different habitat (Heller, 1990). Life spans of terrestrial gastropods range from several months to 19 years. Although many terrestrial gastropods are long-lived, over 50% of the existing records are of short-lived species.

Among genera characterized by short-lived species, one pattern that emerges concerns shell morphology. When each genus is classified into one of three categories according to shell reduction (shell fully calcified, capable of fully housing the retracted animal; shell partly decalcified and semi-transparent, or not capable of housing the retracted animal; shell internalized or absent), longevity tends to be associated with shell reduction. That is, most semi-transparent or shell-less terrestrial gastropod genera are short-lived, whereas most fully calcified ones are long-lived (Heller, 1990). This pattern may be explained in adaptive terms, in that shell absence affects age-specific mortality via growth rates, or that shell-less gastropods more often may utilize transient food resources. The difficulty in accepting either of these two adaptive explanations stems from the near ubiquity of the relationship between shell loss and a short life span. Almost every slug species, over a wide range of habitats in the sea and on the land, is short-lived. The correlation may be also explained in non-adaptive terms: shell and longevity co-vary, so that an initial, adaptive change in the shell engenders a secondary, automatic change in the life span. If this non-adaptive explanation is indeed valid, then the short life span of many gastropods may be a by-product of selection on the shell rather than an independently selected trait. One major difficulty in accepting this non-adaptive explanation is that it lacks evidence at the genetic level. Such an argument would imply ubiquity in the ecology of entire groups of shell-less gastropods, which is difficult to accept (Heller, 1990).

A second pattern among short-lived terrestrial gastropods concerns microenvironments exposed to high solar radiation and high temperatures. Over 35% of the genera with short-lived species have well-calcified shells. The combined traits of occupancy of exposed habitat and possession of fully calcified shells are more frequent among genera with short-lived members than among the genera with long-lived ones (Heller, 1990). The relationship between longevity and exposure may perhaps be explained in adaptive terms in that ionizing radiation could perhaps increase the rate of ageing and reduce the average life span of animals. A short life could perhaps thus be enforced in gastropods dwelling on the tips of vegetation, where they are subject to heavier radiation than those dwelling, for example, beneath stones or in forest leaf litter (Heller, 1990). An exception to this generalization concerns gastropods that inhabit environments that, in addition to being hot, are also extremely unpredictable, such as deserts. In such places, where the conditions for growth and reproduction occur infrequently and unpredictably, short longevity would be an inappropriate life strategy as there would be a high vulnerability to local extinction. Gastropods of these habitats tend to have longer life spans than their close relatives in more favourable conditions. Accordingly, short-lived terrestrial gastropods are restricted to environments in which weather is predictable.

A third pattern among short-lived terrestrial gastropods concerns size. Short life spans tend to occur more frequently among minute gastropods (Heller, 1990; Lazarodou-Dimitriadou and Sgardelis, 1995; see, however, Heller *et al.*, 1997).

Acknowledgements

Annette and Bruno Baur (of Basel) originally were planned to co-write this chapter. They helped me considerably in its general layout, and kindly criticized an advanced draft.

References

Alexander, R.D. (1974) The evolution of social behaviour. *Annual Review of Ecology and Systematics* 4, 325–383.

Arad, Z. and Avivi, T. (1993) Population dynamics of the snail *Theba pisana* (Helicidae) in the sand dunes of northern Israel. *Israel Journal of Zoology* 39, 245–254.

Arad, Z., Goldenberg, S. and Heller, J. (1992) Intraspecific variation in resistance to desiccation and climatic gradients in the distribution of the land snail *Xeropicta vestalis. Journal of Zoology* 226, 643–656.

Auffenberg, K. and Auffenberg, G. (1984) Nest building of *Cochlostyla pithogaster* (Ferussac) (Pulmonata: Bradybaenidae). *American Malacological Bulletin* 3, 1.

Baker, G.H. (1988) Population dynamics of the white snail, *Cernuella virgata* (Mollusca: Helicidae), in a pasture-cereal rotation in South Australia. In: Stahle, P.P. (ed.) *Proceedings of the Fifth Australasian Conference on Grassland Invertebrate Ecology.* D. & D. Printing, Melbourne, pp. 177–183.

Baker, G.H. (1991) Production of eggs and young snails by adult *Theba pisana* (Müller) and *Cernuella virgata* (da Costa) (Mollusca: Helicidae) in laboratory cultures and field populations. *Australian Journal of Zoology* 39, 673–679.

Baker, G.H. and Hawke, B.G. (1991) Fecundity of *Cochlicella acuta* (Müller) (Mollusca: Helicidae) in laboratory cultures. *Invertebrate Reproduction and Development* 20, 243–247.

Baker, G.H. and Vogelzang, B.K. (1988) Life history, population dynamics and polymorphism of *Theba pisana* (Mollusca: Helicidae) in Australia. *Journal of Applied Ecology* 25, 867–887.

Baker, G.H., Hawke, B.G. and Vogelzang, B.K. (1991) Life history and population dynamics of *Cochlicella acuta* (Müller) (Gastropoda: Helicidae) in a pasture–cereal rotation. *Journal of Molluscan Studies* 57, 259–266.

Baur, A. (1994) Within- and between-clutch variation in size and nutrient content of eggs of the land snail *Arianta arbustorum* (L.). *Functional Ecology* 8, 581–586.

Baur, B. (1984) Early maturity and breeding in *Arianta arbustorum* (L.) (Pulmonata: Helicidae). *Journal of Molluscan Studies* 50, 241–242.

Baur, B. (1987a) The minute land snail *Punctum pygmaeum* (Draparnaud) can reproduce in the absence of a mate. *Journal of Molluscan Studies* 53, 113–115.

Baur, B. (1987b) Can cannibalistic hatchlings of the land snail *Arianta arbustorum* distinguish between sib and non-sib eggs? *Behaviour* 103, 259–265.

Baur, B. (1987c) Effects of early feeding experience and age on the cannibalistic propensity of the land snail *Arianta arbustorum. Canadian Journal of Zoology* 65, 3068–3070.

Baur, B. (1988a) Egg-species recognition in cannibalistic hatchlings of the land snails *Arianta arbustorum* and *Helix pomatia. Experientia* 44, 276–277.

Baur, B. (1988b) Do the risks of egg cannibalism and desiccation influence the choice of oviposition sites in the land snail *Arianta arbustorum? Journal of Zoology* 216, 495–502.

Baur, B. (1989) Growth and reproduction of the minute snail *Punctum pygmaeum* (Draparnaud). *Journal of Molluscan Studies* 55, 383–387.

Baur, B. (1990a) Egg cannibalism in hatchlings of the land snail *Helix pomatia*: nutritional advantage may outweigh lack of kin recognition. *Malacological Review* 23, 103–105.

Baur, B. (1990b) Possible benefits of egg cannibalism in the land snail *Arianta arbustorum* (L.). *Functional Ecology* 4, 679–684.

Baur, B. (1992) Cannibalism in gastropods. In: Elgar, M.A. and Crespi, B.J. (eds) *Cannibalism, Ecology and Evolution Among Diverse Taxa.* Oxford University Press, Oxford, pp. 102–128.

Baur, B. (1993) Intraclutch egg cannibalism by hatchlings of the land snail *Arianta arbustorum*: non-random consumption of eggs? *Ethology, Ecology and Evolution* 5, 329–336.

Baur, B. (1994a) Parental care in terrestrial gastropods. *Experientia* 50, 5–14.

Baur, B. (1994b) Inter-population differences in propensity for egg cannibalism in hatchlings of the land snail *Arianta arbustorum*. *Animal Behaviour* 48, 851–860.

Baur, B. (1994c) Multiple paternity and individual variation in sperm precedence in the simultaneously hermaphroditic land snail *Arianta arbustorum*. *Behaviour, Ecology and Sociobiology* 35, 413–421.

Baur, B. and Baur, A. (1986) Proximate factors influencing egg cannibalism in the land snail *Arianta arbustorum* (Pulmonata, Helicidae). *Oecologia* 70, 283–287.

Baur, B. and Baur, A. (1992) Effect of courtship and repeated copulation on egg production in the simultaneously hermaphroditic land snail *Arianta arbustorum*. *Invertebrate Reproduction and Development* 21, 201–206.

Baur, B. and Raboud, C. (1988) Life history of the land snail *Arianta arbustorum* along an altitudinal gradient. *Journal of Animal Ecology* 57, 71–87.

Bayne, C.J. (1968) Survival of the embryos of the grey field slug *Agriolimax reticulatus*, following desiccation of the egg. *Malacologia* 9, 391–401.

Ben-Yehuda, O. (1995) Factors limiting the distribution of *Trochoidea simulata* into Mediterranean regions. PhD thesis, Hebrew University, Jerusalem (Hebrew, English abstract).

Bequaert, J (1950) Studies of the Achatininae, a group of African land snails. *Bulletin of the Museum of Comparative Zoology at Harvard* 105, 1–216.

Berry, A.J. (1964) The reproduction of the minute cyclophorid snail *Opisthostoma* (*Plecostoma*) *retrovertens* from a Malayan limestone hill. *Proceedings of the Zoological Society of London* 142, 655–663.

Bonnet, J.C., Avipinel, P. and Vrillon, J.L. (1990) *L'Escargot Helix aspersa*. Institut National de la Recherche Agronomique, Paris.

Bulman, K. (1995) Life history of *Cochlodina laminata* – preliminary observations. In: Guerra, A., Rolan, E. and Rocha, F. (eds) *Proceedings of the Twelfth International Malacological Congress*. Instituto de Investigationes Marinas, Vigo, pp. 146–147.

Cain, A.J. (1983) Ecology and ecogenetics of terrestrial molluscan populations. In: Russell-Hunter, W.D. (ed.) *The Mollusca*, Vol. 6, *Ecology*. Academic Press, New York, pp. 597–647.

Cain, A.J. and Cook, L.M. (1989) Persistence and extinction in some *Cepaea* populations. *Biological Journal of the Linnaean Society* 38, 183–190.

Cain, A.J., Cook, L.M. and Currey, J.D. (1990) Population size and morph frequency in a long-term study of *Cepaea nemoralis*. *Proceedings of the Royal Society of London, Series B* 240, 231–250.

Calow, P. (1978) The evolution of life cycle strategies in freshwater gastropods. *Malacologia* 17, 351–364.

Calow, P. (1983) Life-cycle patterns and evolution. In: Russell-Hunter, W.D. (ed.) *The Mollusca*, Vol. 6, *Ecology*. Academic Press, New York, pp. 649–678.

Carrick, R. (1938) The life history and development of *Agriolimax agrestis* L. the gray field slug. *Transactions of the Royal Society of Edinburgh* 29, 563–597.

Carter, M.A. and Ashdown, M. (1984) Experimental studies on the effects of density, size, and shell colour and banding phenotypes on the fecundity of *Cepaea nemoralis*. *Malacologia* 25, 291–302.

Charnov, E.L. and Schaffer, W.M. (1973) Life history consequences of natural selection: Cole's result revisited. *American Naturalist* 107, 791–793.

Chatfield, J.E. (1968) The life history of the helicid snail *Monacha cantiana* (Montagu) with reference also to *M. cartusiana* (Müller). *Proceedings of the Malacological Society of London* 38, 233–245.

Chen, X. (1993) Comparision of inbreeding and outbreeding in hermaphroditic *Arianta arbustorum* (L.) (land snail). *Heredity* 71, 456–461.

Chen, X. (1994) Self-fertilization and cross-fertilization in the land snail *Arianta arbustorum* (Mollusca, Pulmonata: Helicidae). *Journal of Zoology* 232, 465–471.

Chen, X. and Baur, B. (1993) The effect of multiple mating on female reproductive success in the simultaneously hermaphroditic snail *Arianta arbustorum*. *Canadian Journal of Zoology* 71, 2431–2436.

Clarke, B. and Murray, J. (1969) Ecological genetics and speciation in land snails of the genus *Partula*. *Biological Journal of the Linnean Society* 1, 31–42.

Clutton-Brock, T.H. (1991) *The Evolution of Parental Care*. Princeton University Press, Princeton, New Jersey.

Cowie, R.H. (1980) Precocious breeding of *Theba pisana* (Müller) Pulmonata: Helicidae). *Journal of Conchology* 30, 238.

Cowie, R.H. (1984) The life cycle and productivity of the land snail *Theba pisana* (Mollusca: Helicidae). *Journal of Animal Ecology* 53, 311–325.

Creek, G.A. (1951) The reproductive system and embryology of the snail *Pomatias elegans* (Müller). *Proceedings of the Zoological Society of London* 121, 599–640.

Creek, G.A. (1953) The morphology of *Acme fusca* (Montagu) with special reference to the genital system. *Proceedings of the Malacological Society of London* 29, 228–240.

Davies, S.M. (1987) *Arion flagellus* Collinge and *A. lusitanicus* Mabille in the British Isles: a morphological, biological and taxonomical investigation. *Journal of Conchology* 32, 339–354.

De Winter, A.J and Gittenberger, E. (1998) The land snail fauna of a square kilometer of rainforest in south-western Cameroon: high species richness, low abundance and seasonal fluctuations. *Malacologia* 40, 231–250.

Eickwort, K.R. (1973) Cannibalism and kin selection in *Labidomera clivicollis* (Coleoptera: Chrysomelidae). *American Naturalist* 107, 452–453.

Elwell, A.S and Ulmer, M.J (1971) Notes on the biology of *Anguispira alternata* (Stylommatophora: Endodontidae). *Malacologia* 11, 199–215.

Fearnley, R.H. (1996) Heterogenic copulatory behaviour produces non-random mating in laboratory trials in the land snail *Helix aspersa* Müller. *Journal of Molluscan Studies* 62, 159–164.

Foltz, D.W., Ochman, H., Jones, J.S., Evangelisti, M. and Selander, R.K. (1982) Genetic population structure and breeding systems in arionid slugs (Mollusca: Pulmonata). *Biological Journal of the Linnean Society* 17, 225–242.

Foltz, D.W., Ochman, H. and Selander, R.K. (1984) Genetic diversity and breeding systems in terrestrial slugs. *Malacologia* 25, 593–606.

Frettter, V. and Graham, A. (1994) *British Prosobranch Molluscs*. Ray Society, London.

Frömming, E. (1954) *Biologie der Mitteleuropäischen Landgastropoden*. Duncker & Humblot, Berlin.

Gadgil, M. and Bossert, W. (1970) Life history consequences of natural selection. *American Naturalist* 104, 1–24.

Heatwole, H. and Heatwole, A. (1978) Ecology of the Puerto Rican camaenid tree-snails. *Malacologia* 17, 241–316.

Heller, J. (1982) Natural history of *Theba pisana* in Israel. *Journal of Zoology* 196, 475–487.

Heller, J. (1988) The biogeography of the land snails of Israel. In: Yom Tov, Y. and Tchernov, E. (eds) *The Zoogeography of Israel*. Dr Junk Publishers, Dordrecht, pp. 325–354.

Heller, J. (1990) Longevity in molluscs. *Malacologia* 31, 259–295.

Heller, J. (1993) Hermaphroditism in molluscs. *Biological Journal of the Linnaean Society* 48, 19–42.

Heller, J. and Dolev, A. (1994) Biology and population dynamics of a crevice-dwelling landsnail, *Cristataria genezarethana* (Clausiliidae). *Journal of Molluscan Studies* 60, 33–46.

Heller, J. and Ittiel, H. (1990) Natural history and population dynamics of the land snail *Helix texta* in Israel (Pulmonata: Helicidae). *Journal of Molluscan Studies* 56, 189–204.

Heller, J. and Volokita, M. (1981) Gene regulation of shell banding in a land snail from Israel. *Biological Journal of the Linnaean Sociey* 16, 261–277.

Heller, J., Pimstein, R. and Vaginsky, E. (1991) Cave-dwelling *Cecilioides genezarethanensis* (Pulmonata, Ferussaciidae) from Israel. *Journal of Molluscan Studies* 57, 289–300.

Heller, J., Sivan, N. and Hodgson, A.N. (1997) Reproductive biology and population dynamics of an ovoviviparous landsnail, *Lauria cylindracea* (Pupillidae). *Journal of Zoology* 243, 263–280.

Hoffmann, R.J. (1983) The mating system of the terrestrial slug *Deroceras laevae*. *Evolution* 37, 423–425.

Jarne, P., Vianey-Liaoud, M. and Delay, B. (1993) Selfing and outcrossing in hermaphrodite freshwater gastropods (Basommatophora): where, when and why. *Biological Journal of the Linnean Society* 49, 99–125.

Johnson, M.S. (1982) Polymorphism for direction of coil in *Partula suturalis*: behavioural isolation and positive frequency dependent selection. *Heredity* 49, 145–151.

Johnson, M.S. and Black, R. (1991) Growth, survivorship, and population size in the land snail *Rhagada convicta* Cox, 1870 (Pulmonata: Camaenidae) from a semiarid environment in western Australia. *Journal of Molluscan Studies* 57, 367–374.

Johnson, M.S., Clark, E.B. and Murray, J. (1977) Genetic variation and reproductive isolation in *Partula*. *Evolution* 31, 116–126.

Johnson, M.S., Clark, E.B. and Murray, J. (1986) High genetic similarities and low heterozygosities in land snails of the genus *Samoana* from the Society Islands. *Malacologia* 27, 97–106.

Kasinathan, R. (1975) Some studies of five species of cyclophorid snails from peninsular India. *Proceedings of the Malacological Society of London* 41, 379–394.

Kerney, M.P. and Cameron, R.A.D. (1979) *A Field Guide to the Land Snails of Britain and North-west Europe*. Collins, London.

Kobayashi, S.R. and Hadfield, M.G. (1996) An experimental study of growth and reproduction in the Hawaiian snails *Achatinella mustelina* and *Partulina redfieldi* (Achatinellinae). *Pacific Science* 50, 339–354.

Lazaridou-Dimitriadou, M. (1981) Contribution á l'étude biologique et écologique des escargots *Cernuella virgata* (Da Costa) et *Xeropicta arenosa* Ziegler (Gastropoda, Pulmonata) vivant sur les microdunes de Potidea, Chalkidiki (Grèce due nord). V. *Conveno Nationale della Societa Malacologica Italiana, Pavia*, pp. 73–83.

Lazaridou-Dimitriadou, M. (1995) The life cycle, demographic analysis, growth and secondary production of the snail *Helicella* (Xerothracia) *pappi* (Schütt, 1962) (Gastropoda Pulmonata) in E. Macedonia (Greece). *Malacologia* 37, 1–11.

Lazaridou-Dimitriadou, M. and Sgardelis, S. (1995) Biological strategies and population dynamics of the northern Greek terrestrial gastropods. In: Guerra, A., Rolan, E. and Rocha, F. (eds) *Proceedings of the 12th International Malacological Congress*, Instituto de Investigationes Marinas, Vigo, pp. 206–207.

Lipton, C. and Murray, J. (1979) Courtship of land snails of the genus *Partula*. *Malacologia* 19, 129–146.

Livshits, G.M. (1985) Ecology of the terrestrial snail *Brephulopsis bidens* (Pulmonata: Enidae): mortality, burrowing and migratory activity. *Malacologia* 26, 213–223.

MacArthur, R.H. and Wilson, E.O. (1967) *The Theory of Island Biogeography*. Princeton University Press, Princeton, New Jersey.

Madec, L. and Daguzan, J. (1993) Geographic variation in reproductive traits of *Helix aspersa* Müller studied under laboratory conditions. *Malacologia* 35, 99–117.

McCracken, G.F. and Brussard, P.F. (1980) Self-fertilization in the white-lipped snail, *Triodopsis albolabris*. *Biological Journal of the Linnaean Society* 14, 429–434.

Mordan, P.B. (1978) The life cycle of *Aegopinella nitidula* (Draparnaud) (Pulmonata: Zonitidae) at Monks Wood. *Journal of Conchology* 29, 247–252.

Murray, J. (1964) Multiple mating and effective population size in *Cepaea nemoralis*. *Evolution* 18, 283–291.

Murray, J. and Clarke, B. (1966) The inheritance of polymorphic shell characters in *Partula* (Gastropoda). *Genetics* 54, 1261–1277.

Murray, J. and Clarke, B. (1976) Supergenes in polymorphic land snails. 1. *Partula taeniata*. *Heredity* 37, 253–269.

Oosterhoff, L.M. (1977) Variation in growth rate as an ecological factor in the land snail *Cepaea nemoralis* (L.). *Netherlands Journal of Zoology* 27, 1–132.

Owen, D.F. (1965) A population study of an equatorial land snail, *Limicolaria martensiana*. *Proceedings of the Zoological Society of London* 144, 361–382.

Parker, G.A (1983) Mate quality and mating decisions. In: Bateson, P. (ed.) *Mate Choice*. Cambridge University Press, Cambridge, pp. 141–166.

Partridge, L. (1983) Non-random mating and offspring fitness. In: Bateson, P. (ed.) *Mate Choice*. Cambridge University Press, Cambridge, pp. 227–256.

Partridge, L. and Harvey, P.H. (1988) The ecological context of life history evolution. *Science* 241, 1449–1455.

Patterson, C.M (1970) Self-fertilization in the land snail family Succineidae. *Journal de Conchyologie* 108, 61–62.

Peake, J. (1978) Distribution and ecology of the Stylommatophora. In: Fretter, V. and Peake, J. (eds) *The Pulmonates*. Academic Press, London, pp. 429–526.

Piechocki, A. (1982) Life cycle and breeding biology of *Vestia elata* (Rossm.) (Gastropoda, Clausiliidae). *Malacologia* 22, 219–223.

Plummer, J.M. (1975) Observations on the reproduction, growth and longevity of a laboratory colony of *Archachatina* (*Calachatina*) *marginata* (Swainson) subspecies *ovum*. *Proceedings of the Malacological Society of London* 41, 395–413.

Pokryszko, B.M. (1990) The Vertiginidae of Poland (Gastropoda: Pulmonata: Pupilloidea) – a systematic monograph. *Annales Zoologie, Warsaw* 43, 134–257.

Pollard, E. (1975) Aspects of the ecology of *Helix pomatia* L. *Journal of Animal Ecology* 44, 305–329.

Potts, D.C. (1975) Persistence and extinction of local populations of the garden snail *Helix aspersa* in unfavourable environments. *Oecologia* 21, 313–334.

Raut, S.K. and Chose, K.C. (1982) Viability of sperms in aestivating *Achatina fulica* Bowdich and *Macrochlamys indica* Godwin-Austin. *Journal of Molluscan Studies* 48, 84–86.

Reise, H. and Backeljau, T. (1995) Behavioural and genetic studies on the mating systems of *Deroceras rodnae* (Gastropoda: Agriolimacidae). In: Guerra, A., Rolan, E. and Rocha, F. (eds) *Proceedings of the Twelfth International Malacological Congress*. Instituto de Investigationes Marinas, Vigo, pp. 398–399.

Reznik, D.N., Byrga, H. and Endler, J.A. (1990) Experimentally induced life history evolution in a natural population. *Nature* 346, 357–359.

Rollo, C.D. (1983) Consequences of competition on the reproduction and mortality of three species of terrestrial slugs. *Researches on Population Ecology* 25, 20–43.

Runham, N.W. and Hunter, P.J. (1970) *Terrestrial Slugs*. Hutchinson, London.

Schaffer, W.M. (1974) Optimal reproductive effort in fluctuating environments. *American Naturalist* 108, 783–790.

Schaffer, W.M. (1979) Equivalence of maximizing reproductive strategies. *Proceedings of the National Academy of Sciences USA* 76, 3567–3569.

Schemske, D.W. and Lande, R. (1985) The evolution of self-fertilization and inbreeding depression in plants. 2. Empirical observations. *Evolution* 39, 41–52.

Selander, R.K. and Kaufman, D.W. (1973) Self-fertilization and genetic structure in a colonising land snail. *Proceedings of the National Academy of Sciences USA* 70, 1186–1190.

Shachak, M. and Phillips, N.J.A. (1971) *Some Ecological Studies on the Orientation in Space and Time for the Activities of Two Species of Desert Snails,* Sphincterochila boissieri and Helicella seetzeni. Midrashat Sde Boker, Sde Boker.

Shachak, M., Orr, Y. and Steinberg, Y. (1975) Field observations on the natural history of *Sphincterochila* (S.) *zonata* (Bourguignat, 1853) (= S. *boissieri* Charpentier, 1847). *Argamon* 5, 20–46.

Shen, J. (1995) Cannibalism in the terrestrial slug *Deroceras laeve*. *Nautilus* 109, 41–42.

Solem, A. (1972) *Tekoulina*, a new viviparous tornatellinid land snail from Rarotonga, Cook Islands. *Proceedings of the Malacological Society of London* 40, 93–114.

Solem, A. (1976) *Endodontid Land Snails from Pacific Islands (Mollusca: Pulmonata: Sigmurethra). Part 1. Family Endodontidae.* Field Museum of Natural History, Chicago, Illinois.

Solem, A. (1978) Classification of the land Mollusca. In: Fretter, V. and Peake, J. (eds) *The Pulmonates*, Vol. 2A, *Systematics, Evolution and Ecology.* Academic Press, London, pp. 49–98.

South, A. (1965) Biology and ecology of *Agriolimax reticulatus* (Müll) and other slugs: spatial distribution. *Journal of Animal Ecology* 34, 403–417.

South, A. (1989) A comparision of the life cycles of the slugs *Deroceras reticulatum* (Müller) and *Arion intermedius* Normand on permanent pasture. *Journal of Molluscan Studies* 55, 9–22.

South, A. (1992) *Terrestrial Slugs.* Chapman & Hall, London.

Staikou, A. and Lazaridou-Dimitriadou, M. (1990) Aspects of the life cycle, population dynamics, growth and secondary production of the snail *Monacha cartusiana* (Müller, 1774) (Gastropoda Pulmonata) in Greece. *Malacologia* 31, 353–362.

Staikou, A. Lazaridou-Dimitriadou, M. and Farmkis, N. (1988) Aspects of the life cycle, population dynamics, growth and secondary production of the edible snail *Helix lucorum* Linnaeus, 1758 (Gastropoda, Pulmonata) in Greece. *Journal of Molluscan Studies* 54, 139–155.

Staikou, A., Lazaridou-Dimitriadou, M. and Pana, E. (1990) The life cycle, population dynamics, growth and secondary production of the snail *Bradybaena fruticum* (Müller, 1774) (Gastropoda, Pulmonata) in northern Greece. *Journal of Molluscan Studies* 56, 137–146.

Stanlen, R. (1917) On the calcareous eggs of terrestrial Mollusca. *Journal of Conchology* 15, 154–164.

Stearns, S.C. (1976) Life history tactics: a review of the ideas. *Quarterly Review of Biology* 51, 3–47.

Stearns, S.C. (1992) *The Evolution of Life Histories.* Oxford University Press, Oxford.

Stearns, S.C. and Crandall, R.E. (1981) Quantitative predictions of delay maturity. *Evolution* 35, 445–463.

Taylor, J.W. (1894–1921) *Monograph of the Land and Freshwater Mollusca of the British Isles.* 3 volumes + 3 parts (unfinished). Taylor Brothers, Leeds.

Tomiyama, K. (1993) Growth and maturation pattern in the giant African snail *Achatina fulica* (Férussac) (Stylommatophora: Achatinidae). *Venus* 52, 87–100.

Tomiyama, K. (1994) Courtship behaviour of the giant African snail *Achatina fulica* (Stylommatophora: Achatinidae). *Journal of Molluscan Studies* 60, 47–54.

Tomiyama, K. (1996) Mate choice criteria in a protandrous simultaneously hermaphroditic land snail *Achatina fulica. Journal of Molluscan Studies* 62, 101–112.

Tompa, A. (1976) A comparative study of the ultrastructure and mineralogy of calcified land snail eggs. *Journal of Morphology* 150, 861–888.

Tompa, A. (1979a) Studies on the reproductive biology of gastropods: part 1. The systematic distribution of egg retention in the subclass Pulmonata (Gastropoda). *Journal of the Malacological Society of Australia* 4, 113–120.

Tompa, A. (1979b) Oviparity, egg retention and ovoviviparity in pulmonates. *Journal of Molluscan Studies* 45, 155–160.

Tompa, A. (1984) Land snails (Stylommatophora). In: Tompa, A.S., Verdonk, N.H. and van den Biggelaar, J.A.M. (eds) *The Mollusca*, Vol. 7, *Reproduction.* Academic Press, New York, pp. 47–140.

Tuskes, P.M. (1981) Population structure and biology of *Liguus* tree snails on Lignumvitae Key, Florida. *Nautilus* 95, 162–169.

Uminski, T. (1975) Life cycles of some Vitrinidae from Poland. *Annales de Zoologie* 23, 17–33.

Vaught, K.C. (1989) *A Classification of the Living Mollusca.* American Malacologists Inc., Melbourne, Florida.

Ward, D. and Slotow, R. (1992) The effects of water availability on the life history of the desert snail, *Trochoidea seetzeni. Oecologia* 90, 572–580.

Wolda, H. (1963) Natural populations of the polymorphic landsnail *Cepaea nemoralis* (L.). *Archives Néerlandaises de Zoologie* 15, 381–471.

Wolda, H. (1965) The effect of drought on egg production in *Cepaea nemoralis. Archives Néerlandaises de Zoologie* 16, 387–399.

Wolda, H. (1967) The effect of temperaure on reproduction in some morphs of the landsnail *Cepaea nemoralis* (L.). *Evolution* 21, 117–129.

Wolda, H. and Kreulen, D.A. (1973) Ecology of some experimental populations of the landsnail *Cepaea nemoralis* (L). 2. Production and survival of eggs and juveniles. *Netherlands Journal of Zoology* 23, 168–188.

Yom Tov, Y. (1971) The biology of two desert snails *Trochoidea* (*Xerocrassa*) *seetzeni* and *Sphincterochila boissieri. Israel Journal of Zoology* 20, 231–248.

Yom Tov, Y. (1972) Field experiments on the effect of population density and slope direction on the reproduction of the desert snail *Trochoidea* (*Xerocrassa*) *seetzeni. Journal of Animal Ecology* 41, 17–22.

Yom Tov, Y. (1983) Life history tactics in two species of desert snails. *Journal of Arid Environments* 6, 39–41.

13 Behavioural Ecology: On Doing the Right Thing, in the Right Place at the Right Time

A. Cook

School of Environmental Studies, University of Ulster, Coleraine, Northern Ireland BT52 1SA, UK

Introduction

The title of this chapter is taken from Rollo *et al.* (1983a) who contend that the success of terrestrial gastropods is built on a series of behavioural decisions which depend upon the integration of internal physiological factors with external factors such as climate. These behaviours are performed within a narrow envelope of appropriate conditions.

Slugs and snails are unlikely land animals. Most rely on adhesive locomotion, they possess a permeable, wet integument and (in the case of snails) carry a relatively heavy, unsupported shell. Rollo and Wellington (1977) put it thus 'A bag of cold water that cannot even move unless it leaks should not be able to survive outside a bog'.

Terrestrial gastropods can be considered to exist in two different states – moving and roosting. The moving animal lays down a water-laden mucus on which it moves, exposes its integument to a potentially drying atmosphere and increases water losses through the pallial cavity because of the necessity for gas exchange. Shell-less species are the most vulnerable and, in these, most water is lost through the skin rather than in the urine (Deyrup-Olsen and Martin, 1982). In return for exposing themselves to these potential hazards, terrestrial gastropods gain access to food, mates and water, and can disperse. A roosting terrestrial gastropod deploys a variety of passive mechanisms for water conservation, including the direct protection of its wet surfaces from drying conditions, avoidance of temperature extremes, the creation of more favourable microclimates and decreases in gas exchange. Despite the contrast in the conditions of these two states, some of the most vulnerable gastropods (slugs) can maintain remarkably constant water content. Lyth (1983) measured the water content of a range of slug species collected in the field and found that it remained relatively stable. This was true in comparisons between animals of the same species collected at the same time from the same site and between individuals of the same species collected at different times. Thus for *Deroceras reticulatum* (Müller) (Agriolimacidae), the most common species at the sites, the water content of the animals only varied from 86.4 to 90%, despite the soil moisture content varying from 15 to 57% on different occasions. Furthermore, the coefficient of variation between field-collected individuals was comparable with that measured for successive observations on individuals kept in standardized laboratory cultures (Lyth, 1982). The inevitable conclusion from these observations is that these animals move between hydrating and dehydrating conditions in such a way as to keep a stable water content in the field while at the same time fulfilling reproductive and nutritional imperatives. Such sensitive control of movement will rely on behavioural choices reflecting proactive strategies depending on endogenous components and reactive mechanisms responding to prevailing ecological factors.

The behavioural choices that are made are those concerned with:

- where to be active;
- when to be active;
- what to do when active;
- where to be inactive; and
- how to get to roosting sites.

These behavioural choices are underpinned with physiological or ecological causation and overlain with functional consequences, many of which are associated with water balance.

This chapter, therefore, is concerned with the description and analysis of those behaviours of terrestrial gastropods that contribute to their abilities to survive in an unstable environment while maintaining their ability to function effectively.

Where to be Active

Habitat selection

Many factors contribute to the local distribution of species. The physical features of the habitat are central factors associated with the choice of habitat (Peake, 1978). Inorganic ions (Burch, 1955; Hermida *et al.* 1995), pH (Cameron, 1973; Nekola and Smith, 1999), drainage (Stephenson, 1967; Paul, 1978), soil texture (Stephenson, 1966, 1975; Hermida *et al.* 1995), local temperature variations (Jones *et al.*, 1977; Cameron, 1970; Cain and Curry, 1963), altitude (Cowie *et al.*, 1995) and topographic features associated with shelter (South, 1965) have all been implicated in determining species distribution. Other factors are associated with food preferences (Cook and Radford, 1988; Negovetic and Jokela, 2000), while still others are determined by behavioural interactions with other species and conspecifics (Rollo and Wellington, 1979). Lominicki (1964) suggested that populations of *Helix pomatia* Linnaeus (Helicidae) were controlled by the availability of shelters, giving individuals which were familiar with the area an advantage over interlopers. Baker (1988) found the distribution of *Theba pisana* (Müller) (Helicidae) in pasture to vary with season, and suggested that this species responds to olfactory cues from food plants as their quality varies through the year. Kasigwa (1999b), on the other hand, related diurnal changes in the vertical distribution of *Sitala jenynsi* (Pfeiffer) (Ariophantidae) to local changes in relative humidity.

Most terrestrial gastropods are general herbivores, and specific associations between gastropods and plant species have been reported (Karlin, 1965; South, 1965; Beyer and Saari, 1977). These, however, may have more to do with the nature of the microclimate than with highly specific feeding preferences. Indeed, where correlations between habitat and species have been recorded, further examination normally shows that a third factor such as ground cover, soil moisture, etc., accounts for the observed distribution. Beyer and Saari (1978) noted a strong association between *Arion subfuscus* (Draparnaud) (Arionidae) and timothy grass (*Phleum pratense* Linnaeus) (Gramineae) in wooded areas of New York state, but concluded that since this plant was not a dominant food item, the association was related to the shelter provided. Furthermore, a detailed examination of the relationship between the distribution of species and local environmental factors in an area of Hawaii (Cowie *et al.*, 1995) has shown that the most probable link is with altitude and lava type rather than associated vegetation changes. South (1965) has suggested that the distribution of *D. reticulatum* in grassland is a result of the animals' choice of egg-laying sites and differential egg mortality. This will leave most juveniles in areas selected by the adults for oviposition, the distribution preserved by homing (Newell, 1968) and differential mortality in drying conditions (Bayne, 1969).

Studies of dispersal of individuals have shown that long-term movements can be restricted by the availability of suitable sites. Thus the dispersal of *Chondrina clienta* (Westerlund) (Chondrinidae), a species associated with rock, is restricted by the distances between suitable rocks, being greatest on limestone pavement (264 cm year^{-1}) and least on a more structurally diverse stone wall (88 cm year^{-1}) (Baur and Baur, 1995). Similarly, the long-term movements of *Arianta arbustorum* (Linnaeus) (Helicidae) have been observed to be restricted to favourable areas: within a pattern of otherwise random movements, dispersal was blocked at the junction of favourable vegetation and an area of drier mown meadow (Baur and Baur, 1993). Similarly, in a study of the dispersion of *S. jenynsi*, Kasigwa (1999a) found movements to be restricted by factors such as discontinuous vegetation cover and dry weather, but promoted by rainy weather and continuous cover. In addition, dispersal is restricted by the availability of food plants (Baker, 1988; Baur, 1993; Kasigwa, 1999a).

Sites of activity

Species vary in the sites over which they are active during their excursions. Cameron (1978) and Cain and Cowie (1978) both relate the sites of activity of gastropod snails to the shape of the shell, with tall-shelled species preferring vertical hard surfaces and those with more flattened shells preferring bare soil. Species with globular shells were found on live vegetation. Cook and Radford (1988) related different sites of activity to feeding preferences of sympatric limacid species. Work on the microdistribution of other sympatric species came to a variety of conclusions. Conroy (1980), for example, concluded that the differential distribution of *Arion lusitanicus* Mabille, *Arion ater* (Linnaeus) and *A. subfuscus* was related to the aspect of the site.

Competition

It is axiomatic that closely related species living in the same place avoid interspecific competition by diverging in some characteristic that impinges directly on the utilization of resources in short supply. Such partial niche separation has been demonstrated in sympatric limacid species (Cook and Radford, 1988) on the basis of feeding preferences and site occupancy, and, in sympatric *Cepaea* Held (Helicidae) species, minor dietary differences have also been detected (Carter *et al.*, 1979). The sympatric rock-dwelling species *C. clienta* and *Balea perversa* (Linnaeus) (Clausiliidae) similarly differ in their food preferences, which may be associated with their ability to handle the different secondary products of their host lichens (Baur *et al.*, 1994). *H. pomatia* and *Cantareus aspersus* (Müller) (Helicidae) display different activity patterns (Blanc *et al.*, 1989). The observations of Asami (1993) have shown that *Mesodon normalis* (Pilsbry) (Polygyridae) and *Triodopsis albolabris* (Say) (Polygyridae), which live in leaf litter in the Appalachian forests, are also separated by different activity patterns in which the former was crepuscular and the latter strictly nocturnal. These differences persist when either species is observed in allopatry, and therefore do not arise from interspecific interaction, but have evolutionary origins. Getz (1959) has demonstrated different feeding preferences in three sympatric slug species, *Arion circumscriptus* Johnston, *Deroceras laeve* (Müller) and *D. reticulatum*. Cameron (1978) has shown that different species occupying the same area are actually active over different parts of that site. Hatziioanou *et al.* (1994) have shown that 83% of the species at a site in Greece can be distinguished on the basis of their resting sites. In all the above examples, niche separation between sympatric species is incomplete and competition for resources will occur.

Interaction with other animals will affect where some species may be active. Rollo and Wellington (1979) examined the distribution of slugs in closed arenas with limited daytime resting sites and in which the food was concentrated at one end. They found that an aggressive species such as *Limax maximus* Linnaeus (Limacidae) displaced *A. ater* and *Ariolimax columbianus* (Gould) (Arionidae) from the shelters nearest the food. Indeed, in some experiments, *L. maximus* either eliminated these other species from the arenas completely or those that survived were badly wounded. In experiments using smaller arenas and single large home sites, Cook (1981b) showed in monocultures that *Limacus flavus* (Linnaeus) and *Limacus pseudoflavus* (Evans) (Limacidae) always occupied the home site, but this tendency was reduced in *L. maximus* and *Lehmannia marginata* (Müller) (Limacidae). In mixed-species groups, the home site occupancy of *L. flavus* and *L. pseudoflavus* was unchanged. *L. marginata* showed an increased occupancy when mixed with *L. flavus* or *L. pseudoflavus*, whereas mixture with *L. maximus* made no significant difference. *L. maximus* showed a decreased occupancy when mixed with non-aggressive species and an increased occupancy when mixed with *L. marginata*. Within the home site, *L. flavus* and *L. pseudoflavus* form tightly packed huddles (Cook, 1981a,b) (Fig. 13.4A) and the total area occupied by these species is relatively small. Aggressive species do not huddle in summer when these experiments were conducted. In the home site, *L. maximus* comes to rest when it makes contact with another individual, and *L. marginata* actively avoids other individuals. The effective area occupied by these aggressive species is, therefore, comparatively large. The results of Cook (1981b) and Rollo and Wellington (1979) may not, therefore, be as contradictory as they may first appear when the actual area occupied by the slugs in the home is considered.

Aggressive species tend to reduce the resources available to non-aggressive species (Rollo, 1983a,b) through changing the behaviour of the latter. In combination, *A. ater* and *A. columbianus* had no detectable effect on each other. *L. maximus*, on the other hand, disrupted the time allocated to feeding, mating and locomotor behaviours in these non-aggressive species, but its own time budget was not modified. These changes in time budget and their consequent effects on resource acquisition had a cumulative effect on the reproduction and mortality of the non-aggressive species.

The interaction between species may not be a simple aggressive one. There is an accumulating body of evidence that the deposition of mucus has subtle effects on the growth and behaviour of individuals of both conspecifics and other species. Cameron and Carter (1979) demonstrated reduced activity levels and growth rates in *Cepaea nemoralis* (Linnaeus), *Cepaea hortensis* (Müller) and *C. aspersus* in containers pre-treated with the mucus of conspecifics. Further experiments showed that the mucus of one *Cepaea* species led to a reduction in the activity of the other, but that the mucus of *C. aspersus* had no effect on either *Cepaea* species. The mechanism involved, therefore, cannot be due to simple fouling or resource depletion since the effect differed between species. Similar differential effects were found by Dan and Bailey (1982), where mucus from adult *C. aspersus* affected the activity of *C. nemoralis*, *Trichia striolata* (Pfeiffer) (Hygromiidae) and juvenile *C. aspersus*, but had no effect on adults of *C. aspersus* or *Candidula intersecta* (Poiret) (Hygromiidae). Tattersfield (1981) concluded that mucus conditioning of the environment was the most likely cause of variations in the shell size of *C. intersecta*, *Cochlicella acuta* (Müller) (Hygromiidae) and *Helicella itala* (Linnaeus) (Hygromiidae) that occurred in dunes at different densities. Jess and Marks (1995) observed the effects of crowding in *C. aspersus* cultures, which could be alleviated by cleaning the containers, and concluded that the early effects of high density persisted even after the animals were returned to lower stocking densities. The frequency of cleaning had little effect until containers were changed every 2 days. They concluded that the effects of population density were rapid and persistent.

Laboratory experiments with *C. clienta* and *B. perversa* (Baur, 1988, 1990; Baur and Baur, 1990) demonstrate that both inter- and intraspecific competition influence the growth rate of juveniles and the ultimate size of the adults. This competition effect was postulated to have been mediated via the mucus. While such interactions can be observed in the laboratory, Baur (1993) has also established that they may not necessarily be reflected in the field.

The field observations of the population dynamics of *Helix texta* Mousson by Heller and Ittiel (1990) and of *Cristataria genezarenthana* (Tristram) (Clausiliidae) by Heller and Dolev (1994) led to the conclusion that the final growth phase, to adulthood, was inhibited by the presence of already mature animals, probably through their deposition of mucus. The evidence for this is compelling since very little recruitment to the adult population of *H. texta* was observed until about 50% of the adults in the study area had been eaten by wild boar, *Sus scrofa* Linnaeus: only after this catastrophic event did the smaller snails in the population grow to adulthood (Heller and Ittiel, 1990).

Most of the aforementioned experiments point to some residue in gastropod mucus which adversely affects the growth rate by reducing the activity in sensitive species. Such an effect might account for some of the inhibitory effects which adults exert on the growth of juvenile gastropods. If this is not a direct toxic effect, however, then it must be assumed that there is an advantage to both the producer of the effect and the individuals which respond. The advantage to the producer is clear; it reduces competition for space and food. However, the advantage to the individual that responds, by reducing its growth and delaying its maturation, is not readily apparent since an individual which does not respond in this way gains an advantage over its peers. Cook (1989) argues that the reported effects of mucus on activity and growth could be an example of a 'war of attrition' (Parker, 1978) in which adults produce substances which inhibit the activity of juveniles. The appropriate response in such a 'game' would be for individuals to vary their activity and growth rates randomly, but this has not been observed directly. Growth rates, however, can be extremely variable, as are the sizes at which individuals become sexually mature (Daguzan, 1982; Prior, 1983b). Bull *et al.* (1992) report that they could not repeat the results of Smallridge and Kirby (1988) in which the mucus of *T. pisana* influenced the growth and survival of *Cernuella virgata* (da Costa) (Hygromiidae). Such failures to replicate results, even by the same authors from year to year (e.g. Bull *et al.*, 1992), might be expected if the biological strategy adopted were one of random variability.

The response of juveniles to the inhibitory effects of the mucus of conspecific adults could also involve kin selection where the probability of the adults and the juveniles being closely related is high. This would occur in species that occupy persistent home ranges in which they lay their eggs and in which the tendency of the juveniles to migrate is low.

The reactions to mucus are not always inhibitory on activity and growth. Landaur and Chapnick (1981) have shown that the mucus of stressed conspecifics is aversive to *Lehmannia valentiana* (de Férussac). Thus, mucus might perform a variety of functions in gastropod populations, ranging from the inhibition of growth in conspecific juveniles to liberating alarm substances.

In conclusion, it is clear that the effects of competition on growth rates and adult size must be mediated via a range of mechanisms including the secretion and response to mucus and direct aggressive interactions resulting in exclusion from food and home sites. While these features can be demonstrated in laboratory experiments, the interpretation of the more complex interactions that occur in the field remains to be resolved.

When to be Active

In most climates, there is a regular variation in conditions through the period of a day and a year. The activity of terrestrial gastropods is adapted to exploit these variations, enabling concealment from unfavourable conditions and exposure to favourable ones.

Seasonal activity

In population studies of terrestrial gastropods, many species have been found to exhibit a decline in abundance in both the summer and winter. In some cases, this probably represents a genuine decline in numbers, but in others it is best interpreted as substantial proportions of the population becoming inactive and therefore not being sampled (Voss *et al.*, 1999). Uminski (1983) attributed the summer decline in numbers of *Vitrina pellucida* (Müller) (Vitrinidae), an annual species, to their migration away from the study site and becoming inactive. Heller and Ittiel (1990) have shown that for *H. texta* in the Mediterranean region of Israel, activity is concentrated into the winter rainy season (October to December) and that these gastropods spend the whole of the hot summer in aestivation. Similarly Dallas *et al.* (1991) found the activity of *Trigonephrus haughtoni* Connolly (Acavidae), in the Namib Desert, to be restricted to the few hours in which dew may persist, giving an acceptable combination of high humidity (higher than 85%) and low temperature (less than 11 °C).

In temperate regions, we might expect nocturnal activity to occupy a greater part of the dark hours. This is only true to a limited extent. Ford and Cook (1994), working with L. pseudoflavus, showed that there was a trend towards increasing activity in the longer nights of spring and autumn. This did not, however, continue into the winter, despite the conditions in the laboratory being unchanged. The animals they used, however, were freshly collected from the field, and no allowance was made for the effects of recent experience. In general, animals which had been kept in laboratory culture tended to spend more time in locomotory activity than those freshly brought in from the field, but no greater time was spent feeding. Rollo (1991) and Ford (1986) have shown that the duration of the excursions of D. reticulatum and L. pseudoflavus, respectively, is dependent on the availability of food. Airey (1987) has reported that the distances travelled by D. reticulatum in a night increase with food deprivation. These observations indicate that the duration and extent of activity is determined partly by the demand for food. Thus in temperate regions, winter low temperatures would lead to reduced nutritional requirements, and since, in most species, the reproductive cycle has been completed, there is a low motivation to become active. This is seen both as a relatively low proportion of the population becoming active and as a low duration of activity for those that do emerge.

The variation of activity with season may merely reflect responses to prevailing environmental factors such as day length and temperature. The entry into a state of dormancy either in winter (hibernation) or in summer (aestivation), however, may have more profound causations. Bailey (1981, 1983) proposed that the onset of hibernation in C. aspersus is under the control of a circannual rhythm entrained by day length. H. pomatia hibernate in response to decreasing day length and low temperatures (Jeppesen and Nygard, 1976; Jeppesen, 1977) and, since the tendency to hibernate varies through the year, there is some measure of internal control. Further, Edelstam and Palmer (1950) report a marked homing behaviour related to winter quarters in H. pomatia. Cooke (1895) presents anecdotal evidence that gastropods hibernate in locations other than their normal day time resting sites. Internal, pre-programmed mechanisms cannot be the only ones involved in the determination of the onset of dormancy, since laboratory experiments have shown that factors such as dehydration are implicated in the onset of dormancy in Helix lucorum Linnaeus and the role of photoperiod appears to be a minor one. Nevertheless, in the field, H. lucorum will hibernate prior to the onset of conditions that trigger dormancy in the laboratory and will remain inactive after favourable conditions return (Lazaridou-Dimitriadou and Saunders, 1986). Similarly, the activity of H. texta (Heller and Ittiel, 1990) and C. genezarenthana (Heller and Dolev, 1994) reaches a peak before the period of highest rainfall, and these animals become quiescent before the end of the rainy season.

Some snails produce an epiphragm that seals the shell aperture during prolonged inactivity, and epiphragm production has been cited as evidence for a pre-programmed onset of a period of quiescence. Klein-Rollais and Daguzan (1990) suggest that the autumn increase in water content of C. aspersus is a preparation for the greater mucus production involved in epiphragm production. None the less, the epiphragm may be built up during inactivity since A. arbustorum secretes a series of epiphragms within the body whorl of the shell (Terhivuo, 1978). Many species, including Sphincterochila boissieri de Charpentier (Sphincterochilidae) (Yom-Tov, 1971), secrete a new epiphragm after every bout of activity and, therefore, in these species the epiphragm is not associated specifically with these seasonal periods of enforced rest. The epiphragm itself, therefore, may be a result of inactivity rather than a pre-programmed preparation for it.

The differential mortality of snails adopting different orientations during hibernation tends to lend strength to the view that these rest periods are more opportunistic than pre-programmed. A. arbustorum, for instance, showed a lower mortality after hibernation when in an 'apex down' position. These individuals also remained inactive for longer (Terhivuo, 1978). Baur and Baur (1991) report higher survival among C. clienta and B. perversa individuals that formed part of a hibernation aggregation than those individuals that hibernated singly.

There are then probably two elements in the control of the onset of dormancy in the form of either hibernation or aestivation. First, there is a pre-programmed element that is controlled by a circannual rhythm entrained by day length and which involves physiological preparations and an increased tendency to adopt traditional winter roosting sites. Secondly, this preparation may be overridden by events and animals may be forced into a dormant state by prolonged dehydration, low temperatures, etc.

Bailey (1983) suggested that the termination of inactivity is delayed until the reduction of a hibernating substance has occurred. Lazaridou-Dimitriadou and Saunders (1986) found that the duration of hibernation is influenced by previous brief periods of dormancy, and some minimum period of dormancy may be required before arousal. The same authors also observed that H. lucorum terminated hibernation by rejecting the epiphragm without necessarily resuming activity. Thus, breaking dormancy need not coincide with the resumption of activity. Even after any necessary minimum dormant period, animals emerging from hibernation may be active only when conditions are favourable and may indeed aestivate immediately after breaking of hibernation. The conditions necessary for the resumption of activity post-hibernation vary with species, some emerging only after rain or dew fall (Yom-Tov, 1971), others requiring mechanical disturbance (Herreid and Rokitka, 1976). It will be of adaptive advantage if gastropods do not become active in the winter just because there is a short-term rise in temperature. Premature activity does, however, appear to occur, and the termination of hibernation may be under less rigid control than it is in some other invertebrate groups.

It appears, therefore, that hibernation and aestivation in shelled terrestrial gastropods may only have a very weak endogenous component,
being largely controlled both in its onset and its termination by the prevailing weather conditions. There is little information on physiological
dormancy in slug forms, although the activity of Agriolimacidae,
Arionidae, Milacidae and Limacidae is generally recognized as being
severely limited by low temperatures and low humidities. *Arion intermedius* Normand has the ability to aestivate during dry periods in cells
constructed of mucus, soil and moss (Barker, 1999). Cooke (1895) reports
that slugs may spend the winter encysted in a mucous covering, but this
may not necessarily constitute hibernation.

Circadian activity

During periods of activity, terrestrial gastropods must acquire the
necessary resources for both their immediate and long-term needs. Given
the obvious perils of activity, therefore, these animals tend to be inactive
unless resources are required.

The components of gastropod activity in the field are: (i) emergence –
animals become active and leave their roosts; (ii) excursion – animals
make an excursion during which feeding, drinking and courtship are
interspersed with periods of rest; and (iii) roosting – animals locate a
suitable resting site, often by homing to previously occupied sites, and
come to rest. The factors which control these various elements of activity
will control the observed field activity (Fig. 13.1).

Attempts to relate activity to simple environmental variables under
laboratory conditions are almost certain to succeed. Dainton (1954a,b)
and Dainton and Wright (1985) showed that the activity of *A. ater* and
other species was initiated by small changes in temperature. Similarly,
changes in wind speed (Richter, 1976), mechanical disturbance (Ross,
1979), relative humidity (Machin, 1975; Burton, 1983; Hess and Prior,
1985; Takeda and Ozaki, 1986), light (Karlin, 1961) and soil moisture
(Young and Port, 1991; Speiser and Hochstrasser, 1998) can all be shown
to individually influence activity. The absolute values of these variables
as well as the length of exposure to them are important influences on
activity (Lewis, 1969a; Munden and Bailey, 1989).

Some species show well-ordered patterns of diurnal activity, and it is
unlikely that these patterns are controlled directly by simple, individual
cues such as temperature or humidity (Ford and Cook, 1987). Lewis
(1969a,b) attempted to relate the activity of *A. ater* to the variation of
physical factors such as temperature and light. He concluded that an
endogenous rhythm existed which was controlled by light cycles but that
temperature variations had little effect on activity.

In the field, environmental factors act together and interact with
endogenous elements such as hydration, nutritional status and activity
rhythms (Rollo *et al.*, 1983a). In addition, different colour morphs and the

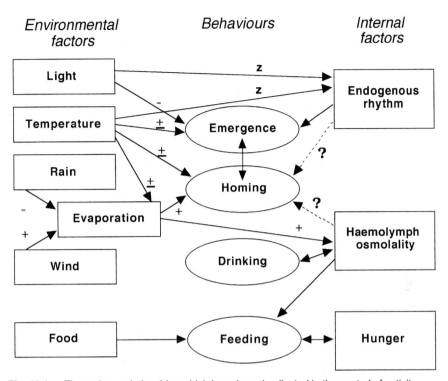

Fig. 13.1. The various relationships which have been implicated in the control of activity in terrestrial gastropods. Solid lines represent relationships for which there is clear evidence, dotted lines (?) refer to speculative relationships and z represents the zeitgebers implicated in the entrainment of the circadian rhythm.

extent of physiological adaptation may influence the extent of activity in some species (Abdel-rehim, 1983).

Field studies of the effect of weather on activity are of two types. The first seeks to relate those nights on which high activity levels occur to potential causal factors. The second attempts to relate the minute to minute changes in the activity of populations of animals to causal environmental and physiological factors. Dundee and Rogers (1977) used discriminant function analysis to relate prevailing weather to gastropod activity. They concluded that there was a significant quantitative difference between factors such as temperature, soil moisture and wind speed on nights when gastropods were active, compared with nights when they were inactive. Detailed examination of some of the factors involved, however, shows that some of these do not have a direct causal relationship. For instance, wind speed at 0.5 m above the ground was a significant factor, whereas that at the surface was not. In addition, soil moisture 75 mm below the ground was positively related, whereas soil moisture at the surface was negatively related to nights on which activity occurred. Clearly, while the relationship between activity and weather is an important aspect of the control of behaviour, it is not a simple one.

Table 13.1. Climatic variables limiting the activity of slugs as determined by threshold models

Species	Limacus pseudo-flavus (Evans) (Limacidae) (Ford, 1986)	Limax maximus Linnaeus (Limacidae) (Rollo, 1982)	Deroceras reticulatum (Müller) (Agriolimacidae) (Rollo, 1991)	Deroceras reticulatum (Müller) (Agrio-limacidae) (Young and Port, 1989)
Surface temperature	< 3.3°C	6.8–18.7°C	7.8–19°C	5.1–15.1°C
Evaporation	> 0.82 mmHg (VPD)	0.53 mm^3 min^{-1} (direct measure)	0.66 mm^3 min^{-1} (direct measure)	26.2% (w/w) (soil moisture)
Relative humidity	> 89.5%	Not included	Not included	81.5%
Wind speed	< 8 m s^{-1}	Not included	< 5 m s^{-1}	0.30 m s^{-1}

Where appropriate, values are taken as those between which 50% of the population were observed to be active.

Young and Port (1989) related weather data to high- and low-activity nights and used a limit model to predict the activity of *D. reticulatum* in an adjacent field plot (i.e. they determined the range of weather factors within which activity would be greatest). They identified values of wind speed, temperature, humidity and soil moisture content that limited activity (Table 13.1). This model tended to err towards under-predicting the numbers of nights on which *D. reticulatum* will be active, but correctly predicted 'high' activity in the test plot on 17 out of 20 nights. This study has the benefit of giving a practical method of predicting periods of crop vulnerability.

There have been a number of studies in which gastropod activity (measured as the proportion of the population or numbers active) and weather variables have been measured over a series of diurnal cycles (Webley, 1964; Crawford-Sidebotham, 1972; Bailey, 1975; Rollo, 1982, 1991; Ford, 1986; Cook and Ford, 1989). The subsequent analysis of the relationships between activity and the factors which control it has been attempted in two ways, regression modelling and threshold modelling. A summary of these methods and the type of data from which the models are derived are shown in Fig. 13.2.

Simple regression models have been used to identify individual factors whose variation was significantly correlated with the levels of activity. Multiple regression using these variables (and sometimes their transformations and products) have then been employed to construct a regression model relating activity to weather. Table 13.2 shows the results of various attempts at this form of modelling. Regression models normally account for a significant proportion of the variance in the data but seldom accurately describe or predict the precise timing of the activity or its magnitude.

In all attempts to relate activity to environmental change over a diurnal cycle, factors associated with time, day length and light intensity were the dominant features of regression models. This indicates the primacy of an activity rhythm in the determination of locomotor behaviour. One of the criticisms of regression modelling is that it assumes that an unfavourable value of one factor can be compensated by an

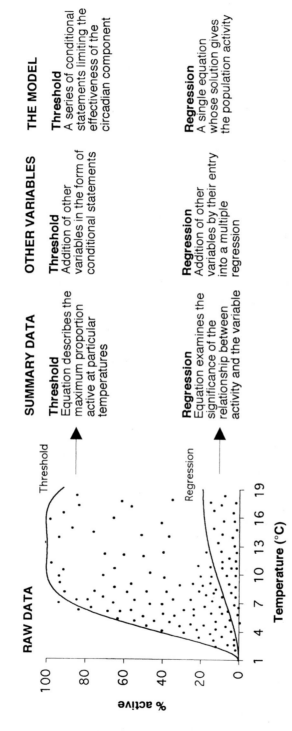

Fig. 13.2. A summary of the processes involved in the production of regression and threshold models of activity. Raw data after Ford (1986).

optimum value of another; for example, that a gastropod species might be active at extremely low temperatures provided that the humidity was high enough. This is not necessarily true. Nevertheless, the models do highlight environmental factors expected to be associated with activity and can raise new research questions. Ford (1986) showed that for models developed to describe *L. pseudoflavus* activity in each of 8 months, temperature was a significant component in only 3 months, humidity (RH% or vapour pressure deficit) in 4 months, wind speed in 3 months and rainfall in 2 months. Barometric pressure was a significant factor in 4 months (Table 13.3). Whether species such as *L. pseudoflavus* can perceive barometric pressure is not known, and it may be that the inclusion of this factor is an anomaly, or that it is acting as a surrogate for a less obvious, unmeasured variable.

The second approach to modelling activity was developed by Rollo (1978). This takes a more biological view and supposes that activity is controlled by a rhythm, and then the expression of that rhythm as activity

Table 13.2. A summary of the findings of regression modelling (such models highlight important weather factors controlling activity but rarely accurately model the form of the activity)

Authority	Species	Time scale (frequency of observations)	Most significant weather factor	Variance accounted for by multiple regression
Crawford-Sidebotham (1972)	Various slugs	1 per night	Temperature and vapour pressure deficit	27–91% depending on species
Bailey (1975)	*Cantareus aspersus* (Müller) (Helicidae)	24 h	Soil moisture Surface temperature	62%
Rollo (1982)	*Limax maximus* Linnaeus (Limacidae)	1 h	Temperature at home Soil moisture	81%
Ford (1986)	*Limacus pseudoflavus* (Evans) (Limacidae)	30 min		60%
Rollo (1991)	*Deroceras reticulatum* (Müller) (Agriolimacidae)	1 h		87%

Table 13.3. The frequency (%) with which weather factors limited activity in *Limacus pseudoflavus* (Evans) (Limacidae)

Month	Circadian rhythm	Wind speed	Temperature	Humidity	Light
March	9	16	39	35	0
April	14	24	42	14	7
May	92	0	3	2	4
June	71	0	0	8	22
July	64	6	10	1	21
August	37	5	48	2	8
September	60	10	24	5	0
November	9	41	40	15	0
Year	42	13	28	10	8

Limiting factors were examined in each 15-min interval between 20:15 and 6:45 for between seven and 12 consecutive nights in each month (72 nights in all). From data in Ford (1986).

is determined by the value of various environmental factors. These factors present a series of thresholds rather than integrate together as implied by a regression model. These threshold or limit models are therefore based on the examination of data to determine the limits within which activity occurs. They have a compelling biological simplicity that conceals a mathematical complexity. Rollo (1982, 1991) fitted equations to the outer limit of scattergrams of a variable such as temperature plotted against the proportion of the population active. The logic behind this is that beyond the outer envelope, the variable in question must have restricted activity, whereas within the envelope other factors were limiting. The fitted line then describes the proportion of a population which might be expected to be active at a particular temperature provided no other factor was limiting (Fig. 13.2). Rollo (1982) described the diurnal rhythm itself by fitting a polynomial expression to transformations of time, phase, scotoperiod, shelter temperature and some of their products 'when weather factors were not inhibitory'. Ford (1986) followed a similar protocol for his two-dimensional threshold model, except that the circadian rhythm function was determined solely from time-based variables and included only those nights on which over 80% of the *L. pseudoflavus* became active. The process of constructing a limit model is illustrated in Fig. 13.2.

In an extension of simple threshold modelling, Ford (1986) and Cook and Ford (1989) presumed that the threshold for a particular variable varied with the time of night. Three-dimensional surfaces were fitted to plots of environmental variable, time and the maximum proportion of the population active at each time–variable intersection using Chebyshev polynomial expansions. Such a procedure allows the changing conditions during *L. pseudoflavus* excursions to influence the model. From these models, it is clear, for instance, that low relative humidities exert a greater influence on activity later in the night when the animals may have lost water than earlier when in a state of full hydration (Cook and Ford, 1989).

Comparison of the various modelling techniques is not simple since two elements need to be resolved: that is the statistical fit of the predicted activity to that measured and the qualitative relationship between the form of the predicted and observed fits. Ford (1986) found that the various models accounted for roughly the same proportion of the variance (two-dimensional, 62%; three-dimensional, 61%; regression, 60%). Rollo *et al.* (1983a) too found a similar statistical fit in different models, but found a generally higher proportion of variance explained: regression modelling accounted for a maximum of 87% of the variance; threshold modelling accounted for 83%. Both Ford and Rollo reported that regression modelling consistently underestimated activity when observed activity was high and overestimated it when it was low. Thus the form of the curves developed from threshold modelling often fitted reality better than those derived from regression modelling despite accounting for similar amounts of variance (Fig. 13.3). Thus limit or threshold models are more accurate in their description of the timing of the onset and cessation of activity and the timing and magnitude of its peak (Ford, 1986).

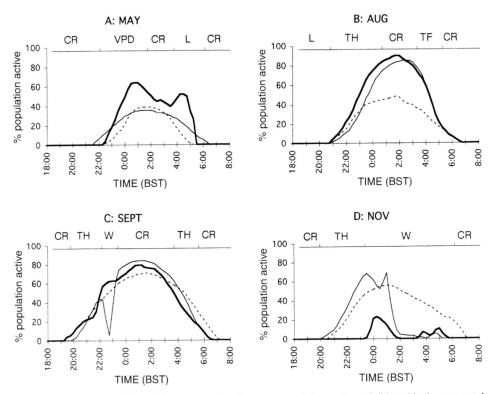

Fig. 13.3. Examples of the activity observed in *Limacus pseudoflavus* (Evans) (Limacidae) ——— on 4 days and the predicted activity using threshold modelling ——— and regression modelling ------. The bar at the top of each graph refers to the element in the threshold model limiting activity for the time period indicated. Note that the amplitude of activity is rarely predicted accurately by the regression model. The threshold model sometimes predicts rapid fluctuations in activity (e.g. C: September in response to a change in the wind in this case) which do not materialize in the field. (Unpublished data from work by A. Cook and D.J.G. Ford.) CR, circadian rhythm; VPD, vapour pressure deficit; L, Light; TF, temperature at the food; TH, temperature in the home; W, wind.

Detailed comparisons of threshold models are not possible because different authors used different environmental variables. Thus Rollo (1982, 1991) measured evaporation directly using an evaporimeter, while Ford (1986) measured vapour pressure deficit using a ventilated psychrometer. Further, the inclusion of a variable in a threshold model developed from field data is only possible if that variable reached limiting values during the observation periods. Thus, the summer temperatures in Northern Ireland never reached the upper limits described by Rollo in Canada and, therefore, Ford's (1986) models of activity for *L. pseudoflavus* contain no upper temperature limits while those of Rollo (1982) for *L. maximus* do. Some gross comparisons between the various attempts at relating weather to activity are presented in Table 13.1. Table 13.3 shows the frequency with which various factors limited the activity of *L. pseudoflavus* during the course of a year in Northern Ireland. The circadian rhythm function of the model and light intensity clearly dominate the control of activity during the short nights of winter. Temperature is a major influence in early spring and autumn, as are humidity and wind speed. This is much as would be expected in a mild, wet climate like Northern Ireland. However, the details of the model would change if the same species were examined in different climates. For this reason, the approach of Rollo *et al.* (1983a) in which some limiting values of environmental variables were established experimentally might provide a clearer way forward in the predictive modelling of terrestrial gastropod activity.

A further level of sophistication in modelling activity comes with the incorporation of motivational factors such as the tendency to home, to mate, to lay eggs, to seek water, etc. (Rollo *et al.*, 1983a). Such integrated models of the relationship between internal and external influences on behaviour are possible and appear to have predictive value. Further, they open the way to comparisons between species that may deploy different strategies in responding to the unstable terrestrial environment. They will probably serve to highlight outstanding questions of the control of activity and how mechanisms and strategies may vary between species.

In conclusion, it is clear that the work on the formulation of predictive models of terrestrial gastropod activity is converging on the view that the appropriate model is composed of a circadian component which promotes the initiation of activity and, if no other factor intervenes, subsequent homing. The expression of this rhythm may be influenced by internal motivational factors, and by external climatic ones. Once an animal is active, the activity rhythm is modified by the initiation of homing by a range of exogenous factors including low or high temperatures, low humidity and high wind speed.

The development of models which are capable of predicting gastropod activity have value in that they highlight those factors which most influence behaviour and point the way towards techniques for predicting periods of maximum damage to crops or the most efficacious time to apply control treatments. Such models, however, do not illuminate the mechanisms underlying the behaviour that they seek to describe. It is clear from those models developed to date that normal activity in the field is dependent on an activity rhythm. The form of this rhythm varies with species and, within the same species, with the environmental conditions. Many species, for example, tend to be crepuscular (*C. aspersus, H. lucorum* and *H. pomatia*) with a bimodal activity pattern (Bailey, 1975; Bailey and Lazaridou-Dimitriadou, 1986; Blanc *et al.*, 1989). There is some suggestion that *Arion lusitanicus* Mabille shows a trimodal activity pattern, with observation of dawn and dusk peaks and an implied nocturnal peak (Grimm *et al.*, 2000). Other species tend to utilize both the light and dark periods equally (e.g. *Punctum pygmaeum* (Draparnaud) (Punctidae), Baur and Baur, 1988). In contrast, other species such as *Deroceras caruanae* (Pollonera) (Morton, 1979) tend to be nocturnal, with a unimodal activity pattern. *A. columbianus* is apparently unusual in displaying a unimodal peak of feeding centred around midday in spring and autumn and being crepuscular in summer (Richter, 1976).

Superimposed upon this regular activity is an opportunistic element that varies from species to species. *D. reticulatum*, *L. marginata*, *Geomalacus maculosus* Allman (Arionidae) (Platts and Speight, 1988) and *C. aspersus* (Klein-Rollais and Daguzan, 1990) frequently are active during the daylight hours if the prevailing conditions are both humid and overcast. Cameron (1970) studied the activity of three helicid species under laboratory conditions and found that the extent of the daytime activity varied with temperature. At higher temperatures, *A. arbustorum* maintained a high level of daytime activity (above 50%) compared with *C. nemoralis* and *C. hortensis* in which daytime activity fell to about 40% of the total.

There have been several attempts to analyse the endogenous and exogenous components controlling activity. The studies of Bailey (1975), Rollo (1982, 1991), Wareing and Bailey (1985), Bailey and Lazaridou-Dimitriadou (1986), Ford and Cook (1987, 1988, 1994), Blanc *et al.* (1989), Staikou *et al.* (1989), Flari and Lazaridou-Dimitriadoa (1995a) and Hommay *et al.* (1998), have sought to determine the nature of the factors which control the activity rhythm. All these studies are consistent with there being an endogenous rhythm controlling the onset of activity. In *L. pseudoflavus* (Ford and Cook, 1987) and *C. aspersus* (Bailey and Lazaridou-Dimitriadou, 1986), the rhythm free-runs in constant dark conditions with a period of between 24 and 25 h. Secondary rhythms with a period of about 12 h persist in *H. lucorum* and reflect its normal crepuscular activity pattern. *L. pseudoflavus* with its natural unimodal activity pattern lacks a secondary activity period. In experimental conditions in which no dark refuge is provided, then *L. pseudoflavus* and *L. maximus* may become active at dawn and appear to display a bimodal activity rhythm (Sokolove *et al.*, 1977; Ford and Cook, 1994). This seldom appears in the field and may be the animals' response to increased illumination in an experimental arena without a home site (Rollo, 1982; Ford and Cook, 1987).

The oscillator governing activity has an endogenous rhythm in excess of 24 h (Bailey and Lazaridou-Dimitriadou 1986; Ford and Cook, 1987), but is entrained continuously by 'zeitgebers'. It has been shown that both light and temperature act as zeitgebers since cycles of either will entrain the activity rhythm to non-24 h periods (e.g. a 19 h light:dark cycle, a 23 h 13°C:24°C cycle and a 23 h light and temperature cycle). The strongest entrainment occurs with combined temperature and light cycles (Ford and Cook, 1987).

Further insight into the nature of endogenous rhythms can be gained by exposing animals to light pulses simulating the complete diurnal cycle. Phase response curves may be constructed which display the time shift of the activity period relative to the timing of a single light pulse. Ford and Cook (1988) have shown such a response to be similar to that shown in other nocturnal animals. The responses of *L. pseudoflavus*, however, are only exhibited when pulses are of long duration (30 min, as opposed to <1 min for some insects) and the responses to the pulse are delayed. This is interpreted as the oscillators controlling the behaviour being coupled to the stimuli that entrain them with a time delay of several hours. This is seen further when twin pulses are used to simulate dawn and dusk. This technique results in some unpredictable entrainment when the second pulse falls within the delay in the response to the first. Experiments with *H. lucorum* in which light pulses were inserted into the normally dark night period have shown that its bimodal activity pattern persists but the dawn peak is shifted by the timing of the extra light pulse. Flari and Lazaridou-Dimitriadou (1995b) speculate that a minimum period of 10 h darkness is required to retain a normal diurnal rhythm.

Thus, activity can be viewed as being under triple control. It is initiated endogenously by a rhythm which is entrained by light and temperature cycles. The prevailing weather conditions may limit the expression of the rhythm. Finally, internal factors such as nutritional, reproductive and osmotic status determine whether an excursion takes place and the extent of that excursion. Nutritional status and weather can completely override the rhythm in some species. The modification of activity in response to food deprivation may also involve the amount of activity rather than its timing. *D. reticulatum*, for example, is more active when starved. This may reflect increases in food-searching behaviour or an increased tendency to migrate (Airey, 1987). Overcrowding in laboratory conditions has also been shown to influence the expression of a feeding rhythm in *L. valentiana* (Hommay *et al.*, 1998). Rollo *et al.* (1983b) suggest that *A. ater* is a 'risk taker', frequently departing from the safest strategy and becoming active during the day in less than optimal conditions, whereas *L. maximus* is conservative and seldom leaves the safest environmental envelope. Certainly, casual observations of *L. marginata*, *D. reticulatum* and *L. maximus* being active during dull days would indicate that species vary considerably in their dependence on endogenous rhythmicity to determine activity periods. Differences in the activity patterns between species may ultimately be attributable to the balance of influence between the endogenous activity rhythm, environmental responses and motivational status (Fig. 13.1).

What to Do When Active

Townsend (1975) viewed the efficient allocation of time to the various elements of the behavioural repertoire as central to the success of a species. The proportions of time allocated to various activities are therefore indicative of the significance of that activity or its consequences. The motivation for activity is undoubtedly resource acquisition, be it food, water or mates. The details of the behaviours that a gastropod performs while active are assumed to be dictated by the deficits (or occasionally, in the case of water, surpluses) of these resources.

Locomotion

Locomotor behaviour potentially serves three functions – orientation towards resources such as food, mates, water, etc., orientation back to a refuge and the shedding of excess water (Kerkut and Taylor, 1956). The extent of crawling in laboratory experiments, however, far exceeds that required by considerations of resources and shelter. Ford and Cook (1994) showed that *L. pseudoflavus* spent about 2 h crawling per night in most months of the year in arenas just $23 \times 13 \times 5$ cm and with food placed in the opposite corner to the home. *C. aspersus* normally exhibits 2–3 rest intervals per night, constituting between 42 and 58% of the total recorded activity (Bailey, 1989a). In laboratory studies, much of the time spent crawling is not associated directly with feeding and drinking. In the field, excursions are often more than simple journeys to and from food even when outside the reproductive season and when water availability is not a problem (Cook, 1980). Clearly, if time allocation is indicative of the priority an animal places on a particular need, then actively moving around serves more complex functions than simple travelling between home and food sources. These functions have not been elaborated experimentally. Dispersal would be an obvious function, but many animals which have been observed are conservative in their dispersal strategies (Rollo *et al.*, 1983b). Nevertheless, of 72 *L. pseudoflavus* excursions filmed by Cook (1980) over the course of 9 days, only 40% involved animals leaving and entering the home site. The remaining excursions were of animals crossing the field of view without reference to the home site. There was only one regular home site in the area observed. These additional excursions are of uncertain function, but one interpretation is that *L. pseudoflavus* individuals were searching for new resources in terms of both food and home sites. For these large limacids, opportunistic dispersal is a possibility, but homing appears to be the norm. The search for alternative food sources may be particularly important in laboratory experiments in which the diet normally is restricted. While exploration might be an appropriate function for animals that can rely on memory for orientation, this may not be the case for animals with limited central nervous functions. Stephens and McGaugh (1972) showed that *C. aspersus* could learn to avoid noxious chemicals but that such training lasted no longer than a week. Croll and Chase (1977) demonstrated that *Achatina fulica* Bowdich (Achatinidae) retained information concerning food odours for up to 120 days. Their experiments involved the choice in a Y-maze between food odours to which the animals had and had not been pre-conditioned, and probably mimic field circumstances more closely than those experiments involving pairing food stimuli with noxious stimuli. However, despite detailed investigations, no gastropod has been shown to be capable of retaining information that is sufficiently complex to encompass the orientation to home or conspecifics without the use of specific chemical information (Mpitsos and Lukowiak, 1985).

Feeding clearly is the most significant event during most terrestrial gastropod excursions. Following experimental observations of animal activity under circumstances in which the availability of food within and outside the home was varied, Ford (1986) concluded that for *L. pseudoflavus* surface activity was primarily a foraging response. Similar results were obtained by Rollo (1978) by placing food in *D. reticulatum* shelters.

Drinking

The control of water is central to the success of terrestrial gastropods as land animals. They lose water from their integument as if it were a surface of free water (Machin, 1975) and deposit it freely with the mucus as they move across the substrate. They are capable of sustaining relatively high rates of evaporative water loss (Dainton, 1954a,b; Machin, 1975; Burton, 1983; Riddle, 1983). Despite this apparently profligate attitude towards water, terrestrial gastropods have been shown to be capable of maintaining a remarkably constant water content (Lyth, 1982, 1983; Klein-Rollais and Daguzan, 1990). Water in gastropods is held in at least three accessible compartments: outside the body cavity in the pallial cavity (Blinn, 1964) and oesophageal crop (Machin, 1975); in the haemolymph; and in tissue cells. Although water may be held in the superficial mucus, it is unlikely that this water can be reabsorbed since it has a relatively high osmolality and gastropods cannot take up water through the foot against a concentration gradient (Prior, 1989). Water loss from the integument through evaporation will manifest itself as a change in the osmolality of the haemolymph, but this may be replenished in the short term from stores in the pallial cavity and oesophageal crop. The mobilization of such reserves might account for the exponential change in haemolymph osmolality during dehydration (Prior *et al.*, 1983). It has been demonstrated elegantly by Prior (Prior *et al.*, 1983; Prior, 1984, 1985) that the change in haemolymph osmolality has a number of physiological and behavioural consequences. Experiments involving the reduction of body weight to between 60 and 70% of the initial body weight by evaporation results in *L. maximus* ceasing other behaviours and starting water-seeking and drinking behaviours. Banta *et al.* (1990), following experiments with *L. maximus* and involving the use of combinations of dehydration and pharmacological agents to elicit drinking, conclude that changes in osmolality are mediated through changes in ionic concentration rather than through direct osmotic effects. There is some evidence that some central nervous events are stimulated by peripheral osmoreceptors

(Rósza, 1963), but no link has been established between their activity and behavioural events.

Drinking is achieved by the bulk flow of water directly through the integument of the foot sole (Prior and Uglem, 1984). During this form of drinking, which has been termed 'contact rehydration', the foot is flattened on to the wet substrate. Such drinking continues until the body weight is restored. There is considerable variability between individuals, but Prior (1984) has shown that drinking continues until specific rehydration occurs ('the rehydration set point'). It has been possible to define this set point in terms of a percentage of the initial body weight (between 80 and 100% for *L. maximus* and between 65 and 90% for *L. valentiana*). The field observations of Matanock and Welsford (1995) show that the time that *L. maximus* spend in rain water pools is inversely correlated with haemolymph osmolality. Further, the haemolymph osmolality of animals entering water was about 180 mOsmol kg^{-1} H$_2$O, whereas that of animals leaving the pools was about 104 mOsmol kg^{-1} H$_2$O. These values compare with Prior's (1984) experimentally deter- mined values of about 140 mOsmol kg^{-1} H$_2$O for hydrated animals and the 200 mOsmol kg^{-1} H$_2$O required to initiate contact rehydration. Given the differences between carefully prepared laboratory experiments and the uncertain history of animals in the field, these values are remarkably similar.

It is clear that behavioural changes are mediated via changes in haemolymph osmolality and that these are not laboratory abstractions. It is therefore of significance that haemolymph osmolality has also been implicated in the control of feeding (Prior, 1983a; Phifer and Prior, 1985), locomotor activity (Kerkut and Taylor, 1956; Hess and Prior, 1985), heart rate (Grega and Prior, 1986; Biannic *et al.*, 1994) and a variety of nerve cell functions (Hughes and Kerkut, 1956; Grega and Prior, 1986).

Water loss through mucus secretion is probably isosmotic and therefore will be apparent through a change in volume. The regulation of body volume has been poorly examined in entire gastropods, and thus the role of behaviour in such regulation is unclear. In *in vitro* preparations, however, it appears that regulation of increased pressures within the body cavity is achieved by heightened mucus secretion (Deyrup-Olsen and Martin, 1982; Martin *et al.*, 1990). The consequences of lowered blood volumes remain to be investigated.

Feeding

Details of feeding behaviours are addressed elsewhere in this volume (Speiser, Chapter 6). It appears that feeding bouts are rhythmic events. Periodogram analyses have shown that the commencement of feeding occurs with a repetition period which is significantly different from that of the bouts of locomotion within which it appears (Ford and Cook, 1987). This rhythm may be controlled endogenously but an equally likely cause might be the time taken to digest previous meals. The amount of time devoted to feeding during an excursion has been recorded by Rollo *et al.* (1983b), Ford and Cook (1987, 1994), Bailey (1989a) and Hommay *et al.* (1998).

Rest

Ford and Cook (1994) divided the activity of *L. pseudoflavus* into crawling before a meal, the meal itself, crawling after the meal and an after-meal rest. Table 13.4 shows the division between these categories of behaviour averaged over the course of a year. Considerable time is spent at rest (about 45% of the time in these feeding-associated behaviours). Newell (1968) also noted that activity and feeding were interspersed with rest in *D. reticulatum*. Research results on the activity of some species may be distorted by the failure to recognize rest as an 'activity'. Sokolove *et al.* (1977) examined the activity of *L. maximus* in activity wheels, and Bailey (1975) and Bailey and Lazaridou-Dimitriadou (1986) determined the frequency of crossing infrared beams to measure the activity of *C. aspersus* and *H. lucorum*. Subsequently, Bailey (1989a) recorded the behaviour of *C. aspersus* using time-lapse cinephotography and noted that this species normally exhibits 2–3 rest intervals per night, constituting between 42 and 58% of the total recorded activity.

The function of these rest periods is not clear. Resting after feeding might be interpreted anthropomorphically as a time when digestion of a recent meal is given physiological priority. In addition, contact rehydration normally is associated with rest (Prior, 1985).

Where to be Inactive

The selection of roosting sites by terrestrial gastropods is influenced by a wide variety of factors. Shelled species may aestivate in exposed or elevated positions to minimize temperature rises during the day (Yom-Tov, 1971; Jaremovic and Rollo, 1979; Tilling, 1986) (Fig. 13.4B). Cook and Freeman (1986) related the heating properties of shells of the mangrove forest species *Littoraria pallescens* (Philippi) (Littorinidae) to their choice of roosting sites. Pallant (1969) observed that *D. reticulatum* selected

Table 13.4. The mean time (min) spent in various activities associated with feeding during excursions by *Limacus pseudoflavus* (Evans) (Limacidae) freshly collected from the field

Event	Pre-feed crawl	Feeding	Post-feed crawl	Post-feed rest
Mean time (min)	27	46	18	75

For methods, see Ford and Cook (1994). Each mean is the result of observation on six animals for six continuous days and repeated for ten samples over the course of a year.

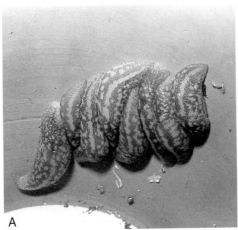

A   B

Fig. 13.4.   (A) A huddle of *Limacus flavus* (Linnaeus) (Limacidae) in a flower pot. The eight animals are tightly packed, flank to flank, reducing the effective evaporative area by an average of 17% per flank in contact (Cook, 1981a). (B) A mixed-species aggregation of snails on a fence post in Cyprus. Shelled species aestivating in elevated positions avoid the high temperatures associated with heat conduction from the ground.

roosting sites close to possible food sources, and much of Rollo's work previously mentioned would imply that *L. maximus* also determines its roosting sites by their proximity to food (Rollo and Wellington, 1979).

There is some evidence that the selection of hibernating sites is not always opportunistic since Edelstam and Palmer (1950) and Pollard (1975) report the return of animals to traditional roosting sites. Further, the detailed position within a hibernation site may influence the animal's survival (Baur and Baur, 1991; Terhivuo, 1978).

How to Get to Roosting Sites

Homing

At least some species of terrestrial gastropods demonstrate an ability to return to previously occupied, favourable sites. Step (1945) reports 'A few years since I noticed that a *Richardia*, that stood in a large pot in the front of the house, had been much eaten, and a large thick shelled *aspersa* [*Cantareus aspersa*] was found clinging to the shady side of the pot. Pencilling my initials on his shell I hurled the snail as far as I could. Next morning he was again attached to the pot though to regain it he had to cross a very broad road and a low wall.'

Both Edelstam and Palmer (1950) and Pollard (1975) observed seasonal migratory patterns in *H. pomatia*, indicating a return to traditional hibernating sites. Displaced *A. fulica* also show an ability to return to the area from which they were taken (Southwick and Southwick, 1969).

Bailey (1989a), using time-lapse video-recording techniques in artificial arenas, demonstrated that adult C. aspersus consistently return to roosting sites which they shared with several conspecifics. Laboratory observations, however, must be interpreted with caution, since Bailey (1989b) has also showed that the same species in the field showed little evidence of returning to the same site with any precision. In one area, there was evidence of a return to a general area; in another, animals moved less far than a random walk would predict; whilst in a third, the distance moved over time conformed closely to that predicted by a random walk. Only in animals that had been displaced from their original home range was the movement observed greater than that predicted by the random walk. Staikou et al. (1989) working with H. lucorum failed to demonstrate a consistent occupation of roosts during 24 h periods. Similarly, Baur and Baur (1993) have demonstrated that the movements of A. arbustorum conform to a random movement model within the confines of favoured vegetation. Tomiyama and Nakane (1993) observed movements of A. fulica in the field. In a 2-month period, old adults moved an average of about 1.5 m, young adults 3.6 m and juveniles 8.3 m from their previous location per day. This is consistent with the older animals having relatively small, stable home ranges and the juveniles having large home ranges which change. Thus, although shelled gastropods have a demonstrable capacity to home, this is not necessarily exploited in the field. It seems probable, therefore, that homing is an opportunistic behaviour in snails, being brought into play when animals fail to find suitable roosts during their excursions. Some animals may, therefore, never demonstrate their homing ability in the field, while other individuals of the same species may display collective homing (Chelazzi, 1990) or homing to individual locations.

The advantages of homing are clear in hostile environments but less so where many locations provide suitable resting sites. Thus, homing should be viewed as a reserve mechanism which is only deployed when animals fail to find acceptable sites. This would account for the variability sometimes seen in snails which are offered some protection by their shell, as well as the regularity of the return to roosting sites by shell-less species, slugs, which are much more vulnerable to desiccation during the day. Slugs can thus be assumed to have much more rigorous criteria defining an acceptable daytime resting site. While slugs may come to rest in occasional sites when conditions are equable throughout the area, this is comparatively rare in laboratory experiments: most will home to the sites provided. Newell (1968) first observed that D. reticulatum returned regularly to specific locations on the soil as rest sites. Since that time, many workers (e.g. Gelperin, 1974; Dundee et al., 1975; Cook, 1979b; Rollo et al., 1983a) have examined homing in slugs from a number of gastropod families and none has reported a consistent failure to demonstrate the phenomenon. Cook has reported field observations of the regular homing of L. pseudoflavus in a paved yard. For A. lusitanicus, however, dispersal after release in grassland may take precedence over any tendency to home (Grimm *et al.*, 2000).

If homing is defined as the consistent return to the same site by an individual, then it is clear that very few terrestrial gastropods home. Many species, however, do return to sites that have been occupied by the same species on the previous night. Thus 'homing' can be viewed as existing at a species level rather than at the level of individuals.

Experiments on homing fall into three categories:

1. Displacement experiments. Animals are moved outside their normal sites of activity. Where this has been conducted, a proportion of animals may return successfully (e.g. Southwick and Southwick, 1969). This type of result merely indicates that animals can detect the home from a distance. That detection could depend on natural features of the home site (odour, vision, etc.) or on the home having been labelled previously. Displacement experiments, then, serve to eliminate the outward journey as a source of information for the return trip.

2. Field observations. These demonstrate how an animal behaves in the field and throw little light on the mechanisms underlying that behaviour (Cook, 1980, 1992). However, field observations are useful in demonstrating that laboratory results are not irrelevant abstractions, that is phenomena that are only demonstrable when the majority of distractions have been eliminated.

3. Laboratory experiments. These can be used to display the range of mechanisms available to animals. They do not, however, show that these mechanisms are used in the field nor that they are the most important.

The mechanism underlying gastropod homing is the deposition, while at rest, of substances in the daytime resting site which members of the same species find attractive (Gelperin, 1974; Chase *et al.*, 1978; Cook, 1979a). This olfactory beacon effectively labels suitable resting sites, and terrestrial gastropods use distant chemoreception mechanisms to 'home' on to this beacon. Presumably, the more suitable a site is the more animals stay there and the greater the deposition of attractive substances. Clear evidence supporting the chemoreception of the beacon is seen in the observation that *L. pseudoflavus* leaves home downwind and then subsequently homes upwind (Cook, 1980). The individual need never lose olfactory contact with the home. The nature of the beacon is unknown. It might be deposited in the faecal material left in the home by previous occupants or the attractive principal in the home might have been secreted with the mucus. Difficulties in separating faecal materials from the mucus have resulted in this element of the mechanism remaining unresolved. It is curious, however, that terrestrial gastropods do deposit substantial volumes of faecal material in their daytime resting sites.

Sometimes animals stray upwind and, therefore, lose contact with home. Under these circumstances, trail following has been observed

(Cook, 1980). The following of previously laid trails is widespread in the animal kingdom. Gastropods are pre-adapted to use trails as an orienting mechanism since a mucus trail is laid as an inevitable consequence of adhesive locomotion. It has been shown that trail following is of central importance in some activities of terrestrial gastropods, such as the detection of prey and mates (Cook, 1994).

In experiments designed to test which information source animals would follow in situations in which there was a choice, Cook (1992) and Chelazzi et al. (1988) concluded that trail following was of relatively low priority in the context of homeward navigation. Chelazzi et al. (1988) have shown that trail following is relatively rare in L. flavus (only ~13% of crossings of trails resulted in trail following). These experiments involved the second (tracker) animal approaching the original (marker) trail from the side. As a consequence, most of the contact between the marker trail and the tracker animal was at angles in excess of 45°. Nevertheless, Chelazzi et al. (1988) report a significant deviation of the trail of the tracker animal towards that of the marker. In an open field experimental design, the frequency of trail following by L. pseudoflavus was substantially higher (30%) and this may reflect the lower mean angle of approach between the trails (Cook, 1977, 1992). This frequency of trail following is similar to that observed after the analysis of movements of the same species in the field (Cook, 1994). The implication of the finding that L. pseudoflavus normally home upwind (Cook, 1980) is that anemotaxis is a mechanism associated with homing. Chase and Croll (1981) report a series of experiments using A. fulica which demonstrates the ability of this species to orient with respect to odours using anemotaxis, trail following and distant chemoreception. Anemotaxis has been examined systematically in few species, although it has also been observed in C. aspersus (Farkas and Shorey, 1976).

It is probable that terrestrial gastropods in general do not rely solely on one homing mechanism but that there is considerable redundancy in the sensory cues and their responses to them.

Species specificity of trail following

Cook (1977) demonstrated that L. pseudoflavus would follow the trails of conspecifics but not those of Tandonia budapestensis (Hazay) (Milacidae) or D. reticulatum. It did follow trails of L. flavus, but at a very much reduced frequency. This relationship between L. pseudoflavus and L. flavus is also seen in other circumstances such as their formation of common aggregations (Cook, 1981b) and may indicate that they are sibling species. Chase et al. (1978) showed that A. fulica followed trails of conspecifics but not those of Eobania vermiculata (Müller) (Helicidae).

It would be expected that species specificity in trail following might be more pronounced in contexts other than homing since suitable homes might be similar for a variety of different species. Suitable mates, for instance, would be species-specific. In the carnivorous snail *Euglandina rosea* (de Férussac) (Oleacinidae), it has been shown that trail following is an integral component of both courtship (Cook, 1985d) and prey location (Cook, 1985c). This species displays a variable tendency to follow prey trails, which is related to the palatability of the trail layer. Further, individuals will follow the trails of conspecifics but not their own (Cook, 1983, 1985b).

Directionality in trail following

One of the key factors in the usefulness of trail following is the direction in which trails are followed. Species from a variety of gastropod lineages (Crisp, 1969; Wells and Buckley, 1972; Townsend, 1974; Cook and Cook, 1975; Trott and Dimock, 1978; McFarlane, 1980), including *D. reticulatum* (Wareing, 1986), exhibit trail following and respond to the direction in which the original trail was laid. It is not a universal phenomenon, however, and no evidence for the detection of polarity has been reported in studies of *L. pseudoflavus* (Cook, 1977, 1992) and *E. rosea* (Cook, 1985b), for example. The differences among species in their response to polarity in the trail may be a function of the context in which trail following was observed and the experimental protocols used. Cook (1985b), studying *E. rosea*, for instance, reported a consistent positive response to the trail direction until other, non-trail sensory cues were removed. In general, where the direction of trail following has been observed in *D. reticulatum*, the tracker animal has followed in the same direction as the marker (Wareing, 1986). The end product of most trail-following events in this species is courtship, and such directional trail following should be viewed in this context. Where trail following has been observed in species with restricted mating seasons, for example large limacids, then directional trail following has not been observed regularly.

Three possible mechanisms have been suggested for directional trail following. First, the directional cue might lay in the structure of the trail itself. Denny and Gosline (1980) have observed thread-like fibres in mucus and these, if detectable, could serve as directional cues. Bretz and Dimock (1983) report similar structures. Secondly the directional cues may lie in the chemical composition of the trail. If a cue is short lived, then inevitably a gradient will be established as the trail-following substance(s) decay with time. Ushadevi and Krishnamoorthy (1980) established that the trail-following cues in *Mariaella dussumieri* Gray (Ariophantidae) decayed on exposure to light and were volatile and thermolabile. Bousefield *et al.* (1981) similarly speculate that the directionality in following observed in the aquatic basommatophoran *Biomphalaria glabrata* (Say) (Planorbidae) is a function of the active substances diffusing from the trail. While trail following, the head waving movements of *L. pseudoflavus* tend to decrease (Cook, 1985a) and the ocular tentacles are held across the trail rather than along it. The olfactory epithelium of ocular cephalic tentacles has been shown not to produce mucus (Chase, 1986), and this is consistent with the sensitivity to mucus being located at the tips of the cephalic tentacles. Gradients, therefore, do not seem likely cues unless they are detected when the animal first approaches the trail from the side, when the cephalic tentacles are held along the length of the original trail. A second source of chemical information in the trail may reside in a left–right asymmetry. Such a difference between the left and the right side of the trail would be detectable as the animal progressed along the trail, would be detectable independent of the angle at which the trail was approached and would be consistent with the asymmetric design of the gastropod body. There is no direct evidence to support these possibilities. The third mechanism might involve no directional information in the trail itself at all and the directional cue could be determined by distant chemoreception of the cue source. Some or all of these mechanisms may be used in examples of trail following in the context of homing, prey finding or reproduction. Distinguishing between them will await appropriate experimental protocols.

Distant chemoreception is mediated via the olfactory epithelia of the cephalic tentacles (Suzuki, 1967; Gelperin, 1974). Removal of these organs prevents homing but not trail following (Cook, 1985a; Chase and Croll, 1981). The functional areas of the cephalic tentacles are held well apart and in front of the body. Time-lapse video-recording or photography of terrestrial gastropods in the field show that these animals indulge in considerable head waving and that the cephalic tentacles pass through a large volume of air immediately in front of the animal. These features indicate the suitability of the cephalic tentacles for sampling the air upwind of the body, and the separation between the sensory surfaces would allow gradients to be detected. Unilateral removal of an cephalic tentacle in *A. fulica* prevents orientation to a distant odour source, indicating that comparisons between the responses from the two tentacles are involved in orientation (Chase and Croll, 1981).

The inferior tentacles are also implicated in trail following. Their removal significantly decreases the accuracy with which trails are followed, and it seems that they may be used to detect the trail immediately under the head (Chase and Croll, 1981). The details of the functioning of the inferior tentacles vary with species. *A. fulica* inferior tentacles make no contact with the substrate (Chase and Croll, 1981) and therefore contact chemoreception cannot be involved. In contrast, those of *L. pseudoflavus* make regular contact with the substrate (Cook, 1985a). The functional significance of an inferior tentacle actually making contact with the substrate is not clear. Contact results in immediate retraction and therefore a loss of appropriate sensory input while the tentacle is cleaned. Presumably, however, molecules of low volatility might serve as sensory cues. Not making contact allows a continuous monitoring of the substrate but requires the cues to be volatile in nature.

A series of laboratory experiments over the last 20 years has shown clearly that the mechanisms used by terrestrial gastropods to home include both distance and contact (or close) chemoreception. In summary, many species detect their communal homes by the chemoreception of substances previously deposited in the home site by conspecifics. Sometimes, especially if contact with the home is lost, mucous trails may be followed. There is no convincing evidence that other mechanisms are involved.

Functional consequences of homing

The regular occupation of daytime resting sites is important for two reasons. First, gastropods aggregating in a group create a humid microclimate within the group, thus making conditions more favourable to their survival when the ambient air is dry. This group effect may account for the mixed-species aggregations sometimes found at aestivation sites (e.g. Pomeroy, 1968) (Fig. 13.4B). Secondly, it has been shown for slugs that aggregating within daytime resting sites serves to conceal parts of the body from the atmosphere and thus reduce the evaporative surface area of individuals (Cook, 1981a; Prior, 1985; Waite, 1988) (Fig. 13.4A). Cook (1981a) showed, for example, that for each flank of a L. pseudoflavus which is in contact with another there is a saving of approximately 17% of the normal water loss for each flank covered. Cooke (1895) states that several individuals of C. aspersus confined to a tube will raise the temperature by one or two degrees, but the methodology for these observations is not clear. Nevertheless, there is a possibility that the metabolism of an aggregation of gastropods in the confined space of a daytime resting site may significantly affect the temperature.

Examination of interspecific aggregation and huddling in slugs has shown that, although different species will home to common sites, they do not mix freely within those sites. L. flavus and L. pseudoflavus will form mixed huddles, but within those huddles there is a preference for the neighbour being a conspecific. L. maximus tends not to huddle except in winter. It is an aggressive species, actively defending daytime resting sites and excluding other species (Rollo and Wellington, 1979) when suitable sites are a limited resource. L. marginata exhibits little tendency to huddle within the home (Cook, 1981b).

Although aggregations of snails have been observed in sheltered areas during hibernation and in exposed areas during aestivation, there is as yet no information concerning the microclimate within the aggregations or any potential physiological benefits to the aggregated snails.

Conclusion

Inevitably, drawing the threads of a discussion together is easy at a gross level but becomes progressively more difficult as one examines the details more closely. At a gross level, Rollo's contention that gastropods survive in a hostile, changing environment by doing the right thing, at the right place, at the right time is a truism. All animals survive in this way. In the case of gastropods, however, the right thing to do is to do nothing, the place to do it is in a place of concealment and the time to do it is as often as possible.

Hibernation and aestivation, however initiated and concluded, are strategies that avoid exposure to the harshest environmental extremes. The gastropod version of an extreme condition, however, is one that would rarely limit the activities of other terrestrial animal groups, particularly arthropods, the only other invertebrate phylum to have exploited the terrestrial environment successfully. The limitation of most activity to the night or to dull, humid conditions reduces the problem of desiccation. Coupled with this restricted activity is the necessity to regain a safe haven during the daylight hours of inactivity, which leads inevitably to the development of a range of homing mechanisms.

These generalizations, however, fail to recognize the flexibility of many strategies and the variation between the behaviours of even closely related species. The failure to replicate some experiments, the frequent mismatch between laboratory results and field observations and the extreme variability in gastropod growth and behaviour all point to a group of animals whose over-riding strategy is to be pragmatic. If a suitable resting site becomes available, then an individual does not home; if the weather suddenly becomes cold or dry, then an individual will enter a dormant phase; when dormancy is broken, then activity is not immediately inevitable; if olfactory contact is lost with a home site, then trail following ensues. One only needs to compare Rollo's accounts of the behaviour of *L. maximus* with Cook's of *L. pseudoflavus* to appreciate that the former is a lion while the latter is a lamb. Rollo's work (Rollo *et al.*, 1983b) is central here. He graphically illustrates the variation in risk-taking strategies between species that at first sight appear to be ecologically similar. It appears that most of the mechanisms associated with activity which can be discerned from laboratory experiments are held in reserve in the field and only deployed when an individual finds itself in the wrong place at the wrong time.

References

Abdel-rehim, A.H. (1983) The effects of temperature and humidity on the nocturnal activity of different shell colour morphs of the land snail *Arianta arbustorum*. *Biological Journal of the Linnaean Society* 20, 385–395.

Airey, W.J. (1987) The influence of food deprivation on the activity of slugs. *Journal of Molluscan Studies* 53, 37–45.

Asami, T. (1993) Divergence of activity patterns in coexisting species of land snails. *Malacologia* 35, 399–406.

Bailey, S.E.R. (1975) The seasonal and daily patterns of locomotor activity in the snail *Helix aspersa* Müller, and their relationship to environmental variables. *Proceedings of the Malacological Society of London* 41, 415–428.

Bailey, S.E.R. (1981) Circannual and circadian rhythms in the snail *Helix aspersa* Müller and the photoperiodic control of annual activity and reproduction. *Journal of Comparative Physiology* 142, 89–94.

Bailey, S.E.R. (1983) The photoperiodic control of hibernation and reproduction in the land snail *Helix aspersa* Müller. *Journal of Molluscan Studies, Supplement*, 12A, 2–5.

Bailey, S.E.R. (1989a) Daily cycles of feeding and locomotion in *Helix aspersa*. *Haliotis* 19, 23–31.

Bailey, S.E.R. (1989b) Foraging behaviour of terrestrial gastropods: integrating field and laboratory studies. *Journal of Molluscan Studies* 55, 263–272.

Bailey, S.E.R. and Lazaridou-Dimitriadou, M. (1986) Circadian components in the daily activity of *Helix lucorum* L. from northern Greece. *Journal of Molluscan Studies* 52, 190–192.

Baker, G.H. (1988) Dispersal of *Theba pisana* (Mollusca: Helicidae). *Journal of Applied Ecology* 25, 889–900.

Banta, P.A., Welsford, I.G. and Prior, D.J. (1990) Water orientation behaviour in the terrestrial gastropod *Limax maximus*: the effects of dehydration and arginine vasotocin. *Physiological Zoology* 63, 683–690.

Barker, G.M. (1999) *Naturalised Terrestrial Stylommatophora (Mollusca: Gastropoda)*. Fauna of New Zealand 38. Manaaki Whenua Press, Lincoln.

Baur, A. (1990) Intra- and inter-specific influences on age at first reproduction and fecundity in the land snail *Balea perversa*. *Oikos* 57, 333–337.

Baur, A. (1993) Effects of food availability and intra- and inter-specific interactions on the dispersal tendency in the land snail *Chondrina clienta*. *Journal of Zoology* 230, 87–100.

Baur, A. and Baur, B. (1988) Individual movement patterns of the minute land snail *Punctum pygmaeum* (Draparnaud) (Pulmonata: Endodontidae). *Veliger* 30, 372–376.

Baur, A. and Baur, B. (1991) The effect of hibernation position on winter survival of the rock dwelling land snails *Chondrina clienta* and *Balea perversa* on Öland, Sweden. *Journal of Molluscan Studies* 57, 331–336.

Baur, A. and Baur, B. (1993) Daily movement patterns and dispersal in the land snail *Arianta arbustorum*. *Malacologia* 35, 89–98.

Baur, A., Baur, B. and Froberg, L. (1994) Herbivory on calcicolous lichens: different food preferences and growth rates in two co-existing land snails. *Oecologia* 98, 313–319.

Baur, B. (1988) Micro-geographical variation in shell size of the land snail *Chondrina clienta*. *Biological Journal of the Linnaean Society* 35, 247–259.

Baur, B. and Baur, A. (1990) Experimental evidence for intra- and interspecific competition in two species of rock dwelling land snails. *Journal of Animal Ecology* 59, 301–305.

Baur, B. and Baur, A. (1995) Habitat-related dispersal in the rock-dwelling land snail *Chondrina clienta*. *Ecography* 18, 123–130.

Bayne, C.J. (1969) Survival of the embryos of the grey field slug, *Agriolimax reticulatus*, following desiccation of the egg. *Malacologia* 9, 391–401.

Beyer, W.N. and Saari, D.M. (1977) Effect of tree species on the distribution of slugs. *Journal of Animal Ecology* 46, 697–702.

Beyer, W.N. and Saari, D.M. (1978) Activity and ecological distribution of the slug *Arion subfuscus* (Draparnaud) (Stylommatophora, Arionidae). *American Midland Naturalist* 100, 359–367.

Biannic, M., Coillot, J.P. and Daguzan, J. (1994) Heart-rate in relation to dehydration in the snail *Helix aspersa* Müller (Gastropoda, Pulmonata). *Comparative Biochemistry and Physiology* 108, 65–67.

Blanc, A., Buisson, B. and Pupier, R. (1989) Evolution en laboratoire du rythme specifique d'activité de deux mollusques gasteropodes (*Helix pomatia* L. et *Helix aspersa* Müller) en situation de cohabitation sous differentes photoperiodes. *Haliotis* 19, 11–21.

Blinn, W.C. (1964) Water in the mantle cavity of land snails. *Physiological Zoology* 37, 329–337.

Bousefield, J.D., Tait, A.I., Thomas, J.D. and Towner-Jones, D. (1981) Behavioural studies on the nature of the stimuli responsible for triggering mucus trail tracking by *Biomphalaria glabrata*. *Malacological Review* 14, 49–64.

Bretz, D.D. and Dimock, R.V. (1983) Behaviourally important characteristics of the mucous trail of the marine gastropod *Ilynassa obseleta* (Say). *Journal of Experimental Marine Biology and Ecology* 143, 181–191.

Bull, C.M., Baker, G.H., Lawson, L.M. and Steed, M.A. (1992) Investigations of the role of mucus and faeces in interspecific interactions of two land snails. *Journal of Molluscan Studies* 58, 433–442.

Burch, J.B. (1955) Some ecological factors of the soil affecting the distribution and abundance of land snails in Eastern Virginia. *Nautilus* 69, 62–69.

Burton, R.F. (1983) Ionic regulation and water balance. In: Saleuddin, A.S.M. and Wilbur, K.M. (eds) *The Mollusca*, Vol. 5, *Physiology*, Part 2. Academic Press, New York, pp. 291–352.

Cain, A.J. and Cowie, R.H. (1978) Activity of different species of land snails on surfaces of different inclinations. *Journal of Conchology* 29, 267–272.

Cain, A.J. and Curry, J.D. (1963) Area affects in *Cepaea*. *Philosophical Transactions of the Royal Society, Biological Sciences* 246, 1–81.

Cameron, R.A.D. (1970) The effect of temperature on the activity of three species of helicid snail (Mollusca: Gastropoda). *Journal of Zoology* 162, 303–315.

Cameron, R.A.D. (1973) Some woodland Mollusc faunas from south east England. *Malacologia* 14, 355–370.

Cameron, R.A.D. (1978) Differences in the sites of activity of coexisting species of land mollusc. *Journal of Conchology* 29, 273–278.

Cameron, R.A.D. and Carter, M.A. (1979) Intra- and interspecific effects of population density on growth and activity in some helicid snails (Gastropoda: Pulmonata). *Journal of Animal Ecology* 48, 237–246.

Carter, M.A., Jeffrey, R.C.V. and Williamson, P. (1979) Food overlap in co-existing populations of the land snails *Cepaea nemoralis* (L.) and *Cepaea hortensis* (Müll.). *Biological Journal of the Linnaean Society* 11, 169–176.

Chase, R. (1986) Lessons from snail tentacles. *Chemical Senses* 11, 411–426.

Chase, R. and Croll, R.P. (1981) Tentacular function in snail olfactory orientation. *Journal of Comparative Physiology* 143, 357–362.

Chase, R., Pryer, K., Baker, R. and Madison, D. (1978) Responses to conspecific chemical stimuli in the terrestrial snail *Achatina fulica* (Pulmonata: Sigmurethra). *Behavioural Biology* 22, 302–315.

Chelazzi, G. (1990) Eco-ethological aspects of homing behaviour in molluscs. *Ethology, Ecology and Evolution* 2, 11–26.

Chelazzi, G., leVoci, G. and Parpagnoli, D. (1988) Relative importance of airborne odours and trails in the group homing of *Limacus flavus* (Linnaeus) (Gastropoda, Pulmonata). *Journal of Molluscan Studies* 54, 173–180.

Conroy, B.A. (1980) Co-existence of two closely related species of *Arion* in natural habitats. *Journal of Conchology* 30, 189–200.

Cook, A. (1977) Mucus trail following by the slug *Limax grossui. Animal Behaviour* 25, 774–781.

Cook, A. (1979a) Homing by the slug *Limax pseudoflavus. Animal Behaviour* 27, 545–552.

Cook, A. (1979b) Homing in the Gastropoda. *Malacologia* 18, 315–318.

Cook, A. (1980) Field studies of homing in the pulmonate slug *Limax pseudoflavus* Evans. *Journal of Molluscan Studies* 46, 100–105.

Cook, A. (1981a) Huddling and the control of water loss by the slug *Limax pseudoflavus. Animal Behaviour* 29, 289–298.

Cook, A. (1981b) A comparative study of aggregation in pulmonate slugs (Genus *Limax*). *Journal of Animal Ecology* 50, 703–713.

Cook, A. (1983) Feeding by the carnivorous snail *Euglandina rosea* Férussac. *Journal of Molluscan Studies*, Supplement 12A, 32–35.

Cook, A. (1985a) Tentacular function in trail following by the pulmonate slug, *Limax pseudoflavus* Evans. *Journal of Molluscan Studies* 51, 240–247.

Cook, A. (1985b) Functional aspects of trail following in the carnivorous snail *Euglandina rosea* Férussac. *Malacologia* 26, 173–181.

Cook, A. (1985c) The organisation of feeding in the carnivorous snail *Euglandina rosea* Férussac. *Malacologia* 26, 173–81.

Cook, A. (1985d) The courtship of *Euglandina rosea* Férussac. *Journal of Molluscan Studies* 51, 211–214.

Cook, A. (1989) Crowding effects on the growth of juvenile slugs (*Limax pseudoflavus*). In: Henderson, I.F. (ed.) *Slugs and Snails in World Agriculture.* British Crop Protection Council Monograph No. 41, pp. 193–200.

Cook, A. (1992) The function of trail following in the pulmonate slug *Limax pseudoflavus. Animal Behaviour* 43, 813–821.

Cook, A., (1994) Trail following in slugs – the stimulus, its reception and the behavioural response. *Ethology, Ecology and Evolution* 6, 55–64.

Cook, A. and Ford D.J.G. (1989) The control of activity of the pulmonate slug *Limax pseudoflavus* by weather. In: Henderson, I.F. (ed.) *Slugs and Snails in World Agriculture.* British Crop Protection Council Monograph No. 41, pp. 337–342.

Cook, A. and Radford, D.J. (1988) The comparative ecology of four sympatric limacid slug species in Northern Ireland. *Malacologia* 28, 131–146.

Cook, L.M. and Freeman P.M. (1986) Heating properties of morphs of the mangrove snail *Littoraria pallescens. Biological Journal of the Linnaean Society* 29, 295–300.

Cook, S.B. and Cook, C.B. (1975) Directionality in the trail following response of the pulmonate limpet *Siphonaria alternata. Marine Behaviour and Physiology* 3, 147–155.

Cooke, A.H. (1895) *The Cambridge Natural History*, Vol. III, *Molluscs*. Harmer, S.F. and Shipley (eds) A.E. Macmillan, London.

Cowie, R.H., Nishida, G.M., Basset, Y. and Gon, S.M. (1995) Patterns of land snail distribution in a montane habitat on the island of Hawaii. *Malacologia* 36, 155–169.

Crawford-Sidebotham, T.J. (1972) The influence of weather upon the activity of slugs. *Oecologia* 9, 141–154.

Crisp, M. (1969) Studies on the behaviour of *Nassarius obseletus* (Say) (Mollusca: Gastropoda). *Biological Bulletin of the Marine Laboratory, Woods Hole* 139, 355–373.

Croll, R.P. (1983) Gastropod chemoreception. *Biological Reviews* 58, 293–319.

Croll, R.P. and Chase, R. (1977) A long term memory for food odours in the land snail (*Achatina fulica*). *Behavioural Biology* 19, 261–268.

Daguzan, J. (1982) Contribution a l'elevage de l'escargot Petit-gris: *Helix aspersa* Müller (Mollusque gasteropode pulmone stylommatophore). II Evolution de la population juvenile de l'eclosion a l'age de 12 semaines, en bâtiment et en conditions d'elevage controlées. *Annales de Zootechnie* 31, 87–110.

Dainton, B.H. (1954a) The activity of slugs: I. The induction of activity by changing temperatures. *Journal of Experimental Biology* 31, 165–187.

Dainton, B.H. (1954b) The activity of slugs: II. The effect of light and air currents. *Journal of Experimental Biology* 31, 188–197.

Dainton, B.H. and Wright, J. (1985) Falling temperature stimulates activity in the slug *Arion ater*. *Journal of Experimental Biology* 118, 439–443.

Dallas, H.F., Curtis, B.A. and Ward, D. (1991) Water exchange, temperature tolerance, oxygen consumption and activity of the Namib desert snail, *Trigonephrus* sp. *Journal of Molluscan Studies* 57, 359–366.

Dan, N. and Bailey, S.E.R. (1982) Growth, mortality, and feeding rates of the snail *Helix aspersa* at different population densities in the laboratory, and the depression of activity of helicid snails by other individuals, or their mucus. *Journal of Molluscan Studies* 48, 257–265.

Denny, M.W. and Gosline, J.M. (1980) The physical properties of the pedal mucus of the terrestrial slug, *Ariolimax columbianus*. *Journal of Experimental Biology* 88, 375–393.

Deyrup-Olsen, I. and Martin, A.W. (1982) Surface exudation in terrestrial slugs. *Comparative Biochemistry and Physiology* 72C, 45–51.

Dundee, D.S. and Rogers, J.S. (1977) Factors affecting molluscan activity. *Bulletin of the American Malacological Union* 43, 61–63.

Dundee, D.S., Tizzard, M. and Traub, M. (1975) Aggregative behaviour in veronicellid slugs. *Nautilus* 89, 69–72.

Edelstam, C. and Palmer, C. (1950) Homing behaviour in gastropods. *Oikos* 2, 259–270.

Farkas, S.R. and Shorey, H.H. (1976) Anemotaxis and odour trail following by the terrestrial snail *Helix aspersa*. *Animal Behaviour* 24, 686–689.

Flari, V. and Lazaridou-Dimitriadou, M. (1995a) The locomotor-activity rhythm of the edible snail, *Helix lucorum* L., in symmetrical skeleton photoperiod regimes. *Animal Behaviour* 509, 635–644.

Flari, V. and Lazaridou-Dimitriadou, M. (1995b) The impact of nocturnal light pulses on the activity pattern of terrestrial snails (*Helix lucorum*) entrained to a photoperiod of a 12 h light 12 h dark. *Canadian Journal of Zoology* 73, 1214–1220.

Ford, D.J.G. (1986) Rhythmic activity of the pulmonate slug *Limax pseudoflavus* (Evans). PhD thesis. University of Ulster, Northern Ireland.

Ford, D.J.G. and Cook, A. (1987) The effects of temperature and light on the circadian activity of the pulmonate slug, *Limax pseudoflavus* Evans. *Animal Behaviour* 35, 1754–1765.

Ford, D.J.G. and Cook, A. (1988) Responses to pulsed light stimuli in a pulmonate slug (*Limax pseudoflavus*). *Journal of Zoology* 214, 663–672.

Ford, D.J.G. and Cook, A. (1994) The modulation of rhythmic behaviour in the pulmonate slug *Limax pseudoflavus* by season and photoperiod. *Journal of Zoology* 232, 419–434.

Gelperin, A. (1974) Olfactory basis of homing behaviour in the giant garden slug, *Limax maximus*. *Proceedings of the National Academy of Sciences USA* 71, 966–970.

Getz, L.L. (1959) Notes on the ecology of slugs: *Arion circumscriptus*, *Deroceras reticulatum* and *D. laeve*. *American Midland Naturalist* 61, 485–498.

Grega, D.S. and Prior, D.J. (1986) Modification of cardiac activity in response to dehydration in the terrestrial slug *Limax maximus*. *Journal of Experimental Zoology* 237, 185–190.

Grimm, B., Paill, W. and Kaiser, H. (2000) Daily activity of the pest slug *Arion lusitanicus* Mabille. *Journal of Molluscan Studies* 66, 125–128.

Hatziioanou, M., Eleutheriadis, E. and Lazaridou-Dimitriadou, M. (1994) Food preferences and dietary overlap by terrestrial snails in Logos area (Edessa, Macedonia, northern Greece). *Journal of Molluscan Studies* 60, 331–333.

Heller, J. and Dolev, A. (1994) Biology and population dynamics of a crevice living land snail, *Cristata genezarethana* (Clausiliidae). *Journal of Molluscan Studies* 60, 33–46.

Heller, J. and Ittiel, H. (1990) Natural history and population dynamics of the land snail *Helix texta* in Israel (Pulmonata: Helicidae). *Journal of Molluscan Studies* 56, 189–204.

Hermida, J., Ondina, P. and Outeiro, A. (1995) Influence of soil characteristics on the distribution of terrestrial gastropods in northwest Spain. *European Journal of Soil Biology* 31, 29–38.

Herreid C.F., II, and Rokitka, M.A. (1976) Environmental stimuli for arousal from dormancy in the land snail *Otala lactea* (Müller). *Physiological Zoology* 49, 181–190.

Hess, S.D. and Prior, D.J. (1985) Locomotor activity of the terrestrial slug *Limax maximus*: response to progressive dehydration. *Journal of Experimental Biology* 116, 323–330.

Hommay, G., Jacky, G. and Ritz, M.F. (1998) Feeding activity of *Limax valentianus* Férussac: nocturnal rhythm and alimentary competition. *Journal of Molluscan Studies* 64, 137–146.

Hommay, G., Lorvelec, O. and Jacky, G. (1998) Daily activity rhythm and use of shelter in the slugs *Deroceras reticulatum* and *Arion distinctus* under laboratory conditions. *Annals of Applied Biology* 132, 167–185.

Hughes, G.M. and Kerkut, G.A. (1956) Electrical activity in a slug ganglion in relation to the concentration of Locke solution. *Journal of Experimental Biology* 33, 282–294.

Jaremovic, R. and Rollo, C.D. (1979) Tree climbing by the snail *Cepaea nemoralis*: a possible method for regulating temperature and hydration. *Canadian Journal of Zoology* 57, 1010–1014.

Jeppesen, L.L. (1977) Photoperiodic control of hibernation in *Helix pomatia* L. (Gastropoda: Pulmonata). *Behavioural Processes* 2, 373–382.

Jeppesen, L.L. and Nygard, K. (1976) The influence of photoperiod, temperature and internal factors on the hibernation of *Helix pomatia* (L.) (Gastropoda: Pulmonata). *Videnskabelige Meddelesler fra Dansk Naturhistorik Forening* 139, 7–20.

Jess, S. and Marks, R.J. (1995) Population density effects on growth in culture of the edible snail *Helix aspersa* var. *maxima. Journal of Molluscan Studies* 61, 313–323.

Jones, J.S., Leith, B.H. and Rawlings, P. (1977) Polymorphism in *Cepaea*: a problem with many solutions? *Annual Review of Ecology and Systematics* 8, 109–143.

Karlin, E.J. (1961) Temperature and light as factors affecting the locomotor activity of slugs. *Nautilus* 74, 125–130.

Karlin, E.J. (1965) Ecological relationships between vegetation and the distribution of land snails in Montana, Colorado and New Mexico. *American Midland Naturalist* 65, 60–66.

Karowe, D.N., Pearce, T.A. and Spaller, W.R. (1993) Chemical communication in freshwater snails: behavioural responses of *Physa parkeri* to mucous trails of *P. parkeri* (Gastropoda: Pulmonata) and *Campeloma decisum* (Gastropoda: Prosobranchia). *Malacological Review* 26, 9–14.

Kasigwa, P.F. (1999a) Dispersian factors in the arboreal snail *Sitala jenynsi* (Gastropoda: Ariophantidae). *South African Journal of Zoology* 34, 145–153.

Kasigwa, P.F. (1999b) Snail arboreality: the microdistribution of *Sitala jenynsi* (Gastropods: Ariophantidae). *South African Journal of Zoology* 34, 154–162.

Kerkut, G.A. and Taylor, B.J.R. (1956) The sensitivity of the pedal ganglion of the slug to osmotic pressure changes. *Journal of Experimental Biology* 33, 493–501.

Klein-Rollais, D. and Daguzan, J. (1990) Variation of water content in *Helix aspersa* Müller in a natural environment. *Journal of Molluscan Studies* 56, 9–15.

Landaur, M.R. and Chapnick, S.D. (1981) Responses of terrestrial slugs to secretions of stressed conspecifics. *Psychological Reports* 49, 617–618.

Lazaridou-Dimitriadou, M. and Saunders, D.S. (1986) The influence of humidity, photoperiod and temperature on the dormancy and activity of *Helix lucorum* L. (Gastropoda Pulmonata). *Journal of Molluscan Studies* 52, 180–189.

Lewis, R.D. (1969a) Studies on the locomotor activity of the slug *Arion ater* (Linnaeus). I. Humidity, temperature and light reactions. *Malacologia* 7, 295–306.

Lewis, R.D. (1969b) Studies on the locomotor activity of the slug *Arion ater* (Linnaeus). II. Locomotor activity rhythms. *Malacologia* 7, 307–312.

Lominicki, A. (1964) Some results of experimental introduction of new individuals into a natural population of the roman snail *Helix pomatia* L. *Bulletin de L'Academie Polonaise des Science* 12, 301–304.

Lyth, M. (1982) Water content of slugs (Gastropoda: Pulmonata) maintained in standardised culture conditions. *Journal of Molluscan Studies* 48, 214–217.

Lyth, M. (1983) Water contents of slugs (Gastropoda: Pulmonata) in natural habitats and the influence of culture conditions on water content stability in *Arion ater* Linné. *Journal of Molluscan Studies* 49, 179–184.

Machin, J. (1975) Water relationships. In: Fretter, V. and Peake, J. (eds) *Pulmonates*, Vol. 1. Academic Press, London, pp. 105–163.

Martin A.W., Deyrup-Olsen, I. and Stewart, D.M. (1990) Regulation of body volume by the peripheral nervous system of the terrestrial slug *Ariolimax columbianus*. *Journal of Experimental Zoology* 253, 121–131.

Matanock, H.L. and Welsford, I.G. (1995) Contact rehydration in the terrestrial slug *Limax maximus* L in the field. *Canadian Journal of Zoology* 73, 607–609.

McFarlane, I.D. (1980) Trail following and trail searching behaviour in homing of the intertidal gastropod mollusc, *Onchidium verraculatum*. *Marine Behaviour and Physiology* 7, 95–108.

Morton, B. (1979) The diurnal rhythm and the cycle of feeding and digestion in the slug *Deroceras caruanae*. *Journal of Zoology* 187, 135–152.

Mpitsos, G.J. and Lukowiak, K. (1985) Learning in gastropod molluscs. In: Willows, A.O.D. (ed.) *The Mollusca*, Vol. 8, *Neurobiology and Behaviour*, Part I. Academic Press, London, pp. 96–267.

Munden, S.K. and Bailey, S.E.R. (1989) The effects of environmental factors on slug behaviour. In: Henderson, I.F. (ed.) *Slugs and Snails in World Agriculture*. British Crop Protection Council Monograph No. 41, pp. 349–354.

Negovetic, S. and Jokela, J. (2000) Food choice behaviour may promote habitat specificity in mixed populations of clonal and sexual *Potamopyrgus antipodarum*. *Animal Behaviour* 60, 435–441.

Nekola, J.C. and Smith, T.M. (1999) Terrestrial gastropod richness patterns in Wisconsin carbonate cliff communities. *Malacologia* 41, 253–269.

Newell, P.F. (1966) The nocturnal behaviour of slugs. *Medical and Biological Illustration* 16, 146–159.

Newell, P.F. (1968) The measurement of light and temperature as factors controlling the surface activity of the slug *Agriolimax reticulatus* (Müller). In: Wadsworth, R.M. (ed.) *The Measurement of Environmental Factors in Terrestrial Ecology*. Blackwell, Oxford, pp.141–146.

Paine, R.T. (1963) Food recognition and predation on opisthobranchs by *Navanax inermis*. *Veliger* 6, 1–9.

Pallant, D. (1969) Daytime resting sites of *Agriolimax reticulatus* (Müller) in woodland. *Journal of Conchology* 27, 9–10.

Parker, G.A. (1978) Searching for mates. In: Krebs, J.R. and Davies, N.B. (eds) *Behavioural Ecology: an Evolutionary Approach*. Blackwell, Oxford, pp. 214–244.

Paul, C.R.C. (1978) The ecology of mollusca in ancient woodland. I. The analysis of distribution and experiments in Hayley Wood, Cambridgeshire. *Journal of Conchology* 29, 281–294.

Peake, J. (1978) Distribution and ecology of the stylommatophora. In: Fretter, V. and Peake, J. (eds) *Pulmonates*, Vol. 2A. Academic Press, New York, pp. 429–526.

Phifer, C.B. and Prior, D.J. (1985) Body hydration and haemolymph osmolality affect feeding and its neural correlate in the terrestrial gastropod *Limax maximus*. *Journal of Experimental Biology* 118, 405–421.

Platts, E.A. and Speight, M.C.D. (1988) The taxonomy and distribution of the Kerry slug *Geomalacus maculosus* Allman 1843 (Mollusca: Arionidae) with a discussion of its status as a threatened species. *Irish Naturalists' Journal* 22, 417–430.

Pollard, E. (1975) Aspects of the ecology of *Helix pomatia* L. *Journal of Animal Ecology* 44, 305–329.

Pomeroy, D.E. (1968) Dormancy in the land snail, *Helicella virgata* (Pulmonata: Helicidae). *Australian Journal of Zoology* 16, 857–869.

Prior, D.J. (1983a) Hydration induced modulation of feeding responsiveness in terrestrial slugs. *Journal of Experimental Zoology* 227, 15–22.

Prior, D.J. (1983b) The relationship between age and body size of individuals in isolated clutches of the terrestrial slug, *Limax maximus* Linnaeus. *Journal of Experimental Zoology* 225, 321–324.

Prior, D.J (1984) Analysis of contact-rehydration in terrestrial gastropods: osmotic control of drinking behaviour. *Journal of Experimental Biology* 111, 63–73.

Prior, D.J. (1985) Water regulatory behaviour in terrestrial gastropods. *Biological Reviews* 60, 403–424.

Prior, D.J. (1989) Contact-rehydration in slugs – a water regulatory behaviour. In: Henderson, I.F. (ed.) *Slugs and Snails in World Agriculture*. British Crop Protection Council Monograph No. 41, pp. 217–223.

Prior, D.J. and Uglem, G.L. (1984) Analysis of contact re-hydration in terrestrial gastropods: absorption of ^{14}C insulin through the foot. *Journal of Experimental Biology* 111, 75–80.

Prior, D.J., Hume, M. Varga, D. and Hess, S.D. (1983) Physiological and behavioural aspects of water balance and respiratory function in the terrestrial slug *Limax maximus*. *Journal of Experimental Biology* 104, 111–127.

Richter, K.O. (1976) The foraging ecology of the banana slug *Ariolimax columbianus* Gould (Arionidae). PhD thesis, University of Washington, Washington, DC.

Riddle, W.A. (1983) Physiological ecology of land snails and slugs. In: Russell-Hunter, W.D. and Wilbur, K.M. (eds) *The Mollusca*, Vol. 6, *Ecology*. Academic Press, New York, pp, 431–461.

Rollo, C.D. (1978) The behavioural ecology of terrestrial slugs. PhD thesis, University of British Columbia, Vancouver.

Rollo, C.D. (1982) The regulation of activity in populations of the terrestrial snail *Limax maximus* (Gastropoda; Limacidae). *Researches on Population Ecology* 24, 1–32.

Rollo, C.D. (1983a) Consequences of competition on the reproduction and mortality of three species of terrestrial slug. *Researches on Population Ecology* 25, 20–43.

Rollo, C.D. (1983b) Consequences of competition on the time budgets, growth and distribution of three species of slug. *Researches on Population Ecology* 25, 44–68.

Rollo, C.D. (1991) Endogenous and exogenous regulation of activity in *Deroceras reticulatum*, a weather sensitive terrestrial slug. *Malacologia* 33, 199–220.

Rollo, C.D. and Wellington, W.G. (1977) Why slugs squabble. *Natural History* 89, 46–51.

Rollo, C.D. and Wellington, W.G. (1979) Intra- and inter specific agonistic behaviour among terrestrial slugs (Pulmonata: Stylommatophora). *Canadian Journal of Zoology* 57, 846–855.

Rollo, C.D., Vertinsky, I.B., Wellington W.G., Thompson, W.A. and Kwan, Y. (1983a) Description and testing of a comprehensive simulation model of the ecology of terrestrial gastropods in unstable environments. *Researches on Population Ecology* 25, 150–179.

Rollo, C.D., Vertinsky, I.B., Wellington, W.G. and Kenetkar, V.K. (1983b) Alternative risk-taking styles: the case of time budgeting strategies of terrestrial gastropods. *Researches on Population Ecology* 25, 321–335.

Ross, R.J. (1979) The effects of mechanical disturbances on the behaviour of inactive terrestrial snails. *Journal of Molluscan Studies* 45, 35–38.

Rózsa, K.S. (1963) A reflex mechanism changing the activity in gastropods upon osmotic effects. *Acta Biologica, Supplement* 5, 44.

Schmidt-Nielsen, K., Taylor, C.R. and Shkolnik, A. (1971) Desert snails: problems of heat, water and food. *Journal of Experimental Biology* 55, 385–398.

Smallridge, M.A. and Kirby, G.C. (1988) Competitive interactions between the land snails *Theba pisana* (Müller) and *Cernuella virgata* (da Costa) from South Australia. *Journal of Molluscan Studies* 54, 251–258.

Sokolove, P.G., Beiswanger, C.M., Prior, D.J. and Gelperin, A. (1977) A circadian rhythm in the locomotor behaviour of the giant garden slug *Limax maximus*. *Journal of Experimental Biology* 66, 47–64.

South, A. (1965) Biology and ecology of *Agriolimax reticulatus* (Müll) and other slugs: spatial distribution. *Journal of Animal Ecology* 34, 403–417.

Southwick, C.H. and Southwick, H.M. (1969) Population density and preferential return in the giant African snail *Achatina fulica*. *American Zoologist* 9, 566.

Speiser, B. and Hochstrasser, M. (1998) Slug damage in relation to watering regime. *Agriculture Ecosystems and Environment* 70, 273–275.

Staikou, A., Lazaridou-Dimitriadou, M. and Kattoulas, M.E. (1989) Behavioural patterns of the edible snail *Helix lucorum* L. in the different seasons of the year in the field. *Haliotis* 19, 129–136.

Step, E. (1945) *Shell Life*. Warne, London.

Stephens, G.J. and McGaugh, J.L. (1972) Biological factors related to learning in the land snail (*Helix aspersa* Müller). *Animal Behaviour* 20, 309–315.

Stephenson, J.W. (1966) Notes on the rearing and behaviour in soil of *Milax budapestensis* (Hazay). *Journal of Conchology* 26, 141–145.

Stephenson, J.W. (1967) The distribution of slugs in a potato crop. *Journal of Applied Ecology* 4, 129–135.

Stephenson, J.W. (1975) Laboratory observations on the different distribution of *Agriolimax reticulatus* (Müll.) in different aggregate fractions of garden loam. *Plant Pathology* 24, 12–15.

Suzuki, N. (1967) Behavioural and electrical responses of the land snail *Ezohelix flexibilis* (Fulton), to odours. *Journal of the Faculty of Science, Hakkaido University* 16, 174–185

Takeda, N. and Ozaki, T. (1986) Induction of locomotor behaviour in the giant African snail, *Achatina fulica*. *Comparative Biochemistry and Physiology* 83, 77–82.

Tattersfield, P. (1981) Density and environmental effects on shell size in some sand dune snail populations. *Biological Journal of the Linnaean Society* 16, 71–81.

Terhivuo, J. (1978). Growth, reproduction and hibernation of *Arianta arbustorum* (L.) (Gastropoda, Helicidae) in southern Finland. *Annalea Zoologici Fennici* 15, 8–16.

Tilling, S.M. (1986) Activity and climbing behaviour: a comparison between two closely related land snail species, *Cepaea nemoralis* (L.) and *C. hortensis* (Müller). *Journal of Molluscan Studies* 52, 1–5.

Tomiyama, K. and Nakane, M. (1993) Dispersal patterns of the giant African snail, *Achatina fulica* (Férussac) (Stylommatophora: Achatinidae), equipped with a radio transmitter. *Journal of Molluscan Studies* 59, 315–322.

Townsend, C.R. (1974) Mucus trail following by the snail *Biomphalaria glabrata* (Say). *Animal Behaviour* 22, 170–177.

Townsend, C.R. (1975) Strategic aspects of time allocation in the ecology of a freshwater pulmonate snail. *Oecologia* 19, 105–115.

Trott, T.J. and Dimock, R.V. (1978) Intraspecific trail following by the mud snail *Ilynassa obseleta*. *Marine Behaviour and Physiology* 5, 91–101.

Uminski, T. (1983) Vitrinidae (Mollusca, Gastropoda) of Poland. Their density and related problems. *Annales Zoologici* 37, 290–311.

Ushadevi, S.V. and Krishnamoorthy, R.V. (1980) Do slugs have silver track pheromone? *Indian Journal of Experimental Biology* 18, 1502–1504.

Voss, M.C., Hoppe, H.H. and Ulber, B. (1998) Estimation of slug activity and slug abundance. *Journal of Plant Diseases and Protection* 105, 314–321.

Waite, T.A. (1988) Huddling and postural adjustments to desiccating conditions in *Deroceras reticulatum* (Müller). *Journal of Molluscan Studies* 54, 249–250.

Wareing, D.R. (1986) Directional trail following in *Deroceras reticulatum* (Müller). *Journal of Molluscan Studies* 52, 256–258.

Wareing, D.R. and Bailey, S.E.R. (1985) The effects of steady and cycling temperatures on the activity of the slug *Deroceras reticulatum*. *Journal of Molluscan Studies* 51, 257–266.

Webley, D. (1964) Slug activity in relation to weather. *Annals of Applied Biology* 53, 407–414.

Wells, M.J. and Buckley, S.K.L. (1972) Snails and trails. *Animal Behaviour* 20, 345–355.

Yom Tov, Y. (1971) The biology of two desert snails *Trochoidea* (*Xerocrassa*) *seetzeni* and *Sphincterochila boissieri*. *Israel Journal of Zoology* 20, 231–248.

Young, A.G. and Port, G.R. (1989) The effect of microclimate on slug activity in the field. In: Henderson, I.F. (ed.) *Slugs and Snails in World Agriculture*. British Crop Protection Council Monograph No. 41, pp. 263–269.

Young, A.G. and Port, G.R. (1991) The influence of soil moisture content on the activity of *Deroceras reticulatum* (Müller). *Journal of Molluscan Studies* 57, 138–140.

14 Soil Biology and Ecotoxicology

R. Dallinger, B. Berger, R. Triebskorn-Köhler[1]
and H. Köhler[2]

Institut für Zoologie und Limnologie, Abteilung Ökophysiologie, Universität Innsbruck, Technikerstraße 25, A-6020 Innsbruck, Austria; [1]Steinbeis-Transfer Center Ecotoxicology and Ecophysiology, Kreuzlinger Strasse 1, D-72108 Rottenburg, Germany; and [2]Animal Physiological Ecology, Zoological Institute, Konrad-Adenaver Strasse 20, D-72072 Tübingen, Germany

Terrestrial Gastropods and Their Impact on the Cycling of Nutrients and Trace Elements

During their evolution, gastropods repeatedly have achieved terrestriality (Barker, Chapter 1, this volume). Indeed, they are among the most successful invertebrates in terrestrial ecosystems, including the most inhospital arctic (Kerney and Cameron, 1979), high alpine (Baur and Raboud, 1988) and arid desert (Crawford, 1981; Arad *et al.*, 1993) areas. In the central alpine region of Europe, for instance, more than 20 species of gastropods live at altitudes over 2000 m above sea level, and still four species occur at altitudes of 3000 m or above (Kerney *et al.*, 1983). This is remarkable, since terrestrial gastropods depend on sufficient amounts of water in order to maintain their activity (Stöver, 1973; Ward and Slotow, 1992). Further, they have developed efficient mechanisms of coping with freezing (Riddle, 1981; Riddle and Miller, 1988), starvation and desiccation (Machin, 1967; Schmidt-Nielsen *et al.*, 1971; Burla and Gosteli, 1993). Nevertheless, gastropods pay a high price for their occupancy of dry and cold terrestrial environments, which is reflected, for instance, in slow growth rates and high stochastic mortality in species living in bleak habitats (Baur, 1984, 1988; Lazaridou-Dimitriadou and Kattoulas, 1991).

In spite of such biological and physiological limitations, and considering the known negative relationship between individual size of a species and its population density (see, for instance, Perry and Arthur, 1991), gastropods in certain woodland and forest habitats can reach very high population densities (e.g. Mason, 1970; Reichardt *et al.*, 1985). Relatively high population densities have been recorded even for alpine (e.g. Baur, 1993) and desert-dwelling species (e.g. Hermony *et al.*, 1992) living under particularly adverse climatic conditions. From an ecophysiological point of view, most terrestrial gastropods are detrivorous animals, feeding on decaying litter along with bacteria, fungi and algae thriving on the surface of the dead plant material. On the other hand, some species, particularly those in ruderal systems of the northern hemisphere, are regarded as primary consumers exhibiting high rates of food digestion and relatively high assimilation efficiencies (Wieser, 1978; Lazaridou-Dimitriadou and Kattoulas, 1991).

In addition to such general patterns of feeding, many species of terrestrial gastropods, particularly those of snail form, strongly depend on calcium as a major macronutrient constituent of their body. This element is the main component of virtually all gastropod shells and opercula, and in many cases also of the eggs. Calcium is also present, either as carbonate or as pyrophosphate, within storage vesicles of connective tissue cells and of basophil cells in the digestive gland of these animals (Simkiss and Mason, 1983). Most of the calcium absorbed by terrestrial gastropods may enter the animal's body via the epithelium of the intestine (Dexheimer, 1963; Beeby and Richmond, 1988), although some authors have suggested that a certain fraction of the element may also be taken up across the integument (Simkiss and Wilbur, 1977; Ireland, 1982). The calcium can be mobilized readily from the intracellular storage sites during periods of increased demand, such as reproduction (Tompa and Wilbur, 1977), hibernation (Dexheimer, 1963) and shell repair (Abolins-Krogis, 1968). Thus, these animals are able to fix calcium by a process of intracellular and extracellular biomineralization (Simkiss, 1976), hence probably contributing significantly to the retention of calcium in the upper layers of soil ecosystems.

Against this background, it has to be expected that gastropods have a strong impact on nutrient cycling by diverting fluxes and changing availabilities of macronutrients in terrestrial ecosystems. In spite of this likely importance, only a few quantitative studies are available from which we can make an assessment of how much terrestrial gastropods can contribute to nutrient cycling. Most studies relate to temperate ecosystems, and information is particularly available for arid and tropical terrestrial ecosystems. If only ingestion rates are considered, it seems that the role of gastropods as detritivores or herbivores in most indigenous temperate ecosystems is rather modest (e.g. Mason, 1970). However, the sample of studies is not large, and they do not cover adequately the range of species richness, functional diversity and population densities that are known to occur in temperate woodlands, forests and grasslands. In some instances, consumption rates in indigenous ecosystems can be considerable (Mason, 1970; Jennings and Barkham, 1979). Furthermore, the importance of terrestrial gastropods for nutrient cycling in these temperate ecosystems probably lies more in their ability to alter the resource base for microbial-driven decomposition processes. Their feeding fragments coarse pieces of plant material, providing for more rapid access by microorganisms. The deposition of mucus and faeces provides 'resource hot spots' in the litter and soil, and, starting from these, microorganisms may

spread and accelerate decomposition processes of the surrounding substrate (Herlitzius and Herlitzius, 1977). Thus high microbial biomass and
activity in the mucus and faecal material contributes to high carbon and
nutrient turnover. Theenhaus and Scheu (1996a), for example, demonstrated by means of experimental microcosms with *Arion ater* (Linnaeus)
(Arionidae) that beech (*Fagus* Linnaeus) (Fagaceae) litter decomposition
was enhanced by the amendment that resulted by deposition of mucus
and faeces by the gastropods. These faecal and mucus 'hot spots' also
increase the heterogeneity of the soil system, which might be of considerable importance for maintenance of a diverse soil microflora and fauna
(Schaefer, 1991).

The importance of gastropods in macronutrient cycling may be
expected to be higher in cold or moisture-stressed environments where
the rate of nutrient turnover is lower than in more equable temperate
regions. In desert ecosystems, for example, terrestrial gastropods often
constitute the predominate standing biomass in the detritivore and
herbivore guilds. *Euchondrus albulus* (Mousson) (Enidae) and closely
related species in the Negev Highlands feed on endolithic lichens, along
with some of the rocky substrate on which lichens thrive, and deposit
their faeces on the soil under the rocks. In this way, these gastropods
transfer considerable amounts of nitrogen from the lichens to the soil,
thus facilitating establishment and production of higher plants in this arid
region (Jones and Shachak, 1990, 1994).

At the same time, some species have been reported to be very selective
in their preferences for habitat types (e.g. Chang and Emlen, 1993) and
their choice of food (e.g. Baur *et al.*, 1994; Hanley *et al.*, 1995; Linhart and
Thompson, 1995). As a consequence, herbivorous terrestrial gastropods
can exert considerable selective pressure on plant communities and
thus profoundly influence their species composition (e.g. Barker, 1991b;
Speiser and Rowell-Rahier, 1991; Speiser, Chapter 6, this volume).
Thompson *et al.* (1993) have shown that selective herbivory by gastropods
led to a reduction in soil ammonia nitrogen, and to an increased availability of phosphate in the soil, thus affecting plant community composition.

Closely connected to the cycling of macronutrients is the cycling
of essential and non-essential trace elements (van Hook *et al.*, 1977;
Zöttl, 1985; Kratz, 1991). It has been shown repeatedly that terrestrial
gastropods possess an exceptional affinity for certain trace elements,
concentrating metals such as copper, zinc, cadmium and lead in their soft
tissues (Coughtrey and Martin, 1976; Ireland, 1979, 1981) or shells (Beeby
and Richmond, 1989; Mulvey *et al.*, 1996), often far above environmental
levels (Dallinger, 1993). This is particularly evident in gastropods from
metal-polluted environments (Martin and Coughtrey, 1982; Greville and
Morgan, 1989a; Berger and Dallinger, 1993; Rabitsch, 1996), but can also
be observed, at least with respect to certain metals, in gastropods in apparently uncontaminated habitats (Knutti *et al.*, 1988). It is also significant
that the uptake and storage of certain trace elements by these animals
often are connected intimately with the uptake and storage of calcium

(Beeby and Richmond, 1988; Beeby, 1991). The physiological basis for this observation lies in the fact that a series of trace elements in gastropod tissues follows the pathway of calcium along a process of biomineralizat-ion within cellular vesicles (Simkiss, 1981; Simkiss et al., 1982). Certain trace elements, such as cadmium, can also be immobilized by detoxifica-tion and binding to metallothioneins (see below) (Dallinger and Berger, 1993; Dallinger, 1996).

The increasing exploitation of natural resources by human activities during the past few centuries has adversely affected the global balance of trace elements (Nriagu, 1990), causing a gradual increase of concentra-tions and availabilities of metals in soil ecosystems (van Hook et al., 1977; Beyer et al., 1985; Zöttl, 1985). Processes of acidification additionally may increase the availability of most trace elements to invertebrates in the soil (van Straalen and Bergema, 1995). The role of terrestrial gastropods in this process of continuous accumulation is not yet clear. However, given the significant contribution of these animals to nutrient cycling (see above), and considering their central position in terrestrial food chains (Laskowski and Hopkin, 1996a), it can be suggested that gastropods must also have a strong impact on the turnover of trace elements. It has to be expected, for instance, that an increasing proportion of essential and non-essential trace elements is retained in upper soil layers due to the presence of terrestrial gastropods which may act as biomineralization sinks and 'buffers' for certain metals (Beeby, 1985), hence changing their biological availability (Dallinger, 1993). At the same time, the activity of terrestrial gastropods may also be responsible for an acceleration of metal cycling in the soil ecosystem as a whole.

Interactions with Other Soil Biota

As ecological key components of terrestrial foodwebs, gastropods may also contribute significantly to the transfer of pollutants from plants to primary consumers (Carter, 1983) and terrestrial predators (Martin and Coughtrey, 1976; Reichardt et al., 1985). Unfortunately, only a few examples exist in which the role of terrestrial gastropods has truly been analysed for the transfer of pollutants in terrestrial habitats. One reason for this lack of information might be that the quantification of food chain transfer of chemicals by terrestrial gastropods is made difficult by the ability of many species to reduce their activity or to aestivate under unfavourable climatic or ecological conditions (Laskowski and Hopkin, 1996a). It has been shown clearly, however, that in transferring nutrients and pollutants through terrestrial food chains, gastropods interact with a variety of other soil organisms, thereby significantly influencing pollutant fluxes and availabilities.

Soil particles, and especially the living and decaying plant and fungal material, exhibit a high capacity for retaining certain trace elements and organic pollutants on their surface or within their organic structure

(Somers, 1978; Boháč *et al.*, 1990; Calderbank, 1994; Alberti *et al.*, 1996). Terrestrial gastropods feeding on this material will take up pollutants in relation to their biological availability in the food source, being able to accumulate metals far above the element levels encountered in their food source (Williamson, 1979; Carter, 1983; Dallinger, 1993; Gomot and Pihan, 1997). The interspecific differences in the capacity of plants to accumulate certain trace elements are well documented (Hunter *et al.*, 1987a). Similarly, dead plant matter and decaying litter, as well as micro-organisms thriving on this material, may accumulate trace elements to different degrees. Some species of macrofungi are able to accumulate high concentrations of trace elements in their fruiting bodies (Stijve and Besson, 1977; Kojo and Lodenius, 1989) on which some species of gastro-pods seem to feed normally (Beyer *et al.*, 1985). It is speculated that under such circumstances, terrestrial gastropods influence metal fluxes through soil ecosystems simply by their preferential selection of certain types of food (see above).

The absorption of pollutants in the gastropod's gut is also influenced by interactions with microorganisms from the substrate or bacteria colo-nizing the animal's alimentary tract. Valuable examples of this have been provided in connection with the accumulation of certain trace elements. The uptake of zinc by *Cantareus aspersus* (Müller) (Helicidae), for instance, depends on microorganisms in the alimentary tract and the abundance of bacteria in the soil substrate (Simkiss and Watkins, 1990, 1991). Furthermore, sulphate-reducing bacteria in the gastropod oesophageal crop can facilitate the absorption of copper (Simkiss, 1985). Another kind of interaction between gastropods and bacteria is exemplified by the inoculating effect which gastropod faeces exert on the soil substrate, leading to increasing bacterial and fungal biomasses, and hence to rising decomposition rates (Theenhaus and Scheu, 1996b). As a consequence, organic matter becomes more accessible to other soil organisms, and this probably increases the availability of nutrients and pollutant metals. Their mobilization from the substrate may also be enhanced due to interactions of terrestrial gastropods with other soil invertebrates such as earthworms (Thompson *et al.*, 1993).

Terrestrial gastropods are likely to play an important role in directly transferring nutrients and certain pollutants to higher trophic levels of ter-restrial food chains, with these animals often serving as prey or hosts for a variety of other animals (e.g. Reichardt *et al.*, 1985). It has been shown, for instance, that a decline in gastropod abundance due to processes of soil acidification can adversely affect populations of forest passerines which need calcium for their eggshells (Graveland and Vanderwal, 1996).

Concerning trace elements, it has been argued that the availability of certain metals for potential predators may depend on the chemical form in which the metal is immobilized in the gastropod's body (Dallinger, 1993; Laskowski and Hopkin, 1996a), even if the gastropod itself may be considered rather tolerant towards the metal (Greville and Morgan, 1991). It can be assumed, for instance, that zinc and lead, which are accumulated in cellular granules along with calcium, may be less available to predatory animals than cadmium, which is mainly bound to chelating proteins such as metallothioneins (Dallinger and Wieser, 1984b; Dallinger *et al.*, 1993a; Berger *et al.*, 1994). In this context, the question arises of whether terrestrial gastropods can contribute to biomagnification of metals along terrestrial food chains. However, it has been stated (Laskowsi, 1991) that biomagnification of trace elements normally does not occur in terrestrial ecosystems, at least for metals. On the other hand, it should be considered that even low transfer rates for trace elements between trophic levels may suffice to exert adverse effects on predators due to metal toxicity (Dallinger, 1993).

Environmental Contamination and Sequestration

Trace elements

Contamination of terrestrial environments by trace elements leads to an increasing uptake of metals by soil invertebrates (Strojan, 1978a; Beyer *et al.*, 1985; Hunter *et al.*, 1987b), and hence to adverse effects of toxic elements on the terrestrial invertebrate fauna (van Straalen and Bergema, 1995), or on the functioning of the soil ecosystem as a whole (Strojan, 1978b). The first studies on terrestrial gastropods as metal-accumulating animals in polluted terrestrial habitats were carried out more than 20 years ago (Meincke and Schaller, 1974; Coughtrey and Martin, 1976). Since that time, many studies on the role of terrestrial gastropods in metal accumulation have been performed (for an overview, see, for instance, Dallinger, 1993), and it has become evident that gastropods belong to those invertebrate species which exhibit the highest capacities for metal accumulation throughout the animal kingdom.

The process of metal accumulation by terrestrial gastropods is determined by some important prerequisites. First, the animals are confronted – as are many other organisms – with trace elements exerting essential and non-essential biological functions (see, for instance, Underwood, 1977). Secondly, the metabolism of many trace elements in terrestrial gastropods is pre-determined by the metabolic pathway of calcium, which is a major constituent of the animal's body (see above). Thirdly, metal accumulation in terrestrial gastropods is conditioned by their obligate need to avoid excessive water loss. Probably as a consequence of such constraints, terrestrial gastropods have developed strategies of metal sequestration at both the cellular and molecular levels (Dallinger, 1993).

The uptake of most trace elements by terrestrial gastropods follows the route of food uptake via the alimentary tract (Hopkin, 1989; Dallinger, 1993; Triebskorn and Köhler, 1996a); moreover, it has to be assumed that terrestrial gastropods do not possess efficient mechanisms allowing them selectively to exclude certain trace metals from being absorbed through the wall of their gut. Available concentrations of trace elements in soil organic matter on which the animals feed are, therefore, the primary extrinsic factors which determine the rates of metal uptake (Dallinger, 1993). For some elements, an additional pathway of entry into the animal's body seems to be the skin (Ryder and Bowen, 1977; Ireland, 1982).

For most trace elements, the main site of accumulation in the gastropod's body is the digestive gland (or midgut gland) (Dallinger, 1993). This organ can be regarded as central to the animal's metabolism, serving as a site of enzyme synthesis, nutrient absorption and decomposition, as well as metal storage and detoxification (Janssen, 1985). In individuals of *C. aspersus* from different contaminated field sites, for example, significant concentrations of cadmium, zinc and lead were detected in the digestive gland (Coughtrey and Martin, 1976; Cooke *et al.*, 1979), whereas only minor amounts of these metals were found in other tissues. Similar results were reported for *Helix pomatia* Linnaeus (Helicidae), *Arianta arbustorum* (Linnaeus) (Helicidae) and *A. ater* after exposure to metals in the laboratory (Schötti and Seiler, 1970; Ireland, 1981; Dallinger and Wieser, 1984a; Marigómez *et al.*, 1986a; Berger and Dallinger, 1989a). A few studies involving the tropical species *Achatina fulica* (Bowdich) (Achatinidae) have confirmed the importance of the digestive gland as a major site of metal accumulation (Ireland, 1988; Ireland and Marigómez, 1992). Other important tissues for accumulation of certain trace elements include, to various degrees, the foot sole, the mantle and the intestine (Coughtrey and Martin, 1976; Dallinger and Wieser, 1984a; Berger and Dallinger, 1989a).

The metabolic pathway of a trace element within the animal's tissues depends mainly on the physicochemical properties of the metal involved. Following a proposal by Nieboer and Richardson (1980), metals can be classified according to their tendency to form ionic or covalent compounds, and hence their preferences in ligand binding. According to this scheme, metals of class A include the alkali and earth alkali elements together with aluminium. In forming chemical compounds, these metals exhibit a clear preference for oxygen-carrying ligands. Metals of class B comprise, for the most part, non-essential heavy metals such as silver, gold, lead (IV) and bismuth. They prefer sulphur-bearing ligands in their chemical compounds. The third category (class C) contains the so-called borderline metals, most of them being essential trace elements such as iron, manganese, zinc and copper(II), or non-essential metals such as cadmium and lead(II). These elements can behave, depending on their chemical environment, like metals either of class A or class B (for more detailed information, see Nieboer and Richardson, 1980).

Class A also includes calcium, potassium and magnesium, which can be regarded as essential macronutrients. The important role of calcium in nutrient metabolism of terrestrial gastropods has already been stressed (Ireland, 1991). As already mentioned above, many trace elements in gastropod tissues follow the pathway of calcium.

Once assimilated, calcium serves a variety of tasks, being an essential constituent of many of the animal's structures and molecules. Sufficient amounts of calcium should, therefore, always be available to meet the gastropod's varying requirements. On the other hand, the intracellular activity of free calcium ions must be kept as low as possible, since this element exerts toxic effects towards cellular structures at elevated concentrations (Taylor *et al.*, 1988). Probably as a consequence of this metabolic dilemma, calcium in gastropod tissues (but not only there; for a general view of this topic, see Brown, 1982) becomes precipitated by a process of biomineralization within intracellular, membrane-surrounded granules, thus forming an energetically convenient and inactivated kind of storage and/or detoxification product (Beeby, 1991). There exist several cell types in gastropod tissues which contain calcium granules. Simkiss and Mason (1983) have distinguished between calcium-containing storage cells and detoxification cells. According to this classification, the most important storage cells are the connective tissue calcium cells which are found in connective tissues of the mantle, foot, visceral complex, and other organs. They contain granules of amorphous calcium carbonate which seems to be readily available upon metabolic demand (Simkiss and Mason, 1983).

More interesting with regard to detoxification processes are the basophilic calcium cells, as well as the digestive and the excretory cells within the gastropod's digestive gland. The basophilic calcium cells contain vesicles with granular, concentrically structured deposits of calcium salts (Abolins-Krogis, 1970; Simkiss, 1976). As shown in *C. aspersus*, the calcium in these granules is present in an amorphous form of pyrophosphate (Simkiss and Mason, 1983). Detailed analyses of such granules have revealed that they contain about 18% water, 5% organic matter and 76% inorganic material, with predominant amounts of Ca^{2+}, Mg^{2+} and $P_2O_7^{4-}$, and minor proportions of the trace elements zinc and manganese (Howard *et al.*, 1981). Through a series of studies, it became clear that these calcium granules can also be regarded as intracellular detoxification sites for trace elements (Simkiss, 1981; Mason and Simkiss, 1982; Simkiss *et al.*, 1982; Almedros and Porcel, 1992a). Cytological studies have shown that the calcium granules of the basophilic cells are formed in close connection to the Golgi system (Simkiss and Mason, 1983), showing elevated activities of acidic phosphatase (Almedros and Porcel, 1992b). It seems that at least in some gastropod species, these granules can be released into the tubular lumen of the digestive gland (Mason and Simkiss, 1982), hence providing a route of trace element liberation.

Trace elements following the calcium pathway into detoxifying granules of basophilic cells can belong to all of the three classes (A, B and borderline) defined by Nieboer and Richardson (1980), although metals of class A and borderline metals are, due to their binding preferences, expected to show a more pronounced affinity for such granules than elements of class B. Traces of aluminium, a metal of class A, were observed in calcium granules of *C. aspersus* (Almedros and Porcel, 1992a) but, in contrast to these findings, Brooks and White (1995) have shown that most of the aluminium assimilated by this species can be detected in excretory granules of the excretory cells (see below). Other metals of class A which become incorporated into calcium granules of the basophilic cells are barium, strontium and beryllium (Simkiss, 1981). Moreover, a variety of borderline metals are also precipitated in this manner, and include manganese (Howard *et al.*, 1981; Mason and Simkiss, 1982; Greaves *et al.*, 1984; Taylor *et al.*, 1988), zinc (Howard *et al.*, 1981; Dallinger, 1993; Triebskorn and Köhler, 1996a), copper (Ireland, 1979; Marigómez *et al.*, 1986a) and possibly lead (Triebskorn and Köhler, 1996a). In contrast to this, few metals of class B have been reported to occur in granules of the basophilic calcium cells (e.g. silver) (Simkiss, 1981; Almedros and Porcel, 1992a).

The digestive gland of terrestrial gastropods contains, besides the basophilic cells, additional cell types involved in metal detoxification: the digestive cells and the excretory cells. The excretory cells of *C. aspersus*, for instance, characteristically contain large, yellowish excretory granules. The predominant elemental constituents of these granules are sulphur, phosphorus and calcium. Only recently has the important role of these granules in the detoxification of aluminium been discovered (Brooks and White, 1995). The digestive cells have a function in the sequestration of lead (Dallinger, 1993).

Apart from cellular compartmentalization, certain trace elements in gastropod tissues can be sequestered by complexation to metallothioneins (Dallinger, 1995, 1996). These are low-molecular weight, metal-binding proteins with a high affinity for certain borderline metals and some trace elements of class B such as cadmium, zinc and copper (I) (Kägi and Schäffer, 1988). Other outstanding features of these proteins are their high cysteine content and their lack of aromatic amino acids. The cysteine residues of the polypeptide chain are highly conserved in their sequential position, being arranged in several Cys-X-Cys motifs, with 'X' designating any other amino acid except cysteine. The metals are bound to the protein backbone by means of several bridging sulphur atoms of the cysteine residues, forming tetrahedrally coordinated metal–thiolate clusters located in two separate protein domains (Kägi and Schäffer, 1988). So far, metallothioneins have been discovered in species from nearly all phyla of the animal kingdom (Kägi, 1993). The metallothioneins from molluscs, and especially from terrestrial gastropods, differ in some of their structural and biochemical features from those of most species in other phyla (Dallinger *et al.*, 1993b; Berger *et al.*, 1995a; Dallinger, 1996). In spite of these differences, gastropod metallothioneins probably perform functions similar to those of most other metallothioneins, being involved in both metal detoxification and trace element regulation (Dallinger, 1996).

The first indication of the presence of a water-soluble, cadmium-binding protein in gastropods was provided by Cooke and colleagues for *C. aspersus* (Cooke *et al.*, 1979); due to its high molecular weight and because of a lack of information about its amino acid composition, it is not clear whether this protein can be regarded as a metallothionein. True

metallothioneins have been fractionated by gel permeation chromatography (Dallinger and Wieser, 1984b) and characterized further by amino
acid composition and sequencing from *H. pomatia* (Dallinger *et al.*,
1989a, 1993b), *A. arbustorum* (Dallinger *et al.*, 1989a; Berger *et al.*, 1995a),
A. ater (Ireland, 1981) and *Arion lusitanicus* Mabille (Dallinger *et al.*,
1989b; Janssen and Dallinger, 1991). As shown by amino acid sequencing,
terrestrial gastropod metallothioneins consist of 63–66 amino acids, 18 of
which are cysteine residues (Dallinger, 1996; Berger *et al.*, 1997). The
molecular weight of the proteins is about 6.4–6.6 kDa.

A fascinating aspect of the biological activity of metallothioneins in
gastropod tissues is their involvement in metal detoxification on the one
hand, and trace element regulation on the other. It was argued that the
function of metallothionein in terrestrial gastropods may have shifted
from trace element regulation towards a more efficient detoxification
of cadmium (Dallinger and Berger, 1993). This applies mainly to the
animals's digestive gland. This organ is by far the most efficient of all
gastropod tissues in accumulating exceptionally high amounts of
cadmium (Dallinger and Wieser, 1984b). As shown by quantitative
chromatographic fractionation of the digestive gland in *H. pomatia*,
85–95% of the cadmium is tightly bound to an inducible, cadmium-
specific metallothionein isoform (Dallinger *et al.*, 1993b; Dallinger, 1996).
The inducibility of metallothionein synthesis by cadmium exposure
(Janssen and Dallinger, 1991; Berger *et al.*, 1995a), along with the
subsequent sequestration of this metal by the induced isoform (Dallinger,
1996) and the long-term persistence of cadmium in the gastropod's
digestive gland (Dallinger and Wieser, 1984a), can therefore be regarded
as a true detoxification process. The exceptional efficiency of this kind of
detoxification also lies in the fact that the trace elements zinc and copper
are additionally present in the animal's digestive gland, but are bound
by different molecules that do not seem to interfere with the cadmium-
specific metallothionein isoform (Berger *et al.*, 1993, 1994; Dallinger
et al., 1993a). Typical gel chromatographic elution patterns showing
metallothionein induction and cadmium sequestration in the digestive
gland of uncontaminated and cadmium-exposed individuals of *H.
pomatia* are exemplified in Fig. 14.1.

Cadmium sequestration by metallothionein binding in the digestive
gland may confer to the animals a high degree of tolerance against this
toxic metal, at least in terms of individual survival, as demonstrated for
C. aspersus (Russell *et al.*, 1981) and *H. pomatia* (Berger *et al.*, 1993).
However, exposed animals exhibit reduced growth rates at even moderate
concentrations of cadmium (Russell *et al.*, 1981).

Recently, a copper-specific metallothionein isoform was isolated and
characterized from the mantle tissue of *H. pomatia* (Berger *et al.*, 1997).
In contrast to the cadmium-specific isoform of the digestive gland, the
copper-binding protein in the mantle cannot be induced at all by
cadmium. Moreover, this isoform does not bind the metal, neither after
cadmium feeding nor by direct administration of cadmium by injection

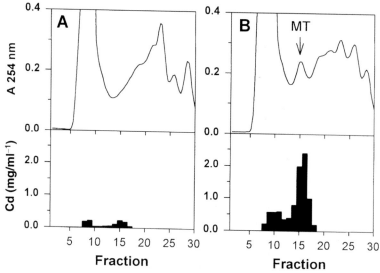

Fig. 14.1. Typical elution patterns after gel permeation chromatography (Sephacryl S-200, Pharmacia Biochemical) of pooled (n = 3–4) digestive gland supernatants from *Helix pomatia* (Linnaeus) (Helicidae) for uncontaminated individuals (A) and individuals exposed for 40 days to a dietary cadmium concentration of 133.5 ± 41.4 $\mu g\ g^{-1}$ dry weight (B). Note the very low concentrations of cadmium in animals on an uncontaminated diet, and the high cadmium concentrations and the presence of induced metallothionein (MT) in animals exposed to contaminated food.

into the mantle tissue. The function of this strictly copper-specific metallothionein isoform must, therefore, be related to the metabolism of the essential trace element copper, probably in connenction with the synthesis of haemocyanin (Dallinger *et al.*, 1997).

It is concluded that the role of metallothioneins in tissues of terrestrial gastropods certainly is not restricted to metal detoxification. However, at least some species synthesize organ- and metal-specific metallothionein isoforms by virtue of which they become able to detoxify the non-essential cadmium efficiently, and at the same time maintain the homeo-static regulation of the essential trace element copper, possibly for the benefit of haemocyanin synthesis. So far, this clear sharing of tasks among metallothionein isoforms in handling of essential and non-essential trace elements is unique to gastropods among the animal kingdom (Dallinger *et al.*, 1998). This evident diversification of metallothionein isoforms is probably also one of the main keys to a better understanding of the exceptional tolerance of terrestrial gastropods towards exposure to cadmium and copper in the environment.

Organic pollutants

In contrast to the bulk of detailed research on metal relationships, relatively few studies have been published dealing with the uptake and metabolism of organic pollutants in terrestrial gastropods. The only exception in this respect are studies focusing on molluscicides (see, for example, Triebskorn and Künast, 1990; Triebskorn and Schweizer, 1990; Triebskorn, 1991).

Some studies dealing with the accidental uptake and accumulation of DDT and related compounds by terrestrial gastropods were published 20–30 years ago. As an example of this kind of research, a thorough examination on the uptake of DDT by *Cepaea hortensis* (Müller) (Helicidae) has demonstrated that after a single meal. DDT residues were accumulated and retained in the digestive gland (Dindal and Wurzinger, 1971). Considerable concentrations of DDT were also detected in the ovotestis of this species, while most of the other organs accumulated the pesticide to only moderate degrees. Moreover, it was reported that concentration factors (referred to the soil substrate on a dry weight basis) for DDT residues are generally lower in terrestrial snails (with values varying from 0.1 to 1.0) than in slugs (with values in the range 2.2–17.9). It was suggested that terrestrial gastropods, as non-target organisms for DDT, may serve as a source for the transfer of this pesticide to vertebrate predators (Dindal and Wurzinger, 1971), thus contributing to the biomagnification of DDT in terrestrial food chains.

Many other studies on the uptake of organic chemicals by terrestrial gastropods were interested primarily in the efficiency of dietary absorption of chemicals and their efficacy as molluscicides, with pestiferous species being target organisms (see, for instance, Briggs and Henderson, 1987). However, a chapter dealing with the accumulation and action of molluscicides is included in a related volume (see Henderson and Triebskorn, 2001), and is, therefore, not a subject of this article.

Terrestrial Molluscs as Biological Indicators of Environmental Pollution

According to a very general definition, biological indicators can be regarded as species which indicate the condition or state of the environment in which they live (Spellerberg, 1991). This broad definition includes the so-called 'bioindicators of accumulation'. These are animal or plant species which indicate, by virtue of their capacity to accumulate certain environmental pollutants, the presence of these toxicants at concentrations possibly hazardous to the biota of the respective ecosystem. The use of terrestrial gastropods as biological indicators so far has been restricted mainly to metal pollution, by employing these animals as accumulation indicators of contaminant trace elements in terrestrial ecosystems (see, for example, Martin and Coughtrey, 1982).

As discussed above, many species of terrestrial gastropods have been shown to concentrate trace elements in their tissues owing to the efficient mechanisms of sequestration which they possess. Accordingly, the final concentration of a metal reached in the gastropod's body is the result of a process of accumulation involving metal uptake, storage and elimination (Dallinger, 1993). The idea behind the concept of biological indication in this respect is based on an expected quantitative or semi-quantitative relationship between the metal concentration in the animal's tissues and the metal levels detected in the environment (Williamson, 1979). This relationship ideally can be expressed by a simple equation (Dallinger, 1994):

$$y = a.x + b$$

where y is the metal concentration in the gastropod's body or tissue, x is the metal concentration in the environment (or substrate), a is the biological concentration factor and b is the background burden of the metal in the animal's body or tissue.

The biological concentration factor (a) in this linear equation predicts by how much a terrestrial gastropod will concentrate a metal in relation to environmental levels. Considering the animal as a whole, a terrestrial gastropod may therefore be classified according to its biological concentration factor with respect to a certain trace element. Species which accumulate a metal far above environmental concentrations (with $a > 2$) have been defined as 'macroconcentrators', whereas species accumulating a trace metal in equal proportion to environmental levels ($a = 1-2$) or below ($a < 1$) have been classified as 'microconcentrators' and 'deconcentrators', respectively (Boháč and Pospíšil, 1989; Dallinger, 1993). It appears, for instance, that some helicid species such as *H. pomatia, C. aspersus, Cepaea nemoralis* (Linnaeus), *A. arbustorum* and arionid species can be regarded as 'macroconcentrators' for cadmium and copper, while they mostly behave as 'microconcentrators' with respect to zinc and lead (Dallinger, 1993; Laskowski and Hopkin, 1996a).

Whereas the concept of quantitative bioindication has been used quite successfully for soil invertebrates such as isopods (Wieser *et al.*, 1976; Hopkin *et al.*, 1986; Dallinger *et al.*, 1992), the linear relationship explained above might be applied to terrestrial gastropods only with some reservations. The reason for this lies in the fact that in contrast to terrestrial isopods which normally feed on a rather homogeneous substrate, the foods utilized by terrestrial gastropods are remarkably varied (Martin and Coughrey, 1976; Reichardt *et al.*, 1985; Hopkin, 1989; Speiser and Rowell-Rahier, 1991). Moreover, metal concentrations in the litter substrate on which terrestrial gastropods feed can display a high degree of variability (Carter, 1983), probably due to different states of decomposition. Additional factors which adversely interfere with the concept of quantitative bioindication in terrestrial gastropods are variabilities of metal concentrations in the animal tissues depending on their body size (Coughtrey and Martin, 1977; Williamson, 1980; Greville and Morgan, 1990; Berger and Dallinger, 1989b; Dallinger and Berger, 1992), species-specific peculiarities in metal accumulation (Greville and Morgan, 1990; Gomot and Pihan, 1997), seasonal factors (Williamson, 1979; Ireland, 1981, 1984; Berger and Dallinger, 1989b, 1993; Greville and

Morgan, 1989b) and prevailing temperature (Meincke and Schaller, 1974). Some studies also suggest that the bioindication approach probably does not work at all for certain metals such as copper (Berger and Dallinger, 1993).

In spite of such adversities, terrestrial gastropods have been employed as biological indicators of environmental metal contamination with a fair degree of success (Martin and Coughtrey, 1982; Berger and Dallinger, 1993). In this respect, it appears that snails are probably more appropriate for the purpose of biomonitoring than slugs. This is explained by the fact that even closely related species of slugs can display large intra- and interspecific variabilities in their metal body burdens (Greville and Morgan, 1989a; Berger and Dallinger, 1993), thus rendering a correct interpretation of the data more difficult. In spite of these circumstances, however, some slugs such as *A. ater* have been shown to be suitable animals for biomonitoring in metal-contaminated soil ecosystems (see, for instance, Popham and D'Auria, 1980).

One major difficulty often encountered in using terrestrial gastropods as biological indicators of environmental pollution is the apparent lack of a significant correlation between metal concentrations in the gastropod tissues and concentrations of the respective metals in plants or leaf litter on which the animals probably feed. This absence of correlation may be explained by the technical problem of identifying the food item the animals had consumed before they were collected, and the selectivity in choice of food displayed by many species of terrestrial gastropods. Accurate identification of the food source is important considering the variability of metal concentrations in different plant species or in litter substrate of mixed composition (Martin and Coughtrey, 1982; Hunter *et al.*, 1987a). Jones and Hopkin (1991) have shown that there may exist better correlations in cadmium concentrations between the gastropods *C. aspersus*, *C. hortensis* and *Monacha cantiana* (Montagu) (Hygromiidae) and the isopods *Porcellio scaber* (Latreille) and *Oniscus asellus* (Linnaeus) (Oniscidae) from the same sites in the field, than between snails and their nettle (*Urtica dioica* Linnaeus) (Urticaceae) diet. Like-wise, body tissue burdens of cadmium and lead have been demonstrated to be remarkably similar between the gastropod species *A. arbustorum* and the isopod *P. scaber* collected at the same sampling points in the urban area of Innsbruck, Austria (Dallinger *et al.*, 1989c; Berger, 1990; Dallinger and Berger, 1992).

A semi-quantitative approach to the use of terrestrial gastropods as bioindicators of metal contamination has been presented by Berger and Dallinger (1993). Concentrations of cadmium, lead, copper and zinc were first measured in individuals of *A. arbustorum* collected at different urban and rural sites of known metal contamination levels near Innsbruck (Austria). The results from this reference data set were then compared with all available data on metal concentrations in terrestrial gastropod species from many European sites. In this way, three distinct levels (classes) of environmental contamination could be established for cadmium, zinc and lead, corresponding to unpolluted reference sites (class 1), areas moderately contaminated by traffic exhausts or urban emissions (class 2) and heavily polluted sites due to mining activity or metal-working industry (class 3), respectively (Berger and Dallinger, 1993) (Fig. 14.2). This system has been applied successfully to moderately contaminated areas in urban environments (Dallinger and Berger, 1992). The sensitivity of the biomonitoring approach with terrestrial gastropods might be improved further by using the digestive gland tissues only, instead of whole animal preparations, for the metal analyses, as recently proposed by Rabitsch (1996).

Using Biomarkers in Terrestrial Gastropods

'Biomarker' is a relatively recent concept and currently under much discussion (see, for instance, van Gestel and van Brummelen, 1996). Sanders (1990) suggested a biomarker concept based on the measurement of molecular, biochemical, cellular and physiological parameters in animals at the subindividual or individual levels in response to a pollutant chemical as screening tools in environmental toxicology. Biomarkers are – in contrast to bioindicators – not species, but subindividual parameters. Their potential use in terrestrial environmental toxicology presently is being researched by many investigators (see, for instance, Kammenga *et al.*, 1996). A commonly used biomarker approach, for example, is that based on the cytochrome P450 monooxygenase system. Surprisingly, this concept so far has not been applied successfully to terrestrial gastropods. Instead, three other biomarker approaches have been followed in these animals: cellular biomarkers, and biochemical biomarkers with metallothioneins on the one hand, and stress proteins on the other.

Cellular biomarkers in terrestrial gastropods

As summarized by Braunbeck and Storch (1989), histological and ultrastructural alterations in certain organs of animals can be used as biomarkers to detect individual responses to toxicant exposure, with the ultimate purpose of evaluating the toxic potential of a given chemical in the environment (see also Storch, 1988; Triebskorn *et al.*, 1991). In a study with mercury-exposed individuals of *A. ater*, for example, Marigómez and colleagues proposed the use of this species as a sentinel organism assessing cellular biomarkers of exposure to metallic pollutants (Marigómez *et al.*, 1996).

As in the field of human medicine, light and electron microscopy are used to detect cellular and subcellular symptoms of injury resulting from poisoning by organic xenobiotics or toxic elements (Cajaraville *et al.*, 1989; Marigómez *et al.*, 1990; Triebskorn *et al.*, 1991; Triebskorn, 1995). Both techniques can be applied to determine symptoms preceding cell

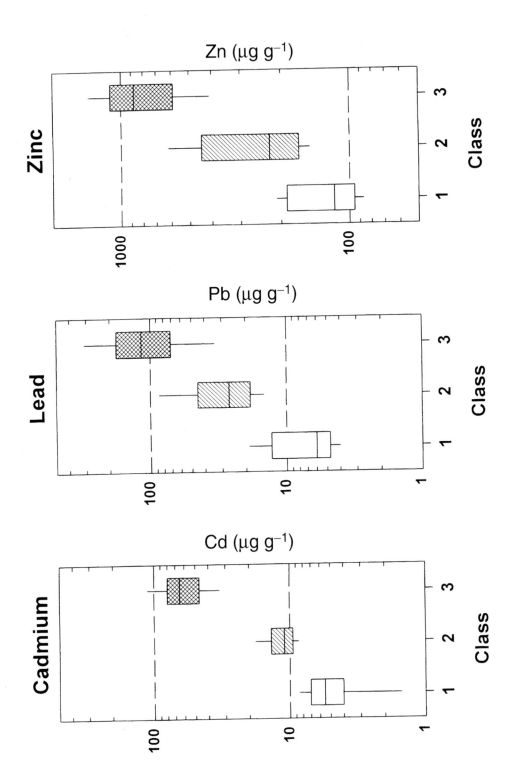

death (Sparks, 1972; Bowen, 1981), or to disclose sublethal reaction (or compensation) in animals tolerating elevated concentrations of toxicants induced through processes of biotransformation or detoxification (Köhler and Triebskorn, 1996; Triebskorn and Köhler, 1996b; Triebskorn et al., 1996).

It has to be noted, however, that interpretation of cytotoxicity as a tool for environmental diagnosis is influenced by a variety of endogenous and exogenous factors, such as the nutritional and developmental state and gender of the animal, and prevailing environmental temperature and humidity, as shown for many species of terrestrial invertebrates (Carstens and Storch, 1980; Alberti and Storch, 1983; Storch, 1984; Neumann, 1985; Hryniewiecka-Szyfter and Storch, 1986; Triebskorn and Schweizer, 1990; Fischer and Molnár, 1992). Therefore, cytotoxicity refers to a cellular status quo that results from several factors which include the possible toxic impact of a xenobiotic substance. Moreover, cellular reactions in animals exposed to pollutants often differ not qualitatively, but quantitatively from those occurring in the 'normal' metabolic state. These subcellular effects observed after toxic impacts often reflect modifications of established metabolic pathways intensified by pollutants (Moore, 1985; Köhler and Triebskorn, 1998), rather than being the development of completely new metabolic pathways and related structures. A further consideration in interpreting cellular diagnosis is the fact that cells in a particular organ often do not react simultaneously, with the consequence that 'hot spots' of cellular injury are distributed unevenly throughout the organ.

In order to overcome such disadvantages in histological and ultrastructural diagnosis, the 'control status', that is the diversity and plasticity of cellular reactions and structures occurring in an organ under normal conditions, has to be well known before evaluating the cytotoxicological state of an organ after chemical exposure. This already indicates the comparative character of biomarkers. Furthermore, a series of tissue samples from each of several individuals per treatment is needed, in order to cope with individual and organ-specific variations in controls and contaminated animals. Additionally, a combined approach involving both light and electron microscopy is indispensable. Light microscopic histological studies permit a gross examination of the target organ and allow the researcher to select areas suspected to have developed ultrastructural damage. These tissue areas can, in turn, be analysed in more detail by electron microscopy.

Fig. 14.2. (Opposite) Diagrams (modified after Berger and Dallinger, 1993) showing three classes (x-axes) of environmental contamination with cadmium (left), lead (middle) and zinc (right) as reflected by metal concentrations ($\mu g\ g^{-1}$ dry weight) in whole animals of several species of terrestrial gastropods sampled at a range of sites, expressed on a logarithmic scale (y-axes). Boxes represent ranges (25 and 75% percentiles) of metal concentrations, with medians (horizontal lines within the boxes) and 10 and 90% percentiles (vertical lines) for each class, corresponding to uncontaminated environments (class 1), moderately contaminated environments (class 2), and heavily polluted environments (class 3), respectively.

In terrestrial gastropods, the digestive gland has been found to be the major site of metal accumulation (see above), and is the organ in which cellular alterations are most evident after exposure to organic and inorganic pollutants (Marigómez et al., 1986b; Récio et al., 1988; Triebskorn, 1989; Triebskorn and Künast, 1990; Felder, 1992; Bradley and Runham, 1996; Marigómez and Dussart, 1996; Triebskorn et al., 1996). Consequently, ultrastructural changes in basophilic and digestive cells, the two major cell types in the digestive gland of terrestrial gastropods, can be used successfully as biomarkers in toxicant-exposed animals (Triebskorn and Köhler, 1996a) (Table 14.1).

In the basophilic cells, alterations of the granular endoplasmic reticulum are the most prominent reactions to toxicant exposure. Symptoms such as dilatation, vesiculation and degranulation of the endoplasmic reticulum, as well as formation of circular arrays can be related to induced processes of biotransformation or to an activation of the animal's metabolism (Triebskorn and Köhler, 1992) (Table 14.1). Furthermore, an increase in the number of electron-dense calcium granules (spherites) can be related directly to metal pollution (Hopkin, 1989; Triebskorn and Köhler, 1996a). Bradley and Runham (1996) have described alterations in calcium granules of basophilic cells in the digestive gland of C. aspersus after exposure to manganese, and a necrosis of basophilic cells after exposure to zinc. Marigómez and Dussart (1996) have suggested a shift in 'responsibility' for metal detoxification from digestive to basophilic cells in the digestive gland of arionid species exposed to copper, zinc, mercury and lead.

In the digestive cells, the 'status of reaction' to exposure to metals or pesticides is reflected mainly by an increase in cellular vacuolization and lysosomal instability (Triebskorn, 1989; Triebskorn and Künast, 1990) (Table 14.1). Moreover, mitochondria are swollen and stocks of storage products become reduced. Above a certain threshold of pollution (representing the 'status of destruction'), cell membranes are ruptured, the lysosomal system breaks down and cell death occurs, each readily apparent in electron microscopy. Such a type of necrosis has been described, for instance, in the digestive gland of C. aspersus (Bradley and Runham, 1996) and Deroceras reticulatum (Müller) (Agriolimacidae) (Triebskorn and Köhler, 1996a) after exposure to zinc.

Apart from the digestive gland, mucocytes of the gastropod's skin and intestine have also been found to exhibit prominent alterations after exposure to environmetal pollutants (Triebskorn and Ebert, 1989; Triebskorn and Schweizer, 1990). Characteristic ultrastructural symptoms reflecting toxicant-induced reactions of these cells (which are related to an increased mucus production and extrusion) include a dilatation of the endoplasmic reticulum and an increase in the number and dilatation of Golgi cisternae and vesicles. Again, the membrane system of the cells breaks down above a certain threshold of pollutant exposure, and the whole cells are extruded from the epithelium. Such symptoms obviously represent the 'status of destruction' (see above). Bradley and Runham

Table 14.1. Ultrastructural changes in the basophilic and digestive cells of the digestive gland of *Deroceras reticulatum* (Müller) (Agriolimacidae) following exposure to cadmium over a period of 21 days in a contaminated lettuce (*Lactuca sativa* Linnaeus: Asteraceae) leaf diet.

	Cadmium concentration in the food (dry weight)			
	Control 0.7 µg g⁻¹	59 µg g⁻¹	358 µg g⁻¹	751 µg g⁻¹
Basophilic cells:				
Cytoplasm	Homogeneous	100% of cytoplasm inhomogeneous, partly highly aggregated, partly very electron-lucent	90% of cytoplasm inhomogeneous, 10% of cytoplasm reduced in extent	50% of cytoplasm electron-lucent, 50% reduced
Nuclei	OK	80% of nuclei very electron-lucent, 5% karyolysis, in 5% of nuclei increased mitosis	80% of nuclei electron-lucent, in 10% of nuclei karyolysis	90% of nuclei electron-lucent, in 10% of nuclei karyolysis
Mitochondria	OK	OK	60% of mitochondria dilated, with reduced cristae, 20% of mitochondria destroyed	60% of mitochondria dilated, with reduced cristae, 20% of mitochondria destroyed
Rough endoplasmic reticulum	OK	80% of RER degranulated, vesiculated or dilated	80% of RER with short cisternae, vesiculated or dilated, 5% of RER with myelin bodies	80% of membranes destroyed, forming myelin bodies
Golgi apparatus	OK	Activated secretion	Dilated or compressed cisternae, increased in number	50% of the cisternae destroyed
Calcium granules	Concentrically structured	Strongly mineralized, with dark appearance	Large vacuoles with light contents	Concentrically structured, vacuoles with dark contents
Lipid content	High	Absent	Absent	Absent, lipofuscin appearing
Glycogen content	Medium	Medium	Medium	High
Digestive cells:				
Cell apices	Mostly OK	Elongated pinocytotic channels, 5% of cell apices destroyed	50% of cell apices with elongated pinocytotic channels, 50% destroyed	95% of cell apices destroyed
Microvilli	Few damaged	80–90% of microvilli reduced in number, 10% shortened	80–90% of microvilli reduced in number, 10% shortened	100% of microvilli reduced in number
Cytoplasm	Homogeneous	100% of cytoplasm inhomogeneous, electron-lucent	90% of cytoplasm inhomogeneous, electron-lucent, 10% of cytoplasm reduced	50% of cytoplasm electron-lucent, 50% reduced
Nuclei	OK	OK	10% of nuclei dark	10% of nuclei dark, with structural changes in the nucleolus
Mitochondria	OK	OK	50% of mitochondria dilated	80–90% of mitochondria dilated, with membrane whorls
Vacuoles	OK	Increased fusion and very light contents	Increased fusion, enlarged in appearance, with partly destroyed membranes	Increased fusion, enlarged in appearance, with partly destroyed membranes
Lipid content	High	Medium	Very low	Absent
Glycogen content	Medium	Absent	High	Absent

For details on tissue preparation and transmission electron microscopy, see Triebskorn and Köhler (1996a).

(1996) have described, moreover, a kind of general necrosis in the foot sole of *C. aspersus* after exposure to copper.

Finally, the epithelial cells of the oesophageal crop have also been shown to represent sensible tools for monitoring environmental contamination. Several authors have demonstrated ultrastructural alterations to occur in these cells after pesticide application (Bourne *et al.*, 1991; Triebskorn, 1989; Triebskorn and Künast, 1990; Triebskorn *et al.*, 1990). The most prominent reaction of epithelial crop cells to pollutant exposure is a reduction in intracellular energy stores such as lipids and glycogen, probably due to the energy-consuming synthesis of biotransformation enzymes, metallothioneins or stress proteins (Triebskorn and Köhler, 1996a).

Gastropod metallothioneins as biomarkers

The strong inducibility of metallothionein synthesis by certain trace metals and their subsequent binding to the induced protein (see above) make this molecule a potential biomarker for environmental metal pollution (Dallinger, 1994, 1996; Berger *et al.*, 1995b). The fact that metallothionein synthesis in many animal species additionally can be promoted by non-metallic inducers such as organic chemicals and other stress factors (see, for instance, Kägi and Schäffer, 1988) further increases the importance of this protein for biomonitoring purposes (Dallinger *et al.*, 2000).

So far, the most important parameter reflecting the metallothionein status in an animal is, of course, its concentration in different tissues (see, for example, Roesijadi, 1993; Berger *et al.*, 1995b). Methods allowing metallothionein quantification at the level of the expressed protein can be regarded as direct approaches. They are based mainly on saturation assays in which metals such as cadmium (Bartsch *et al.*, 1990), silver (Scheuhammer and Cherian, 1986) and mercury (Dutton *et al.*, 1993) are added to the metallothionein to saturate its binding sites completely by replacing other metal ions having a lower affinity for the protein molecule. An indispensable prerequisite for this kind of quantification is the knowledge of the metallothionein's metal-binding stoichiometry (e.g. Dallinger *et al.*, 2001). Apart from direct approaches, there exist some indirect methods for assessing metallothionein synthesis which are based mainly on molecular techniques such as quantification of metallothionein mRNA (Unger and Roesijadi, 1993).

A cadmium saturation method (the so-called Cd-Chelex assay) has been used to measure metallothionein concentrations in terrestrial gastropods (Berger *et al.*, 1995b). This approach allows not only the quantification of the cadmium-binding metallothionein pool, but also the assessment of the native protein's saturation with this toxic metal. In this way, the metallothionein status in the digestive gland of metal-exposed individuals of *H. pomatia* was assessed in terms of the protein's concentration and its saturation with cadmium, and related to critical threshold values observed in animals that were moribund after long-term exposure. It appeared that the mortality of metal-exposed *H. pomatia* significantly increased above concentrations of (Cd)-metallothionein in the digestive gland of about 1000 μg g^{-1} (fresh weight), and at a cadmium saturation of the (Cd)-metallothionein pool of more than 60% (Berger *et al.*, 1995b).

The Cd-Chelex assay recently has been modified in order also to detect concentrations of copper-binding metallothionein and of total metallothionein (Dallinger *et al.*, 2000) in the digestive gland and the mantle of some species of helicid species. Figure 14.3 depicts an application of this method, showing significant differences in the status of different metallothionein pools in the digestive gland of uncontaminated and metal-exposed *H. pomatia* individuals.

The extent to which the metallothionein approach in terrestrial gastropods can be applied to field conditions as a biomarker for environmental pollution remains to be established. One possible approach in the future will be to correlate metallothionein concentrations in terrestrial

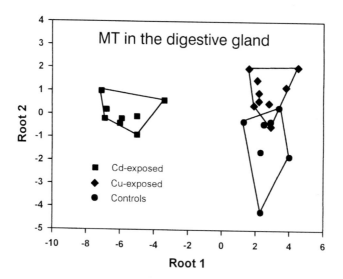

P-levels

	Controls	Cd-exp.	Cu-exp.
Controls	--	0.01	0.01
Cd-exp.	0.01	--	0.01
Cu-exp.	0.01	0.01	--

Factor structure matrix

	Root 1	Root 2
Total-MT	-0.47	-0.81
Cd,Zn-MT	-0.79	-0.18
Cu-MT	-0.11	-0.92

Fig. 14.3. Top: Graphic representation of a discriminant function of the total metallothionein concentrations, (Cd, Zn)-metallothionein and (Cu)-metallothionein in the digestive gland of *Helix pomatia* (Linnaeus) (Helicidae) after different exposure treatments (controls, cadmium-exposed and copper-exposed). Bottom: tables showing the significance levels (*P*-levels) of differences between treatments (left) and the factor structure matrix (right) for concentrations of total metallothionein (Total-MT), (Cd, Zn)-metallothionein (Cd, Zn-MT) and (Cu)-metallothionein (Cu-MT), respectively.

gastropods with sublethal parameters of toxicity. Such parameters include, for example, growth and fertility, as recently studied in metal-contaminated individuals of the species *C. aspersus* (Laskowski and Hopkin, 1996b; Gomot, 1997).

Stress proteins of terrestrial gastropods as biomarkers

First discovered in the species group of the fruit fly, *Drosophila melano-gaster* Meigen (Drosophilidae) in response to increased temperature (Tissières *et al.*, 1974), a set of inducible proteins, originally called heat shock proteins, since has been found to occur ubiquitously in animal tissues after exposure to a variety of chemicals, but also in response to viral infections and both minor and massive damage to tissues (e.g. Schlesinger *et al.*, 1982, 1990; Nover, 1984). The term 'heat shock proteins' has been replaced by the expression 'stress proteins' for this class of molecules (Gething and Sambrook, 1992). Despite their apparent non-specificity in induction, the suitability of stress proteins as molecular biomarkers for chemical exposure and cellular stress has become more and more evident during the past few years (Peakall and Walker, 1994; Sanders and Dyer, 1994).

According to their molecular weight, eukaryotic stress proteins are classified into different protein families. Invertebrate animals possess: (i) a heterogeneous group of low molecular weight stress proteins with molecular masses of about 15–40 kDa; (ii) a group of so-called stress-60 proteins (hsp60, chaperonin, cpn60 and tcp1; mol. mass: 58–60 kDa); (iii) proteins of the prominent stress-70 family (hsp70 and BiP; mol. mass: 66–78 kDa); (iv) groups of stress-90 (hsp90; mol. mass: 83–90 kDa); and (v) high molecular weight stress proteins (mol. mass: 100–110 kDa) (Sanders, 1993). Additionally, a small protein of 7 kDa, called ubiquitin, which is involved in non-lysosomal protein degradation, as well as a variety of associated proteins assisting the aforementioned protein groups in their physiological function, are usually assigned to the stress proteins.

The physiological rationale for using stress proteins as biomarkers is based on the mode of induction of hsp70 which so far represents the best investigated family of stress proteins. Hsp70 is involved in intracellular protein folding and membrane translocation (Pelham, 1986; Morimoto *et al.*, 1990). Thus, whenever a stressor leads to an increase in denatured or malfolded protein molecules in the cell, the elevated presence of uncoiled polypeptide chains protruding from those proteins promotes a cascade of induction (see Gething and Sambrook, 1992), resulting in an enhanced synthesis of hsp70 ('abnormal protein hypothesis'; see Eding-ton *et al.*, 1989; Sorger and Nelson, 1989; Craig and Gross, 1991). Based on this background, it is now well accepted that the main advantage of the hsp70 proteins as biomarkers is their ability to integrate effectively a variety of adverse effects on protein integrity collectively termed proteo-toxicity (Sanders, 1993). The increased level of stress proteins indicates the presence of a stressor, but usually does not give any information about its physical or chemical nature (Peakall and Walker, 1994). Some of the numerous members of the aforementioned stress protein families, however, are thought to be stressor-specific (Sanders and Dyer, 1994).

The use of stress proteins as biomarkers for environmental hazards has gained increased prominence during the last decade. Most of the research in this field, however, has been directed towards animals from aquatic environments (e.g. Sanders, 1993), with only a limited number of studies focusing on soil invertebrates (Köhler et al., 1992, 1996a). The supposed ubiquity and structural conservatism of at least the three major families of stress proteins (hsp90, hsp70 and hsp60) imply their presence also in terrestrial gastropods. However, evidence for this so far has only been provided for hsp70 and hsp60 in A. ater and D. reticulatum.

In individuals of D. reticulatum exposed to a set of environmentally pollutant chemicals, hsp70 seems to be a 'better' biomarker than hsp60 due to its higher sensitivity in response to stressors (Rahman, 1994). Accordingly, it has been shown that poisoning in D. reticulatum and A. ater by carbamate molluscicides leads to an induction of hsp70 proteins (Köhler et al., 1992). Moreover, subchronic exposure of D. reticulatum to elevated concentrations of pentachlorophenol over a period of 2 weeks resulted in a dose-dependent induction of these proteins (Rahman, 1994).

Most of the work on stress proteins as biomarkers in terrestrial gastropods has been carried out with trace elements as inducing agents. Subchronic exposure of D. reticulatum to metal ions (Zn^{2+}, Cd^{2+} and Pb^{2+}) over a period of 3 weeks, for instance, resulted in the induction of hsp70 in a concentration-dependent manner (Köhler et al., 1994, 1996a) (Fig. 14.4). This has been demonstrated at the level of protein expression by means of accumulation of the stress protein itself, as well as at the mRNA level (Köhler, 1996). In the exposure experiments with cadmium and zinc, the levels of hsp70 accumulated after 3 weeks in animal tissues were found to be good predictors of mortality in life cycle exposure experiments. Accordingly, RT–PCR (reverse transcription–polymerase chain reaction) studies have shown an elevated transcription rate of the hsp70 gene, and/or an increased longevity of the hsp70 mRNA after 2 weeks of exposure, to reflect (and possibly to result in) a decline in fecundity and viability of eggs in D. reticulatum under conditions of lifetime exposure (Köhler, 1996; Köhler et al., 1996b).

The molecular stress response in terrestrial invertebrates is especially useful as a biomarker for sublethal effects of chemicals. Like other biochemical biomarkers, the stress protein response is maximal with initial increases in intensity of the stressor, but then decreases above a certain threshold level of exposure. This commonly is explained by a pathological inhibition of gene transcription and protein synthesis at too high intensities of exposure. Such a pattern of response has been demonstrated for the hsp70 levels in the isopod O. asellus (Eckwert et al., 1997) and the nematode Caenorhabditis elegans (Maupas) (Rhabditidae) (Guven et al., 1994). It might also be expected to occur in the tissues of terrestrial

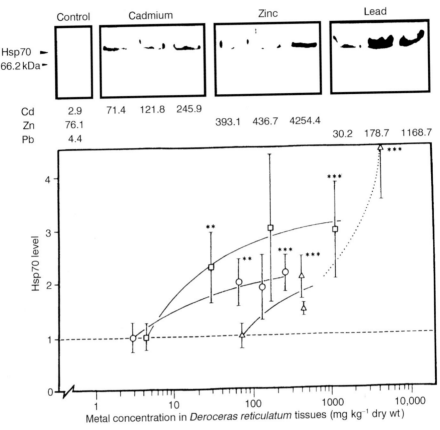

Fig. 14.4. Accumulation of the stress protein hsp70 induced by exposure of *Deroceras reticulatum* (Müller) (Agriolimacidae) to cadmium-, zinc- and lead-enriched food and substrate over a period of 21 days under constant laboratory conditions (temperature: 10°C; photoperiod: 16/8 h light/dark) (data taken from Köhler *et al.*, 1996a). Top: representative western blots of equal amounts of total protein, extracted from whole body homogenates of unaffected (Control) and metal-exposed animals (Cadmium, Zinc, Lead), showing increasingly dense bands of hsp70 reflecting increasing concentrations (in mg kg^{-1} dry weight) of metals in the animals's soft tissues, as specified below the blots. Visualization of hsp70 bands was achieved by monoclonal antibodies raised against hsp70. Bottom: densitometric quantification of hsp70 bands (hsp70 level) generated by western blot analysis with increasing metal concentrations (means and standard deviations) in animal tissues (mg kg^{-1} dry weight). The mean value of the hsp70 band in controls was set, as a reference, to 1.00 (dashed horizontal line). Circles, cadmium; squares, lead; triangles, zinc. For zinc exposure only, the experiment resulting in the highest metal accumulation (up to 1168.7 mg Zn kg^{-1} tissue dry weight; dotted line) was terminated after 9 days instead of 21 because of the high mortality rate in this group. Significance levels (Student's *t*-test, difference from controls) were as follows: **$P \leq 0.01$; ***$P \leq 0.001$.

gastropods. It is concluded, therefore, that a proper interpretation of the effects of pollutant exposure might only be achieved by combining the stress protein response with cellular biomarkers, and, depending on the intensity of exposure, taking into account pathological alterations.

It is not clear yet to what extent the biomarker concept in terrestrial gastropods might contribute to a more efficient impact assessment in polluted terrestrial ecosystems (see Treweek, 1996). Much work in this novel field of ecototoxicology is still in progress.

References

Abolins-Krogis, A. (1968) Shell regeneration in *Helix pomatia* with special reference to the elementary calcifying particles. *Symposia of the Zoological Society of London* 22, 75–92.

Abolins-Krogis, A. (1970) Electron microscope studies on the intracellular origin and formation of calcifying granules and calcium spherites in the hepatopancreas of the snail *Helix pomatia. Zeitschrift für Zellforschung und Mikroskopische Anatomie* 108, 501–515.

Alberti, G. and Storch, V. (1983) Zur Ultrastruktur der Mitteldarmdrüsen von Spinnentieren (Scorpiones, Araneae, Acari) unter verschiedenen Ernährungsbedingungen. *Zoologischer Anzeiger Jena* 211, 145–160.

Alberti, G., Hauk, B., Köhler, H.R. and Storch, V. (1996) *Dekomposition – Qualitative und quantitative Aspekte und deren Beeinflussung durch geogene und anthropogene Belastungsfaktoren.* Ecomed Verlagsgesellschaft AG & Co KG, Landsberg.

Almedros, A. and Porcel, D. (1992a) A structural and microanalytical (EDX) study of calcium granules in the hepatopancreas of *Helix aspersa. Comparative Biochemistry and Physiology* 103A, 757–762.

Almedros, A. and Porcel, D. (1992b) Phosphatase activity in the hepatopancreas of *Helix aspersa. Comparative Biochemistry and Physiology* 103A, 455–460.

Arad, Z., Goldenberg, S. and Heller, J. (1993) Intraspecific variation in resistance to desiccation and climatic gradients in the distribution of the bush-dwelling snail *Trochoidea simulata. Journal of Zoology (London)* 229, 249–265.

Barker, G.M. (1991a) Biology of slugs (Agriolimacidae and Arionidae, Mollusca) in New Zealand hill country pastures. *Oecologia* 85, 581–595.

Barker, G.M. (1991b) Slug density–seedling establishment relationships in a pasture renovated by direct drilling. *Grass and Forage Science* 46, 113–120.

Bartsch, R., Klein, D. and Summer, K.H. (1990) The Cd-Chelex assay: a new sensitive method to determine metallothionein containing zinc and cadmium. *Archives of Toxicology* 64, 177–180.

Baur, A., Baur, B. and Froberg, L. (1994) Herbivory on calcicolous lichens – different food preferences and growth rates in 2 coexisting land snails. *Oecologia* 98, 313–319.

Baur, B. (1984) Shell size and growth rate differences for alpine populations of *Arianta arbustorum* (L.) (Pulmonata: Helicidae). *Revue Suisse de Zoologie* 91, 37–46.

Baur, B. (1988) Population regulation in the land snail *Arianta arbustorum*: density effects on adult size, clutch size and incidence of egg cannibalism. *Oecologia* 77, 390–394.

Baur, B. (1993) Population structure, density, dispersal and neighbourhood size in *Arianta arbustorum* (Linnaeus, 1758) (Pulmonata: Helicidae). *Annalen des Naturhistorischen Museums, Wien* 94/95 (B), 307–321.

Baur, B. and Raboud, C. (1988) Life history of the land snail *Arianta arbustorum* along an altitudinal gradient. *Journal of Animal Ecology* 57, 71–87.

Beeby, A. (1985) The role of *Helix aspersa* as a major herbivore in the transfer of lead through a polluted ecosystem. *Journal of Applied Ecology* 22, 267–275.

Beeby, A. (1991) Toxic metal uptake and essential metal regulation in terrestrial invertebrates: a review. In: Newman, M.C. and McIntosh, A.W. (eds) *Metal Ecotoxicology – Concepts and Applications*. Lewis Publishers, London, pp. 65–89.

Beeby, A. and Richmond, L. (1988) Calcium metabolism in two populations of the snail *Helix aspersa* on a high lead diet. *Archives of Environmental Contamination and Toxicology* 17, 507–511.

Beeby, A. and Richmond, L. (1989) The shell as a site of lead deposition in *Helix aspersa*. *Archives of Environmental Contamination and Toxicology* 18, 623–628.

Berger, B. (1990) Beiträge zum Schwermetallstoffwechsel der terrestrischen Gastropoden *Arianta arbustorum* L. und *Cepaea hortensis* Müller (Fam. Helicidae): Quantitative Aspekte und Cadmium bindende Proteine. PhD thesis, University of Innsbruck, Austria.

Berger, B. and Dallinger, R. (1989a) Accumulation of cadmium and copper by the terrestrial snail *Arianta arbustorum* L.: kinetics and budgets. *Oecologia* 79, 60–65.

Berger, B. and Dallinger, R. (1989b) Factors influencing the contents of heavy metals in terrestrial snails from contaminated urban sites. In: Vernet, J.P. (ed.) *Proceedings of the 7th International Conference on Heavy Metals in the Environment*, Geneva, Vol. II. CEP Consultants Ltd, Edinburgh, pp. 550–553.

Berger, B. and Dallinger, R. (1993) Terrestrial snails as quantitative indicators of environmental metal pollution. *Environmental Monitoring and Assessment* 25, 65–84.

Berger, B., Dallinger, R., Felder, E. and Moser, J. (1993) Budgeting the flow of cadmium and zinc through the terrestrial gastropod, *Helix pomatia* L. In: Dallinger, R. and Rainbow, P.S. (eds) *Ecotoxicology of Metals in Invertebrates*. Lewis Publishers, Boca Raton, Florida, pp. 291–313.

Berger, B., Dallinger, R., Gruber, A. and Moser, J. (1994) Uptake, assimilation and ligand binding of cadmium and zinc in *Helix pomatia* after combined exposure to both metals. In: Eijsackers, H., Heimbach, F. and Donker, M. (eds) *Ecotoxicology of Soil Pollution*. Lewis Publishers, Boca Raton, Florida, pp. 347–354.

Berger, B., Hunziker, P.E., Hauer, C.R., Birchler, N. and Dallinger, R. (1995a) Mass spectrometry and amino acid sequencing of two cadmium-binding metallothionein isoforms from the terrestrial gastropod *Arianta arbustorum*. *Biochemical Journal* 311, 951–957.

Berger, B., Dallinger, R. and Thomaser, A. (1995b) Quantification of metallo-thionein as a biomarker for cadmium exposure in terrestrial gastropods. *Environmental Toxicology and Chemistry* 14, 781–791.

Berger, B., Dallinger, R., Hunziker, P.E. and Gering, E. (1997) Primary structure of a copper-binding metallothionein from mantle tissue of the terrestrial gastropod *Helix pomatia* L. *Biochemical Journal* 311, 951–957.

Beyer, W.N., Pattee, O.H., Sileo, L., Hoffman, D.J. and Mulhern, B.M. (1985) Metal contamination in wildlife living near two zinc smelters. *Environmental Pollution (Series A)* 38, 63–86.

Boháč, J. and Pospísil, J. (1989) Accumulation of heavy metals in invertebrates and its ecological aspects. In: Vernet, J.P. (ed.) *Proceedings of the 7th International Conference on Heavy Metals in the Environment*, Geneva, Vol. II. CEP Consultants Ltd, Edinburgh. p. 354.

Bohác, J., Krivolutskii, D.A. and Antononva, T.B. (1990) The role of fungi in the biogenous migration and in the accumulation of radionuclides. *Agriculture, Ecosystems and Environment* 28, 31–34.

Bourne, N.B., Jones, G.W. and Bowen, I.D. (1991) Endocytosis in the crop of the slug *Deroceras reticulatum* (Müller) and the effects of the ingested molluscicides, metaldehyde and methiocarb. *Journal of Molluscan Studies* 57, 71–80.

Bowen, I.D. (1981) Techniques for demonstrating cell death. In: Bowen, I.D. and Lockshin, R.A. (eds) *Cell Death in Biology and Pathology*. Chapman & Hall, London, pp. 379–444.

Bradley, M.D. and Runham, N.W. (1996) Heavy metal toxicity in the snail *Helix aspersa maxima* reared on commercial farms: cellular pathology. In: Henderson, I.F. (ed.) *Slugs and Snail Pests in Agriculture*. British Crop Protection Council Monograph No. 41, pp. 353–358.

Braunbeck, T. and Storch, V. (1989) Zelle und Umwelt: Wie wirken sich Umweltgifte auf Zellen aus? *Biologie in Unserer Zeit* 19, 127–132.

Briggs, G.G. and Henderson, I.F. (1987) Some factors affecting the toxicity of poisons to the slug *Deroceras reticulatum* (Müller) (Pulmonata: Limacidae). *Crop Protection* 6, 341–346.

Brooks, A.W. and White, K.N. (1995) The localization of aluminium in the digestive gland of the terrestrial snail *Helix aspersa*. *Tissue and Cell* 27, 61–72.

Brown, B.E. (1982) The form and function of metal-containing 'granules' in invertebrate tissues. *Biological Reviews* 57, 621–667.

Burla, H. and Gosteli, M. (1993) Thermal advantage of pale coloured morphs of the snail *Arianta arbustorum* (Helicidae, Pulmonata) in alpine habitats. *Ecography* 16, 345–350.

Cajaraville, M.P., Marigómez, J.A. and Angulo, E. (1989) Ultrastructural study of the short-term toxic effects of naphthaline on the kidney of the marine prosobranch *Littorina littorea*. *Journal of Invertebrate Pathology* 55, 215–224.

Calderbank, A. (1994) The consequences of bound pesticide residues in soil. In: Donker, M.H., Eijsackers, H. and Heimbach, F. (eds) *Ecotoxicology of Soil Organisms*. Lewis Publishers, Boca Raton, Florida, pp. 71–76.

Carstens, S. and Storch, V. (1980) Beeinflussung der Ultrastruktur von Fettkörper und Mitteldarm des Staphyliniden *Atheta fungi* (Grav.) durch Umweltein- flüsse. *Zoologisches Jahrbuch für Anatomie* 103, 73–84.

Carter, A. (1983) Cadmium, copper, and zinc in soil animals and their food in a red clover system. *Canadian Journal of Zoology* 61, 2751–2757.

Chang, H.W. and Emlen, J.M. (1993) Seasonal variation of microhabitat distribu- tion of the polymorphic land snail *Cepaea nemoralis*. *Oecologia* 93, 501–507.

Cooke, M., Jackson, A., Nickless, G. and Roberts, D.J. (1979) Distribution and speciation of cadmium in the terrestrial snail, *Helix aspersa*. *Journal of Environmental Contamination and Toxicology* 23, 445–451.

Coughtrey, P.J. and Martin, M.H. (1976) The distribution of Pb, Zn, Cd and Cu within the pulmonate mollusc *Helix aspersa* Müller. *Oecologia* 23, 315–322.

Coughtrey, P.J. and Martin, M.H. (1977) The distribution and speciation of cadmium in the terrestrial snail, *Helix aspersa* Müller, and its relevance to the monitoring of heavy metal contamination of the environment. *Oecologia* 27, 65–74.

Craig, E.A. and Gross, C.A. (1991) Is hsp70 the cellular thermometer? *Trends in Biological Sciences* 16, 135–140.

Crawford, C.S. (1981) *Biology of Desert Invertebrates*. Springer-Verlag, Berlin.

Dallinger, R. (1993) Strategies of metal detoxification in terrestrial invertebrates. In: Dallinger, R. and Rainbow, P.S. (eds) *Ecotoxicology of Metals in Invertebrates.* Lewis Publishers, Boca Raton, Florida, pp. 245–289.

Dallinger, R. (1994) Invertebrate organisms as biological indicators of heavy metal pollution. *Applied Biochemistry and Biotechnology* 48, 27–31.

Dallinger, R. (1995) Metabolism and toxicity of metals: metallothioneins and metal elimination. In: Cajaraville, M.P. (ed.) *Cell Biology in Environmental Toxicology.* Servicio Editorial, Universidad del Pais Vasco, Bilbao, pp. 171–190.

Dallinger, R. (1996) Metallothionein research in terrestrial invertebrates: synopsis and perspectives. *Comparative Biochemistry and Physiology* 113C, 125–133.

Dallinger, R. and Berger, B. (1992) Bio-monitoring in the urban environment. In: Bonotto, S., Nobili, R. and Revoltella, R.P. (eds) *Biological Indicators for Environmental Monitoring.* Serono Symposia Reviews No. 27. pp. 227–242.

Dallinger, R. and Berger, B. (1993) Function of metallothioneins in terrestrial gastropods. *Science of the Total Environment, Supplement* 1993, 607–615.

Dallinger, R. and Wieser, W. (1984a) Patterns of accumulation, distribution and liberation of Zn, Cu, Cd, and Pb in different organs of the land snail *Helix pomatia* L. *Comparative Biochemistry and Physiology* 79C, 117–124.

Dallinger, R. and Wieser, W. (1984b) Molecular fractionation of Zn, Cu, Cd, and Pb in the midgut gland of *Helix pomatia* L. *Comparative Biochemistry and Physiology* 79C, 125–129.

Dallinger, R., Berger, B. and Bauer-Hilty, A. (1989a) Purification of cadmium-binding proteins from related species of terrestrial Helicidae (Gastropoda, Mollusca): a comparative study. *Molecular and Cellular Biochemistry* 85, 135–145.

Dallinger, R., Janssen, H.H., Bauer-Hilty, A. and Berger, B. (1989b) Characterization of an inducible cadmium-binding protein from hepatopancreas of metal-exposed slugs (Arionidae, Mollusca). *Comparative Biochemistry and Physiology* 92C, 355–360.

Dallinger, R., Birkel, S., Bauer-Hilty, A. and Berger, B. (1989c) Biokartierung der Schwermetallbelastung im Raum Innsbruck. *Veröffentlichungen der Universität Innsbruck,* Bd. 171.

Dallinger, R., Berger, B. and Birkel, S. (1992) Terrestrial isopods: useful biological indicators of urban metal pollution. *Oecologia* 89, 32–41.

Dallinger, R., Berger, B. and Gruber, C. (1993a) Quantitative aspects of zinc and cadmium binding in *Helix pomatia*: differences between an essential and a nonessential trace element. In: Dallinger, R. and Rainbow, P.S. (eds) *Ecotoxicology of Metals in Invertebrates.* Lewis Publishers, Boca Raton, Florida, pp. 315–332.

Dallinger, R., Berger, B., Hunziker, P., Birchler, N., Hauer, C.R. and Kägi, J.H.R. (1993b) Purification and primary structure of snail metallothionein. Similarity of the N-terminal sequence with histones H4 and H2A. *European Journal of Biochemistry* 216, 739–746.

Dallinger, R., Berger, B., Hunziker, P. and Kägi, J.H.R (1997) Metal-specific metallothionein isoforms in cadmium detoxification and copper metabolism of terrestrial snails. *Nature* 388, 237–238.

Dallinger, R., Berger, B., Hunziker, P. and Kägi, J.H.R. (1998) Structure and function of metallothionein isoforms in terrestrial snails. In: Klaassen C. (ed.) *Metallothionein IV.* Birkhäuser Verlag, Basel, pp. 173–178.

Dallinger, R., Berger, B., Gruber, C., Hunziker, P. and Stürzenbaum, S., (2000) Metallothioneins in terrestrial invertebrates: structural aspects, biological signifcance and implications for their use as biomarkers. *Cellular and Molecular Biology* 46 (2), 331–346.

Dallinger, R., Wang, Y., Berger, B., Mackay, E. and Kägi, J.H.R (2001) Spectroscopic characterization of metallothionein from the terrestrial snail, *Helix pomatia*. *Eurpean Journal of Biochemistry* (in press).

Dexheimer, L. (1963) Beiträge zum Kalkstoffwechsel der Weinbergschnecke (*Helix pomatia*). *Zoologisches Jahrbuch* 63, 130–152.

Dindal, D.L. and Wurzinger, K.-H. (1971) Accumulation and excretion of DDT by the terrestrial snail, *Cepaea hortensis*. *Journal of Environmental Contamination and Toxicology* 6, 362–371.

Dutton, M.D., Stephenson, M. and Klaverkamp, J.F. (1993) A mercury saturation assay for measuring metallothionein in fish. *Environmental Toxicology and Chemistry* 12, 1193–1202.

Eckwert, H., Alberti, G. and Köhler, H.R (1997) The induction of stress proteins (hsp) in *Oniscus asellus* (Isopoda) as a molecular marker of multiple heavy metal exposure. I. Principles and toxicological assessment. *Ecotoxicology* 6, 249–262.

Edington, B.V., Wheelan, S.A. and Hightower, L.E. (1989) Inhibition of heat shock (stress) protein induction by deuterium oxide and glycerol: additional support for the abnormal protein hypothesis of induction. *Journal of Cell Physiology* 139, 219–228.

Felder, E. (1992) Ultrastrukturelle Veränderungen in der Mitteldarmdrüse von *Helix pomatia* nach Cadmiumbelastung. Diploma thesis, University of Innsbruck, Austria.

Fischer, E. and Molnár, L. (1992) Environmental aspects of the chloragogenous tissue of earthworms. *Soil Biology and Biochemistry* 24, 1723–1727.

Gething, M.-J. and Sambrook, J. (1992) Protein folding in the cell. *Nature* 355, 33–45.

Gomot, A. (1997) Dose-dependent effects of cadmium on the growth of snails in toxicity bioassays. *Archives of Environmental Contamination and Toxicology* 33, 209–216.

Gomot, A. and Pihan, F. (1997) Comparison of the bioaccumulation capacities of copper and zinc in 2 snail subspecies (Helix). *Ecotoxicology and Environmental Safety* 38, 85–94.

Graveland, J. and Vanderwal, R. (1996) Decline in snail abundance due to soil acidification causes eggshell defects in forest passerines. *Oecologia* 105, 351–360.

Greaves, G.N., Simkiss, K., Taylor, M. and Binsted, N. (1984) The local environment of metal sites in intracellular granules investigated by using X-ray-absorption spectroscopy. *Biochemical Journal* 221, 855–868.

Greville, R.W. and Morgan, A.J. (1989a) Concentrations of metals (Cu, Pb, Cd, Zn, Ca) in six species of British terrestrial gastropods near a disused lead and zinc mine. *Journal of Molluscan Studies* 55, 31–36.

Greville, R. and Morgan, A.J. (1989b) Seasonal changes in metal levels (Cu, Pb, Cd, Zn and Ca) within the grey field slug, *Deroceras reticulatum*, living in a highly polluted habitat. *Environmental Pollution* 59, 287–303.

Greville, R.W. and Morgan, A.J. (1990) The influence of size on the accumulated amounts of metals (Cu, Pb, Cd, Zn and Ca) in six species of slug sampled from a contaminated woodland site. *Journal of Molluscan Studies* 56, 355–362.

Greville, R.W. and Morgan, A.J. (1991) A comparison of (Pb, Cd and Zn) accumulation in terrestrial slugs maintained in microcosms – evidence for metal tolerance. *Environmental Pollution* 74, 115–127.

Guven, K., Duce, J.A. and de Pomerai, D.I. (1994) Evaluation of a stress-inducible transgenic nematode strain for rapid aquatic toxicity testing. *Aquatic Toxicology* 29, 119–137.

Hanley, M.E., Fenner, M. and Edwards, P.J. (1995) The effect of seedling age on the likelihood of herbivory by the slug *Deroceras reticulatum. Functional Ecology* 9, 754–759.

Henderson, I.F. and Triebskorn, R. (2001) Molluscicidal chemicals. In: Barker, G.M. (ed.) *Molluscs as Crop Pests.* CAB International, Wallingford, UK.

Herlitzius, R. and Herlitzius, H. (1977) Streuabbau in Laubwäldern. *Oecologia* 30, 147–171.

Hermony, I., Shachak, M. and Abramsky, Z. (1992) Habitat distribution in the desert snail *Trochoidea seetzenii. Oikos* 64, 516–522.

Hopkin, S.P. (1989) *Ecophysiology of Metals in Terrestrial Invertebrates.* Elsevier Applied Science Publishers, London.

Hopkin, S.P., Hardisty, G.N. and Martin, M.H. (1986) The woodlouse *Porcellio scaber* as a biological indicator of zinc, cadmium, lead and copper pollution. *Environmental Pollution* 11B, 271–290.

Howard, B., Mitchell, P.C.H., Ritchie, A., Simkiss, K. and Taylor, M. (1981) The composition of intracellular granules from the metal-accumulating cells of the common garden snail (*Helix aspersa*). *Biochemical Journal* 194, 507–511.

Hryniewiecka-Szyfter, Z. and Storch, V. (1986) The influence of starvation and different diets on the hindgut of Isopoda (*Mesidotea entomon, Oniscus asellus, Porcellio scaber*). *Protoplasma* 134, 53–59.

Hunter, B.A., Johnson, M.S. and Thompson, D.J. (1987a) Ecotoxicology of copper and cadmium in a contaminated grassland ecosystem. I. Soil and vegetation contamination. *Journal of Applied Ecology* 24, 573–586.

Hunter, B.A., Johnson, M.S. and Thompson, D.J. (1987b) Ecotoxicology of copper and cadmium in a contaminated grassland ecosystem. II. Invertebrates. *Journal of Applied Ecology* 24, 587–599.

Ireland, M.P. (1979) Distribution of essential and toxic metals in the terrestrial slug *Arion ater. Environmental Pollution* 20, 271–278.

Ireland, M.P. (1981) Uptake and distribution of cadmium in the terrestrial slug *Arion ater* (L.). *Comparative Biochemistry and Physiology* 68A, 37–41.

Ireland, M.P. (1982) Sites of water, zinc and calcium uptake and distribution of these metals after cadmium administration in *Arion ater* (Gastropoda: Pulmonata). *Comparative Biochemistry and Physiology* 73A, 217–221.

Ireland, M.P. (1984) Seasonal changes in zinc, manganese, magnesium, copper and calcium content in the digestive gland of the slug *Arion ater. Comparative Biochemistry and Physiology* 78A, 855–858.

Ireland, M.P. (1988) A comparative study of the uptake and distribution of silver in a slug *Arion ater* and a snail *Achatina fulica. Comparative Biochemistry and Physiology* 90C, 189–194.

Ireland, M.P. (1991) The effect of dietary calcium on growth, shell thickness and tissue calcium distribution in the snail *Achatina fulica. Comparative Biochemistry and Physiology* 98A, 111–116.

Ireland, M.P. and Marigómez, J.A. (1992) The influence of dietary calcium on the tissue distribution of Cu, Zn, Mg and P and histological changes in the digestive gland cells of the snail *Achatina fulica* Bowdich. *Journal of Molluscan Studies* 58, 157–168.

Janssen, H.H. (1985) Some histophysiological findings on the midgut gland of the common garden snail, *Arion rufus* (L.) (Syn. *A. ater rufus* (L.), *A. empiricorum* Férussac), Gastropoda: Stylommatophora. *Zoologischer Anzeiger, Jena* 215, 33–51.

Janssen, H.H. and Dallinger, R. (1991) Diversification of cadmium binding proteins due to different levels of contamination in *Arion lusitanicus*. *Archives of Environmental Contamination and Toxicology* 20, 132–137.

Jennings, T.J. and Barkham, J.P. (1979) Litter decomposition by slugs in mixed deciduous woodland. *Holarctic Ecology* 2, 21–29.

Jones, C.G. and Shachak, M. (1990) Fertilization of the desert soil by rock-eating snails. *Nature* 346, 839–841.

Jones, C.G. and Shachak, M. (1994) Desert snails' daily grind. *Natural History* 103, 56–61.

Jones, D.T. and Hopkin, S.P. (1991) Biological monitoring of metal pollution in terrestrial ecosystems. In: Ravera, O. (ed.) *Terrestrial and Aquatic Ecosystems – Perturbation and Recovery*. Ellis Horwood, Chichester, UK, pp. 148–152.

Kägi, J.H.R. (1993) Evolution, structure and chemical activity of class I metallothioneins: an overview. In: Suzuki, K.T., Imura, N. and Kimura, M. (eds) *Metallothionein III*. Birkhäuser Verlag, Basel, pp. 29–55.

Kägi, J.H.R. and Schäffer, A. (1988) Biochemistry of metallothionein. *Biochemistry* 27, 8509–8515.

Kammenga, J.E., Köhler, H.-R., Dallinger, R., Weeks, J.M. and van Gestel, C.A.M. (1996) *Biochemical Fingerprint Techniques as Versatile Tools for the Risk Assessment of Chemicals in Terrestrial Invertebrates*. Third Technical Report. Ministry of the Environment and Energy, National Environmental Research Institute, Silkeborg, Denmark.

Kerney, M.P. and Cameron, R.A.D. (1979) *A Field Guide to the Land Snails of Britain and North-west Europe*. Collins, London.

Kerney, M.P., Cameron, R.A.D. and Jungbluth, J.H. (1983) *Die Landschnecken Nord- und Mitteleuropas*. Verlag Paul Parey, Hamburg.

Knutti, R., Bucher, P., Stengl, M., Stolz, M., Tremp, J., Ulrich, M. and Schlatter, C. (1988) Cadmium in the invertebrate fauna of an unpolluted forest in Switzerland. *Environmental Toxin Series* 2, 171–191.

Köhler, H.-R. (1996) Part B. Detailed report of the contractors. In: Kammenga, J.E. (ed.) *Biochemical Fingerprint Techniques as Versatile Tools for the Risk Assessment of Chemicals in Terrestrial Invertebrates*. National Environmental Research Institute, Silkeborg, Denmark, pp. 15–20.

Köhler, H.-R. and Triebskorn, R. (1996) The validation of cytotoxic effects by a synthetic technique integrating qualitative and quantitative aspects. In: *Adaptation to Stress in Aquatic and Terrestrial Ecosystems*. European Society for Comparative Physiology and Biochemistry, Antwerp, p. 184.

Köhler, H.-R. and Triebskorn, R. (1998) The validation of cytotoxic effects by a synthetic technique integrating qualitative and quantitative aspects. *Biomarkers* 3, 109–127.

Köhler, H.-R., Triebskorn, R., Stöcker, W., Kloetzel, P.-M. and Alberti, G. (1992) The 70 kD heat shock protein in soil invertebrates: a possible tool for monitoring environmental toxicants. *Archives of Environmental Contamination and Toxicology* 22, 334–338.

Köhler, H.-R., Rahman, B. and Rahmann, H. (1994) Assessment of stress situations in the grey garden slug, *Deroceras reticulatum*, caused by heavy metal intoxication: semi-quantification of the 70 kD stress protein (hsp70). *Verhandlungen der Deutschen Zoologischen Gesellschaft* 87, 328.

Köhler, H.-R., Rahman. B., Gräff, S., Berkus, M. and Triebskorn, R. (1996a) Expression of the stress-70 protein family (hsp70) due to heavy metal contamination in the slug, *Deroceras reticulatum*: an approach to monitor sublethal stress conditions. *Chemosphere* 33, 1327–1340.

Köhler, H.-R., Belitz, B. and Eckwert, H. (1996b) The stress-70 (hsp70) response to metals in slugs on the transcriptional and translational level. In: *Adaptation to Stress in Aquatic and Terrestrial Ecosystems*. European Society for Comparative Physiology and Biochemistry, Antwerp, p 72.

Kojo, M.R. and Lodenius, M. (1989) Cadmium and mercury in macrofungi – mechanisms of transport and accumulation. *Angewandte Botanik* 63, 279–292.

Kratz, W. (1991) Cycling of nutrients and pollutants during litter decomposition in pine forests in the Grunewald, Berlin. In: Nakagoshi, N. and Golley, F.B. (eds) *Coniferous Forest Ecology from an International Perspective*. SPB Academic Publishing bv, The Hague, pp. 151–160.

Laskowski, R. (1991) Are the top carnivores endangered by heavy metal biomagnification? *Oikos* 60, 387–390.

Laskowski, R. and Hopkin, S.P. (1996a) Accumulation of Zn, Cu, Pb and Cd in the garden snail (*Helix aspersa*): implications for predators. *Environmental Pollution* 91, 289–297.

Laskowski, R. and Hopkin, S.P. (1996b) Effect of Zn, Cu, Pb, and Cd on fitness in snails (*Helix aspersa*). *Ecotoxicology and Environmental Safety* 34, 59–69.

Lazaridou-Dimitriadou, M. and Kattoulas, M.E. (1991) Energy flux in a natural population of the land snail *Eobania vermiculata* (Müller) (Gastropoda: Pulmonata: Stylommatophora) in Greece. *Canadian Journal of Zoology* 69, 881–891.

Linhart, Y.B. and Thompson, J.D. (1995) Terpene-based selective herbivory by *Helix aspersa* (Mollusca) on *Thymus vulgaris* (Labiatae). *Oecologia* 102, 126–132.

Machin, J. (1967) Structural adaptation for reducing water-loss in three species of terrestrial snail. *Journal of Zoology (London)* 152, 55–65.

Marigómez, J.A. and Dussart, D.G.J. (1996) Cellular basis of the adaptation to metal pollution in sentinel slugs: eco(toxico)logical implications. In: *Adaptation to Stress in Aquatic and Terrestrial Ecosystems*. European Society for Comparative Physiology and Biochemistry, Antwerp, p. 81.

Marigómez, J.A., Angulo, E. and Moya, J. (1986a) Copper treatment of the digestive gland of the slug, *Arion ater* L. 1. Bioassay conduction and histochemical analysis. *Journal of Environmental Contamination and Toxicology* 36, 600–607.

Marigómez, J.A., Angulo, E. and Moya, J. (1986b) Copper treatment of the digestive gland of the slug, *Arion ater* L. 2. Morphometrics and histophysiology. *Journal of Environmental Contamination and Toxicology* 36, 608–615.

Marigómez, J.A., Cajaraville, M.P., Angulo, E. and Moja, J. (1990) Ultrastructural alterations in the renal epithelium of cadmium treated *Littorina littorea* (L.). *Archives of Environmental Contamination and Toxicology* 19, 863–881.

Marigómez, J.A., Soto, M. and Kortabitarte, M. (1996) Tissue-level biomarkers and biological effect of mercury on sentinel slugs, *Arion ater*. *Archives of Environmental Contamination and Toxicology* 31, 54–62.

Martin, M.H. and Coughtrey, P.J. (1976) Comparisons between the levels of lead, zinc and cadmium within a contaminated environment. *Chemosphere* 5, 15–20.

Martin, M.H. and Coughtrey, P.J. (1982) *Biological Monitoring of Heavy Metal Pollution: Land and Air*. Applied Science Publishers, London.

Mason, C.F. (1970) Snail populations, beech litter production and the role of snails in litter decomposition. *Oecologia* 5, 215–239.

Mason, A.Z. and Simkiss, K. (1982) Sites of mineral deposition in metal-accumulating cells. *Experimental Cell Research* 139, 383–391.

Meincke, K.F. and Schaller, K.H. (1974) Über die Brauchbarkeit der Weinbergschnecke (*Helix pomatia* L.) im Freiland als Indikator für die Belastung der Umwelt durch die Elemente Eisen, Zink und Blei. *Oecologia* 15, 393–398.

Moore, M.N. (1985) Cellular responses to pollutants. *Marine Pollution Bulletin* 16, 134–139.

Morimoto, R., Abravaya, K., Mosser, G. and Williams, G.Tt. (1990) Transcription of the human hsp70 gene: *cis*-acting elements and *trans*-acting factors involved in basal, adenovirus E1A, and stress-induced expression. In: Schlesinger, M.J., Santoro, M.G. and Garaci, E. (eds) *Stress Proteins – Induction and Function*. Springer Verlag, Berlin, pp. 1–17.

Mulvey, M., Newman, M.C. and Beeby, A. (1996) Genetic and conchological comparison of snails (*Helix aspersa*) differing in shell deposition of lead. *Journal of Molluscan Studies* 62, 213–223.

Neumann, W. (1985) Veränderungen am Mitteldarm von *Oxidus gracilis* (C.L. Koch, 1847) während einer Häutung (Diplopoda). *Bijdragen tot de Dierkande* 55, 149–158.

Nieboer, E. and Richardson, D.H.S. (1980) The replacement of the nondescript term 'heavy metals' by a biologically and chemically significant classification of metal ions. *Environmental Pollution (Series B)* 1, 3–26.

Nover, L. (1984) *Heat Shock Response in Eucaryotic Cells*. Springer Verlag, Berlin, pp. 1–82.

Nriagu, J.O. (1990) Global metal pollution – poisoning the biosphere? *Environment* 32, 7.

Peakall, D.B. and Walker, C.H. (1994) The role of biomarkers in environmental assessment (3). Vertebrates. *Ecotoxicology* 3, 173–179.

Pelham, H.R.B. (1986) Speculations on the function of the major heat shock and glucose-regulated proteins. *Cell* 46, 959–961.

Perry, R. and Arthur, W. (1991) Shell size and population density in large helicid land snails. *Journal of Applied Ecology* 60, 409–421.

Popham, J.D. and D'Auria, J.M. (1980) *Arion ater* (Mollusca: Pulmonata) as an indicator of terrestrial environmental pollution. *Water, Air and Soil Pollution* 14, 115–124.

Rabitsch, W.B. (1996) Metal accumulation in terrestrial pulmonates at a lead/zinc smelter site in Arnoldstein, Austria. *Journal of Environmental Contamination and Toxicology* 56, 734–741.

Rahman, B. (1994) Induktion von Streßproteinen und konjugierenden Proteinen bei Pulmonaten in Reaktion auf Umweltchemikalien. Diploma Thesis, University of Stuttgart-Hohenheim, Germany.

Récio, A., Marigómez, J.A., Angulo, E. and Moya, J. (1988) Zinc treatment of the digestive gland of the slug *Arion ater* I: cellular distribution of zinc and calcium. *Journal of Environmental Contamination and Toxicology* 41, 858–864.

Reichardt, A., Raboud, C., Burla, H. and Baur, B. (1985) Causes of death and possible regulatory processes in *Arianta arbustorum* (L., 1758) (Pulmonata, Helicidae). *Basteria* 49, 37–46.

Riddle, W.A. (1981) Cold hardiness in the woodland snail, *Anguispira alternata* (Say) (Endodontidae). *Journal of Thermal Biology* 6, 117–120.

Riddle, W.A. and Miller, V.J. (1988) Cold-hardiness in several species of land snail. *Journal of Thermal Biology* 13, 163–167.

Roesijadi, G. (1993) Response of invertebrate metallothioneins and MT genes to metals and implications for environmental toxicology. In: Suzuki, K.T., Imura, N. and Kimura, M. (eds) *Metallothionein III*. Birkhäuser Verlag, Basel, pp. 141–158.

Russell, L.K., DeHaven, J.I. and Botts, R.P. (1981) Toxic effects of cadmium on the garden snail (*Helix aspersa*). *Journal of Environmental Contamination and Toxicology* 26, 634–640.

Ryder, T. and Bowen, I.D. (1977) The slug foot as a site of uptake of copper molluscicide. *Journal of Invertebrate Pathology* 30, 381–386.

Sanders, B.M. (1990) Stress proteins: potential as multitiered biomarkers. In: McCarthy, J.F. and Shughart, L.R. (eds) *Biomarkers of Environmental Contamination*. CRC Press, Boca Raton, Florida, pp. 165–192.

Sanders, B.M. (1993) Stress proteins in aquatic organisms: an environmental perspective. *Critical Reviews of Toxicology* 23, 49–75.

Sanders, B.M. and Dyer, S.D. (1994) Cellular stress response. *Environmental Toxicology and Chemistry* 13, 1209–1210.

Schaefer, M. (1991) The animal community: diversity and resources. In: Röding, E. and Ulrich, M. (eds) *Temperate Deciduous Forests (Ecosystems of the World IV)*. Elsevier, Amsterdam, pp. 51–120.

Scheuhammer, A.M. and Cherian, M.G. (1986) Quantification of metallothionein by a silver-saturation method. *Toxicology and Applied Pharmacology* 82, 417–425.

Schlesinger, M.J., Ashburner, M. and Tissières, A. (1982) *Heat Shock – from Bacteria to Man*. Cold Spring Harbor Laboratory, Cold Spring Harbor, New York.

Schlesinger, M.J., Santoro, M.G. and Garaci, E. (1990) *Stress Proteins. Induction and Function*. Springer Verlag, Berlin.

Schmidt-Nielsen, K., Taylor, C.R. and Shkolnik, A. (1971) Desert snails: problems of heat, water and food. *Journal of Experimental Biology* 55, 385–398.

Schötti, G. and Seiler, H.G. (1970) Uptake and localisation of radioactive zinc in the visceral complex of the land pulmonate *Arion rufus*. *Experientia* 26, 1212–1213.

Simkiss, K. (1976) *Intracellular and Extracellular Routes in Biomineralization*. Symposia of the Society of Experimental Biology No XXX: Calcium in Biological Systems. Cambridge University Press, Cambridge, pp. 423–444.

Simkiss, K. (1981) Calcium, pyrophosphate and cellular pollution. *Trends in Biological Sciences* 6, 3–5.

Simkiss, K. (1985) Prokaryote–eukaryote interactions in trace element metabolism. *Desulfovibrio* sp. in *Helix aspersa*. *Experientia* 41, 1195–1197.

Simkiss, K. and Mason, A.Z. (1983) Metal ions: metabolic and toxic effects. In: Hochachka, P.W. (ed.) *The Mollusca*, Vol. 2, *Environmental Biochemistry and Physiology*. Academic Press, New York, pp. 102–164.

Simkiss, K. and Watkins, B. (1990) The influence of gut microorganisms on zinc uptake in *Helix aspersa*. *Environmental Pollution* 66, 263–271.

Simkiss, K. and Watkins, B. (1991) Differences in zinc uptake between snails (*Helix aspersa* (Muller) from metal- and bacteria-polluted sites. *Functional Ecology* 5, 787–794.

Simkiss, K. and Wilbur, K.M. (1977) The molluscan epidermis and its secretions. In: Spearmen, R.I.C. (ed.) *Comparative Biology of Skin*. Academic Press, London, pp. 35–76.

Simkiss, K., Jenkins, K.G.A., McLellan, J. and Wheeler, E. (1982) Methods of metal incorporation into intracellular granules. *Experientia* 38, 333–335.

Somers, G.F. (1978) The role of plant residues in the retention of cadmium in ecosystems. *Environmental Pollution* 17, 287–295.

Sorger, P.K. and Nelson, H. (1989) Trimerization of a yeast transcriptional activator via a coiled-cell-motif. *Cell* 59, 807–813.

Sparks, A.K. (1972) *Invertebrate Pathology, Noncommmunicable Diseases*. Academic Press, New York.

Speiser, B. and Rowell-Rahier, M. (1991) Effects of food availability, nutritional value, and alkaloids on food choice in the generalist herbivore *Arianta arbustorum* (Gastropoda: Helicidae). *Oikos* 62, 306–318.

Spellerberg, I.F. (1991) *Monitoring Ecological Change*. Cambridge University Press, Cambridge.

Stijve, T. and Besson, R. (1977) Mercury, cadmium, lead and selenium content of mushroom species belonging to the genus *Agaricus*. *Chemosphere* 2, 151–158.

Stöver, H. (1973) Über den Wasser- und Elektrolythaushalt von *Arianta arbustorum* (L.). *Journal of Comparative Physiology* 83, 51–61.

Storch, V. (1984) The influence of nutritional stress on the ultrastructure of the hepatopancreas of terrestrial isopods. *Symposia of the Zoological Society of London* 53, 167–184.

Storch, V. (1988) Cell and environment: a link between morphology and ecology. In: Iturrondobeitia, J.C. (ed.) *Biologia Ambiental I*. Servicio Editorial de la Universad del Pais Vasco, Bilboa. pp. 179–191.

Strojan, C.L. (1978a) The impact of zinc smelter emissions on forest litter arthropods. *Oikos* 31, 41–46.

Strojan, C.L. (1978b) Forest leaf litter decomposition in the vicinity of a zinc smelter. *Oecologia* 32, 203–212.

Taylor, M.G., Simkiss, K., Greaves, G.N. and Harries, J. (1988) Corrosion of intracellular granules and cell death. *Proceedings of the Royal Society of London, Series B* 234, 463–476.

Theenhaus, A. and Scheu, S. (1996a) The influence of slug (*Arion rufus*) mucus and cast material addition on microbial biomass, respiration, and nutrient cycling in beech leaf-litter. *Biology and Fertility of Soils* 23, 80–85.

Theenhaus, A. and Scheu, S. (1996b) Successional changes in microbial biomass, activity and nutrient status in faecal material of the slug *Arion rufus* (Gastropoda) deposited after feeding on different plant materials. *Soil Biology and Biochemistry* 28, 569–577.

Thompson, L., Thomas, C.D., Radley, J.M.A., Williamson, S. and Lawton, J.H. (1993) The effect of earthworms and snails in a simple plant community. *Oecologia* 95, 171–178.

Tissières, A., Mitchell, H.K. and Tracy, U.M. (1974) Protein synthesis in the salivary glands of *D. melanogaster*. Relation to chromosome puffs. *Journal of Molecular Biology* 84, 389–398.

Tompa, A.S. and Wilbur, K.M. (1977) Calcium mobilization during reproduction in snail *Helix pomatia*. *Nature* 270, 53–54.

Treweek, J. (1996) Ecology and environmental impact assessment. *Journal of Applied Ecology* 33, 191–199.

Triebskorn, R. (1989) Ultrastructural changes in the digestive tract of *Deroceras reticulatum* (Müller) induced by a carbamate molluscicide and by metaldehyde. *Malacologia* 31, 141–156.

Triebskorn, R. (1991) Cytological changes in the digestive system of slugs induced by molluscicides. *Journal of Medical Application of Malacology* 3, 113–123.

Triebskorn, R. (1995) Tracing molluscicides and cellular reactions induced by them in slugs' tissues. In: Cajaraville, M.P. (ed.) *Cell Biology in Environmental Toxicology*. Universidad del Pais Vasco, Bilbao. pp. 193–220.

Triebskorn, R. and Ebert, D. (1989) The importance of mucus production in slugs' reaction to molluscicides and the impact of molluscicides on the mucus producing system. In: Henderson, I.F. (ed.) *Slugs and Snail Pests in Agriculture*. British Crop Protection Council Monograph No. 41, pp. 373–379.

Triebskorn, R. and Köhler, H.-R. (1992) Plasticity of the endoplasmatic reticulum in three cell types of slugs poisoned by molluscicides. *Protoplasma* 169, 120–129.

Triebskorn, R. and Köhler, H.-R. (1996a) The impact of heavy metals on the grey garden slug, *Deroceras reticulatum* (Müller): metal storage, cellular effects and semi-quantitative evaluation of metal toxicity. *Environmental Pollution* 93, 327–343.

Triebskorn, R. and Köhler, H.-R. (1996b) Cellular adaptations in slugs to environmental pollution. In: *Adaptation to Stress in Aquatic and Terrestrial Ecosystems*. European Society for Comparative Physiology and Biochemistry, Antwerp, p. 90.

Triebskorn, R. and Künast, C. (1990) Ultrastructural changes in the digestive system of *Deroceras reticulatum* (Mollusca; Gastropoda) induced by lethal and sublethal concentrations of the carbamate molluscicide Cloethocarb. *Malacologia* 32, 89–106.

Triebskorn, R. and Schweizer, H. (1990) *Influence du Molluscicide Métaldehyde sur les Mucocytes du Tractus Digestif de la Petite Limace Grise (Deroceras reticulatum Müller)*. Proceedings of the 2nd ANPP Conference. int. ravag. agric., pp. 183–190.

Triebskorn, R., Künast, C., Huber, R. and Brem, G. (1990) Tracing a ^{14}C-labeled carbamate molluscicide through the digestive system of *Deroceras reticulatum* (Müller). *Pesticide Science* 28, 321–330.

Triebskorn, R., Köhler, H.-R., Zahn, T., Vogt, G., Ludwig, M., Rumpf, S., Kratzmann, M., Alberti, G. and Storch, V. (1991) Invertebrate cells as targets for hazardous substances. *Zeitschrift für Angewandet Entomologie* 78, 277–287.

Triebskorn, R., Henderson, I.F., Martin, A. and Köhler, H.-R. (1996) Slugs as target and non-target organisms for environmental chemicals. In: Henderson, I.F. (ed.) *Slugs and Snail Pests in Agriculture*. British Crop Protection Council Monograph No. 41, pp. 65–72.

Underwood, E.J. (1977) *Trace Elements in Human and Animal Nutrition*, 4th edn. Academic Press, New York.

Unger, M.E. and Roesijadi, G. (1993) Sensitive assay for molluscan metallo-thionein induction based on ribonuclease protection and molecular titration of metallothionein and actin mRNAs. *Molecular Marine Biology and Biotechnology* 2, 319–324.

Van Gestel, C.A.M. and van Brummelen, T.C. (1996) Incorporation of the biomarker concept in ecotoxicology calls for a redefinition of terms. *Ecotoxicology* 5, 217–225.

Van Hook, R.I., Harris, W.F. and Henderson, G.S. (1977) Cadmium, lead, and zinc distributions and cycling in a mixed deciduous forest. *AMBIO* 6, 281–286.

Van Straalen, N.M. and Bergema, W.F. (1995) Ecological risks of increased bioavailability of metals under soil acidification. *Pedobiologia* 39, 1–9.

Ward, D. and Slotow, R. (1992) The effects of water availability on the life span of the desert snail, *Trochoidea seetzeni*. *Oecologia* 90, 572–580.

Wieser, W. (1978) Consumer strategies of terrestrial gastropods and isopods. *Oecologia* 36, 191–201.

Wieser, W., Busch, G. and Büchel, L. (1976) Isopods as indicators of the copper content of soil and litter. *Oecologia* 23, 107–114.

Williamson, P. (1979) Comparison of metal levels in invertebrate detrivores and their natural diets. Concentration factors reassessed. *Oecologia* 44, 75–79.

Williamson, P. (1980) Variables affecting body burdens of lead, zinc, and cadmium in a roadside population of the snail *Cepaea hortensis* Müller. *Oecologia* 44, 213–220.

Zöttl, H.W. (1985) Heavy metal levels and cycling in forest ecosystems. *Experientia* 41, 1104–1113.

Index

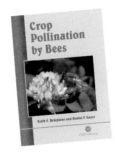
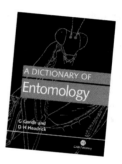